ANNUAL REVIEW OF PHYTOPATHOLOGY

ANNUAL REVIEW OF PHYTOPATHOLOGY

KENNETH F. BAKER, *Editor*
The University of California, Berkeley

GEORGE A. ZENTMYER, *Associate Editor*
The University of California, Riverside

ELLIS B. COWLING, *Associate Editor*
North Carolina State University

VOLUME 13

1975

ANNUAL REVIEWS INC.   4139 EL CAMINO WAY   PALO ALTO, CALIFORNIA 94306

ANNUAL REVIEWS INC.
Palo Alto, California, USA

International Standard Book Number: 0-8243-1313-5
Library of Congress Catalog Card Number: 63-8847

REPRINTS

The conspicuous number aligned in the margin with the title of each article in this volume is a key for use in ordering reprints. Available reprints are priced at the uniform rate of $1 each postpaid. Effective January 1, 1975, the minimum acceptable reprint order is 10 reprints and/or $10.00, prepaid. A quantity discount is available.

PRINTED AND BOUND IN THE UNITED STATES OF AMERICA

PREFACE

Annual Review of Phytopathology, the eleventh member of the Annual Review family that now numbers twenty one, was started in 1963. It is strongly international in content; approximately 43% of the papers in the first 13 volumes of this series were written by authors from 23 foreign countries. Great Britain, Canada, Australia, Germany, Japan, the Netherlands, and Israel have provided the most authors from outside the United States.

The first 13 volumes averaged 475 pages and included 254 papers by 344 authors on the variety of subjects indicated in Volume 6 (pages 453–56), Volume 11 (pages 556–59), and Volume 13 (pages 408–12).

A recent analysis (*Current Contents* 6 (2):5–9, 1975) of 50 world botanical journals found that, in the number of botanical articles cited, *"Annual Review of Phytopathology* . . . ranked second . . . , despite . . . extremely low self-citation rates. That says a great deal about the importance and the nature of review material." The journal was ranked first in the average citation rate by all journals of the articles we published ("overall impact") and in "botanical impact" (citation by botanical journals). In another analysis (*Current Contents* 6(4):5–9, 1975), of 41 journals that published 101 highly cited botanical papers in 1961–1972, *Annual Review of Phytopathology* ranked tenth in frequency as the source journal.

Plant pathologists should be aware of the uniqueness of Annual Reviews Inc. It is something quite different from the many commercial publishers, who may also provide good journals in which to publish and may be good sources of literature reviews. Annual Reviews Inc., a nonprofit organization, is closely involved in the scientific community, which provides its Board of Directors, Editorial Committees, and editors, authors, and readers.

The content of *Annual Review of Phytopathology* is determined by an Editorial Committee selected from members of the American Phytopathological Society, rather than by the staff of Annual Reviews Inc. The profession is thus provided a journal responsive to its interests and needs without itself having to publish it. Furthermore, special sales arrangements are often made by Annual Reviews Inc. through various professional societies for the benefit of the societies and their members. We trust that our own list of subscribers will grow and enable us to continue to provide this service to the profession.

We welcome R. M. Lister as the new member of the Editorial Committee replacing T. O. Diener, who concludes his five-year tenure, and we thank W. F. Rochow and A. F. Sherf, guest committeemen who helped plan this volume.

THE EDITORIAL COMMITTEE

CONTENTS

SOME RELATED ARTICLES IN OTHER ANNUAL REVIEWS

From the *Annual Review of Biochemistry,* Volume 43, 1974
Fungal Sex Hormones, G. W. Gooday, 35–50
Biochemistry of Bacterial Cell Envelopes, V. Braun and K. Hantke, 89–122
Bacterial Transport, W. Boos, 123–46
The Electron Microscopy of DNA, H. B. Younghusband and R. B. Inman, 605–20
The Selectivity of Transcription, M. J. Chamberlin, 721–76

From the *Annual Review of Ecology and Systematics,* Volume 5, 1974
The Evolution of Weeds, H. G. Baker, 1–24
Closed Ecological Systems, F. B. Taub, 139–60
The Measurement of Species Diversity, R. K. Peet, 285–308

From the *Annual Review of Entomology,* Volume 20, 1975
Federal and State Pesticide Regulations and Legislation, E. Deck, 119–32

From the *Annual Review of Genetics,* Volume 8, 1974
Biochemical Genetics of Bacteria, J. S. Gots and C. E. Benson, 79–102
Genetics of Amino Acid Transport in Bacteria, Y. S. Halpern, 103–34
On the Origin of RNA Tumor Viruses, H. M. Temin, 155–78
Fungal Genetics, D. G. Catcheside, 279–300
Genetics of DNA Tumor Viruses, W. Eckhart, 301–18

From the *Annual Review of Microbiology,* Volume 28, 1974
Viroids: the Smallest Known Agents of Infectious Disease, T. O. Diener, 23–40
Bacterial Photosynthesis, W. W. Parson, 41–60
Fungal Cell Wall Glycoproteins and Peptidopolysaccharides, J. E. Gander, 103–20
Regulation of Enzyme Activity During Differentiation in Dictyostelium discoideum, K. A. Killick and B. E. Wright, 139–66
Inclusion Bodies of Prokaryotes, J. M. Shively, 167–88
Bacterial Genetics Excluding E. coli, M. Levinthal, 219–30
Mycoplasmas and Rickettsiae in Relation to Plant Diseases, K. Maramorosch, 301–24
Cyclic AMP in Prokaryotes, H. V. Rickenberg, 353–70
Microscopes, Microscopy, and Microbiology, R. Barer, 371–90

From the *Annual Review of Plant Physiology,* Volume 26, 1975
Plasmodesmata, A. W. Robards, 13–30
Carbohydrates, Proteins, Cell Surfaces, and the Biochemistry of Pathogenesis, P. Albersheim and A. J. Anderson-Prouty, 31–52
The "Washing" or "Aging" Phenomenon in Plant Tissues, R. F. M. Van Steveninck, 237–58

ANNUAL REVIEWS INC. is a nonprofit corporation established to promote the advancement of the sciences. Beginning in 1932 with the *Annual Review of Biochemistry*, the Company has pursued as its principal function the publication of high quality, reasonably priced Annual Review volumes. The volumes are organized by Editors and Editorial Committees who invite qualified authors to contribute critical articles reviewing significant developments within each major discipline.

Annual Reviews Inc. is administered by a Board of Directors whose members serve without compensation.

Annual Reviews are published in the following sciences: Anthropology, Astronomy and Astrophysics, Biochemistry, Biophysics and Bioengineering, Earth and Planetary Sciences, Ecology and Systematics, Entomology, Fluid Mechanics, Genetics, Materials Science, Medicine, Microbiology, Nuclear Science, Pharmacology, Physical Chemistry, Physiology, Phytopathology, Plant Physiology, Psychology, and Sociology. The *Annual Review of Energy* will begin publication in 1976. In addition, two special volumes have been published by Annual Reviews Inc.: *History of Entomology* (1973) and *The Excitement and Fascination of Science* (1965).

Oil painting by Deane Keller

James S. Horsfall

FUNGI AND FUNGICIDES   ❖3608
The Story of a Nonconformist

James G. Horsfall
The Connecticut Agricultural Experiment Station, New Haven, Connecticut 06504

The Editorial Committee has granted me the high honor of addressing you, my friends, in the pages of the *Annual Review of Phytopathology.* Had they not done so, I would never have had the courage or the necessary immodesty to do it somewhere else.

They told me to be autobiographical, philosophical, or both. Being a cornfield philosopher and a ham at heart, I cheerfully accept the challenge. I do hope you like it.

Although I have wandered around extensively over the field of plant pathology, I have specialized in fungicides, and so I shall discuss that. Being by nature and nurture a nonconformist, I shall discuss that too.

THE PHILOSOPHY OF A NONCONFORMIST

After fifty years, man and boy, I still think that nonconformity is a great asset to a scientist. We must be curious to see if what we see is what we seem to see. We must analyze it, open it up, turn it over, look underneath it, and look behind. The conformist is simply not programmed for this.

Perhaps the harshest lesson a young scientist has to learn is that while nonconformity is a great asset in science, it is counterproductive in living. It can be a joy in science. It can be a heartache in living.

The herd instinct is strong in the human animal. An old aphorism describes it. "As one banana said to the other, you wouldn't have been stepped on if you had not got away from the bunch." Society has learned from psychologists to call the isolated banana an introvert.

Apparently I didn't inherit the herd instinct. As a child I was the stereotype of an introvert. I had to learn the hard way that isolated bananas can be stepped on. I found myself in the position of the man who woke up all bandaged in the hospital. He said, "I had the right of way." The nurse gently said, "But the other driver had a truck."

I had been playing a dangerous game. It could not continue. Therefore, I became a synthetic extrovert. I couldn't go all the way, to be sure. And so, I learned to set up side shows. Mycology was crowded, so I got into fungicides. Bordeaux mixture and elemental sulfur were heavily worked. Whereupon, I got into organics. When the emphasis was on field testing, I turned to mechanisms of action. When this got crowded, I returned to epidemiology and disease assessment, a field I have fiddled with for years.

THE PLEASURES OF PLANT PATHOLOGY

Lest the above sound too lugubrious, let me say, "Cheer up." The heartaches of nonconformity aside, I have had great fun—a great life in plant pathology. Plant pathology has been kind to me. It has taken me all over the world to make friends everywhere. I take this means to thank you all for your hospitality.

J. C. Walker introduced me once as a person who had fun with fungicides. I must say I was tempted to title this piece Fun With Fungi and Fungicides, but I finally decided to be a conformist and label it simply Fungi and Fungicides.

Roland Thaxter, the first Connecticut plant pathologist to have fun with fungicides (15), later in his life teased the plant pathologists by dubbing them "squirt gun botanists." I am sure he had his tongue in cheek because 18 years later he became a charter member of the American Phytopathological Society. Following Thaxter, I often call myself a squirt gun botanist. If we can't laugh at ourselves, we are stodgy indeed.

A PHILOSOPHY OF SCIENCE

My philosophy of science derives by alchemical osmosis from the policy of my institution. Its policy derives in turn from the title of a book, *How Crops Grow,* written in 1869 by S. W. Johnson, founder of the experiment station where I write.

At that time his experiment station was only a gleam in his eye, but he knew that the *quid pro quo* must be the payoff in the fields and granaries of the nation.

His thesis, frequently expounded, was that theoretical and useful science must march together. His title, *How Crops Grow,* is "clear, curt, and complete." The *how* denotes the theoretical, the *crops* denote the useful, *grow* denotes the mechanisms. I wish I could condense as much as well.

My philosophy of research matches Johnson's completely. It was expressed in the preface of my first book on fungicides (6). "It [the book] attempts to develop the underlying theory on which the practice is based, and by which the practice may be improved."

The modern terms, pure and applied or basic and applied, are only analogue terms that are not really equivalent to Johnson's theoretical and practical. One can do theoretical research on corn or wheat that is intellectually just as deep as that on *Chlorella.* Its snob value is lower but its scientific value is as high and it *can* get one a Nobel prize or elected to the National Academy of Sciences.

A very slight shift in the wording of Johnson's title gives us How to Grow Crops. There is no theory in this, and the research under it tends to be shallow and inadequate. In the early days many experiment stations and US Department of Agriculture tended to use this phrase as policy, however.

Those in academe who consider *crops* a dirty word have rephrased Johnson's title to How Plants Grow. This rephrasing has been heavily underscored since 1950.

The rephrasing of Johnson's title to How Plants Grow knocks out the useful and eventually knocks out the societal support. In the late sixties the policy generated by this title came under intense scrutiny by congressmen. Senator Allott of Colorado writing in *Science* in 1968 (1) said that, for some time he had been warning members of the scientific community that unless they explained to the taxpayer what he was buying with his research dollars there would be a severe cutback in funds allocated for research. And there was!

The National Science Foundation listened to Senator Allott and the rising tide of antiscience sentiment and established RANN—Research Applied to National Needs. They decided a century after Johnson that the voice of taxpayers is persuasive, and that the theoretical and useful should march together.

In my philosophical talk to the Phytopathological Society in the summer of 1969 (7), "Are We Smart Outside?," I urged us to attune our ears to Senator Allott and to return to a better balance between theoretical and useful. Some said I was a renegade who had deserted basic research, but like Mark Twain's premature obituary, this was "slightly exaggerated," however.

It is fitting that this epistle will appear in the late summer of 1975 because it was in the late summer of 1925, fifty years ago, that I shook the cotton field dust from my feet in Marianna, Arkansas, put on my shoes, and took the train to Ithaca, New York and Cornell University to work with the great H. H. Whetzel. My only recollection of the trip was that as the train from St. Louis clattered its way eastward across Indiana, a fellow traveler asked my destination. I told him Ithaca, New York. "That's where Ithaca guns are made," said he. I had never heard of Ithaca guns and he had never heard of Cornell University.

A half century of plant pathology has gone by. In 1925 mycology had us in its grip. In fact, I minored in mycology, having been forced to take all the courses anyway. It shortened my graduate student days somewhat, but perhaps I should have stayed longer and minored in organic chemistry. Who knows? It would have helped the theoretical side of my squirt gun botany.

In my student days Whetzel was dusting the apple orchards of New York State, Greaney the wheat in western Canada. Viruses were still "filterable viruses"; they had hardly escaped from the phrase, "contagium vivum fluidum." Mycoplasmas were a light year away. Fungicides were mercuries, lime sulfur, and Bordeaux mixture—that holy water of plant pathology. Farlow had called it *Eau Benite*.

Some of the giants of American plant pathology were Duggar, Edgerton, Jones, Melhus, R. E. Smith, "Erwinia" F. Smith, Stakman, and Whetzel. The latter pair always arranged a well-orchestrated squabble to attract members to the business meetings of the Society. Charles Walker, Joyce Riker, Max Gardner, and James Dickson were on their way up. Cap Weston from Mt. Olympus enlivened the banquets by needling the plant pathologists. The only sally that I can still remember was his speaking of the Boyce Thompson Institute in Rompers-On-Hudson. Abroad, some of the giants were Appel, W. Brown, Butler, Gäumann, Hemmi, Miyabe, McAlpine, Sorauer, and Westerdijk.

Those were experimentally simple days in 1925. We had never heard of low-volume spraying, captan, zineb, chromatography, NMR, electron microscopes, and the double-crossed helix. LSD meant least significant difference. The causal organism was called a pathogene, not pathogen. I have never found out why the final "e" was dropped.

The Coolidge boom was on. The flappers reigned. The Charleston was the dance. Harold Cook and I built a heterodyne radio from a kit. About the only stations we could tune in were KDKA in Pittsburg and WEAF in New York, and they crackled and fried as we listened.

The Model T Ford was still alive but gasping for breath. It had two more years to go before it died.

Surely plant pathology has moved ahead with exciting developments. We have grown out of the swaddling clothes of descriptive mycology. We are now concerned with mechanisms of action. We now enlist recruits from the ranks of those who fixed their mothers' alarm clocks, not from the ranks of the collectors of stamps and butterflies.

ANTECEDENTS

There must be something earlier than 1925. Yes, of course there was. My grandfather Horsfall belonged to a family of ship owners in Liverpool in England. But he, being a nonconformist, too, got crossways with his father and was banished to the US as a remittance man with £1000 per year. He eventually settled in eastern Arkansas because the bird hunting was good in the Mississippi River flyway. Nonconformity cost me only a delectable job or two. It cost him a fortune. Needless to say, we pulled few plums from the family pie.

When I was born, my father, poor as a churchmouse, was working in a tiny village in a tiny independent fruit experiment station in southwest Missouri. By arranging to be so born, I automatically inherited an interest in science.

My father saw no future in it, however, and departed to become a schoolman in his home state, Arkansas. He was an agricultural schoolman. We were so far down in southern Arkansas that you could chase a stray steer into the Mississippi River to the east or into Louisiana to the south. My father maintained his interest in horticultural research, however, eventually obtaining a Master's degree at the University of Missouri.

My father was his own type of nonconformist, but to that extent, he was a great mentor in science. He was wont to say, "Son, what is often accepted as fact is often not so." I can recall sitting on the front porch one Sunday afternoon. Seeing a cow in the pasture he asked me what color it was. "Black," said I. "How do you know?"

said he. "I can see," said I. "You can see only this side. How do you know it is black on the other side?" I had to make an experiment. I had to walk down and verify that it was black on the other side. Luckily, it was.

I recognized my first plant disease when my scientist father said our pear tree was dying of fireblight. We were advised to prune out the blighted shoots. After a few years of such Draconian treatment we had only a stub left. The treatment was a success but the tree died.

He sent me off to the University of Arkansas, where he had graduated. I wanted to be an engineer, having shown an aptitude to repair Model T cars. At the University I discovered, though, that engineering is the job of a mathematician not a grease monkey. My grades in mathematics were all right, but I didn't like math.

Enter Dwight Isely, an entomologist. He loved science and he stimulated me to love science. He provided me with pin money to impale his *Chrysomelids* on pins in Schmidt boxes. This bored me, but allowed me to take my girl to the movies. It was more exciting when he sent me two summers to Marianna, Arkansas to chase cotton boll weevils on horseback. I couldn't see the boll weevils from my horse, but I could see the yellow infested "squares." I never learned why the triangular flower bud of cotton is called a square.

Dwight Isely was pioneering the use of insect counts to determine when to dust the cotton. This is now called integrated control. He had the idea; all I did was chase the bugs. You can imagine my surprise 48 years later to hear from the stage of the auditorium of the National Academy of Sciences that I was the first "scout" to do so (Smith 14, p. 53).

Pioneer or not, no entomology department would give me a graduate scholarship. My nonconformity was definitely nonproductive here. H. R. Rosen and V. H. Young found me a place at Cornell in plant pathology, and that is how it came about that I left the cotton fields, never to return.

FUN WITH FUNGICIDES

By February 1929 graduate school was over and I was out in the cold world. By great good luck a few days ahead of time I was offered a job at the Agricultural Experiment Station in Geneva, New York. The station was larger, the town was larger, and the future was larger than my father had had a generation earlier at that little experiment station in the Missouri Ozarks.

The ink was still shining wetly on my diploma when two greenhouse growers of tomato seedlings came in to see this callow PhD. "Doctor," they said, (I had difficulty answering to that august title) "Can you soak tomato seeds in a copper sulfate solution and control damping-off?" Sensing that they already knew the answer, I responded that I thought so, and proceeded to test it experimentally. To my intense delight my first experiment was successful, and thus did I discover fun with fungicides. I was on my way to becoming a squirt gun botanist.

I was so pleased that I gave a paper (5) on the results at the Annual Meeting of the Phytopathological Society. And there on the front row sat the great L. R. Jones. I was so awed I could hardly speak, rare as that occasion is. Nevertheless, here was proof I needed. Fungicides interested one of the top men. Maybe it would provide a ladder for me.

From time to time I studied the base of Jones' success. It was clear that Jones carried water on both shoulders. He could encourage theoretical epidemiology on the one hand and cabbage breeding on the other. This was several years ahead of my exposure to the same principles in Connecticut.

To control damping-off was my practical aim. Where lay the theoretical? I stumbled along in search of some basics in damping-off. To describe the dynamics of damping-off and its control I needed four new terms: pre-emergence damping-off, post-emergence damping-off, inoculum potential, and pasteurizing soil. These found an audience. Plant pathologists were indeed interested in the theoretical as well as in squirting Bordeaux.

My pattern was complete. I would try to do both theoretical and applied research, and do both on crops and diseases that were important in my state.

Bordeaux Mixture

Like many other pathologists of the day, I began by squirting Bordeaux mixture. In the late twenties I used canning crop tomatoes that were subject to foliage blights. Bordeaux didn't work very well. The dry thirties came along. I continued to spray but the disease was scanty. This turned out to be a boon in disguise.

Suddenly I could see that Bordeaux mixture was deleterious to tomatoes. What was going on? Basic research was needed. I needed to understand. We found that stomates were closed by the spray, the middle lamellas were hardened, the leaf cuticles were weakened by the alkaline spray, and the plants were dwarfed. To find all this out was great fun.

For potatoes at that time, Bordeaux mixture was still king. "Spare the Bordeaux and spoil the potatoes," was the hue and cry of plant pathologists and entomologists. "It stimulated potatoes," they said. Being a nonconformist, I didn't believe it. Since it was so deleterious to tomatoes, the cousin of potatoes, how could it be stimulatory to potatoes? It must be deleterious to potatoes, too, but this was hidden by the bug and blight control.

Since the lime in Bordeaux mixture was a potent source of plant injury, we tried to substitute red and yellow cuprous oxides. Cuprous oxide was free of lime but could not control the insects on potatoes as well as Bordeaux mixture and, hence, could not succeed on potatoes.

Nevertheless, in the mid-forties I went to one of Harry Young's fungicide forays in Ohio. I spoke up to say that Bordeaux mixture on potato was a dead horse that had not yet fallen over. This was heresy and the local aficionados almost excommunicated me that night.

About the same time, a reviewer of my 1945 book wrote that, "not all will agree with the author's apparent attitude that organic fungicides will soon replace the inorganics, and that 'Bordeaux mixture and elemental sulfurs will be turned out to pasture to spend their last years in leisure for a good job well done.'"

Bordeaux mixture wasn't dead yet, of course. I was just an ebullient nonconformist.

Organic Fungicides

My philosophy about fungicides had changed a few years earlier with a personal near-tragedy. My small daughter had had serious inner ear infections from time to time. In 1937 she was miraculously healed from a dangerous infection with a new synthetic organic compound called sulfanilamide.

My mind was made up. I was bound to find useful organic fungicides for agriculture. My colleagues were pessimistic. Farmers wouldn't buy an organic compound at $1.50 a pound when they could buy Bordeaux for 6 cents. Can you imagine 6 cents for copper sulfate? I believed they would if the compound performed.

TETRACHLOROQUINONE With the help of Walter C. O'Kane of the Crop Protection Institute we persuaded the US Rubber Company (now Uniroyal) to try. They were as ignorant then of plant pathology as we were of rubber chemistry. Upon questioning we said that the then current dogma was that copper in Bordeaux mixture killed by an oxidizing action. They said that copper acts as an oxidant in rubber as well, and causes rubber to crack and deteriorate. Why not, therefore, try tetrachloroquinone, an organic pro-oxidant? In April 1938 we did. It worked, and in a couple of years farmers of New York State were clamoring to buy it at $1.50 per pound for treating pea seed. The price myth was exploded.

I never published it. US Rubber wouldn't release the chemistry, and until they did I wouldn't write a paper.

Eric Sharvelle was in our laboratory then. Along with Cunningham (2), he published it under a code number.

Unfortunately, we couldn't make it work in a squirt gun. It was not destined to be the nemesis of Bordeaux mixture. It hydrolyzed in the sun and the dew. 2,3-Dichloro-1,4-naphthoquinone did not deteriorate on the leaf, and it found commercial adoption for some foliage diseases, but still not for potatoes.

ETHYLENEBISDITHIOCARBAMATE I remember sitting in a cheap restaurant on 42nd Street just outside Grand Central Station in New York City one day in the fall of 1939 with my friend Donald F. Murphy of Rohm and Haas Company, talking about our joint work on cuprous oxide. I felt it was time for a change. I said to Murphy, "Let us try to develop organic fungicides. Sulfur is a fungicide. Let us try organic sulfur compounds."

Murphy persuaded his company. In January 1940, they sent us 100 samples to try. He-175 was among them. This went on to have a new code, D-14, and eventually was published in 1943 as disodium ethylenebisdithiocarbamate, or nabam. A potato fungicide was born, and Bordeaux mixture was in trouble.

Dimond and Heuberger were with us then; we published it together (3). The new compound was water soluble. It should not have protected potato foliage, but it did. Heuberger decided to stick it on better with the zinc analogue of Bordeaux mixture ($CaOH_2$ plus $ZnSO_4$). His mixture worked (4).

The company chemists were sure that the zinc salt was formed in the spray tank. If so, their patent was threatened. Before our eyes they dramatically made Heuberger's mixture in a beaker. A precipitate formed. They poured it on a filter paper.

Clear water came through. Presto! The insoluble zinc salt was on the paper, they said. They recommended that the research be dropped.

I was born in Missouri. I had "to be shown." We needed a little basic research. We needed to understand the mechanism. I went home, made Heuberger's mixture, poured it on the paper. A clear liquid came through, just as before. I put my trusty fungus spores on a slide with the precipitate, on another slide with the filtrate. The filtrate killed the spores. The precipitate did not. The zinc salt was not being formed in the tank; it was zinc hydroxide instead.

To learn this was more fun. I understood nature a little better. The patent would not be invalidated, and the study went on. It was only several years later that the explanation hit me. Since Heuberger's precipitate was zinc hydroxide, I guessed that it was less soluble than the zinc salt of my fungicide. I further guessed that the carbon dioxide from the sprayed leaf would convert the zinc hydroxide to zinc carbonate the first night, and that this would be more soluble than the zinc salt. This would permit the zinc salt to be formed.

Whereupon, I made another batch of Heuberger's mix and bubbled CO_2 through it. Presto! The toxicity was in the filter cake, not in the water that went through. This was practical physical chemistry.

The plant was a chemical factory. It produced the zinc salt in situ. I had understood a little more about nature.

Zinc ethylenebisdithiocarbamate plus a good insecticide rapidly replaced Bordeaux mixture on potatoes. And the dead horse did fall over. This was still more fun.

No longer did the cry go out to potato farmers, "Spare the Bordeaux and spoil the potatoes." The new fungicide replaced Bordeaux, not so much because it controlled *Phytophthora infestans* any better than Bordeaux, but because it did not dwarf the plants as much.

The work led us down a short trail. Zinc ethylenebisdithiocarbamate and its progeny, maneb, controls *Alternaria solani* much better than Bordeaux. The mechanism was a mystery for many years, but Lukens & Horsfall (11) showed that it liberates just enough volatile material to kill *Alternaria* spores when they are germinating at high humidity without free water. We understood nature a little better.

Chemotherapy

Before I went to New Haven in the summer of 1939 I had heard how the Dutch elm disease was marching down the streets of that well-named Elm City. In my naïveté, I imagined myself as Sir Galahad. It would be I who would find the Holy Grail, the control of disease in those stately and elegant elms.

Zentmyer came and shared the enthusiasm, quietly of course! We would solve it with chemotherapy. That was a new and untried procedure for elms. The fungus lay inside the tree. We would inject our medicine and search it out as sulfanilamide had searched out the infection pocket in my daughter's inner ear.

We filled beer bottles with our magic elixirs and attached them to the trees with rubber tubes. Since we bought the bottles from the Cremo Beer Company, we claimed we were doing Cremotherapy research.

We could never make it work usefully. Zentmyer got discouraged before I did, and returned to the sunny clime and chocolate-colored smog of southern California.

Dimond came back from Nebraska and struggled with it, but as of this writing the elm disease has done its work on most of New Haven's elms. We simply could not solve the problem in time. We found large numbers of useful compounds. The trees felt better after injection but died as soon as the dosing was stopped. Apparently the tree degraded the compound and the titer in the tree diminished too far.

Benomyl looks better. We hope it lives up to its promise. I think that our compounds failed for two reasons, (a) the tree degraded the compound and (b) the elm had no phagocytes. The compounds were fungistatic, not fungicidal. Penicillin is bacteriostatic, too, not bactericidal; it reduces the infection to a reasonable level. The phagocytes clean up the last stragglers and the patient recovers.

We are convinced that our compounds reduced the infection in the elm to a low level, too, but there were no phagocytes to clean up the last stragglers; the elm degraded the compound and did not recover from the disease.

In any event we had great fun and great excitement, even though we found only experimentally workable compounds. We still have hope.

FUN WITH FUNGI

H. M. Fitzpatrick was a warmhearted teacher of mycology when I was at Cornell, and besides he ate lunch with us graduate students in the "Domecon" cafeteria. He pitched pennies with us (and always won). He added his penny winnings to his private funds, took us on picnics and provided steaks, which we could never afford in the cafeteria.

I became so enamored with Fitzpatrick's fungi that I minored in mycology and helped publish three papers on fungi.

Probably my favorite fungus, however was *Alternaria solani.* On hearing that I was going to my first job at the Geneva Experiment Station, Charles Chupp asked me to control *Septoria* on tomatoes. *Septoria* was in my plots the first year in 1929, but it slowly faded away and was gradually replaced by *Alternaria solani.* I wish I knew why.

The Prodigal Son Principle

Alternaria solani provided me with a couple of examples of the prodigal son principle of research. Here my nonconformity shows again. At heart, I had always been fascinated with the points on the regression line that did not fit the expected curve. They are the prodigal sons. As the Bible says, I would kill the fatted calf for them.

This is why I have seldom worried much about standard deviation, analysis of variance, Chi square, T test, goodness of fit, etc. I like the points that don't fit, not the odds that the others do. Sometimes statistics are needed, however, to show that the prodigal son exists.

The first prodigal son in the *A. solani* case was the very rare "bull" plant that I saw occasionally in farmers' tomato fields. There was no statistical evidence for the one bull plant in 10,000, just your eye. They were sterile, produced no fruit, and were essentially immune to *Alternaria solani.* They stood starkly alone as islands of green in a sea of brown and defoliated plants. They were prodigal sons. Let us kill the fatted calf, celebrate, and cerebrate. Why were they immune?

We were able to produce bull plants by pinching off the blossoms. Obviously bull plants are not genetic freaks. We concluded that the fruits remove something from the leaves that makes them resistant (9). Came the war; we dropped the subject. Later John Rowell in his PhD thesis (12) showed that the substance pulled from the leaves by the fruits is a hexose sugar.

The Umbrella Principle

Rowell's discovery led me to speculate, as discoveries often do. The elucidation of the role of sugar in alternarial blight suggested a general principle. Perhaps sugar is related to resistance in other diseases. Let us call this an umbrella and attempt to see what we could bring under it. With Al Dimond's able assistance we raised such an umbrella (8).

Potato leaves with leaf roll are high in sugar and low in *Alternaria.* By the same token tomato leaves sprayed with maleic hydrazide are high in sugar and low in *Alternaria;* the roots are low in sugar and high in *Fusarium.* Roots of barley plants with yellow dwarf are low in sugar and often high in *Fusarium.* Old tomato leaves are low in sugar and high in *Alternaria.* A search of the literature reveals many low sugar diseases. One is pink root of onion.

This is just observation. What is the mechanism? Horton & Keen (10) showed that sugar inhibits the destructive enzymes of the onion pink-root fungus, and Sands & Lukens (13) have shown the same for *Alternaria* in tomato.

One of my favorite critics, reviewing a book of mine, discussed my concern with the umbrella principle in these words: "He leaps from crag to crag with the nimbleness of a mountain goat." I liked that.

Occam's Razor

I believe, too, in using Occam's razor in my research. We scientists love to think that processes are complex, and we set up wide arrays of assumptions to account for an observed phenomenon.

The Reverend William Occam (or Ockham) was an English cleric of the fourteenth century who proclaimed a class of logic holding that assumptions should be sliced to a minimum. I confess that I surely failed the Reverend Mr. Occam in trying to explain a bull tomato plant. I tried everything but the obvious. Rowell took the simplest case—a fruit is composed of big quantities of carbohydrates—therefore sugar must be the substance moved out of the leaves. Rowell didn't know that he was slicing through the brambles with Occam's razor, but he was.

FUN WITH WORDS

Style

Being a nonconformist, I have always tried to say it differently. It is fun and it gets me into trouble occasionally, but I still like it.

One critic calls it "a breezy and compelling style," while another, perhaps more tolerantly says that the "use of homely analogy and exposition to enlighten some of the more knotty technical problems makes for easy reading and comprehension." I hold that scientific writing should be comprehensible and easy to read. I could never abide the stodgy stilted style of much scientific writing. The English language is an elegant medium for saying exactly what one wants to say—no need to use any of the standard circumlocutions. Say it differently!

I guess it was this propensity that got me into the editing business, beginning with the college student magazine and continuing through a three-volume treatise into the editorship of the *Annual Review of Phytopathology.*

On account of the latter, the current editor, Baker, has asked me to set down here the history of that review.

Origin of the Annual Review of Phytopathology

The story can be traced back to 1951. In that year the council of the American Phytopathological Society rescinded the rule against printing theoretical papers in Phytopathology (*Phytopathology* 42:230. 1952). No longer were the pages limited to "original research" with laboratory and field gadgets. One could now do theoretical research with pen and paper. Einstein was not an experimental physicist. His tools were his intellect, a slide rule, pencil, and paper.

In 1956 (*Phytopathology* 47:320. 1957) the Society set up a Standing Committee on Phytopathological Reviews to solicit and edit reviews to be published in the journal. Members were Stevens, Shay, and Sill. The Committee reported in 1957 (*Phytopathology* 48:116–17 1958) that Annual Reviews Inc. had appointed Horsfall to represent plant pathology on the editorial committee for the *Annual Review of Plant Physiology.* They proposed that his name be added to the Committee on Reviews.

The Committee continued in 1958 (*Phytopathology* 49:162, 169–70. 1959). Stevens complained about the difficulty of obtaining reviews. The members of the Society had clearly not overcome their longstanding feeling that the only worthwhile research was to continue to fragment knowledge, not to put it together again. A review by Gäumann on fusaric acid was accepted and printed, however.

In 1959 (*Phytopathology* 50:243–251. 1960) the committee continued to complain about the difficulty of procuring reviews for *Phytopathology.* Stevens, its dynamic chairman, published his "Bibliography of Reviews" as a supplement to *Phytopathology.*

That year the committee structure was drastically altered for reasons that were not and are not clear to me. Horsfall was made chairman. The other members were Boyce, H. J. Jensen, Ross, and Sill.

In 1960 the Society, seeing a wind change, tacked again and appointed an additional committee—a Special Committee to Study Publications and Public Relations, with Horsfall as chairman, and including Dimock, Hayden, Hewitt, McCallan, P. R. Miller, Ross, Snyder, and Zentmyer (*Phytopathology* 51:45, 53–4 1961).

This committee met during the convention of its appointment at Green Lake, Wisconsin (it should have been called Dry Lake, Wisconsin), and recommended among other things "that the Society should look into the possibility of publishing an annual volume that would attempt to synthesize the knowledge extant in various fields."

We sent to the council in early March 1961 a proposed policy for a new Society publication to be called *Perspectives in Plant Pathology.*

This proposal uncovered an arrangement being worked out by a publishing house in Holland to publish commercially a series, *Recent Advances in Plant Pathology.* This information in turn generated a proposal from Annual Reviews Inc. that they would publish at their own expense an *Annual Review of Phytopathology.*

They requested the American Phytopathological Society to appoint an ad hoc advisory committee. This was done (*Phytopathology* 52:464, 482–83. 1962). Horsfall (chairman), Holton, Kelman, Pound, and Snyder. This group met on call of Annual Reviews Inc. in Palo Alto, California in January 1962.

Ludwig was added, and the group was appointed as the first editorial committee to set up the first volume. In late 1962 K. F. Baker was appointed Associate Editor. The rest of the history is in the volumes of the Annual Review.

THE MEN IN MY LIFE

It goes without saying that no story there would be to tell, had there not been a series of great colleagues who participated in the cooperative research. They did the research, and I did the cooperating. In more or less chronological order they were Z. I. Kertesz, R. O. Magie, Ross Suit, Eric Sharvelle, J. W. Heuberger, Neely Turner, A. E. Dimond, G. A. Zentmyer, G. A. Gries, R. A. Barratt, Saul Rich, Richard Chapman, David Davis, R. J. Lukens, and Paul Waggoner. They honed the experimental designs to a sharp edge. They picked the flaws and improved the philosophy. To them a thousand thanks.

THE END

And thus my tale is told. I come to the end of my epistle. If you have read this far, you are a friend indeed, and I am grateful. Please forgive me for my immodesty. When it comes your turn to do the prefatory chapter for the *Annual Review of Phytopathology,* you will understand.

Literature Cited

1. Allott, G. 1968. Research funds: friends in the Senate. *Science* 162:214–15
2. Cunningham, H. S., Sharvelle, E. G. 1940. Organic seed protectants for lima beans. *Phytopathology* 30:4–5
3. Dimond, A. E., Heuberger, J. W., Horsfall, J. G. 1943. A water soluble protectant fungicide with tenacity. *Phytopathology* 33:1095–97
4. Heuberger, J. W., Manns, T. F. 1943. Effect of zinc sulphate-lime on protective value of organic and copper fungicides against early blight of potato. *Phytopathology* 33:1113
5. Horsfall, J. G. 1931. Seed treatment for damping-off of tomatoes. *Phytopathology* 21:105
6. Horsfall, J. G. 1945. *Fungicides and Their Action.* Waltham, Mass.: Chronica Botanica Co. 239 pp.
7. Horsfall, J. G. 1969. Relevance: Are We Smart Outside? *Phytopathol. News* 3(12):5–9
8. Horsfall, J. G., Dimond, A. E. 1957. Interactions of tissue sugar, growth substances, and disease susceptibility. *Z. Pflanzenkr. Pflanzenschutz* 64:415–21
9. Horsfall, J. G., Heuberger, J. W. 1942.

Causes, effects and control of defoliation on tomatoes. *Conn. Agric. Exp. Stn. Bull.* 456:182–223
10. Horton, J. C., Keen, N. T. 1966. Sugar repression of endopolygalacturonase and cellulase synthesis during pathogenesis by *Pyrenochaeta terrestris* as a resistance mechanism in onion pink root. *Phytopathology* 56:908–16
11. Lukens, R. J., Horsfall, J. G. 1971. Spore germination and appressorial formation, a new assay for fungicides. *Phytopathology* 61:130
12. Rowell, J. B. 1953. Leaf blight of tomato and potato plants. *R. I. Agric. Exp. Stn. Bull.* 320. 29 pp.
13. Sands, D. L., Lukens, R. J. 1974. Effect of glucose and adenosine phosphates on production of extracellular carbohydrases of *Alternaria solani. Plant Physiol.* 54:666–69
14. Smith, E. H. 1972. Implementing pest control strategies. In *Pest Control Strategies for the Future. Nat. Acad. Sci.*, ed. R. L. Metcalf et al, 44–68. Washington DC: NAS. 376 pp.
15. Thaxter, R. 1890. Fungicides. *Conn. Agric. Exp. Stn. Bull.* 102. 7 pp.

SOME HIGHLIGHTS IN PLANT PATHOLOGY IN THE UNITED STATES

❖3609

J. C. Walker

Department of Plant Pathology, University of Wisconsin, Madison, Wisconsin 53706

As this chapter is published, 122 years have elapsed since De Bary (12), at the age of 22, published his epochal paper in Berlin, in which he refuted the age-old theory that fungi parasitic on plants were not distinct organisms, but resulted from tissue breakdown of what we now call the host plant due to adverse environment or malnutrition. Good evidence of this modern viewpoint had already been presented in France by Prévost (45), who published results of his study of wheat bunt in 1807. His views were ignored by Ré (46) in Italy in 1817, and by Unger (69) in Austria in 1833. Berkeley (6) in England had inclined to the same point of view as De Bary in writing on potato late blight in 1848. Tulasne (68) in France in 1853 was also in accord in his description of the powdery mildew fungus on grape..Final proofs of our modern point of view came slowly but steadily during the next few decades.

Midway in this 122-year period, I began my graduate studies in plant pathology in 1914 under Professor L. R. Jones at the University of Wisconsin. He laid much stress in his lectures upon the importance of the historical background of modern plant pathology, as did Professor H. H. Whetzel at Cornell University. In fact, we had a semester seminar on the subject nearly every other year. I mention this for the benefit of beginners in this science who may take time to read this chapter, since it may appear to them to place too much emphasis on old stuff.

Six years after De Bary's paper, a former United States Congressman from Illinois spoke before the Wisconsin Agricultural Society at the Milwaukee, Wisconsin State Fair. He put together some very profound facts in these two sentences: "Every blade of grass is a study; and to produce two where there was but one is both a profit and a pleasure. And not grass alone, but soils, seeds and seasons; hedges, ditches and fences; draining, droughts and irrigations; plowing, hoeing and harrowing; reaping, mowing and threshing;. . . saving crops, pests of crops; diseases of crops, and what will prevent or cure them;. . . the thousands of things of which these are specimens, each a world of study within itself."

15

This man was Abraham Lincoln, who was elected the following year to the presidency of the United States. In 1862 he signed the Morrill Act, which provided land grants to each state, the proceeds of which were to provide instruction in agriculture and mechanical arts. The Section of Mycology, which had been created in the Division of Botany in the United States Department of Agriculture in 1876, was renamed the Division of Vegetable Pathology in 1877. In the same year, the US Congress passed the Hatch Act, which made an annual appropriation of $15,000 to each state to found and support an agricultural experiment station. These stations were usually set up in association with agricultural education in one of the state colleges or universities. This support of agricultural research resulted in the appointment of a station botanist in many of the new experiment stations, and many of these men initiated programs of research on plant diseases in their respective states. Plant pathology received attention soon after 1890 in Kansas under W. A. Kellerman, in Indiana under J. C. Arthur, in North Dakota under H. L. Bolley, in Iowa under L. H. Pammel, in New Jersey under B. D. Halsted, in Alabama under G. F. Atkinson, in New York under F. C. Stewart, in Delaware under F. D. Chester, in Ohio under A. D. Selby, in Vermont under L. R. Jones, in Connecticut under R. T. Thaxter, and in Massachusetts under G. E. Stone.

Fifty years earlier, late blight of potato was epidemic in western Europe and in the northeastern quarter of the United States. The fungus, *Phytophthora infestans* (Mont.) De Bary, was undoubtedly transported originally from Central or South America to Europe. A little later, the grape powdery mildew organism, *Uncinula necator* (Schw.) Burr., was introduced from America to Europe; it was fortunately brought under control by the use of sulfur dust. At this time, the *Phylloxera* root louse was increasingly destructive on French grape cultivars. Because American grapes were much more resistant, they were transported commonly to Europe for use as root stocks. This led eventually to the introduction of the downy mildew organism, *Plasmopara viticola* (Berk. & Curt.) Berl. & DeT., from the eastern United States. The latter disease, while common, was not as destructive on American cultivars as it became in European vineyards, where it was first recognized in 1878. Sulfur dust, which was quite effective against powdery mildew, had no protective value against downy mildew. The fortunate discovery of the protective value of lime-copper mixtures led to the perfection of Bordeaux mixture, described by Millardet (40) in 1885. By this date the importance of greater knowledge of plant diseases and means for their control was recognized and received greater attention in both Europe and America.

Another disease that rapidly increased in importance in the eastern United States was peach yellows. Erwin F. Smith, a graduate of Michigan State University at East Lansing and one of the early plant pathologists in the US Department of Agriculture, studied this disease during this decade. He published a preliminary report in 1888 (55), and in 1891 a more complete account of his study on this disease (56). In 1889, he presented a portion of this work in a doctoral dissertation at the University of Michigan. It was a coincidence that his major professor asked L. R. Jones, a senior majoring in botany, to sit in as an auditor on Smith's final oral examination. This, no doubt, stimulated Jones' interest in plant pathology. It started a close lifetime friendship between the two men. The next year Jones began his professional career at the University of Vermont.

Early in this century, the increasing importance of plant diseases revealed the need for more attention to training of agricultural students in this field. During the first decade, departments or sections were set up for such training at the University of California under R. E. Smith, at the University of Minnesota under E. M. Freeman and E. C. Stakman, at Cornell University under H. H. Whetzel, and at the University of Wisconsin under L. R. Jones. At each of these institutions, graduate training in plant pathology was also stressed and the PhD degree soon became the usual attainment of those seeking positions as specialists in this field.

The American Phytopathological Society was founded in 1908, with 130 charter members. The first volume of its official journal, *Phytopathology*, appeared in 1911. Since that date the Society has gradually increased to a membership of about 2800. In 1958, it celebrated its fiftieth anniversary by special symposia at its annual meeting at Indianapolis. The papers presented at this meeting were published in a separate volume (26). In 1973, it was the host Society for the Second International Congress of Plant Pathology at the University of Minnesota. In 1972 it dedicated a permanent headquarters, built in conjunction with the American Society of Cereal Chemists at St. Paul, Minnesota. The first international review of plant pathology was initiated under the editorship of James G. Horsfall about 1961; the first volume of the *Annual Review of Phytopathology* was published in 1963 by Annual Reviews Inc., Palo Alto, California.

As early as 1911 the need was evident for carrying the results of agricultural research to the rural public. In that year, M. F. Barrus was appointed as Extension Specialist in Plant Pathology for New York State at Cornell University. At that time, there were no county agents, but Professor Barrus lectured at Farmers' Institutes and Grange Meetings. On May 8, 1914, Congress passed the Smith-Lever Act which established the Cooperative Extension Service and appropriated $4,100,000 payable to states over an eight-year period. R. E. Vaughan was appointed Extension Plant Pathologist in Wisconsin in July, 1915. By 1972, 47 states had at least one such position and many had three or more. An excellent account of the Cooperative Extension Service with special reference to plant pathology from its beginning to the present is provided by Sherf (53).

BACTERIAL DISEASES

Smith was able to transmit the peach yellows causal agent only by grafting. Kunkel (36) did not report insect transmission of this disease until 1933. Smith was unable to culture the infective agent, as Arthur (2) had done in 1885 with the fire-blight bacterium at the New York (Geneva) Experiment Station, verifying Burrill's (9) earlier contentions. I was told by L. R. Jones that Smith decided that, before he could make further progress with peach yellows, he must first become thoroughly familiar with bacterial diseases of plants. Thus began his brilliant career as a plant bacteriologist, which lasted the rest of his life. His monumental researches, published in great detail (63), are an outstanding highlight of plant pathology in the

United States, and gave this country the leadership in this area which it has since held. In the decade of 1890–1900, Smith published extensively on three bacterial vascular diseases, i.e. black rot of crucifers (59), bacterial wilt of cucurbits (57), and bacterial wilt of solanaceous plants (58). Wakker (71) published in 1883 a complete report of his studies on the bacterial vascular disease of hyacinths in Holland. Except for Wakker, European plant pathologists tended to cast doubt upon the importance of bacteria as plant pathogens. This culminated in a paper in 1897 by Alfred Fischer, Professor of Botany in the Königlichen Gymnasium at Leipzig, Germany (21). He denied the reliability of the results of Smith and others who claimed to have seen bacteria in plant cells. Smith (60, 61) replied to Fischer in 1899, pointing out his complete lack of knowledge of the subject. Fischer (22) replied the same year, and Smith followed with two rebuttal papers, the last in 1901 (62) published in German. This appeared to settle the controversy, and the role of bacteria as plant pathogens has expanded ever since.

The crown gall disease increased in importance. Numerous investigators in Europe and America between 1892 and 1904 attempted to isolate a causal organism without success. Smith and associates (64) took up a study of the disease in 1904, and by 1907 demonstrated a causal bacterium which they named *Bacterium tumefaciens,* and is now referred to as *Agrobacterium tumefaciens* (E. F. Sm. & Towns.) Conn. Smith (64) described the bacteria as intracellular, but Riker (48) corrected this by showing that they are intercellular in the host tissue. The resemblance of this disease in certain aspects to animal cancer stimulated much fundamental work in Smith's and Riker's laboratories, and later in other centers.

VIRUS DISEASES

As Smith became engrossed in his study of peach yellows, the tobacco mosaic disease received an increasing amount of attention. Smith believed that the two diseases had much in common insofar as the causal agents were concerned. Mayer (38) in Holland tried without success all known methods to isolate a causative bacterium. However, the causal entity was much more easily transmitted than that of peach yellows. In 1892, Ivanowski (27) in Russia passed infectious tobacco juice through a bacteria-proof filter and thereby distinguished the pathogen from hitherto known bacteria. At the turn of this century, virus diseases were recognized as distinct from other infectious diseases of plants.

During the next 70 years, a tremendous amount of work has resulted in great advances in our understanding of virus diseases of plants. There is space here only to cite some of the highlights in this area over these seven decades. Allard (1) of the US Department of Agriculture determined certain physical and chemical properties of the tobacco mosaic virus. He stated: "That this pathogenic entity is highly infectious and is in some manner reproduced within the plant are established facts. If these facts are interpreted according to those fundamental principles upon which all our scientific conceptions in pathology and biology are based, that infectious diseases are associated with parasitism and that self-reproduction is a characteristic of living things alone, it must be admitted that the pathogenic agents responsible for the mosaic disease of tobacco must be parasites."

In 1895, Takata in Japan associated rice dwarf with a leaf hopper (44). Allard (1) reported in 1914 the transfer of tobacco mosaic virus by the tobacco aphid and the peach aphid. Because later workers failed to confirm this, it is now generally assumed that he had a mixture of tobacco and cucumber mosaic viruses, the latter now known to be readily transmitted by aphids, while tobacco mosaic is not. Boncquet & Hartung (8) in 1915 proved the transmission of sugar beet curly top virus by leaf hoppers. Bennett (5) in 1940 reported virus transmission by dodder. A compilation of viruses transmitted by various species of *Cuscuta* is given by Fulton (23). Transmission of the wheat-streak mosaic virus by the mite, *Aceria tulipae* Keifer, was reported by Slykhuis (54) in 1953. Grogan and associates (24) in 1958 reported the first case of transmission of a virus by a fungus, *Olpidium* and the big vein virus of lettuce. Hewitt and associates in the same year (25) were the first to demonstrate transmission of a virus by a nematode in the case of fanleaf disease of grape.

Early in this century investigators, using current cytological techniques, noted unusual intracellular bodies in tobacco tissue infected by the mosaic virus. Ivanowski (28) in Russia first described them in 1903. In 1921, Kunkel (35) described similar bodies in corn tissue infected with a mosaic virus. Plant pathologists were startled at the American Phytopathological Society meetings in Boston in 1921 when Nelson (42) of Michigan State University described bodies of a different nature in the phloem of mosaic-infected bean, clover, tomato, and potato plants and propounded a protozoan theory of causation. This idea was promptly refuted by other workers who showed that the "Nelson bodies" were normal, elongate nuclei, more or less suggestive of protozoa, commonly found in sieve tubes of many species of higher plants, and which had been well illustrated some years earlier in Haberlandt's textbook of botany.

In the same year (1921) Duggar & Karrer (17), published results of a thorough study of the size of the infective particles of tobacco mosaic virus. Duggar (16) stated in 1926 that he had become convinced that the "infectious agency is a particle of almost inconceivably small size, certainly too small to represent an organism with the usual characteristics." He also said: "Pasteur, as we all know, is commonly supposed to have disproved the idea of the spontaneous generation of life. As students of living beings and of their development, however, we know that he disproved no such thing, but rather he proved that known forms of life did not originate in a brief space of time in a sterilized nutrient fluid. This fact is of incalculable value and has prevented many false theories. Nevertheless, the proof which he offered would not operate to deter investigations that may take us nearer to an understanding of lower orders of life whether or not that may be synonymous with a higher order of chemical substances." How modern the last sentence sounds in the light of some very profound experiments performed today!

The next important breakthrough was in 1935 when Stanley (67) of the Rockefeller Institute in Princeton, New Jersey, announced that he had isolated a crystalline protein that possessed the properties of the tobacco-mosaic virus. He stated: "Tobacco-mosaic virus is regarded as an autocatylytic protein which, for the present, may be assumed to require the presence of living cells for multiplication." While Stanley considered the tobacco-mosaic virus to be a globulin, this was promptly corrected at the Rothamsted Experiment Station in England by Bawden and Pirie, who showed it to be a nucleoprotein. During the next three and a half decades, much has been learned about viruses of both plants and animals. Particles of various sizes and shapes have been associated with virus diseases with the aid of the electron microscope. In all plant viruses studied for many years, the nucleic acid was consistently found to be of the ribonucleic (RNA) type. Only recently, Shepherd and associates (50–52) at the University of California at Davis showed the virus of cauliflower mosaic to be of the deoxyribonucleic (DNA) type, as are some viruses affecting animals. Recently (1971) Diener (13, 14) of the US Department of Agriculture described a single-stranded free RNA as the causal entity of potato spindle-tuber virus. He suggested this as a new class of pathogens to which he gave the name "viroid." Likewise, Japanese workers recently described particles in certain diseases regarded as virus in nature to resemble the mycoplasma found in connection with certain diseases of animals. The latter have been grown in culture, but this has not been accomplished with the plant pathogenic agents. It is apparent that great strides have been made in the study of virus diseases of plants since 1900, and a goodly amount of this has been done in the United States.

DISEASE CONTROL

A forward surge in disease control was initiated by the development of Bordeaux mixture in 1885. Its use was taken up rapidly in the United States as a protective fungicide for several hitherto poorly controlled diseases. These included potato late blight, apple scab, and several other foliage blights. By 1900 it was realized that for some crops the phytotoxicity of this fungicide often outweighed its protective value. Some of the effects noted were: blemishes on the fruit, as in apple; foliage injury, as in peach; retarded growth, as in cucumber; excessive transpiration resulting in drought injury, as in ginseng and some vegetables; blossom drop resulting in delayed production, as in tomato. As indicated earlier, the use of sulfur dust for powdery mildews was an established practice by 1860, but its usefulness for other types of disease was practically nil. The mixture of lime and sulfur in various forms had been tested without much success until 1908, when two important breakthroughs were reported. Cordley (10) reported success in apple scab control with boiled lime and sulfur. This in many areas replaced Bordeaux mixture on apples. When this mixture is boiled, the chief constituents are calcium thiosulfate and calcium polysulfides. A large part of the polysulfides is converted to finely divided free sulfur after it is sprayed on the leaf. It was still too toxic to tender peach foliage. Scott (49) of US Department of Agriculture, working chiefly in South Carolina, also developed in 1908 a mixture he called self-boiled lime sulfur. Water was added in small amounts to burnt lime (CaO) and as heat was generated, sulfur was added slowly with vigorous mixing. When boiling ceased, more water was added, the final mixture contained 8 pounds of finely ground sulfur, 8 pounds of burnt lime, and 50 gallons of water. Both boiled and self-boiled lime sulfur were eventually replaced by improved sulfur sprays such as flotation sulfur, wettable sulfur, Kolofog®, and Kolospray®.

The next important development in foliage sprays was the development of organic fungicidal sprays. The first of these was patented by Tisdale and Williams of the duPont Company in 1934; these were derivatives of dithiocarbamic acid. The first of these was thiram, which was most satisfactory as a seed treatment. Metal derivatives such as ferbam (containing iron) and ziram (containing zinc) were more successful as foliage sprays. Dimond and associates (15) at the Connecticut Experiment Station in 1943 reported the fungicidal value of sodium ethylene bisdithiocarbamate (nabam). A zinc and a manganese derivative of nabam, known respectively as zineb and maneb are widely used. The above discoveries stimulated search for more effective spray materials, and many organic compounds have come into use. In general, the organic compounds are less phytotoxic than the earlier inorganic compounds. All those in use on deciduous fruit trees up to 1972 are discussed by Lewis & Hickey (37). All foliage fungicides developed, however, until about 1970 were locally protective, but not systemic. When benomyl was released by the duPont Company, it proved to be another breakthrough because it is successfully systemic. At this writing, its full potential has not been completely explored, but it and newer compounds with similar properties bid fair to revolutionize control of plant foliage and vascular diseases, as well as some subterranean maladies. The status of this and several other systemic fungicides was reviewed by Erwin (19) in 1973.

While the events related in the last two paragraphs were occurring, a similar line of progress was evident in the use of seed, bulb, and tuber disinfectants. Although copper sulfate was effective against certain grain smuts, it was also quite phytotoxic. The addition of lime to copper sulfate powder was common in the United States by 1862. Loose smuts of wheat and barley were not controlled until Jensen (29) in Denmark developed the hot-water treatment, which became standard for many decades and later was extended to cabbage and tomato seeds. The development of organic mercury compounds was another breakthrough about 1920, but eventually these were ruled out because of their poisonous nature to animals and humans. Other organic compounds replaced them. Disinfestation of seeds which were internally infected, as in loose smuts of wheat and barley, by oxathiin compounds was shown in 1966 (18, 70) to be effective in seed treatment against loose smut of barley.

At the turn of the twentieth century, the importance of soil-borne plant pathogens was beginning to be more fully recognized. In the United States the rapid increase in *Fusarium* wilt diseases of cotton, flax, tomato, cabbage, and other crops were causing severe losses. Root-rot diseases, such as tobacco root rot (*Thielaviopsis basicola* (Berk.) Ferr.), *Pythium* root rot, and *Rhizoctonia* root rot of various crops, and various nematode diseases, required attention. The chemicals succeeding against seed-borne and foliage-infecting organisms were not successful against these soil-borne pathogens. In intensive culture, such as glasshouse crops and in outdoor seed beds, heat was an old remedy. Steam under pressure became most practical and commonly used soon after 1900. A marked improvement was devised about 1960 by Baker & Olsen (3) in which aerated steam is employed.

Among the early chemicals used were chloropicrin and formaldehyde: each had decided disadvantages both in phytotoxicity and in difficulties in application. About 1942, a decided breakthrough was experienced, especially with nematode control, in the discovery of three organic compounds. The first of these is widely known as D-D®, a mixture of 1,3-dichloropropene and 1,2-dichloropropane. The second is ethylene dibromide, commonly known as EDB®. The third is methyl bromide, commonly referred to as MB. DD and EDB have high boiling points and volatilize slowly in the soil, and thus can be applied successfully without sealing the surface. They can be applied in the plow sole or by means of special chisels. They came rapidly into use, especially for nematodes, but have low fungicidal value. MB has a low boiling point and is used as a gas or dissolved in propene dichloride, carbon tetrachloride, or xylene. It is especially useful in cold frames, hot beds, and green-house benches where the soil can be covered by gas-proof plastic tarpaulins for periods long enough to secure maximum effectiveness. MB is quite effective against soil-inhabiting pathogens. Other compounds discovered since 1940 used effectively as soil disinfectants or as protectants for roots or underground stems include Vapam® (sodium N-methyldithiocarbamate), Mylone® (3,5-dimethyl-1,3,5,2H-te-trahydrothiadiazine-2-thione), Vorlex® (chlorinated methylisothiocyanate), Terra-clor® (pentachloronitrobenzene), Dexon® [p-(dimethylamino)benzenediazo sodium sulfonate], and Trizone® (methyl bromide 60%, chloropicrin 30%, 3-bromopropyne and related compounds 9%).

While plant pathologists were busily engaged in programs referred to above, relatively little attention was paid to the introduction and long-distance transfer of important pathogens. The "gifts" of grape powdery mildew and grape downy mildew from America to Europe have already been cited. Our powdery mildew of gooseberry, *Sphaerotheca mors-uvae* (Schw.) Berk. & Curt., was transferred from America to southwestern Russia about 1890, and spread rapidly throughout western Europe where it became a far more destructive disease than in America. Snapdragon rust (*Puccinia antirrhini* Diet. & Holw.), widespread in America, was first found in northern France in 1931, and by 1938 was widespread in Europe and was often very destructive.

What has come to be known as white-pine blister rust (*Cronartium ribicola* J. R. Fischer) was common in western Europe on five-needle pines. The aecial stage was first recorded in 1854. Connection of the aecial stage on pine with the uredial and telial stages on *Ribes* was not established until 1905. With the rapidly growing interest in reforestation of eastern white pine (*Pinus strobus* L.) in the northeastern quarter of the United States the practice was established early in the century of collecting seeds and sending them to Europe for the growing of seedlings which were in turn shipped to America for planting. Because our eastern white pine was very susceptible, it was inevitable that pine seedlings would bring the pathogen with them. Thus, the disease was discovered at Geneva, New York in 1906 but was probably present in Maine since 1897. The fungus spread rapidly throughout the range of eastern white pine; because it was especially severe on young trees and *Ribes* was omnipresent, its effect was devastating.

An even more destructive fungus was introduced from the Orient a few years earlier. The chestnut blight fungus, new to science, was discovered in the New York Zoological Park in 1904 on an introduced plant. It was later found to be widely prevalent in China, where native species of chestnut were more tolerant than our American chestnut (*Castanea dentata* Borkh.), which was an important forest tree extending from New England through the Appalachian Mountains to North Carolina. Pycnospores of the fungus, *Endothia parasitica* (Murr.) P. J. Anderson and H. W. Anderson, are water-borne and distributed by birds, while ascospores discharged into the air are carried by wind currents. Within a few years the chestnut forests were, for all practical purposes, destroyed in spite of all attempts at control.

Citrus canker incited by the bacterium *Xanthomonas citri* (Hasse) Dows., first observed in Florida in 1912, threatened the citrus industry there. It had been introduced from the Orient on nursery stock. It was eradicated in Florida, where prompt and strict state measures were enforced. Canker was found on more than 500 properties in 26 counties. About three million trees were destroyed, but the pathogen was permanently eradicated from the state.

Up until 1912, there was no policy of federal plant quarantine in the United States. California had set up a strict state quarantine in 1881. It took the white pine and chestnut episodes to impel the US Congress to pass the Plant Quarantine Act of 1912, authorizing the Secretary of Agriculture to "condition, regulate, restrict, and prohibit the entry into the United States of nursery stock and other plants and plant products." Individual states followed with regulation of interstate traffic.

The sad commentary which American plant pathologists cannot be proud of is the entry of the Dutch elm disease pathogen (*Ceratocystis ulmi* (Buisman) C. Moreau) some years later. First described in Holland in 1921, its rapid decimation of elm in that country left no doubt as to its potential danger. To be sure, elm trees were successfully excluded by our quarantine laws, but our scientists overlooked the fact that the fungus lived for a long period in dead logs from which it was distributed by elm bark beetles. Its first appearance in Ohio in 1930 was a great shock. It was eventually found that elm logs had been imported for the veneer industry from Europe, unimpeded, for some years previous to the outbreak. Because logs were not alive they were not controlled by plant quarantine laws. Logs had entered through several Atlantic and Gulf of Mexico ports. From some of the ports, they had been transported on open flatcars for several hundred miles. Some 14 distinct terminal areas were exposed, together with some 12,000 miles of railway rights-of-way, whence adult borers could leave the logs to deposit the pathogen in growing elms. This explained the appearance of the disease simultaneously in many widely separated localities. Thus, the story of the elm disease is a low light, rather than a highlight in plant pathology in the United States. As witness to this episode are thousands of large dead elm trees still standing throughout eastern and middle United States. What was once perhaps our most popular ornamental tree is nearing extinction.

The development of disease-resistant varieties has become a major means of disease control in the United States. Plant breeders and plant pathologists have independently or cooperatively kept continuous programs going. The degree of success and the rate of progress in any given case is conditioned in part by the availability of resistant germ plasm and the variability in pathogenicity of the causal entity.

Variability in rusts and smuts was experienced early. Eriksson in Sweden had in 1894 set up several pathogenic forms of the wheat black rust organism, which were indistinguishable morphologically, but were distinctly different in the genera or species of grains that they were able to infect. Barrus (4) found two pathogenic strains of the bean anthracnose organism in 1911, based on their ability to attack certain bean varieties. In 1917, Stakman & Piemeisel (66), and in the following year Melchers & Parker (39), described distinct races within Eriksson's form *Puccinia graminis tritici.* These beginnings of extensive studies of races of this and other species based on host reaction were led by Stakman at the University of Minnesota and supplemented by others, especially at Iowa State University and Kansas State University. The US Department of Agriculture has maintained a central rust laboratory at the University of Minnesota which moved into new quarters in 1972. One basis of variability was clarified when Craigie (11), who had worked with Stakman as a graduate student and returned to the rust laboratory in Winnipeg, discovered heterothallism in this fungus. An immense amount of research has followed on the nature of variation in pathogenicity in obligate parasites, and in facultative saprophytes.

Another monumental work in this area is that initiated in the genus *Fusarium* by Snyder & Hansen at the University of California, Berkeley. This genus contains many important pathogens. Wollenweber in Germany divided the genus into several sections, a system that was followed generally for 30 years. Snyder & Hansen (65) in 1940 pointed out that variability within species and within single-spore lines was great enough to invalidate many species distinctions set up by Wollenweber. They proposed that species be based on consistent morphological differences, and that biotypes within such species be designated as forms and races. As an example, all of the vascular-infecting types were morphologically identical, but differed widely in pathogenicity. According to the Snyder & Hansen system, they all belonged to *Fusarium oxysporum.* Within the species, forms were set up, based on host genera or species. Within forms, races were defined on the basis of selective pathogenicity usually between cultivars of a given host species. There is not space here to enumerate the many works on the genetics of pathogenicity in fungi and bacteria that have followed.

Breeding for resistance in cereals was activated in this country about 1900 and has continued aggressively since. About the same time, breeding for resistance to the vascular fusarium wilts began. This was first done with cotton, cowpea, and watermelon in the southeastern states by Orton (43). This was soon followed by successful work with flax wilt by Bolley (7) in the Red River Valley. Essary (20), began work with tomato wilt in Tennessee before 1910, and Jones (32) initiated work with cabbage yellows in Wisconsin in the latter year. In the last seven decades, many major diseases of annual and biennial crops have been brought under control by this method. Space does not permit going into further details. Many review articles are available, the most recent comprehensive one being by Roane (47) in 1973.

HOST-PARASITE INTERACTIONS

This paper so far has dealt chiefly with various incidents in the discovery and description of diseases, and with the various measures devised for their control. Equally important with these have been the fundamental researches into the nature of host-parasite interactions. This has been a steady, persistent effort carried on in many laboratories and it is difficult indeed to pick specific highlights. I mention only a few that have impressed me over the years, but the reader must understand that I will undoubtedly overlook and omit some very important ones.

I may be excused for starting with my revered professor and leader, L. R. Jones. During his first decade at Vermont he found time among his many duties to carry on a research program in which he wrote a dissertation presented about 1903 to the University of Michigan for his doctor's degree. This had to do with the nature of the host-parasite relation of soft rot of carrot, incited by a new species of bacterium, now known as *Erwinia carotovora* (L. R. Jones) Holland. A preliminary paper in German was published in the *Zentralblatt für Bakteriologie* in 1901 (30), and a more complete paper appeared as a bulletin of the Vermont Experiment Station in 1909 (31). This was one of the first papers published in the United States that attempted to explain how a pectolytic enzyme excreted by a pathogen brought about the breakdown of storage tissue of the host plant.

When Jones moved to Wisconsin in 1910, he was well aware that more emphasis must be placed on research into the basic nature of plant disease, and that the period of merely describing diseases and their causal organisms had passed its peak. In his 20 years of experience with late blight of potato and its response to environment in the northeast, he came to realize that we needed more than casual observations of disease occurrence from year to year in order to understand them more completely. On coming to Wisconsin, he chose soil temperature in relation to plant disease development as a place to start. To carry out more exact experiments, the Wisconsin soil temperature tank was devised, largely through the ingenuity of James Johnson, in which plants could be grown in infested soil at a series of constant soil temperatures. The first studies were with *Fusarium* wilts, e.g. flax wilt, tomato wilt, cabbage yellows, and with tobacco root rot. Papers on *Fusarium* root rot of corn and wheat, potato scab, *Rhizoctonia* disease of potato, onion smut, and other diseases followed. A comprehensive review of these early studies was published in 1926 as a Wisconsin Agricultural Experiment Station Research Bulletin (33). Beginnings were made in the studies of the relation of air temperature and humidity to disease development. These studies have expanded since at many institutions, one of the most successful early developments being at the University of Nebraska.

I mention only a few more programs of basic research in the United States. I have already cited several cases in virus disease research and in variability in the selective pathogenicity of infectious organisms. This latter subject is, of course, intimately connected with the nature and inheritance of disease resistance. For the latter, I may refer the reader to the last chapter in my text book, *Plant Pathology* (72). The success in so many cases of developed varieties with resistance to one or more pathogens has naturally attracted attention to the nature of disease resistance. This is not an easy problem, and I can only mention briefly a few basic facts brought forth.

Work at Wisconsin from 1928 to 1954 on several diseases of onion, on clubroot of crucifers, and tomato wilt were reviewed in 1955 by Stahmann and me (73). Early in this period, resistance in onion to several diseases of the bulb associated with bulb color was studied. The dry outer bulb scales of colored varieties were shown to contain water-soluble simple phenols which were associated with the water-insoluble flavone, quercetin. These diffused into the infection drop and prevented germination of spores of *Colletotrichum circinans* (Berk.) Vogl., incitant of the smudge disease, and of *Diplodia natalensis* Pole-Evans, incitant of *Diplodia* rot. White varieties are susceptible and colored varieties resistant to these two pathogens. On the other hand, the black-mold fungus, *Aspergillus niger* V. Tiegh., was not inhibited by the phenols, and attacked all varieties successfully. *Botrytis allii* Munn, *B. byssoidea* Walker, and *B. squamosa* Walker usually invade the bulb in the neck tissue wounded at harvest. All are sensitive to the phenols, but the effectiveness of the latter is erratic. If the spores reach the wounded neck tissue without contacting the dry outer scales, infection proceeds; otherwise it is prevented. Under some weather conditions, there is little evidence of resistance; in other cases it is pronounced.

Pungency of onion bulbs is due largely to allyl sulfide, a volatile substance. This was shown to be very toxic to the bulb pathogens. However, once the pathogens had penetrated the fleshy scale, where the color compound was visible only in the outer epidermal layer, they proceeded equally well in colored and in white varieties. The success of resistance was due to the preformed phenols which could diffuse readily into the infection drop, while it could not in the living scale tissue. Moreover, although the sulfide was equally toxic or more so, it did not prevent infection and disease progress. These researches led the investigators to emphasize that the presence of a toxic material in the healthy tissue did not assure that resistance would be expressed. The relation of the pathogen to host cell and its progress therein was all-important in determining whether or not the toxic material, preformed or otherwise, was functional in the resistant property. There have been many studies of the nature of host resistance with few tangible results to elucidate the nature of resistance.

As pointed out by Kuć (34) in 1972, "phytoalexins have been the object of continuing debate by plant scientists for more than three decades." Phytoalexins were redefined by Müller (41) in 1956 as "antibiotics which are produced as a result of the interaction to two different metabolic systems, the host and the parasite, and which inhibit the growth of microorganisms pathogenic to plants." These substances are not preformed as were those described above in onion, but occur as the result of invasion. Many such substances have been described abroad and in the United States, as summarized recently by Kuć (34), who points out that "even the repeated association of a metabolite or enzyme with a resistance reaction, however, does not prove it is responsible for resistance, since it is difficult to differentiate between cause and effect." This area is only one of many in plant pathology that are centers of active basic research in this country and abroad.

Occasionally in recent years I have heard it said that plant pathologists are engaging too much in basic research and are neglecting applied aspects of this science. With this I cannot agree. Both phases are equally important and must be pursued with vigor. Some persons, by training and temperament are happier, and thus more successful, in one phase than in the other. This is good, but plant pathology must remain a well-balanced science, and I am sure that this will continue to be the case.

Literature Cited

1. Allard, H. A. 1914. The mosaic disease of tobacco. *US Dep. Agric. Bull.* 40:1–33
2. Arthur, J. C. 1885. Proof that bacteria are the direct cause of the disease in trees known as pear blight. *Bot. Gaz. Chicago* 10:343–45
3. Baker, K. F., Olsen, C. M. 1960. Aerated steam for soil treatment. *Phytopathology* 50:82
4. Barrus, M. F. 1911. Variation of varieties of beans in their susceptibility to anthracnose. *Phytopathology* 1:190–95
5. Bennett, C. W. 1940. Acquisition and transmission of viruses by dodder (*Cuscuta subinclusa*). *Phytopathology* 30:2
6. Berkeley, M. J. 1846. Observations, botanical and physiological, on the potato murrain. *J. R. Hortic. Soc.* 1:9–34 (Reprinted as *Am. Phytopathol. Soc. Phytopathol. Classics* 8:13–40. 1948)
7. Bolley, H. L. 1901. Flax wilt and flax sick soil. *N. D. Agric. Exp. Stn. Bull.* 50:26–60
8. Boncquet, P. A., Hartung, W. J. 1915. The comparative effect upon sugar beets of *Eutettix tenella* Baker from wild plants and from curly top beets. *Phytopathology* 5:348–49
9. Burrill, T. J. 1879. (Remarks made in a discussion) *Trans. Ill. State Hortic. Soc. N.S.* 12:79–80
10. Cordley, A. B. 1908. The lime-sulfur spray as a preventive of apple scab. *Rural New Yorker* 67:202
11. Craigie, J. H. 1931. An experimental investigation of sex in the rust fungi. *Phytopathology* 21:1001–40
12. De Bary, A. 1853. *Untersuchungen über die Brandpilze und die durch sie verursachten Krankheiten der Pflanzen mit Rücksicht auf des Getreide und andere Nutzpflanzen.* Berlin: G. W. F. Müller (Engl. trans. R. M. S. Heffner, D. C. Arny, J. D. Moore in *Am. Phytopathol. Soc. Phytopathol. Classics* 11:1–95. 1969)
13. Diener, T. O. 1971. Potato spindle tuber "virus". IV. A replicating, low molecular weight RNA. *Virology* 45:411–28
14. Diener, T. O., Smith, D. R. 1973. Potato spindle tuber viroid. IX. Molecular-weight determination by gel electrophoresis of formylated RNA. *Virology* 53:359–65
15. Dimond, A. E., Heuberger, J. W., Horsfall, J. G. 1943. A water soluble protectant fungicide with tenacity. *Phytopathology* 33:1095–97
16. Duggar, B. M. 1929. The nature of mosaic diseases. *Proc. Int. Congr. Plant Sci.* (1926) 2:1231–42
17. Duggar, B. M., Karrer, J. L. 1921. The sizes of the infective particles in the mosaic disease of tobacco. *Ann. Mo. Bot. Gard.* 8:343–56
18. Edgington, L. V., Reinbergs, E. 1966. Control of loose smut in barley with systemic fungicides. *Can. J. Plant Sci.* 46:336
19. Erwin, D. C. 1973. Systemic fungicides: disease control, translocation, and mode of action. *Ann. Rev. Phytopathol.* 11:389–422
20. Essary, S. H. 1912. Notes on tomato diseases with results of selection for resistance. *Tenn. Agric. Exp. Stn. Bull.* 95:1–18
21. Fischer, A. 1897. *Vorlesungen über Bakterien.* Jena:Fischer. 186 pp.
22. Fischer, A. 1899. Die Bakterienkrankheiten der Pflanzen. *Zentralbl. Bakteriol. Parasitenk. Infektionskr. Hyg. Abt. 2.* 5:279–87
23. Fulton, R. W. 1964. Transmission of plant viruses by grafting, dodder, seed, and mechanical inoculation. In *Plant Virology,* ed. M. K. Corbett, H. D. Sisler, 39–67. Gainesville, Fla.: Univ. Fla. Press. 527 pp.
24. Grogan, R. G., Zink, F. W., Hewitt, W. B., Kimble, K. A. 1958. The association of *Olpidium* with big-vein disease of lettuce. *Phytopathology* 48:292–97

25. Hewitt, W. B., Raski, D. J., Goheen, A. C. 1958. Nematode vector of soilborne fan-leaf virus of grapevines. *Phytopathology* 48:586–95

26. Holton, C. S. et al., Ed. 1959. *Plant Pathology—Problems and Progress, 1908–1958*. Madison, Wis.: Univ. Wisconsin Press. 588 pp.

27. Ivanowski, D. 1892. Über die Mosaikkrankheiten der Tabakspflanze. *St. Petersb. Acad. Imp. Sci. Bull.* 35, (Ser. 4, 3): 67–70. (Engl. transl. by J. Johnson in *Am. Phytopathol. Soc. Phytopathol. Classics* 7:27–30. 1942)

28. Ivanowski, D. 1903. Über die Mosaikkrankheit der Tabakspflanze. *Z. Pflanzenkr.* 13:1–41

29. Jensen, J. L. 1888. The propagation and prevention of smut in oats and barley. *J. R. Agric. Soc. Ser.* 2, 24:397–415

30. Jones, L. R. 1901. *Bacillus carotovorus* n. sp. die Ursache einer weichen Fäulnis der Möhre. *Zentralbl. Bakteriol. Parasitenk. Infektionskr. Hyg. Abt. 2,* 7:12–21, 61–68

31. Jones, L. R. 1909. Pectinase, the cytolytic enzym produced by *Bacillus carotovorus* and certain other soft-rot organisms. *Vt. Agric. Exp. Stn. Bull.* 147:283–360

32. Jones, L. R., Gilman, J. C. 1915. The control of cabbage yellows through disease resistance. *Wis. Agric. Exp. Stn. Res. Bull.* 38:1–70

33. Jones, L. R., Johnson, J., Dickson, J. G. 1926. Wisconsin studies upon the relation of soil temperature to plant disease. *Wis. Agric. Exp. Stn. Res. Bull.* 71:1–144

34. Kuć, J. 1972. Phytoalexins. *Ann. Rev. Phytopathol.* 10:207–32

35. Kunkel, L. O. 1921. A possible causative agent for the mosaic disease of corn. *Hawaii. Sugar Plant. Assoc. Exp. Stn. Bull., Bot. Ser.* 3:44–58

36. Kunkel, L. O. 1933. Insect transmission of peach yellows. *Contrib. Boyce Thompson Inst.* 5:19–28

37. Lewis, F. H., Hickey, K. D. 1972. Fungicide usage on deciduous fruit trees. *Ann. Rev. Phytopathol.* 10:399–428

38. Mayer, A. 1886. Ueber die Mosaikkrankheit des Tabacs. *Landwirtsch. Vers. Stn.* 32:451–67. (Engl. transl. by J. Johnson in *Am. Phytopathol. Soc. Phytopathol. Classics* 7:11–24. 1942)

39. Melchers, L. E., Parker, J. H. 1918. Another strain of *Puccinia graminis*. *Kans. Agric. Exp. Stn. Circ.* 68:1–4

40. Millardet, P. M. A. 1885. Traitement du mildiou et du rot. *J. Agric. Prat.*

2:513–516. Traitement du mildiou par le mélange de sulphate de cuivre et de chaux. Ibid. 707–10. Sur l'histoire du traitement du mildiou par le sulphate de cuivre. Ibid. 801–5. (Engl. transl. by F. J. Schneiderhan in *Am. Phytopathol. Soc. Phytopathol. Classics* 3:7–25. 1933)

41. Müller, K. O. 1956. Einige einfache Versuche zum Nachweis von Phytoalexinen. *Phytopathol. Z.* 27:237–54

42. Nelson, R. 1922. The occurrence of protozoa in plants affected with mosaic and related diseases. *Mich. Agric. Exp. Stn. Tech. Bull.* 58:1–30

43. Orton, W. A. 1909. The development of farm crops resistant to disease. *US Dep. Agric. Yearb. Agric.* 1908:453–64

44. Ou, S. H. 1972. *Rice Diseases,* p. 6. Kew, England: Common. Mycol. Inst. 368 pp.

45. Prévost, B. 1807. *Mémoire sur la cause immédiate de la carie ou charbon des blés, et de plusieurs autres maladies des plantes, et sur les préservatifs de la carie.* Paris (Engl. Transl. G. W. Keitt in *Am. Phytopathol. Soc. Phytopathol. Classics* 6:19–95. 1939)

46. Ré, F. 1817. *Saggis teorica practico sulle mallattie delle piante.* 2nd ed., Milan: Silvestri. 331 pp. (English transl. of 2nd ed. by M. J. Berkeley in *Gard. Chron.* 9:211, 228–29, 244, 260. 1849)

47. Roane, C. W. 1973. Trends in breeding for disease resistance in crops. *Ann. Rev. Phytopathol.* 11:463–86

48. Riker, A. J. 1923. Some relations of the crowngall organism to its host tissue. *J. Agric. Res.* 25:119–32

49. Scott, W. M. 1908. Self-boiled lime-sulphur mixture as a promising fungicide. *US Dep. Agric. Bur. Plant Indus. Circ.* 1:1–18

50. Shepherd, R. J., Bruening, G. E., Wakeman, R. J. 1970. Double-stranded DNA from cauliflower mosaic virus. *Virology* 41:339–47

51. Shepherd, R. J., Wakeman, R. J. 1971. Observation on the size and morphology of cauliflower mosaic virus deoxyribonucleic acid. *Phytopathology* 61:188–93

52. Shepherd, R. J., Wakeman, R. J., Romanko, R. R. 1968. DNA in cauliflower mosaic virus. *Virology* 36:150–52

53. Sherf, A. F. 1973. The development and future of extension plant pathology in the United States. *Ann. Rev. Phytopathol.* 11:487–512

54. Slykhuis, J. T. 1953. The relation of *Aceria tulipae* Keifer to streak mosaic and

other chlorotic symptoms on wheat. *Phytopathology* 43:484–85

55. Smith, E. F. 1888. Peach yellows: a preliminary report. *US Dep. Agric. Bot. Div. Bull.* 9:1–254

56. Smith, E. F. 1891. Additional evidence on the communicability of peach yellows and peach rosette. *US Dep. Agric. Div. Veg. Pathol. Bull.* 1:1–65

57. Smith, E. F. 1895. *Bacillus tracheiphilus* sp. nov., die Ursache die Verwelkens verschiedener Cucurbitaceen. *Zentral. Bakteriol. Parasitenk. Infektionskr. Hyg. Abt. 2,* 1:364–73

58. Smith, E. F. 1896. A bacterial disease of the tomato, eggplant, and Irish potato (*Bacillus solanacearum* n. sp.) *US Dep. Agric. Div. Physiol. Pathol. Bull.* 12:1–26

59. Smith, E. F. 1897. *Pseudomonas campestris* (Pammel). The cause of a brown rot in cruciferous plants. *Zentralbl. Bakteriol. Parasitenk. Infektionskr. Hyg. Abt. 2,* 3:284–91, 408–15, 478–86

60. Smith, E. F. 1899. Are there bacterial diseases of plants? *Zentralbl. Bakteriol. Parasitenk. Infektionskr. Hyg. Abt. 2,* 5:271–78

61. Smith, E. F. 1899. Dr. Alfred Fischer in the rôle of pathologist. *Zentralbl. Bakteriol. Parasitenk. Infektionskr. Hyg. Abt. 2,* 5:810–17

62. Smith, E. F. 1901. Entgegnung auf Alfred Fischer's "Antwort" in betreff der Existenz von durch Bakterien verursachten Pflanzenkrankheiten. *Zentralbl. Bakteriol. Parasitenk. Infek-tionskr. Hyg. Abt. 2,* 7:88–100, 128–39, 190–99

63. Smith, E. F. 1905–1914. *Bacteria in Relation to Plant Diseases.* Vol. 1:1–285. 1905; Vol. 2:1–367. 1911; Vol. 3:1–309. 1914. Washington DC: Carnegie Inst.

64. Smith, E. F., Brown, N. A., Townsend, C. O. 1911. Crown-gall of plants: its cause and remedy. *US Dep. Agric. Bur. Plant Indust. Bull.* 213:1–215

65. Snyder, W. C., Hansen, H. N. 1940. The species concept in *Fusarium. Am. J. Bot.* 27:64–67

66. Stakman, E. C., Piemeisel, F. J. 1917. A new strain of *Puccinia graminis. Phytopathology* 7:73

67. Stanley, W. M. 1935. Isolation of a crystalline protein possessing the properties of tobacco-mosaic virus. *Science* 81:644–45

68. Tulasne, L. R. 1853. Note sur le champignon qui cause la maladie de la vigne. *C. R. Acad. Sci. Paris* 37:605–9

69. Unger, F. 1833. Die Exantheme der Pflanzen. Vienna: C. Gerold. 422 pp.

70. Von Schmeling, B., Kulka, M. 1966. Systemic fungicidal activity of 1,4-oxanthiin derivatives. *Science* 152:659–60

71. Wakker, J. H. 1883. Vorläufige Mitteilungen über Hyacinthenkrankheiten. *Bot. Cent.* 14:315–17

72. Walker, J. C. 1969. *Plant Pathology.* 3rd ed. New York: McGraw-Hill. 819 pp.

73. Walker, J. C., Stahmann, M. A. 1955. Chemical nature of disease resistance in plants. *Ann. Rev. Plant Physiol.* 6:351–66

PREDICTIVE SYSTEMS: ❖3610
MODERN APPROACHES
TO DISEASE CONTROL

R. A. Krause[1]

Department of Plant Pathology, The Pennsylvania State University, University Park, Pennsylvania 16802

L. B. Massie

New York Agricultural Experiment Station, Geneva, New York 14456

SCOPE

The ultimate goal of the science of plant pathology is a complete understanding of plant disease—host, pathogen, environment interactions—so that disease control can be obtained when necessary and economically warranted. Numerous plant diseases have been controlled without an understanding of the disease cycle, or occasionally without actual knowledge of the causal agent. However, the most efficient and most economical control is usually obtained by a thorough knowledge of disease epidemiology. For some diseases, epidemiological studies have led to the formulation of predictive systems which can forecast the occurrence of infection or disease. This advanced knowledge can be used to great advantage in reducing and timing chemical disease control measures necessary to obtain maximum disease control.

This review is not intended to be a categorical listing and in-depth discussion of all plant disease predictive systems, models, and simulators developed to date; indeed, excellent and up-to-date reviews of this nature are already available (8, 41–43). This review discusses several historical and modern predictive systems or models as examples of systems that can be used in applied, predictive disease control. Discussion focuses on the use of computers, modern systems of data acquisition and transmission, and the rapid analysis of such data as an integrated means of controlling plant disease. Examples are given of how such techniques may be

[1]Present address: Agricultural Advisors Inc., Yuba City, California 95991.

applied to existent empirical disease forecast systems that have proven to be accurate, but have not been widely accepted. Methods are also suggested for improving some historical and modern forecast systems so that they may become more useful on a localized or regional basis.

CLASSIFICATION AND OBJECTIVES OF PREDICTIVE SYSTEMS

It has been stated in other reviews (8, 43) that all predictive systems may be divided into two categories: (*a*) disease prediction and (*b*) infection prediction. This categorization is discussed and followed in this review.

Disease Prediction

Because infection always precedes symptom development, most predictive systems forecast the occurrence of disease symptom expression by monitoring certain environmental conditions known to be conducive to infection of the host plant by the causal organism. Once favorable infection periods have occurred, subsequent symptom expression is based on the assumption that certain environmental conditions favorable to colonization will proceed according to historically established weather trends. Some disease prediction systems take into account the amount of inoculum or disease present at a given time (14–16) or at a given time of plant maturity (22, 23). Other systems (3, 29, 32, 64, 68, 69) are based on the assumption that sufficient inoculum to initiate an epidemic is available whenever the crop is present. Predictive systems based on this assumption immediately incur a degree of inaccuracy. This degree of inaccuracy is reduced as the probability of inoculum being available is increased. It has been difficult to evaluate under field conditions systems that assume the presence of inoculum. For if the predicted disease does not occur, either sufficient inoculum was not available, or the predictive system did not measure and/or interpret the infection periods correctly.

Infection Prediction

Whereas disease predictions are only made *after* biological and meteorological conditions favorable for infection have been fulfilled, systems that predict infection periods must forecast, several days in *advance,* the occurrence of such biological and meteorological conditions. Although conditions required for infection and colonization are known for many plant diseases, the actual forecast of the conditions is very difficult. Conditions favoring infection and colonization, depending on the disease under investigation, are the magnitude, duration, and relationships among relative humidity, temperature, rainfall, dew periods, radiation, air movement, etc. Thus, the probability of accurately predicting conditions favorable for infection several days in advance are indeed low. Infection prediction systems should provide the most favorable means of disease control, if the disease is amenable to chemical control. Based on the fact that most plant disease control chemicals are protective rather than therapeutic, knowledge that an infection period is about to take place is more important for disease control than knowledge that an infection period has already taken place.

Development of Predictive Systems

Both disease and infection prediction systems may be categorized further according to the methods by which they were developed. Several terms such as "empirical," "fundamental," "deductive," "inductive," and "logical" have found their way into the literature.

EMPIRICAL PREDICTIVE SYSTEMS By tradition the term empirical or deductive has been used to describe predictive systems developed by studying and comparing historical records of disease occurrence and concurrent weather conditions in the same or approximate locality. Such systems usually result in the formulation of "rules" or specific meteorological conditions that must be fulfilled before disease development can take place. The fact that empirical systems are usually developed from comparison of historical disease and weather records for a specific area does not always mean that the applicability of the system is limited to the geographical area of its development. An empirically derived predictive system based on the interrelationships of biological and meteorological conditions affecting host-pathogen interactions should be applicable, with little modification, wherever the host and pathogen occur. If the system is not always applicable, then it must be assumed that all important factors affecting the host-pathogen interaction are not understood or taken into account. Therefore, after the development of an empirically derived predictive system, it should be evaluated in geographical locations with varying environmental conditions so that all variations of conditions affecting disease occurrence can be accounted for.

FUNDAMENTAL PREDICTIVE SYSTEMS Predictive systems developed from data obtained experimentally in the laboratory or field regarding the relationships of biological and environmental conditions governing host-pathogen interactions are referred to as derived, fundamental, inductive, or logical systems. The classification of disease prediction systems as fundamental or empirical is indeed an arbitrary process. Often a fundamental system has its origin as an empirical system. As the system was tested and modified to fit various geographic and environmental conditions, certain questions arose which had to be answered under controlled experimental conditions. As these questions were answered and the relationships defined mathematically, the system usually became known as a fundamental system. However, the mathematical definition of such relationships is no more than the *empirical* fitting of regression lines to those relationships. Once the lines have been fitted, the empiricism of the mathematical formulas is often forgotten. Nevertheless, the empiricism involved in such regression analysis is less problematical than the empiricism of nonmathematically defined predictive systems. The empiricism of mathematical systems can be analyzed statistically and modified objectively. The accuracy of a predictive system does not depend upon its method of derivation; accuracy depends upon how well the system has interpreted the biological and meteorological relationships that precede infection or disease development. Pragmatically speaking, the ultimate test of a predictive system is its accuracy in predicting infection or disease occurrence under both experimental and commercial conditions.

PREDICTIVE SYSTEMS: PAST AND PRESENT

Although it is not within the scope of this article to describe all predictive systems developed for plant diseases, some systems are reviewed in order to discuss their development, methodology, and application.

Empirical Disease Prediction Systems

More predictive systems have been developed to forecast the occurrence of potato late blight [*Phytophthora infestans* (Mont.) de B.] than any other plant disease. It is understandable that a disease, which is not easily controlled by genetic resistance and which can completely destroy a major food crop, should receive so much attention. The genesis of late blight prediction systems is excellently described and discussed in great detail in a review article by Miller & O'Brien (42).

The late blight forecasting system developed by Hyre (29–32) was modeled after Cook's system (17, 18). Hyre's system (29) is based on records of daily rainfall and maximum and minimum temperatures. The initial appearance of late blight is forecast 7–14 days after the first occurrence of 10 consecutive blight-favorable days. A day is considered to be blight-favorable when the 5-day average temperature is below 25.5° C and the total rainfall for the last 10-day period is $\geq$ 3.0 cm. Days on which the minimum temperature falls below 7.2° C are considered unfavorable for blight development. This system is known as the "moving graph" system because it is always based on the average temperature and total rainfall for the last 5 and 10 days, respectively.

Hyre's system was developed in northeastern United States (30, 31) and was found to be very accurate in predicting the initial occurrence and subsequent spread of late blight. However, under midwestern conditions Hyre's system was not as successful as under eastern conditions. Because this system assumes the presence of adequate inoculum to initiate an epidemic under favorable conditions, Hyre & Horsfall concluded that the system failed to operate accurately in the midwest because of the inconsistent occurrence of primary inoculum as opposed to the consistent occurrence of primary inoculum on the eastern coast (32). However, Wallin (66–68) has shown that Hyre's system does not work in the midwest because rainfalls conducive to blight development are infrequent and thus, RH is more important for disease development than rainfall. The senior author's experience with the Hyre system indicates that the system works in the eastern United States because of fortuitous relationships among rainfall, long periods of high RH, and blight-favorable temperatures. The actual conditions favorable to blight development are more likely leaf wetness and temperature, with leaf wetness being more closely related to high RH than rainfall. Therefore, it is easy to see how blight could develop in the midwest in the absence of rainfall.

Wallin (63) and Wallin & Waggoner (69) developed a system, modeled after the Beaumount method (3), to forecast the initial occurrence and subsequent spread of potato late blight based on RH and temperature. The Wallin system is based on "severity values" which are arbitrary values assigned to specific relationships between duration of RH periods $\geq$ 90% and the average temperature during those periods (57, 64) (Table 1). With this system the first occurrence of blight is predicted 7–14 days after the accumulation of 18–20 severity values recorded from the time of plant emergence. Thurston studied the system and concluded that it accurately predicted the first occurrence of blight and indicated subsequent infection periods and resultant disease increase. However, the system did not relate well to the rate of disease increase, as measured in his studies (57). Wallin's system is a good example of a predictive system which had its origins as an empirical or deductive system and was later modified by field and laboratory investigations (69) to increase its accuracy and application.

Empirical Infection Prediction Systems

The difficulty of accurately predicting the occurrence of infection periods already has been discussed. As the number of parameters necessary to accurately predict infection increases, the probability of making accurate meteorological forecasts for those parameters decreases. This is an important point because the accuracy of infection prediction increases, up to a point, as the number of biological and meteorological parameters used to make the prediction increases. For example, if a system could accurately predict the occurrence of infection based only on average ambient air temperatures, the total system would be very accurate because 7–14 day meteorological forecasts for average daily temperatures can be very accurate. However, if information on the number and duration of RH periods ≥ 90% and the average temperatures during those periods were needed to accurately power a predictive system, the results would most likely be very unreliable because it is very difficult to predict duration of RH periods even 24–48 hr in advance. Therefore, predictive infection systems depend greatly on the availability of accurate microclimate and macroclimate forecasting systems. Moreover, the relationship of macroclimate to microclimate is indeed difficult and poorly understood (20, 27, 28). Solutions to the above problems will greatly aid predictive disease control.

Disregarding the difficulties just discussed, several researchers have developed systems to predict infection periods by use of synoptic weather maps (6, 7, 50, 65). In the early 1950s Bourke (5, 7) developed a predictive system based on the use of surface synoptic weather charts for northwestern Europe. Correlations were made between periods of favorable late blight infection and surface weather charts during

Table 1 Relationship of temperature and RH periods as used in the Wallin late blight forecasting system

Average temperature range[1]	Severity numbers and Hours of 90%, or above, relative humidity				
	0	1	2	3	4
7.2–11.6°C	15	16–18	19–21	22–24	25+
11.7–15.0°C	12	13–15	16–18	19–21	22+
15.1–26.6°C	9	10–12	13–15	16–18	19+

[1] Average temperature of period when the relative humidity equals or exceeds 90%.

those periods. Two important synoptic situations which gave rise to blight situations were identified: (a) open waves of maritime tropical air, and (b) stagnant or slow-moving depressions giving lengthy periods of wet, overcast weather. When synoptic situations similar to these were seen developing, a spray warning was issued to all growers by newspaper and radio messages.

Wallin (65) also studied the use of synoptic weather maps as an aid to his plant disease prediction system already discussed. His objectives were to develop a means of anticipating conditions favorable for potato late blight for the entire northcentral United States. In this way he could minimize the delay of data acquisition and analysis and even project his forecast into the future. His results indicated that it was possible to use 3-, 5-, and 30-day maps of synoptic (prognostic) weather situations for both surface and high altitudes. However, he stated that such predictions should be used only as a supplement, not a substitute, to predictive systems based on weather data collected in the field.

Predictive systems based on synoptic weather maps are empirical systems suited to a large, but limited, geographical area. Therefore, a system developed for one area is not readily usable in another area. Furthermore, such a system cannot take into account geographical variations that can affect climatic conditions and thus affect infection and disease occurrence.

Fundamental Disease Prediction Systems

Eversmeyer & Burleigh (10, 22–24) have developed predictive disease systems (models) for wheat leaf and stem rust (*Puccinia recondita* Rob. ex Desm. f. sp. *tritici*, and *P. graminis* Pers. f. sp. *tritici*, respectively) based on the use of stepwise multiple regression equations. They have defined 15 biological and meteorological parameters that relate to disease increase after the date of disease prediction (DP). They found that for wheat leaf rust 6 of the parameters gave R^2 (coefficient of determination) values ranging from 0.722 to 0.527 and predicted severities $\pm$ 1, 3, and 12%, for 14-, 21-, and 30-day forecasts, respectively. For wheat stem rust the use of 8 parameters gave R^2 values of 0.745, 0.664, 0.509, and 0.362 for 7-, 14-, 21-, and 30-day forecasts, respectively. This work is extremely interesting and very promising especially when Eversmeyer, Burleigh & Roelfs (24) state that not all meteorological and biological variables and their interactions have been identified. Because this system is based on the use of data representing meteorological and biological relationships that have already occurred, application of protective fungicides subsequent to these events should result in disease control by impeding future disease cycles before economic losses result. A system such as this, aided by rapid data acquisition on a daily or weekly basis, and frequent updating of predicted values to conform to observed values, would be an ideal application of a practical, predictive disease control system.

Another fundamental disease prediction system has been developed by Schrödter & Ullrich (51–53). This system is based on stepwise multiple regression using 5 biometeorological parameters relating potato late blight occurrence and the environment. The parameters used are (a) meteorological influence on sporangium production, (b) meteorological influences on sporangial germination and infection, (c) meteorological influence on colonization, (*d*) the magnitude and duration of high humidity periods, and (*e*) the effect of dry periods on disease development. This system was developed for Germany and is based on daily data collected weekly. The data is analyzed by a computer program which produces output in the form of ratings accumulated from the average time of plant emergence. No disease forecast is made and no control measures are necessary until 150 ratings are accumulated. At this time an alert is given that blight could occur within 15–40 days. From this time on growers are instructed to inspect their fields very closely for the appearance of blight. When 270 ratings are accumulated, blight is forecast to occur within 15 days before or 15 days after this date. At this time, a definite call for control measures is made. This system is termed the "negative prognosis system" (58) because it defines blight-free periods. Thus no control measures are needed until 150 ratings are accumulated, and then only in the event of blight occurrence. General blight control measures are called for only after 270 ratings are accumulated.

In experimental field evaluations, this system has produced regression coefficients of 0.756, or it has accounted for 56% of the variation in disease progress by mathematical analysis of the data collected. The developers insist that the predictive system must be powered by data collected from standard weather stations in the vicinity of the crop. The developers of the system apparently believe that collection of data from within the crop canopy is too involved to warrant the effort. We believe that a great deal of the remaining 44% unexplained variation could be accounted for by collection of data within the canopy or microclimate of the crop. The problem of relating microclimate to macroclimate (20, 28) has already been discussed. The German system, as described here, has been in operation in Germany since 1967 and appears to be giving satisfactory results on an area or regional basis. Since Schrödter & Ullrich's work indicates that the system does respond differently for different varieties (52), it would be interesting to evaluate this system in the United States under varying environmental conditions and with various varieties.

Fundamental Infection Prediction Systems

Currently there are no fundamentally derived systems that predict the occurrence of infection based on forecast weather. An exception to this might be systems that deal with the prediction of occurrence and population levels of vectors which harbor and/or transmit a disease-causing organism. An example of this is the system which predicts the occurrence and severity of Stewart's bacterial wilt [*Erwinia stewartii* (Smith) Dye] of sweet corn and field corn for the following season based on temperatures in December, January, and February (54, 55). This system has proven to be accurate and reliable since its introduction in 1933. This system is discussed in more detail at a later point.

ROLE OF MODERN TECHNOLOGY IN QUANTIFICATION AND PREDICTION OF DISEASE

There are good reasons the number of plant disease forecasting systems are few in relation to the number of plant diseases currently known throughout the world. In the past, phytopathologists did not have calculators, computers, modern statistical techniques, and more sophisticated environmental sensing devices to aid them in predicting disease or infection occurrence. They had to rely on personal experience gained over many years of careful observation of disease development and associated weather conditions. Their environmental data were painstakingly collected from thermographic charts, nonrecording rain gauges, etc, while coincidental, careful observations were made on disease development. Then these data were usually plotted in some manner, and with intuition, good judgment, and excellent visual perception, subjective correlations between weather and disease development were made. The task was time consuming and often frustrating, but such perseverance led to the development of empirical forecasting systems such as those for potato late blight (60, 61), apple scab (44), and Stewart's wilt of corn (4, 54, 55).

These systems are empirical and forecast disease development in subjective terms. Forecasts of apple scab infection periods and Stewart's wilt disease use such terms as "light," "moderate," or "severe." Late blight forecasts deal with "favorable days" or "severity values." When these systems were developed, it was physically and mentally impossible for the researchers, with the equipment and facilities available, to grasp, fully understand, and define all of the complex interrelationships among pathogen, environment, and host. Thus, the quantitative relationships of these gross environmental parameters and corresponding measurements of disease severity were expressed in an empirical and subjective manner.

Measurement of Parameters

Relating disease development to weather conditions requires accurate quantification of environmental parameters as well as disease progress. Whereas early workers had relatively few tools available for use in this respect, the mid-1950s saw a considerable upsurge of interest in equipment and instrumentation for measuring various environmental parameters. Good reviews of the literature in this area through the mid-1960s are contained in Platt & Griffith's text, *Environmental Measurement and Interpretation* (48) and Yang's *Agricultural Meteorology* (70). Similarly, during this period equipment for quantifying fungal spore levels in the air was being developed by Hirst (26) and others (9, 36, 47).

Since the mid-1960s, considerable advances have been made in environmental sensing equipment. Sophisticated electronic devices have been developed to replace bulky, somewhat unreliable mechanical devices: thermocouples and thermistors have replaced bimetallic or mercury thermometers, humidity-sensitive ceramic sensors replaced hair hygrographs and wet-bulb dry-bulb thermometers, and hot wire anemometers replaced relatively insensitive cup anemometers. The net result of such new sensing equipment is the capability to measure weather parameters more accurately and with greater sensitivity in locations closer to diseased plants, for example, measurement of temperature on a leaf surface with a thermocouple as opposed to measurement with a thermograph in a weather shelter somewhat removed from the diseased plant.

DATA RECORDING   The availability of electronic data acquisition equipment has presented phytopathologists the opportunity to circumvent, at least in part, the awesome task of collecting and collating data in preparation for analysis. No longer does the researcher have to spend numerous hours manually transcribing data from recorder charts or autographs into record books or onto computer cards. Delay between data recording and final analysis is now greatly reduced. Modern acquisition equipment can automatically monitor large numbers of various sensing devices at selected intervals and record these data on some medium, essentially ready for processing and analysis. Data may be logged on punched paper or magnetic tape for transport to a computer facility, or the acquisition unit may be directly linked to some electronic processing device (computer, minicomputer, or programmable calculator) where analysis can be performed nearly simultaneously with data collection.

DATA PROCESSING   The development of sophisticated, sensitive, and reliable sensing and acquisition equipment is a boon to research phytopathologists but its importance is somewhat overshadowed by the tremendous capabilities made available through electronic data processing (EDP). Increased availability of computers (analog or digital, maxi or mini) and programmable calculators has all but eliminated hand plotting and hand calculations. The normal activities of summation, averaging, plotting, etc, which took days and weeks to do manually can now be completed in seconds or minutes. Additionally, the capability of handling large data collections and easily carrying out a multitude of complex calculations has given rise to increased utilization of established mathematical and statistical techniques for studying and quantifying relationships of various cause-and-effect processes.

Multiple Regression Analysis

Of the many techniques available for multivariate analysis, perhaps the most applicable to phytopathology is multiple regression analysis (MRA). Although MRA has been used for more than a half century in the field of entomology (19) it was not widely used in phytopathology until Van der Plank (59) emphasized the need for quantification in the study of plant disease epidemics and pointed out the use of regressions and correlations in epidemiological studies. The uses of MRA since then have been varied. MRA has been used to explain progress of epidemics in time (11), to relate the rate of disease increase to environmental factors (1), to predict the level of disease at a given point in time (1), or to relate progress of disease cycles (i.e. sporulation, infection) to fluctuations in environmental parameters (37). Each of these uses for MRA are applicable to the area of disease forecasting, and considerable useful information has been and will continue to be produced utilizing this technique. For a more detailed discussion on the place of MRA in phytopathological research, the authors strongly urge the reading of the excellent and complete review article on this subject by Butt & Royle (12).

Models and Simulations

Models, and simulation by the experimental use of models in an attempt to mimic a real process, have been used for some time in several scientific fields (2, 21, 45). However, modeling and the simulation of disease epidemics by the logical linking of several models representing various aspects of a disease cycle have been introduced only recently to phytopathology. In the late 1960s Waggoner (60, 61), Zadoks (71), and Oort (46) produced models and simulations of several plant diseases and began seriously discussing the use of models and systems analysis in the study of plant disease. By definition, models may be simple or complex, but all are only simplifications of reality (25). Models may be of the deterministic type in which the predicted values are computed exactly, or of the stochastic type in which predicted values depend upon probability distributions (49). EPIDEM (61), a simulator of tomato early blight [*Alternaria solani* (Ell. & G. Martin) Sor.] written by Waggoner was the first plant disease simulator to be widely distributed. It resulted in the generation of considerable interest in the area of plant disease modeling and simulation. In recent years, several other simulators have been written: EPIMAY (62) for southern corn leaf blight (*Helminthosporium maydis* Nisik. & Miyake) and EPIVEN (33) for apple scab [*Venturia inaequalis* (Cke.) Wint.]. These two simulators were essentially modifications of the original EPIDEM. McCoy (40) has developed a simulator for ascochyta blight of chrysanthemum (*Mycosphaerella ligulicola* Baker, Dimock & Davis). A. L. Jones and T. B. Sutton (personal communication) are currently developing a simulator for apple scab which is considerably different from that of EPIVEN. Each of the above simulators are deterministic in that repeated simulations using identical input data always yield identical results.

Massie (37) has developed a simulator for southern corn leaf blight which is of the stochastic type. By relying heavily on MRA to develop predictive equations as models for the various segments of the disease cycle, and utilizing probability distributions and observed response variability, the resulting simulator indicates, through repeated simulations with identical input data, not only the average development of the disease, but also the chance variation in development that might be expected. Jones and Sutton are incorporating stochastic properties into their apple scab simulator.

IMPLEMENTATION OF PREDICTIVE SYSTEMS

Efforts toward quantification of plant disease responses to environmental factors with predictive models and simulators have been enhanced by the rapid development of data acquisition and processing. However, the value of these efforts cannot be fully realized until these systems can be used in the field to make decisions concerning control measures, cropping practices, and loss appraisal. Because forecasting of environmental parameters necessary for accurate prediction of plant infection is still a hope of the future, and because most plant diseases are extremely sensitive to small fluctuations in environmental conditions, the immediate future of long-range forecasting for most plant diseases is not encouraging. But such is not the case for evaluation of current infection status and short-term disease prediction.

With a disease such as potato late blight in which infection progresses rapidly, but only periodically throughout the growing season, current knowledge of favorable conditions for disease increase is very important. By knowing immediately that periods favorable for disease increase have begun, growers can apply fungicides to protect foliage against successive infections and thus against continued disease buildup. Using a combination of telephone communications and EDP to effect rapid information retrieval, processing, and dissemination systems, growers could be given an accurate assessment of infection probability in time to initiate effective control procedures.

Apple scab represents another type of disease in which primary infection occurs on a sporadic schedule, greatly influenced by the temperature and duration of rainy periods. Mills (44) has derived a late-warning system to assess the potential severity of infection during wetting periods. However, because Mills' system apparently does not take into account the variability and consequence of inoculum load, which has been clearly demonstrated by Szkolnik (56), the necessity of *current* quantitative information regarding pathogen development is readily apparent.

The forecasting system for Stewart's wilt of corn (54, 55) has proven relatively reliable for many years. Because forecasts of potential disease severity are based on survival of the pathogen vector, the system allows adequate time for control measures based on variety selection, insecticide application, and time of planting. Because there can be considerable local variation in vector survival, the application of this system must, of necessity, be localized. Utilization of information retrieval, processing, and dissemination systems with this forecasting system could enhance its effectiveness and account for local variation in vector survival.

Full utilization of modern technology to develop or augment forecasting systems appears to be the logical path. However, due to initial expense of equipment, relatively high operation and maintenance costs, and the small number of individuals with the necessary training and experience, progress has been slow. Nevertheless, tools are available for use, and as equipment costs decrease through continued improvements in technology and as support for facilities increases through legislative action or grower underwriting, and as more individuals are trained in the utilization of new equipment and techniques, progress in the area of plant disease prediction will be greatly enhanced.

There is considerable pressure, due to socioeconomic factors, to develop the most efficient crop production systems possible. An integral part of this effort must include development of effective disease and insect control systems. In turn, development and/or implementation of accurate plant disease forecasting systems can result in a significant contribution to both effective and efficient disease control practices.

The most immediate results can perhaps be obtained by implementing and evaluating existing prediction and forecasting systems under both experimental and grower conditions. An example of such an approach can be illustrated clearly in a discussion of Blitecast, a computerized forecasting system for potato late blight first written and implemented in 1972 (34, 35, 38).

Forecasting systems for potato late blight developed by Hyre and Wallin had been tested in Pennsylvania and other northeastern states for more than 13 years but were not widely accepted and utilized by growers for several reasons: (*a*) lack of equipment for acquiring environmental data, (*b*) lack of understanding and use of the systems, (*c*) lack of confidence in the forecast's applicability to a particular locality, and (*d*) delay in receiving forecasts after data collection and submission.

The development of Blitecast which combined both the Wallin and Hyre systems provided the means of performing necessary calculations accurately and quickly, and issuing consistent and understandable recommendations to the grower. In 1972, 12 growers in selected test locations throughout Pennsylvania were provided with a hygrothermograph, rain gauge, and weather shelter. These growers were requested to telephone a central processing location periodically to report their most recently recorded environmental data. The operator at the processing center entered the data into a computer which analyzed the data, calculated the probable disease situation, and returned a forecast and spray recommendation based on a combination of both the Hyre and Wallin forecasting systems (Table 2). The entire operation could be completed during a standard three minute telephone call, therefore allowing one operator to process 15 to 20 forecasts per hour from widely separated locations.

During the first year of testing no blight occurred on the 12 test farms; however, blight did occur on adjacent farms that were not following Blitecast recommendations (34). In all cases, blight occurred within the forecast period. Each cooperating grower applied an average of 6 fewer sprays than normally would have been applied.

Continued testing of this program in 41 locations in four states during 1973 showed similar success in both forecasting and control of potato late blight (35). Where apparent failures of the system occurred, they could be traced to one of three causes: (a) incorrect or incomplete data acquisition, (b) failure to apply recommended spray applications, or (c) poor fungicide application technique. Based on experience over several years of both experimental and commercial application of Blitecast, if all recommended sprays are applied (no less) and optimum coverage is obtained, a late blight epidemic should not occur because the amount of disease in the potato field will be continually restricted to a minimum level. As a result of the excellent success that Blitecast has had, growers' confidence and acceptance of Blitecast has grown steadily. The success experienced by this system is a direct result of using applied technology to provide rapid data retrieval, processing, and information dissemination which in turn provided the accuracy, dependability, and timeli-

Table 2 Adjustable matrix used to relate severity values (Wallin's system) and rain-favorable days (Hyre's system) and generate spray recommendations for Blitecast

		Severity values during last 7 days					
		< 3	3	4	5	6	>6
		Message number					
Total rain-favorable	< 5	−1	−1	0	1	1	2
days during last 7 days	> 4	−1	0	1	2	2	2

Message number	Spray recommendations
−1	No spray
0	Late blight warning
1	7-day spray schedule
2	5-day spray schedule

ness of disease control recommendations. Thus a "late warning system" can be used to forecast and prevent disease occurrence.

A somewhat similar approach to forecasting of Stewart's wilt of corn has been undertaken by Castor et al (13). Utilizing an existing forecasting system developed by Stevens (54), and shown to be valid by Stevens & Haenseler (55) and later modified by Boewe (4), a computer program was developed which facilitated the processing of temperature data from many locations. This has increased the speed and efficiency with which *localized* forecasts can be made. The system was first used in Pennsylvania in 1974 and resulted in a wider application of the forecasting system than ever before. If results of continued testing and application of this system to verify its accuracy under varied local conditions remain favorable, the value of the computerized system to corn growers may be considerable.

The previous two examples point out the advantages of reevaluating and implementing (using rapid communications and EDP) empirical forecasting systems which already exist and appear to be valid. There is no doubt there are other systems that can be utilized in a similar manner, either as they have been developed, or with modification. For example, the Mills' system for forecasting (late-warning) infection of apples by the apple scab fungus *Venturia inaequalis* (44) is an empirical system based on many years of field observations of scab infections and their relationship to temperature and duration of wet periods. Through use of these two criteria, scab infection probability was classified as none, light, moderate, or severe. Although the system has been used widely by fruit growers as an aid to spray scheduling, close observation of infection levels under different wetting conditions has shown that infection levels can vary widely within a given Mills classification, and that occasionally more infections may be seen to occur in a "light" infection period than in a "severe" one (unpublished data). Apparently, the fact that Mills' system does not take into account variation in inoculum levels may make it subject to considerable error in forecasting.

Massie & Szkolnik (39) have developed a formula for calculating available mature *V. inaequalis* ascospores based on current knowledge of accumulated precipitation and degree-days from date of leaf-fall for the previous autumn. Equations were developed by using MRA on 17 years data. When the equations were tested on observations from 3 additional years, they adequately forecast the state of ascospore maturation. It may be possible to utilize this formula along with Mills' system to develop a considerably more reliable late-warning system for primary apple scab infection. Thus, the supplementation of an old empirical system with a fundamental model relating availability of inoculum to environmental conditions may result in a very useful disease prediction system.

THE FUTURE

As stated earlier, it is not important that the progress of disease in relation to environment is predicted using empirical (e.g. potato late blight, apple scab) or fundamental (EPIDEM, EPIMAY) systems. What is important is that the systems be reasonably accurate and amenable to implementation. A simulator that proves

95% accurate under laboratory conditions, but cannot be used under field conditions because of the intricate data required, is less useful than a simple empirical system that can be implemented but is only 80% accurate.

All phytopathologists can contribute toward developing and implementing predictive systems: the laboratory researcher can offer considerable data of use in developing fundamental systems; the field researcher can offer data useful in developing either fundamental or empirical systems; the epidemiologist should be able to coordinate the accumulated data and, utilizing all tools and techniques available, develop and/or modify and implement existing predictive systems; the extension phytopathologist should play an important role in development and implementation of such a program, including education of growers in correct interpretation and use of forecast information.

That most systems currently in operation are actually late-warning disease prediction systems rather than infection prediction is in part due to the fact that we do not have the ability to accurately predict the environmental parameters necessary for infection prediction systems. This situation, however, is not critical. If disease prediction warnings are issued in a timely manner, and if suitable prophylactic or therapeutic chemicals are available to permit effective control of disease development, these late warnings serve as well as an accurate infection prediction system. The necessity for timely disease warnings again emphasizes the important role that automatic data acquisition and EDP can play in the area of disease prediction. The fact that most disease prediction systems are late-warning or after-the-fact systems should indicate the need for development and testing of therapeutic ("after infection," "kick-back") type chemicals that can be used most effectively with this type of information.

The future in plant disease prediction will perhaps be centralized national, regional, or state facilities, equipped with data processing machinery linked to remote data acquisition units. These facilities could monitor environmental parameters and process data for diseases, pests, and growing conditions on many crops. Such a network could make disease predictions, issue warnings, and perhaps make definitive disease control recommendations as they are necessary. This will be possible if enough accurate and reliable prediction systems are developed, tested, and implemented to justify the expense of such facilities. With the availability of trained personnel, equipment, and techniques for undertaking such a task, there should be no reason why it should not come about.

Literature Cited

1. Analytis, S. 1973. Zur Methodik der Analyse von Epidemien dargestellt am Apfelschorf (*Venturia inaequalis* (Cook) Aderh.). *Acta Phytomed.* 1:1–79
2. Bailey, N. T. J. 1967. *The Mathematical Approach to Biology and Medicine.* New York: Wiley. 296 pp.
3. Beaumont, A. 1947. The dependence on the weather of the dates of outbreak of potato blight epidemics. *Trans. Br. Mycol. Soc.* 31:45–53
4. Boewe, G. H. 1950. Stewart's disease: prospect for 1950. *Plant Dis. Reptr.* 34:155
5. Bourke, P. M. A. 1953. Potato blight and the weather: a fresh approach. *Ir. Meteorol. Serv. Tech. Note* No. 12
6. Bourke, P. M. A. 1957. The use of synoptic weather maps in potato blight epidemiology. *Ir. Meteorol. Serv. Tech. Note* No. 23. 35 pp.
7. Bourke, P. M. A. 1959. Meteorology and the timing of fungicide applications against potato blight. *Int. J. Biometeorol.* 3:1–8
8. Bourke, P. M. A. 1970. Use of weather information in the prediction of plant disease epiphytotics. *Ann. Rev. Phytopathol.* 12:345–70
9. Bromfield, K. R., Underwood, J. F., Peet, C. E., Grissinger, E. H., Kingsolver, C. H. 1959. Epidemiology of stem rust of wheat: IV. The use of rods as spore collecting devices in a study on the dissemination of stem rust of wheat urediospores. *Plant Dis. Reptr.* 43: 1160–68
10. Burleigh, J. R., Eversmeyer, M. G., Roelfs, A. P., 1972. Development of linear equations for predicting wheat leaf rust. *Phytopathology* 62:947–53
11. Burleigh, J. R., Romig, R. W., Roelfs, A. P. 1969. Characterization of wheat rust epidemics by numbers of uredia and numbers of urediospores. *Phytopathology* 59:1229–37
12. Butt, D. J., Royle, D. J. 1973. Multiple regression analysis in the epidemiology of plant diseases. *Ecological Studies,* ed. J. Kranz, Vol. 13, *Epidemics of Plant Diseases: Mathematical Analysis and Modeling.* Stuttgart: Springer
13. Castor, L. L., Ayers, J. E., MacNab, A. A., Krause, R. A. 1975. Computerized forecasting system for Stewart's bacterial disease on corn. *Proc. Am. Phytopathol. Soc.* 1:In press (Abstr.)
14. Chester, K. S. 1943. The decisive influence of later winter weather on wheat leaf rust epiphytotics. *Plant Dis. Reptr.* Suppl. 143, 133–44
15. Chester, K. S. 1946. *The Nature and Prevention of the Cereal Rusts as Exemplified in the Leaf Rust of Wheat.* Waltham, Mass.: Chron. Bot. Co. 269 pp.
16. Chester, K. S. 1950. Validity and value of plant disease forecasting. *Plant Dis. Reptr.* Suppl. 190, 5–8
17. Cook, H. T. 1947. Forecasting tomato late blight. *Plant Dis. Reptr.* 31:245–49
18. Cook, H. T. 1949. Forecasting late blight epiphytotics of potatoes and tomatoes. *J. Agric. Res.* 78:545–63
19. Cook, W. C. 1921. Studies on the flight of nocturnal Lepidoptera. *Rept. St. Entomol. Minn.* No. 18:43–56
20. DeWeille, G. A. 1964. Some notes on the use of hygrograms for phytopathological purposes with special reference to their reliability. *Neth. J. Agric. Sci.* 12:229–34
21. DeWit, C. T. 1968. *Theorie en Model.* Wageningen, The Netherlands: Veenman. 13 pp.
22. Dirks, V. A., Romig, R. W. 1970. Linear models applied to variation in numbers of cereal rust urediospores. *Phytopathology* 60:246–51
23. Eversmeyer, M. G., Burleigh, J. R. 1970. A method of predicting epidemic development of wheat leaf rust. *Phytopathology* 60:805–11
24. Eversmeyer, M. G., Burleigh, J. R., Roelfs, A. P. 1973. Equations for predicting wheat stem rust development. *Phytopathology* 63:348–51
25. Goodall, D. W. 1972. Building and testing ecosystem models. In *Mathematical Models in Ecology,* ed. J. N. R. Jeffers, 173–94. Oxford: Blackwell
26. Hirst, J. M. 1952. An automatic volumetric spore trap. *Ann. Appl. Biol.* 39:257–65
27. Hirst, J. M., Stedman, O. J. 1956. The effect of height of observation in forecasting potato blight by Beaumont's method. *Plant Pathol.* 5:135–40
28. Holmes, R. M., Dingle, A. N. 1965. The relationship between the macro- and microclimate. *Agric. Meteorol.* 2: 127–33
29. Hyre, R. A. 1954. Progress in forecasting late blight of potato and tomato. *Plant Dis. Reptr.* 38:245–53
30. Hyre, R. A. 1955. Three methods of forecasting late blight of potato and tomato in northeastern United States. *Am. Potato J.* 32:362–71

31. Hyre, R. A., Bonde, R. 1955. Forecasting late blight of potato in northern Maine. *Am. Potato J.* 32:119–25
32. Hyre, R. A., Horsfall, J. G. 1951. Forecasting potato late blight in Connecticut. *Plant Dis. Reptr.* 35:423–31
33. Kranz, J., Mogk, M., Stumpf, A. 1973. EPIVEN: ein Simulator fur Apfelschorf. *Z. Pflanzenkr.* 80:181–87
34. Krause, R. A., Massie, L. B., 1973. Application and implementation of computerized forecasts of potato late blight. *Phytopathology* 63:205 (Abstr.)
35. Krause, R. A., Massie, L. B., Hyre, R. A. 1975. Blitecast: A computerized forecast of potato late blight. *Plant Dis. Reptr.* 59:95–98
36. Malligo, J. E., Idowine, L. S. 1964. Single stage impaction device for particle sizing biological aerosols. *Appl. Microbiol.* 12:32–36
37. Massie, L. B. 1973. *Modeling and simulation of Southern corn leaf blight disease caused by race T of Helminthosporium maydis Nisik. & Miyake.* PhD thesis. The Pennsylvania State Univ., University Park. 84 pp.
38. Massie, L. B., Krause, R. A. 1973. A computer program for the forecasting of late blight of potato. *Phytopathology* 63:205 (Abstr.)
39. Massie, L. B., Szkolnik, M. 1975. Prediction of ascospore maturity of *Venturia inaequalis* utilizing cumulative degree-days. *Proc. Am. Phytopathol. Soc.* 1: In press (Abstr.)
40. McCoy, R. E. 1971. *Epidemiology of chrysanthemum Ascochyta blight.* PhD dissertation. Cornell Univ. Ithaca. 177 pp.
41. Miller, P. R. 1967. Plant disease epidemics—their analysis and forecasting. In *Papers Presented at the FAO Symp. Crop Losses, Rome, 1967,* 9–37. Rome: FAO. 330 pp.
42. Miller, P. R., O'Brien, M. 1952. Plant disease forecasting. *Bot. Rev.* 18:547–601
43. Miller, P. R., O'Brien, M. 1957. Prediction of plant disease epidemics. *Ann. Rev. Microbiol.* 11:77–110
44. Mills, W. D. 1944. Efficient use of sulphur dusts and sprays during rain to control apple scab. *Cornell Univ. Agric. Exp. Stn., Ext. Bull.* 630
45. Newnham, R. M., Smith, J. H. G. 1964. Development and testing of stand models for Douglas fir and lodgepole pine. *For. Chron.* 40:494
46. Oort, A. J. P. 1968. A model of the early stages of epidemics. *Neth. J. Plant Pathol.* 74:177–80
47. Pady, S. M. 1959. A continuous spore sampler. *Phytopathology* 49:757–60
48. Platt, R. B., Griffiths, J. F. 1964. *Environmental Measurement and Interpretation.* New York: Reinhold. 256 pp.
49. Ross, Q. J. S. 1972. Stochastic model fitting by evolutionary operation. See Ref. 25, 297–308
50. Scarpa, M. J., Raniere, L. C. 1964. The use of consecutive hourly dewpoints in forecasting downy mildew of lima bean. *Plant Dis. Reptr.* 48:77–81
51. Schrödter, H., Ullrich, J. 1965. Untersuchungen zur Biometeorologie und Epidemiologie von *Phytophthora infestans* (Mont.) de By. auf mathematisch-statistischer Grundlage. *Phytopathol. Z.* 54:87–103
52. Schrödter, H., Ullrich, J. 1966. Weitere Untersuchungen zur Biometeorologie und Epidemiologie von *Phytophthora infestans* (Mont.) de By. Ein neues Konzept zur Lösung des Problems der epidemiologischen Prognose. *Phytopathol. Z.* 56:265–78
53. Schrödter, H., Ullrich, J. 1967. Eine mathematisch-statistische Lösung des Problems der Prognose von epidemien mit Hilfe meteorologischer Parameter, Dargestellt am Beispiel der Kartoffelkrautfäule (*Phytophthora infestans*). *Agric. Meteorol.* 4:119–35
54. Stevens, N. E. 1934. Stewart's disease in relation to winter temperatures. *Plant Dis. Reptr.* 18:141–49
55. Stevens, N. E., Haenseler, C. M. 1941. Incidence of bacterial wilt of sweet corn, 1935–1940; forecasts and performance. *Plant Dis. Reptr.* 25:152–57
56. Szkolnik, M. 1969. Maturation and discharge of ascospores of *Venturia inaequalis.* *Plant Dis. Reptr.* 53:534–37
57. Thurston, H. D., Knutson, K. W., Eide, C. J. 1958. The relation of late blight development on potato foliage to temperature and humidity. *Am. Potato J.* 35:397–406
58. Ullrich, J., Schrödter, H. 1966. Das problem der Vorhersage des Auftretens der Kartoffelkrautfäule (*Phytophthora infestans*) und die Möglichkeiten seiner Lösung durch eine "Negativ-prognose." *Nachrichtenbl. Dtsch. Pflanzenschutzdienst, Braunschweig* 18:33–40
59. Van der Plank, J. E. 1963. *Plant Diseases: Epidemics and Control.* New York: Academic. 349 pp.
60. Waggoner, P. E. 1968. Weather and the rise and fall of fungi. In *Biometeorology,*

ed. W. P. Lowry, 45–66. Corvallis: Oregon State Univ. Press

61. Waggoner, P. E., Horsfall, J. G. 1969. EPIDEM: A simulator of plant disease written for a computer. *Conn. Agric. Exp. Stn. Bull.* 698. 80 pp.

62. Waggoner, P. E., Horsfall, J. G., Lukens, R. J. 1972. EPIMAY: A simulator of Southern corn leaf blight. *Conn. Agric. Exp. Stn. Bull.* 729. 84 pp.

63. Wallin, J. R. 1951. Forecasting tomato and potato late blight in the north-central region. *Phytopathology* 41:37 (Abstr.)

64. Wallin, J. R. 1962. Summary of recent progress in predicting late blight epidemics in United States and Canada. *Am. Potato J.* 39:306–12

65. Wallin, J. R., Riley, J. A. 1960. Weather map analysis—An aid in forecasting potato late blight. *Plant Dis. Reptr.* 44:227–34

66. Wallin, J. R., Samson, R. W. 1953. The relation of cumulative amount and frequency of rainfall and mean temperature to late blight in Indiana. *Am. Potato J.* 30:262–70

67. Wallin, J. R., Wade, E. K., Darling, H. M. 1953. The relation of weekly mean temperatures and cumulative rainfall to late blight epiphytotics in Wisconsin. *Am. Potato J.* 30:231–42

68. Wallin, J. R., Waggoner, P. E. 1949. The influence of weekly cumulative rainfall and temperature on potato late blight epiphytotics in Iowa. *Plant Dis. Reptr.* 33:210–18

69. Wallin, J. R., Waggoner, P. E. 1950. The influence of climate on the development and spread of *Phytophthora infestans* in artificially inoculated potato plots. *Plant Dis. Reptr.* Suppl. 190, 19–33

70. Yang, J. Y. 1963. *Agricultural Meteorology.* Milwaukee: Pacemaker. 693 pp.

71. Zadoks, J. C. 1965. Epidemiology of wheat rusts in Europe. *FAO Plant Prot. Bull.* 13:97–108

RACES OF THE PATHOGEN   ❖3611
AND RESISTANCE TO COFFEE RUST

C. J. Rodrigues Jr., A. J. Bettencourt,[1] and L. Rijo

Centro de Investigação das Ferrugens do Cafeeiro, Oeiras, Portugal

INTRODUCTION

Coffee, after petroleum, is the most important product in international world trade (19). Coffee growing is therefore one of the most important agricultural occupations, upon which the economy of more than 50 countries depends.

From the 16 *Hemileia* species described in the literature (72), which occur on rubiaceous plants, only *Hemileia vastatrix* B. & Br. and *H. coffeicola* Maubl. & Rog. are able to infect the genus *Coffea*. For this reason, these two species, and particularly the first one because of its major economic importance, have been the object of detailed investigation (53, 63). All the species of *Hemileia* so far known are hemiforms, the pycnial and aecial stages being unknown (68).

During the last quarter of the 19th century, *H. vastatrix* (coffee leaf rust, orange rust) caused enormous losses on arabica coffee plantations in Ceylon and later seriously checked the growth of this species at low altitude in other Asiatic and African countries. The rust is foliicolous and strictly hypophyllous, forming orange or orange-red pustules varying in size from a few millimeters to more than a centimeter in diameter, according to the age of the pustules and to the intensity of the attack. Reports of the rust on berries and tender shoots have occasionally been made (20, 68), but the writers have never observed such cases. A most impressive description of the damage caused by the rust in Ceylon and other countries is given by Wellman (76). In 1970, coffee rust was reported in Brazil, causing justifiable alarm on the American continent, which produces over 65% of the world's coffee. The rust has also been detected recently in Paraguay and Argentina. At present, with the exception of the other coffee growing countries on the American continent not yet affected, the rust is practically endemic in all other regions of the world where coffee is grown.

Chemical control of the rust, first tried by Marshall Ward (42) in Ceylon with lime sulfur, has been further developed and utilized in India (51), Kenya (7, 33, 54,

---

[1]Instituto do Café de Angola, Luanda, Angola.

73), and more recently in Brazil (38), using mainly copper fungicides. The success of this method is, however, limited by several practical circumstances and is in many cases not economical (64). Thus, resistance in coffee to orange rust has been the object of much search.

This review attempts to give a panoramic view of the efforts made to look for resistance in coffee to orange rust, with particular emphasis on the work carried out by the Centro de Investigação das Ferrugens do Caffeeiro (CIFC). We take this opportunity to express our great appreciation of the invaluable pioneer work carried out, first by Marshall Ward in Ceylon on the biology of *H. vastatrix,* and later by other scientists who devoted themselves to the study of this disease in coffee research stations of Java, India, Kenya, Tanzania, Democratic Republic of Congo, and the Ivory Coast.

SEARCH FOR RUST RESISTANCE IN *COFFEA* SPP.
PHYSIOLOGICAL GROUPS OF COFFEE

The genus *Coffea* has been studied in the past by many taxonomists, but the most recent and perhaps most valid review is by Chevalier (18), who divided the true coffee species into the following four groups: Eucoffea K. Schum., Mascaracoffea Chev., Argocoffea Pierre, and Paracoffea Miq. The three first groups include coffees exclusively native to Africa. Most of the representatives of the fourth group (Paracoffea) are native to India, Indochina, Ceylon, and Malaya.

The Eucoffea section, the most economically important of the genus, is in turn subdivided into the following subgroups: Erythrocoffea, comprising the species *C. arabica* L., *C. canephora* Pierre, *C. congensis* Froehner, and *C. eugenioides* Moore; Pachycoffea, including *C. liberica* Hiern, *C. dewevrei* de Wild. et Dur. and its varieties, *C. klainii* Pierre, *C. abeokutae* Cramer, and *C. oyemensis* Chev.; Nanocoffea, with the species *C. humilis* Chev and others; Melanocoffea, comprising the species *C. carrissoi* Chev., *C. stenophyla* G. Don, etc; and Mozambicoffea, including *C. ligustroides* Moore, *C. racemosa* Lour., *C. salvatrix* Swyn. et Phil., etc.

Coffea arabica, the most largely cultivated species in the world, occupies 65% of the total area used for growing coffee; it produces coffee of the best quality, coming mainly from the Americas. It is apparently the only species with a number of chromosomes $2n = 44$ and is self-fertile, whereas all the others are $2n = 22$ and self-sterile. It is followed by *C. canephora* Pierre, also known as Robusta, Uganda, and Quillou, which occupies 33% of the total area, and gives coffee of lower quality, produced chiefly in Africa. Finally, *C. liberica* and *C. dewevrei* considered by some to produce coffee of even lower quality, take third place, occupying 2% (19) of the total area; they are almost exclusively cultivated in Liberia, St. Thome and Fernando Po islands, and French Guinea. *C. racemosa,* also known as Inhambane's coffee because of being indigenous to Inhambane (Mozambique), is only locally consumed.

All the other species have no economic interest, although in some cases they might be locally used as substitutes for other coffees. Special mention must be made, however, of some coffee species devoid of caffein or nearly so, such as those that form the Mascaracoffea section, native to Madagascar and the Mascarene islands. These species have been lately studied from different angles (26, 43).

Coffea arabica

According to historical data, the primary dispersal center of cultivated *C. arabica* was the Yemen (Arabia Felix) where it was possibly introduced from Ethiopia ca 575 AD (77). Seeds of this species were apparently taken to Java by the Dutch in 1690. The botanical garden in Amsterdam received one single coffee seedling (*C. arabica* L. var. *Typica* Cramer) from that island in 1706 and provided seed from this plant to the botanical garden in Paris in 1713. It is believed that this single progeny was the origin of the coffee seed for all coffee plantations established in Antilles, Central and South America (41). Unselected seed of the same variety was also introduced into India, Ceylon, and other Asiatic and African regions.

Because of the absence of *H. vastatrix* in Yemen (66), the arabicas cultivated there could be selected for quality, yield, and drought resistance, but were never subjected to natural selection towards resistance to the leaf rust. Because of this, the single progeny origin, and the predominant self-fertility of *C. arabica*, it would be not surprising if most of the cultivated plants of this species were genetically alike and rust susceptible. A survey of the primitive coffee population in Brazil and other Latin American countries revealed a very restricted gene pool for the arabica species (12). This is the likely reason attempts to find resistance to *H. vastatrix* among cultivated *C. arabica* were initially a complete failure.

The first *C. arabica* plant showing resistance to *H. vastatrix* was found in 1911, in Mr. Kent's estate, at Doddengooda, Mysore, India, among others fully susceptible. Selfed seed of this plant gave rise to the well-known Kent cultivar which was introduced for large-scale planting in 1918–1920, to replace the Coorg cultivar heavily attacked by the rust. Kent's coffee was imported by several countries, and in spite of later becoming susceptible to the rust, its role was quite important in the improvement of *C. arabica*.

Because the southwestern mountains of Ethiopia and the Boma plateau of Sudan were the most likely center of origin of *C. arabica* (48, 67), there would be a chance that a certain degree of genetic variability would be found among the most primitive types of the species. In 1928 Cramer (20) visited some coffee estates in Ethiopia and noting the vigorous, healthy, appearance of the trees, selected mother plants and sent the seed to Java. This Abyssinian coffee gave good results at the lower planting belt and was apparently more resistant to leaf rust. However, whatever might have been the reason, the Dutch never again produced arabica as they did before the appearance of rust, and in 1969 (41) only 10% of Indonesia's production came from this species cultivated in higher altitudes.

New introductions of coffee from Ethiopia were made into Kenya on different occasions (39) through the good offices of the British consulates in that country, an expedition to the Boma plateau by A. S. Thomas, and through serving officers of the East African Forces during the World War II. Collections such as Harar, Geisha, Amfillo, Dalle, Dalle Mixed, Dilla, Dilla and Alghe, Gimma Mbuni and Gimma Galla Sidamo, were made and named after the places of origin. These new varieties were established and some field observations were made, but a positive knowledge of their real abilities, especially concerning rust resistance, was unknown.

Until the 1950s there was, therefore, almost a total lack of knowledge about the existence of true arabica coffees with rust resistance. This fact and the danger that the eventual introduction of rust into the Latin America would threaten the economic and social stability of coffee growers made Dr. F. Wellman, working at the Instituto Interamericano de Ciencias Agrícolas, Turrialba, Costa Rica, alert the responsible authorities towards that potential menace (74, 75). A Point IV Mission composed of Drs. Wellman and W. Cowgill set off to the eastern hemisphere in 1952 to (a) study the behavior of the disease in the field, (b) collect coffee materials of potential resistance or of value in programs of breeding for resistance, and (c) establish contacts with eastern hemisphere coffee scientists, in order to set up mutually advantageous working relationships to provide a concerted effort toward solving the problem. The mission obtained much information and sent back well over 100 types of coffee new to the Americas. These coffees were propagated in quarantine by the US Department of Agriculture and distributed to leading research stations.

During their trip, Wellman and Cowgill visited the Department of Plant Pathology of the Estação Agronómica Nacional, near Lisbon, Portugal. The department's head, Dr. Branquinho d'Oliveira, was conducting research work on coffee screening for rust resistance and rust race differentiation. Such a location, where coffee was not grown outdoors, was ideal to carry on this type of work on an international basis, as rust samples could be received from abroad without danger of introducing new races of the pathogen. During the progress of the Point IV Mission, samples of the rust and seed of coffee varieties were sent to Lisbon, and through Dr. d'Oliveira's own efforts other materials were collected.

In April 1955, the USA and the Portuguese Governments agreed on the establishment of a project [Foreign Operations Administration (FOA) No. 72–11–004] and provided the necessary funds to build up, at Oeiras, the Centro de Investigação das Ferrugens do Cafeeiro (Agreement FO-PO-5) with the objective to study, under international cooperation, the problem of the coffee rusts.

The first of the valuable conclusions of Dr. d'Oliveira's work was to show that all the Latin America growing arabicas (Typica, Caturra, Mundo Novo, Bourbon, S. Bernardo, etc) were highly susceptible to *H. vastatrix,* thus confirming Wellman's fears about the future of coffee growing in the New World, if the rust were eventually introduced there. These results confirmed, in addition, the genetic homogeneity, at least concerning rust resistance, of *C. arabica* varieties cultivated, and the urgent need to look for new sources of resistance among pure arabicas and other coffee species. According to the system adopted by d'Oliveira and co-workers (22–25) of grouping the coffee seedlings in physiologic groups (named arbitrarily after the letters of the Roman and Greek alphabets) according to their reaction spectra to the tested races, the growing American cultivars were included in group E (Table 1). This group is susceptible to 21 rust races, including the most frequently differentiated race II. Coffee cultivars of this group were the ones in culture in Ceylon, Java, India, Phillipines, etc, when the rust epidemics wiped out the arabica coffee from the lower coffee belt about a century ago.

In the sequence of the work carried out at the CIFC, seeds of *C. arabica* have been received from many different origins. So far, not a single cultivar or selection of this species has ever been found with total resistance or total susceptibility to the known races of *H. vastatrix*. Besides group E, already mentioned, the *C. arabica* accessions tested are included in groups β, D, α, C, γ, J, L, I, and W, ranging from susceptibility to 28 races (group β) to susceptibility to only 3 races (group W) (4, 17, 25, 59, 61).

Within group D, with susceptibility to 11 races, are included Kent's selection from India and its derivatives selected in Tanzania, namely Series 'KP' 'F' 'H' and 'X', and in Kenya, namely selections K.7 and S.L.6. From Ethiopia, only one accession (S.16 Wollamo), collected by Pierre Sylvain at Soddu, district of Wollamo, was grouped as such.

All the other groups, except L, have been defined according to coffee types exclusively collected in Ethiopia, either directly sent to the CIFC by Sylvain, Beche-tel, Lejeune, Siegenthaler, and the 1964 Coffee Mission to Ethiopia, or received through the good offices of the USDA. Group β (Matari) is, within the species, the universal suscept, because it is resistant only to two races that do not infect *C. arabica*. On the other hand, groups α (Dilla & Alghe) and γ (S.12 Kaffa) present resistance respectively to 21 and 19 races. These groups along with group E, the most represented in the Ethiopian sampling, provided the basic germ plasm for group C ($\alpha \times E$) with the coffee types Geisha, S.17 Yrgalem, U.1 Dalecho and Sudan Barbuk; group J ($\gamma \times E$) with the types S.4 Agaro and S.6 Cioiccie; group I ($\alpha \times \gamma$) with S.12 Kaffa; and finally group W (C $\times \gamma$) represented also by S.12 Kaffa. That three different physiologic groups (γ, I, W) have been characterized in the same coffee type is surely a consequence of the random way the seed was collected (65, 67).

Group L associates the resistance of groups C and D. It was never found in material collected in Ethiopia, probably because of the rare occurrence there of group D plants.

Diploid Coffees

After the catastrophic appearance of orange rust in the *C. arabica* plantations, coffee growers started introducing other coffee species such as *C. liberica, C. canephora,* and others, which showed a high tolerance to the disease, although of inferior liquoring characteristics.

C. liberica was, perhaps, one of the first species to be introduced. It was brought to cultivation in old French Equatorial Africa on land where arabica growing had been abandoned, and from here imported by many countries, including Mauritius, Reunion, Madagascar, Java, and India. (77). Besides its apparent resistance to rust, the species possessed other attractive characteristics such as drought tolerance, ability to grow in poor soils, and good vigor. The popularity of this species declined,

Table 1 Races of *H. vastatrix*, differentials, and coffee physiologic groups represented by type clones[a]

Mayne's physiologic races	Physiologic races of *H. vastatrix* and number of each type culture	Postulated rust genotypes	A — 832/1-Híbrido de Timor-1[a]	R — 143/269-Híbrido de Timor-2[b]	1 — H.420/10 b,c	2 — H.420/2 b,c	3 — H.419/20 b,c	G — 33/1-288.23[b] ($S_H3,5$)	H — 34/13-S.353 4/5 ($S_H2,3,5$)	M — 644/18-Kawisari hybrid[b]	β — 849/1-Matari[b]	E — 63/1-Bourbon[b] (S_H5)	a — 128/2-Dilla & Alghe[b] (S_H1)	C — 87/1-Geisha ($S_H1,5$)	D — 32/1-DK 1/6[b] ($S_H2,5$)	Y — 635/2-S.12 Kaffa[b] (S_H4)	J — 110/5-S.4 Agaro ($S_H4,5$)	I — 134/4-S.12 Kaffa ($S_H1,4$)	W — 635/3-S.12 Kaffa ($S_H1,4,5$)	L — 1006/10-K.P. 532, tree 31 ($S_H1,2,5$)	Q — 1621/13-C. congensis Uganda[b]	P — 681/7-C. canephora v. Ugandae[b]	K — 829/1-C. canephora[b]	B — 263/1-C. congensis Uganda[b]	N — 168/12-C. excelsa Longkoi[b]	F — 369/3-C. racemosa[b]
2	I (22)	v 2, 5									S	S			S										S	S
1	II (15)	v 5									S	S													S	S
	III (37)	v 1, 5									S	S	S	S										S	S	S
	IV (32)										MS														S	S
	VI (71)	v 3, 5										S													S	S
	VII (130a)										S														S	S
	VIII (166)	v 2, 3, 5						MS	S		S	S	S	S	S										S	S
4	X (137a)	v 1, 4, 5									S	S	S	S		MS	MS	MS	MS						S	S
	XI (221)										MS												S		S	S

XII (167a)	> 1, 2, 3, 5	
XIII (138a)	1, 2, 3, 5	
XIV (178a)	> 2, 3, 4, 5	
XV (70)	> 4, 5	
XVI (178c)	> 1, 2, 3, 4, 5	
XVII (292)	> 1, 2, 5	
XVIII (92)		
XIX (264)	> 1, 4	
XX (394)		
XXI (256)		
XXII (535)		
XXIII (292a)	> 1, 2, 4, 5	
XXIV (22a)	> 2, 4, 5	
XXV (815)		
XXVI (816)		
XXVII (264a)		
XXVIII (999)		
XXIX (1321)		
XXX (1326)		
XXXI (1302)		
XXXII (256a)		

a Blanks = resistance reactions, S = susceptible, MS = moderately susceptible, MR = moderately resistant.
b Used as differential.
c H. 419 = 1535/33 Mundo Novo X HW 26/13. H. 420 = 1535/33 Mundo Novo X HW 26/14. HW. 26 = 19/1 Caturra X 832/1 Híbrido de Timor.

however, after some years; the main possible causes of this were the difficulty in curing the coffee, its lower prices relative to the arabica type, and its increased susceptibility to leaf rust disease (20).

C. canephora was the other species introduced because of its rust-tolerant qualities and greater vigor and productivity than *C. arabica,* hence its general designation, "Robusta."

The screening for rust resistance in coffee species other than *C. arabica* carried out in the CIFC has been mainly on *C. canephora* because of its economic importance (6). The results have shown that, except in those countries where a natural or experimental selection has never occurred, plants fully resistant (group A) to the rust races are commonly found in a number of accessions, contrary to what happens with *C. arabica.* Simultaneous with the presence of resistant plants, some of the accessions show individuals fully susceptible (group F), or to a lesser extent, with partial resistance to some of the races (groups Q, P, K, and B). This segregation is in agreement with the allogamic nature of these coffees.

The species *C. liberica, C. dewevrei* and its varieties *excelsa* Chev., *neo-Arnoldiana* Chev., *aruwimiensis* de Wild. and *dybowskii* Pierre, *C. abeokutae, C. Klainii, C. eugenioides, C. ligustroides,* and *C. salvatrix* have shown, similarly to *C. canephora,* individuals fully resistant or fully susceptible (6). Selection of resistant plants in the diploid species therefore seems to be much less critical than in *C. arabica.*

All the plants of *C. bengalensis* Roxb. ex Heyne and *C. lebruniana* Germ. et Kesl., one plant of *C. humilis,* and most of the plants of *C. racemosa* so far tested proved to be susceptible to all the races of *H. vastatrix.* The species *C. travancorensis* Wight. and *C. Wightiana* Wall have not been inoculated at the CIFC, but some reports (51) indicate that these species are susceptible, at least, to the most common races.

Interspecific Hybrids

The introduction of new coffee species in the regions devastated by *H. vastatrix* led in some cases to the natural appearance of interspecific hybrids and in others to their artificial production. These hybrids were formed despite the natural barriers of cytological nature between the tetraploid *C. arabica* and the other diploid species. Because of this, most of the *C. arabica* × *Coffea* spp. hybrids were practically useless on account of their triploid condition, except in those cases where the *Coffea* spp. was artificially duplicated or when nature itself did all the work. In the cases of productive interspecific hybrids, propagation by seed gives rise to a very heterogeneous, largely worthless offspring, and maintenance of the valuable types requires vegetative propagation (grafting). These hybrids produce in general many "empty fruits," and abnormal beans. In spite of this, the literature cites some hybrids which, in one way or other, had, or still have, a role in coffee culture.

C. ARABICA × C. LIBERICA HYBRIDS   According to Cramer (20) the oldest interspecific hybrid known in Java was the Kalimas, found in 1886 at the Kalimas estate and resulting from a cross between arabica and liberica. About a decade later, a group of analogous hybrids (Kawisari) appeared. Out of four hybrids, Kawisari

B and D were selected as the most valuable and used for field grafting on old unproductive liberica trees. They were quite resistant to rust and grew in places where pure arabicas failed.

The Kawisari hybrid accession in our collection indicated the presence of seedlings of groups M (with resistance to 29 races) and of group E.

When in 1925–1926 the Central Coffee Research Institute of Balehonnur, in Mysore, India, was established, advantage was taken of some hybrids between *C. liberica* and *C. arabica* which had been previously produced by many growers attracted by the vigorous growth of liberica and apparent tolerance to the rust. Some outstanding plants were then marked and studied. Several mother plants like S.26, S.31, S.71, and S.73 were selected, and their selfed progeny backcrossed to Kent or Coorg gave rise to some tetraploid interspecific selections such as S.288, S.333, S.353, and S.795. According to Narasimhaswamy (50), S.795 has vigorous growth and high resistance to the common races of *H. vastatrix* that affect Coorg and Kent, is a good cropper under widely varying conditions, and has good cup quality.

The selections of this origin which have been screened at the CIFC (17, 61), such as S.288–23, S.353 4/5, S.795 and series B.A., showed plants of group G (with resistance to 25 races), group H (with resistance to 26 races), and group E.

In 1935, a natural hybrid between *C. arabica* and *C. dewevrei,* which was designated as C.387, was found in Brazil. Selected plants of its progeny with good yields were backcrossed several times to Bourbon vermelho or Mundo Novo at the Instituto Agronómico de Campinas (49). Seedlings of C.387 tested in the CIFC presented reaction spectra of groups A, M, and E.

C. ARABICA X C. CANEPHORA HYBRIDS   Probably, some of the oldest representatives of this type were, like the Arabica X Liberica, those naturally or artificially produced in Java. Cramer (20) refers to the Bogor Prada Hybrid, introduced at Bangelan in 1914. The original aim in producing these hybrids was to obtain forms suitable for the lower Arabica belt. Coffee literature also refers to other hybrids of this type such as the "Devamachy hybrids," obtained in India, naturally or under controlled pollination, which have been the object of much research. Other references are found to the hybrids *C. arabica* X *C. canephora* (duplicated) artificially obtained in Brazil (13) and the Ivory Coast (10, 11), respectively known as Icatu and Arabusta. However, perhaps the most conspicuous of these hybrids is the tetraploid arabicoid Híbrido de Timor, a natural cross found in the Portuguese Timor (2, 35, 71).

In the CIFC, Híbrido de Timor and Icatu have been studied. Híbrido de Timor appeared spontaneously in Timor and began to be cultivated in this territory in a private estate by the second half of decade 1940–1949 to replace the local arabica seriously damaged by the rust. Since 1956 (35) its growing expanded to practically the whole island. Its heterogeneous population with plants showing promising yields, regular liquoring quality, and a percentage of caffeine a little below the standards of arabica and robusta coffees, offers good grounds for selection. This hybrid presents already an arabica phenotype, is self-fertile, and bears a number of chromosomes $2n = 44$ (57).

Seeds of Híbrido de Timor have been received in the CIFC several times. The two first plants raised from seed received in 1955, CIFC 832/1 and CIFC 832/2, have been intensively inoculated ever since with hundreds of isolates of *H. vastatrix* from different origins and proved to be resistant to all of them. The screening of several thousands of seedlings of this hybrid received on other occasions indicates the presence of a slight segregation characterized by individuals of group A, group R (resistant to 22 races), and group E, with a predominance of about 95% of group A.

It is interesting to note that the hybrid between Caturra Vermelho CIFC 19/1 and the Híbrido de Timor CIFC 832/1 backcrossed to Mundo Novo has produced some seedlings which, initially included in group A, now give origin to the new groups 1, 2, and 3.

Some plants of Híbrido de Timor seem to possess, in addition, some resistance to the coffee berry disease (CBD), caused by a strain of *Colletotrichum coffeanum* Noack, according to the observations made in Tanzania (9, 30, 31, 70) and the laboratory tests carried in the CIFC and in Angola.

Seed of Híbrido de Timor has been freely provided to 27 countries. Field trials have been set up in Angola, Brazil, Colombia, and Costa Rica, either with the hybrid proper, or with this material already crossed and backcrossed to commercial cultivars.

Icatu, a hybrid produced at the Instituto Agronomico de Campinas in 1950, has already been backcrossed to the arabicas Mundo Novo, Caturra Vermelho, and Bourbon Amarelo. The progenies screened at the CIFC indicate the presence of seedlings belonging to groups A, E, and possibly to new groups (D. Marques, unpublished).

RACES OF THE PATHOGEN

The apparent loss of resistance in coffee cultivars in India caught the attention of Mayne in the early 1930s at the Balehonnur Coffee Station. By means of experimental inoculations not only on detached leaves floating on water but also on leaves attached to coffee trees, he was able to differentiate with local rust samples and host differentials four physiologic races of *H. vastatrix* (44, 47). It was thus explained why varieties considered resistant at one time appeared susceptible later.

The work carried out by the CIFC with 779 rust samples from over 30 different regions of the world and with a vast collection of coffee germ plasm led to the differentiation of a total of 30 physiologic races (17, 60) as shown by their different infection spectra on a set of 17 differential hosts. These hosts are clonal lines of 5 *C. arabica* selections, 6 tetraploid hybrids of *C. arabica* × *Coffea* spp, and 6 *Coffea* spp. selections.

The coffee-rust interactions most commonly found in the CIFC are the *fl t* (association of small chlorotic flecks and minute tumefactions) and the type 4 (large uredosporic pustules), usually represented by R (resistant) and S (susceptible), respectively. However intermediate reactions are also observed in certain host-race combinations, designated as MR (moderately resistant) and MS (moderately susceptible). A more detailed description of these infection types is given elsewhere (24). Special reference must be made to the reaction t, characterized by the formation of swellings or tumefactions, either punctiform at the site of the stomata where the rust penetrated and usually associated with flecks, or somewhat irregularly shaped areas surrounding to a certain extent a type 0 reaction (larger flecks) or, even in certain cases an intermediate type MR.

Table 1 presents the list of the differentials and the physiologic races as well as their respective interactions. Races not mentioned in this table are race V, unfortunately lost, and Mayne's race 3, which was never established in the CIFC.

The geographical distribution of rust races is given in Table 2.

From the races listed, only two of them (races VI and XVIII) are innocuous to *C. arabica* and the tetraploid hybrids, but there are others such as races IV, XI, XIX, XX, XXI, XXVII, and XXXII, which by the scarce number of their hosts attacked, and the intermediate type of reaction induced on them seem to be more linked to the diploid coffees. These races have been, in addition found mostly in areas where these coffees are predominant.

Race II, one of the races with narrower infection spectrum, is of widespread occurrence in the world, having been detected in 30 out of 33 different world areas in Africa, Asia, and America. This race is in addition the most prevalent in our sampling, as it has been isolated from more than 50% of the rust samples received. Its generalized presence is probably a consequence of the genetic homogeneity of the majority of arabica cultivars prevalent throughout the world.

Other races with broader infection spectra have been isolated from areas where the presence of hosts with genes complementary to those of the races present have exerted a directional selection in the pathogen population. This is the case of race I, which soon after the discovery of the Kent's selection in India, overcame its resistance and was easily detected. This race is particularly expanded in Kenya where Kent's derivatives K.7 and S.L.6, susceptible to it, constitute a large proportion of the local arabica cultivated coffees. It represents in our screening about 50% of the 152 rust samples received from that country.

The existence of races apparently bound to certain regions (races VIII, XII, XIV, XXIII, XXV, XXVIII, XXXI in India; races XXII, XXVI, XXIX, XXX in the Portuguese Timor) may be tentatively explained by the presence of certain peculiar hosts, more or less confined to those regions, such as interspecific hybrids, bearing genes for resistance coming from species other than *C. arabica,* which replaced the pure Arabica coffees seriously damaged. The variability of the host genotypes at play has probably accelerated the detection of these races (59).

Considering that Ethiopia is the center of origin of *C. arabica* and most probably of *H. vastatrix* as well, it is probable that host and parasite have undergone a parallel evolution, and that the work of natural selection has given rise to resistant plants on one side and to a number of physiologic races of the pathogen on the other (21). However, the screening of 66 rust samples received from several areas of Ethiopia on different occasions (3, 24) led only to the identification of races I(?), II, III, and XV. The presence in Ethiopia of coffee genotypes other than those matched by the above races, as proved in the CIFC, makes it possible to anticipate the existence

Table 2 Geographical distribution of the differentiated physiologic races of *H. vastatrix*

Origin	Number of tested cultures	Number of times each race was identified													
		I	II	III	IV	V	VI	VII	VIII	X	XI	XII	XIII	XIV	XV
Angola	91	9	36	15				1							15
Argentina	1		1												
Brazil	86		84												2
Cambodia	12		8				1				3				
Cameroon	17		16	1											
Central African Republic	9		2			6									
Ceylon	24	6	14												4
Comores	1		1												
Democratic Republic of Congo	6	1	5												
Ethiopia	69	1	30	37											1
Guinea Republic	3		3												
India	24	3	1	2					4			1		1	
Indonesia	7	3		1											3
Ivory Coast	25		22		3										
Kenya	152	70	76					1							2
Laos	4		4												
Madagascar	12	1	9								2				
Malawi	1		1												
Mauritius	1		1												
Mozambique	14						14								
Nigeria	11		7	4											
Papua	3	1	2												
Philippines	12	2	8						1				1		
Republic of Vietnam	2	1	1												
Rhodesia	6		6												
São Thomé	17	1	13	1											1
South African Republic	13		13												
Swaziland	1		1												
Tanzania	30	6	9	8							1				
Timor (Portuguese)	88	4	47	1											3
Uganda	13	1	7		3										
Zambia	13		13												
Zanzibar	11						11								
Total number	779	110	440	70	9	3	26	2	4	1	6	1	1	1	31

XVI	XVII	XVIII	XIX	XX	XXI	XXII	XXIII	XXIV	XXV	XXVI	XXVII	XXVIII	XXIX	XXX	XXXI	XXXII
											15					
			1								1					
1	2						3	3	1			1		1		
				3												
			1													
	3			1				2								
						20			1	6			4	2		
					1											1
1	5	1	1	4	1	20	3	5	2	6	16	1	4	2	1	1

there of other races. Race X, for instance, not yet identified in material from Ethiopia, but with virulent genes complementary to the genes of plant CIFC 635/3 (S.12 Kaffa), a selection only received from that country, is almost certain to exist there. Failure in finding it might be ascribed to a deficient rust sampling.

Although the number of rust samples received at the CIFC might be looked at as reasonably high, one should bear in mind that most of them come mainly from cultivated or semi-spontaneous coffees. Considering the large number of species in the genus *Coffea,* our rust sampling is therefore quite deficient and does not allow any conclusions on the significance of wild coffees as sources of infection for the cultivated ones.

INHERITANCE OF RESISTANCE

The genetic study of the resistance to *H. vastatrix* in *Coffea* spp. was first attempted in India by Mayne (45, 46), who reported the existence of two factors responsible for resistance of some local selections to the prevalent races. After the characterization in the CIFC of a number of physiological groups of coffee, the studies carried out by Noronha-Wagner & Bettencourt (5, 52) with *C. arabica* (groups β, E, C, α, D, I, W, J, γ, L) and some interspecific hybrids of arabica phenotype (groups G, H), allowed the identification of 5 dominant genes S_H1, S_H2, S_H3, S_H4, and S_H5 which, simple or associated, condition the resistance of the groups to some rust races. Genes S_H2 and S_H3 correspond to the two factors postulated by Mayne (46). Gene S_H3 has been found only in plants of arabica phenotype originally from India; thus, it probably comes from *C. liberica,* one of the ancestral parents of some Indian selections.

Recent work carried out with progenies and hybrids of the coffee seedling CIFC 832/1 (Híbrido de Timor - 1) of group A, and of CIFC 1343/269 (Híbrido de Timor - 2) of group R, seems to indicate that new genes are present in these plants.

Flor's concept of the gene-for-gene relationship between host and pathogen (34) has been successfully applied to the coffee-leaf rust association (5, 52) through the analysis of the reaction spectra of each of 12 coffee clones and their hybrids towards 18 physiologic races of *H. vastatrix.* This analysis revealed the genotypes of 14 rust races and predicted the likely occurrence of others not yet differentiated. Unfortunately, the virulence genes of the referred races cannot be confirmed until the sexual stage of *H. vastatrix* is discovered.

BREEDING FOR RUST RESISTANCE

Coffee improvement for rust resistance has deserved special attention in the *C. arabica* breeding programs of several coffee experimental centers.

In India, the attainment of the selections S.288, S.333, S.795, series B.A., etc, already referred to, was indeed a good landmark in coffee improvement for rust resistance, and this material has been largely utilized in research centers. At present, the breeding program of the Coffee Research Institute of Balehonnur includes the analysis of new progenies of these selections and of hybrids of many different genetic combinations. From 22 of these combinations supplied by the CIFC, the most outstanding are S.2581, S.2582, S.2591, and S.2593, which resulted from crosses between Híbrido de Timor and S.795, S.L.30, Kent's, and S.16 Wollamo, respectively (71).

In Tanzania, at the inception of the Lyamungu Coffee Research Station in 1934, a search for high-yielding trees of good quality was carried out within the Bourbon and Kent varieties. The selection work with the Kent yielded the commercial selections of series 'KP', 'H', 'F', and 'X' (29), which maintained the Kent type of resistance. New progenies of these series and hybrids with different factors for rust resistance were being studied there in 1962 (28, 29). The selection work in Kenya, also started with the Kent variety, produced the cultivars K.7 and SL.6, at present well spread in the low altitude coffee plantations (32). Despite bearing the factor S_H2 the growing of these coffees has to be supplemented with fungicide treatments.

In Brazil, breeding work for rust resistance was initiated in 1953 in the Department of Genetics of the Instituto Agronómico de Campinas (IAC) with 76 coffee selections coming from Tanzania, Kenya, India, Ethiopia, and Sudan through the USDA. This work has already yielded coffee populations bearing separately the genes S_H1, S_H2, S_H3, and S_H4 in the homozygous condition, and with vigor and yield similar to Bourbon Vermelho (13, 14). The screening of the progenies of these coffees, as well as of many hundreds of others, has been made in the CIFC which has maintained a very close cooperation with the IAC since the beginning of the program. The characteristics of the progenies of this selected material, namely 1120 (X.321), 1123 (KP.263), 1128 (KP.532), 1130 (H.1), 1132 (KP.423), 1136 (H.66), and 1151 (K.7) with S_H2 S_H2; 1137 and 1350 (Geisha) with S_H1 S_H1; 1110 (BA.10) and 1116 (BA.16) with S_H3 S_H3; 1125, 1472, 1474, 1475 (S.6 Cioiccie), 1164, and 1471 (S.4 Agaro) with S_H4 S_H4, offer good prospects for its utilization in large-scale planting, as soon as there is enough information on its behavior in different ecological regions of the country (14). F_2 hybrids between these progenies and the best yielding varieties Mundo Novo, Catuai, and Bourbon Amarelo, are also being evaluated. The research work of IAC has been in addition directed towards material derived from C. arabica X C. canephora hybrids, such as Icatu (IAC. H.2460) and Híbrido de Timor. Special emphasis has been also put on the study of the F_2 and F_3 generations of 403 hybrids synthesized in the CIFC, involving a gamut of different genotypes for rust resistance.

Since 1970 the Universidade Federal de Viçosa in Brazil has carried out work with F_2 and F_3 populations of the CIFC hybrids. Outstanding among the results so far obtained is progeny UFV 386 of the F_3 generation of the CIFC hybrid HW.26 (Caturra Vermelho X Híbrido de Timor), whose F_2 generation was selected at the Instituto de Investigaçao Agronómica de Angola (IIAA). This progeny shows good uniformity, vigor, yield, and resistance to all the rust races in 96% of the seedlings.

In Colombia, since 1965 the Breeding Department of Cenicafe in Chinchiná, with the cooperation of the CIFC where all the screening is done, has been selecting for rust resistance with germ plasm received from the USDA. Progenies with vigor and yield close to Bourbon and possessing individually the factors for rust resistance in

the homozygous condition have been selected. The most important of these progenies under field observation in several ecological conditions are: derivatives of H.1,
KP.532, S.286–7, F.840, and F.502, all S_H2 S_H2; Geisha A (S_H1 S_H1), and S.4
Agaro (S_H4 S_H4) (15). The same department is also studying the F_2 of interspecific
hybrids between some commercial varieties such as Bourbon, Caturra,
S. Bernardo, etc and Híbrido de Timor (CIFC 1343). Some progenies of these
hybrids seem to be promising not only for productivity but also rust resistance
(16).

The work in Angola was started in 1960 at the Centro de Estudos da Chianga
from the IIAA. Selections from several origins as well as F_1 and F_2 hybrids developed in the CIFC were supplied and studied. The successful work carried out there
allowed the selection of progenies of the following: Dilla & Alghe-IIAA 859
(S_H1 S_H1); S.12 Kaffa-IIAA 490 (S_H4 S_H4); F_3 of Caturra Vermelho X Geisha-
IIAA 1518 (S_H1 S_H1); F_2 of Caturra X S.795-IIAA 860 (S_H3 –); F_3 of Caturra
X S.4 Agaro-IIAA 1530 (S_H4 S_H4); F_2 of K.7 X Dilla & Alghe-IIAA 1526
(S_H1–, S_H3–); F_2 of S.333 X Dilla & Alghe-IIAA 1517 (S_H1–, S_H3–); F_2
of Caturra Vermelho X Híbrido de Timor-IIAA 857; F_3 of Caturra Vermelho X
Híbrido de Timor-IIAA 1525. All this material with characteristics very near the
local susceptible arabica cultivar is now being evaluated under different ecological
conditions (A. M. Gaspar, A. J. Bettencourt, A. M. Ponte, unpublished).

Other field studies on coffee selection for rust resistance have taken place since
1965 in the Estação Regional do Uíge, an experimental center of the Instituto do
Café de Angola (ICA), using coffee germ plasm provided by the CIFC. The most
promising coffees selected in the course of the last six years are progenies of the
following: S.795 - U 343 and U 344 (S_H3 S_H3); F_2 of Caturra Vermelho X Geisha -
U 289 and U 293 (S_H1 S_H1); F_2 of Caturra Vermelho X S.6 Cioiccie - U 278
(S_H4 S_H4); F_2 of Caturra Vermelho X S.4 Agaro - U 346 (S_H4 S_H4); F_2 of Caturra
Vermelho X Híbrido de Timor - U 189. This latter progeny has been resistant to all
rust races. As to the others, despite the presence of the local prevalent races III (v_1
v_5), I (v_2v_5), VII (v_3v_5), and XV (v_4v_5) of $H.$ vastatrix, which infect plants with the
genes S_H1, S_H2, S_H3, and S_H4, respectively, they have not been very much
affected. On the other hand, the progenies bearing the S_H2 gene suffered a uniform
and strong attack of rust; thus they might be cultivated only with chemical control
(A. J. Bettencourt, D. T. Fernandes, unpublished).

In its program of assistance to the coffee growing countries, the CIFC has continued to make hybrids, aiming at the incorporation of factors for resistance to $H.$
vastatrix in 46 commercial varieties of different origins (3). Some of these hybrids
have been backcrossed to Catuai, one of the most outstanding varieties concerning
yield and reduced size. This latter characteristic is very important because it eases
the fruit picking and the phytosanitary treatments. About 1000 progenies of these
hybrids in the F_1, F_2, BC_1, and BC_2 have been selected in greenhouse conditions
according to rust resistance, vegetative vigor, reduced size, and fruit characteristics.
This material is now being evaluated in Angola, Brazil, Colombia, India, etc, as to
productivity and resistance to the local races of $H.$ vastatrix (13, 69).

HISTOPATHOLOGY AND BIOCHEMISTRY OF RESISTANCE

Histopathology

Histopathological studies of coffee-rust interaction have been carried at the CIFC in compatible and incompatible combinations attempting to shed some light on the phenomena of resistance and susceptibility. Unfortunately, only the compatible combinations (type 4 reaction) have been studied under the electron microscope (40, 55, 58), and in this case the general features observed do not differ basically from those found in other host-rust associations.

The most common reaction type found in the incompatible combinations is the chlorotic fleck (*fl*) usually associated with punctiform tumefactions (*t*). The phenotypic expression *fl t* of the interaction occurs generally 12–15 days after the inoculation. The successive microscopic stages of this reaction type may be summarized as follows.

After the appressorium formation, the fungus penetrates through the stoma and a hypha grows into the substomatical chamber, producing at the advancing tip two thick lateral branches. The hypha and its branches resemble in the whole an anchor. From each septate lateral branch is borne a hypha which directs itself upwards to the subsidiary cell and there forms an encapsulated haustorium. The subsidiary cells thus seem to be the first host cells to be parasitized.

The initial growing hypha proceeds down intercellularly and vertically without much branching, reaching within 2–3 days after the inoculation the second or third layer of the mesophyll. From that time on, the infection process clearly seems to slow down and the further mycelial development becomes very poor, seldom reaching the palisade tissue. Very few haustoria are observed.

Only the cells showing haustoria present granular cytoplasm. It is possible, however, that other cells in their vicinity migh be in some way altered (78), although this alteration could not be detected under the optical microscope.

About 6 days after the inoculation, a progressive increase in the volume of the mesophyll cells and the consequent disappearance of the intercellular spaces begin to be noticed. The first cells to increase in size are those subjacent to the substomatical chamber. Later, all the spongy cells become irregularly shaped and some of them with thicker walls. Their nuclei also increase in volume (1.5 X) and so do the nucleoli (2–3 X). The palisade parenchyma does not show any morphological alterations.

As a result of the abnormal growth of the spongy cells, the lower epidermis of the leaf in the affected area is pushed away and a convex area corresponding to the tumefaction is formed (56). During the process of cell volume increase, the mycelium is apparently crushed between the cells.

The symptoms on coffee leaves designated in Kenya and Tanzania as "weak spots," of unknown etiology, are quite similar to the described *fl t*. Cross sections of material, kindly supplied by Dr. R. T. A. Cook, showed the same type of histological alterations, although the presence of mycelium could not be positively demonstrated on the observed sections.

The occurrence of flecks without external symptoms of tumefaction only very seldom have been noticed under our conditions. However, the few cases observed indicated that the same fungal development and host cell characteristics described for the fl t were found. The only difference was that the volume increase of the cells was not big enough to cause the deformation of the lower epidermis.

In this same case of flecks without visible tumefactions, another microscopic feature was accidentally observed. A band of 4–5 meristematic cells concave to the infection point was formed in the second or third mesophyll layer. The callus tissue then produced by this meristem came to exert pressure against the epidermis whose cells died (37). The fungus development was then practically confined to the penetration area, and a so-called limited zone was formed. In the same cross section, the meristematic tissue could appear as a waving ribbon with several limited zones corresponding to different infections.

The occurrence of reaction type O, frequently surrounded by a somehow irregular tumefaction, is found mainly in coffee species other than $C.$ $arabica$ or in $C.$ $arabica$ hybrids with "blood" of another coffee species. This type is characterized by a profuse mycelial development in the chlorotic area, with cells of normal size and clear evidence of haustoria. Limiting that area, a typical tumefaction zone is often noticed, formed by cells of increased size which act as a kind of barrier to the further development of the mycelium.

Biochemistry

Although a great deal is already known about the genetics of resistance of $C.$ $arabica$ and of some tetraploid interspecific hybrids to the races of $H.$ $vastatrix$, almost nothing is yet known about the possible biochemical mechanisms whereby that resistance is expressed.

Attempts to correlate host resistance with higher amounts of phenols and of phenol oxidase activity in coffee leaf extracts have been made (8, 36), comparing healthy and inoculated seedlings of different physiologic groups, but in every case the results were inconclusive.

Exploratory work has been lately carried out at the CIFC on the possible induction of enhanced levels of antifungal substances following inoculation of $C.$ $arabica$ with the rust (62). $C.$ $arabica$ cultivars differing in a single gene for resistance as well as rust races differing in a single gene for virulence allowed the establishment of compatible and incompatible combinations of $H.$ $vastatrix$ and $C.$ $arabica$. All the incompatible combinations produced diffusates which caused significantly increased inhibition of all races tested. This inhibition was observed in relation to the percentage germination and germ tube length. Conversely, diffusates from compatible interactions did not cause significant increase in inhibition compared with diffusates from uninoculated leaves. Similar results were obtained with other compatible and incompatible interactions of host and pathogen. Previous inoculation of coffee leaves with an avirulent race of $H.$ $vastatrix$ significantly (P = 0.001) increased resistance of these leaves to a compatible race.

One simple explanation of the association between incompatibility and toxicity is that the S_H gene products are involved with "recognition" of the incompatible races, and that after recognition a common reaction which results in enhanced toxicity, proceeds (62). It will be interesting to know which fungitoxic substance(s) are involved in the incompatible combinations and whether they are different in each cultivar-rust race used.

Based on the idea explored by some research workers (27) that common antigens shared by the host and pathogen may determine the nature of disease reactions, Alba and co-workers (1) tried to study the antigenic behavior of saline soluble substances obtained from uredospores of *H. vastatrix* race II and of susceptible and resistant coffee leaves. The results obtained so far showed no common antigens present, indicating, in addition, that the fungal and host antigens had an opposite polarity in immunoelectrophoresis assays.

CONCLUDING REMARKS

Even though *H. vastatrix* is still considered as an enemy of coffee growing in the areas where the rust is endemic, and as a strong menace in those countries where it does not exist, the prospects of its control with resistant varieties are by far more optimistic than ever. In fact, the combined efforts of some coffee research centers and the CIFC are now offering, in relation to the arabicoid coffees, new promising leads. These are essentially the following: (*a*) selections bearing one or two of the known genes for rust resistance in the homozygous condition with yield and other characteristics similar to those of the traditional arabicas; (*b*) hybrids between commercial varieties and selections with known S_H factors; (*c*) interspecific hybrids such as Icatu and Híbrido de Timor, and hybrids between the latter and commercial varieties, some of them with total resistance to *H. vastatrix;* (*d*) a great deal of other basic material with a gamut of genotypes, under evaluation in field trials.

The material indicated in *a* has already been largely distributed to Brazilian farmers in small amounts, as a mixture of genotypes (the so-called Iarana), for field evaluation and comparison with the traditional varieties that demand phytosanitary treatments. The remaining material is still under study, but some of it presents progenies very promising concerning productivity and rust resistance. The large spectrum of resistance of the material in *c* might encourage in the near future its use in areas highly damaged by the rust, and/or where chemical control is not economically suitable.

The sources of resistance determined in greenhouse conditions in the last two decades and utilized as basic material for arabica breeding have been selected only towards vertical or differential resistance. However, field trials established in certain regions with this material seem to indicate as well the presence of horizontal resistance, mainly in the Ethiopian collections. Coffee breeders are therefore urged to look for this type of resistance whose likely utilization might be of great interest for coffee improvement against the rust.

In relation to *C. canephora,* the problem of resistance to *H. vastatrix* is much less critical, due to its higher generalized resistance to the rust and wide variability.

Literature Cited

1. Alba, A. P. C., Namekata, T., Moraes, W. B. C., Oliveira, A. R., Figueiredo, M. B. 1973. Serological studies on coffee rust. *Arq. Inst. Biol.* 40:227–31
2. Bettencourt, A. J. 1973. Considerações gerais sobre o 'Híbrido de Timor.' *Instituto Agronómico,* Circ. #23. Campinas, Brazil
3. Bettencourt, A. J., Lopes, J. 1965. Breeding of *C. arabica* L. for rust resistance. *Centro Inv. Ferrug. Caf. Prog. Rept.* 1960–65:124–34
4. Bettencourt, A. J., Lopes, J. 1968. Preliminary report of the coffee leaf rust (*Hemileia vastatrix*) material received from the FAO Coffee Mission to Ethiopia 1964–65. *FAO Coffee Mission to Ethiopia* 1964–65:124–40. Rome: FAO
5. Bettencourt, A. J., Noronha-Wagner, M. 1971. Genetic factors conditioning resistance of *Coffea arabica* L. to *Hemileia vastatrix* B. & Br. *Agron. Lusit.* 31:285–92
6. Bettencourt, A. J., Rodrigues, C. J. Jr. 1965. Routine screening for resistance to *Hemileia vastatrix* B. & Br. on *Coffea canephora* Pierre and *Coffea* spp. accessions from different regions of the world. *Centro Inv. Ferrug. Caf. Prog. Rept.* 1960–65:100–20
7. Bock, K. R. 1962. Control of coffee leaf rust in Kenya Colony. *Trans. Br. Mycol. Soc.* 45:301–13
8. Bruges, J., Contreiras, J. 1967. Aspects biochimiques de la résistance du caféier à l'*Hemileia vastatrix. Port. Acta Biol.* 10:75–88
9. Cannell, M. G. R. 1971. *Coffee Res. Found. Kenya Ann. Rept.* 1970–71:41–48
10. Capot, J. 1972. L'amélioration du caféier en Côte d'Ivoire. Les hybrides "Arabusta." *Café Cacao Thé* 16:3–18
11. Capot, J., Dupautex, B., Durandeau, A. 1968. L'amélioration du caféier en Côte d'Ivoire. Duplication chromosomique et hybridation. *Café Cacao Thé* 12:114–26
12. Carvalho, A. et al 1969. Coffee (*Coffea arabica* L. and *C. canephora* Pierre ex Froehner). *Outlines of Perennial Crop Breeding in the Tropics,* ed. F. P. Ferwerda, F. Wit, 189–241. Wageningen: Landbowhogeschool. 511 pp.
13. Carvalho, A., Monaco, L. C. 1971. Melhoramento do cafeeiro visando a resistência à ferrugem alaranjada. *Ciênc. Cult. São Paulo* 23:141–46
14. Carvalho, A., Monaco, L. C. 1972. Adaptação e produtividade de cafeeiros portadores de factores para resistência à *Hemileia vastatrix. Ciênc. Cult. São Paulo* 24:924–32
15. Castillo, J., Lopez, S., Torres, E. 1972. Comportamento de introducciones de café com resistência à *Hemileia vastatrix* en Colombia. *Centro Nacion. Invest. Café,* Chinchiná, Colombia. 31 pp.
16. Castillo, J., Lopez, S., Torres, E. 1972. Plan de trabajo sobre resistência a la roya del Cafeto. *Centro Nac. Invest. Café,* Chinchiná, Colombia. 47 pp.
17. *Centro de Investigação das Ferrugens do Caffeeiro.* 1971. Oeiras, Portugal. 29 pp.
18. Chevalier, A. 1947. *Les Caféiers du Globe.* Fascicule III. Paris: Paul Lechevalier. 356 pp.
19. Chouleur, P. 1972. Les produits tropicaux et méditerranéens de la production à la consommation. *Marchés Trop.* 1415:3625–3833
20. Cramer, P. J. S. 1957. A review of literature of coffee research in Indonesia, ed. F. Wellman. *Inter-Am. Inst. Agric. Sci. Misc. Publ,* 15. Turrialba, Costa Rica. 262 pp.
21. D'Oliveira, B. 1951. The centers of origin of cereals and the study of their rusts. *Agron. Lusit.* 23:221–26
22. D'Oliveira, B. 1954–57. As ferrugens do caffeeiro. *Rev. Café Port.* 1(4):5–13; 2(5): 5–12; 2(6):5–13; 2(7):9–17; 2(8):5–22; 4(16):5–15
23. D'Oliveira, B. 1958. Selection of coffee types resistant to the Hemileia leaf rust. *Coffee Tea Ind. Flavor Field* 81:112–20
24. D'Oliveira, B., Rodrigues, C. J. Jr. 1959. Progress report to Ethiopia. *Garcia de Orta* 7:279–92
25. D'Oliveira, B., Rodrigues, C. J. Jr. 1961. O problema dos ferrugens do cafeeiro. *Rev. Café Port.* 8(29):5–50
26. D'Ornano, M., Chassevent, F., Pougneaud, S. 1965. *Composition et caractéristiques chimiques de Coffea sauvages de Madagascar.* I. *Recherches préliminaires sur la teneur en caféine et isolement de la cafamarine.* Presented at 2nd Colloq. Int. Chimie des Cafés, Paris
27. Doubly, J. A., Flor, H. H., Clagett, C. O. 1960. Relation of antigens of *Melampsora lini* and *Linum usitatissimum* to resistance and susceptibility. *Science* 131:229

28. Fernie, L. M. 1962. *Tanganyika Coffee Board Res. Rept.* 20–26
29. Fernie. L. M. 1964. *Handbook on Arabica Coffee in Tanganyika*, ed. J. B. D. Robinson, 1–8. Tanganyika Coffee Board. 182 pp.
30. Fernie, L. M. 1965. *Tanganyika Coffee Board Res. Rept.* 20–24
31. Fernie, L. M. 1967. *Tanganyika Coffee Board Res. Rept.* 22–27
32. Firman, I. D., Hanger, B. F. 1963. Resistance to coffee leaf rust in Kenya. *Coffee Turrialba* 5(18):49–59
33. Firman, I. D., Wallis, J. A. N. 1965. Low-volume spraying to control coffee leaf rust in Kenya. *Ann. Appl. Biol.* 55:123–37
34. Flor, H. H. 1955. Host-parasite interaction in flax-rust, its genetics and other implications. *Phytopathology* 45:680–85
35. Gonçalves, M. M., Daenhardt, E. 1971. A *Hemileia vastatrix* B. et Br. em Timor. Nota sobre a sua importância económica e melhoramento do cafeeiro face à doença. MEAU.- 666. *Missao Estudos Agronómicos Ultramar,* Lisboa, Portugal. 17 pp.
36. Guedes, M. E. M., Rodrigues, C. J. Jr. 1974. Disc electrophoretic patterns of phenoloxidase from leaves of coffee cultivars. *Port. Acta Biol.* 13:169–78
37. Hjodo, H., Uritani, J., Akai, S. 1968. Formation of callus tissue in sweet potato stems in response to infection by an incompatible strain of *Ceratocystis fimbriata. Phytopathology* 58:1032–33
38. Instituto Brasileiro do Café–Gerca. 1972. *Novos resultados de controle químico da ferrugem do cafeeiro no Brasil.* Presented at 64th Ann. Meet. Am. Phytopathol. Soc., Mexico City
39. Jones, P. A. 1957. Notes on the varieties of *Coffea arabica* in Kenya. *Selected Articles on Coffee Culture.* Kenya Coffee Board. 185 pp.
40. Kitajima, E. W. 1972. Ultraestrutura do fungo causador da ferrugem do cafeeiro nos tecidos da hospedeira. *Ciênc. Cult. Sao Paulo* 24:267–72
41. Krug, C. A., De Poerck, R. A. 1968. *World Coffee Survey.* Rome: FAO. 476 pp.
42. Large, E. C. 1940. *The Advance of the Fungi.* New York: Dover Publications Inc. 488 pp.
43. Leroy, J. F. 1961. Coffeae novae madagascariensis. *J. Agric. Trop. Bot. Appl.* 8:1–29
44. Mayne, W. W. 1932. Physiologic specialization of *Hemileia vastatrix* B. & Br. *Nature* 129:150
45. Mayne, W. W. 1935. Annual Report of the Coffee Scientific Officer, 1934–35. *Mysore Coffee Exp. Stn. Bull.* 13. 28 pp.
46. Mayne, W. W. 1936. Annual report of the coffee scientific officer, 1935–36. *Mysore Coffee Exp. Stn. Bull.* 14. 21 pp.
47. Mayne, W. W. 1942. Annual report of the coffee scientific officer, 1941–42. *Mysore Coffee Exp. Stn. Bull.* 24. 21 pp.
48. Meyer, F. G. 1965. Notes on wild *Coffea arabica* from southwestern Ethiopia, with some historical considerations. *Econ. Bot.* 19:136–51
49. Monaco, L. C., Carvalho, A., Antunes, C. S. N. 1967. Aproveitamento de uma combinaçao híbrida interespecífica para fins de melhoramento do cafeeiro. *Fitotec. Latinoam.* 4:113–21
50. Narasimhaswamy, R. L. 1960. Arabica selection S.795. Its origin and performance. A study. *Indian Coffee* 24:197–204
51. Narasimhaswamy, R. L. 1961. Coffee leaf disease (*Hemileia*) in India. *Coffee Turrialba* 3:33–39
52. Noronha-Wagner, M., Bettencourt, A. J. 1967. Genetic study of the resistance of *Coffea* spp. to leaf rust. I. Identification and behaviour of four factors conditioning disease reaction in *Coffea arabica* to twelve physiologic races of *Hemileia vastatrix. Can. J. Bot.* 45:2021–31
53. Ramos, H. C. 1973. Royas del cafeto (*Hemileia* spp.). Bibliografia. *Inst. Interam. Cienc. Agric. Org. Estad. Am. Turrialba,* Costa Rica. 71 pp.
54. Rayner, R. W. 1962. The control of coffee rust in Kenya by fungicides. *Ann. Appl. Biol.* 50:245–61
55. Rijkenberg, F. H. J., Truter, S. J. 1973. Haustoria and intracellular hyphae in the rusts. *Phytopathology* 63:281–86
56. Rijo, L. 1972. Histopathology of the hypersensitive reaction *t* (tumefaction) induced on *Coffea* spp. by *Hemileia vastatrix* Berk. & Br. *Agron. Lusit.* 33:427–31
57. Rijo, L. 1974. Observaçoes cariológicas no cafeeiro 'Híbrido de Timor.' *Port. Acta Biol.* 13:157–68
58. Rijo, L., Sargent, J. 1974. The fine structure of the coffee leaf rust, *Hemileia vastatrix. Can. J. Bot.* 52:1363–67
59. Rodrigues, C. J. Jr. 1972. *Resistance in Coffee to the Orange and Powdery Rusts.* Presented at 64th Ann. Meet. Am. Phytopathol. Soc., Mexico City
60. Rodrigues, C. J. Jr., Bettencourt, A. J., Lopes, J. 1965. Study of the physiologic

specialization of the coffee rust *Hemileia vastatrix* B. & Br. and selection of coffee clones for the establishment of a standard range of differential hosts for this rust. *Centro Inv. Ferrug. Caf. Prog. Rept.* 1960–65:21–27

61. Rodrigues, C. J. Jr., Bettencourt, A. J. 1965. Routine screening for resistance to *Hemileia vastatrix* B. & Br. on *Coffea arabica* L. accessions from different coffee producing regions of the world. *Centro Inv. Ferrug. Caf. Prog. Rept.* 1960–65:47–99

62. Rodrigues, C. J. Jr., Lewis, B. G., Medeiros, E. F. 1974. Relationship between a phytoalexin-like response in coffee leaves (*Coffea arabica* L.) and compatibility with *Hemileia vastatrix* Berk & Br. *Physiol. Plant Pathol.* In press

63. Rossetti, V., Silveira, M. L., Luensinger, B. M., Souza, S. R. 1974. Coffee Rust. Bibliography with abstracts 1953–73. *Inst. Biol.* São Paulo, Brasil 221 pp.

64. Schieber, E. 1972. Economic impact of coffee rust in Latin America. *Ann. Rev. Phytopathol.* 10:491–510

65. Sylvain, P. G. 1955. Some observations on *Coffea arabica* L. in Ethiopia. *Turrialba* 5:37–53

66. Sylvain, P. G. 1956. Le café du Yémen. *L'agron. Trop.* 11:62–73

67. Sylvain, P. G. 1958. Ethiopian coffee. Its significance to world coffee problems. *Econ. Bot.* 12:111–39

68. Thirumalachar, M. J., Narasimhan, M. J. 1947. Studies on the morphology and parasitism of *Hemileia* species on Rubiaceae in Mysore. *Ann. Bot. London* 11:77–89

69. Twentyfifth Annual Detailed Technical Report 1971–72. *Indian Coffee Board Res. Dep.*, 46–54

70. Vermeulen, H. 1966. *Coffee Res. Found. Kenya Ann. Rept.* 1964–65:57–60

71. Vishveshwara, S., Govindarajan, A. G. 1970. Studies on 'Híbrido de Timor' coffee collection. *Indian Coffee* 34:71–78

72. Waller, J. M. 1972. Coffee rust in Latin America. *Pest Article News Summ.* 18:402–8

73. Wallis, J. A. N., Firman, I. D. 1962. Spraying arabica coffee for the control of leaf rust. *East Afr. Agric. For. J.* 28:89–104

74. Wellman, F. L. 1952. Peligro de introducción de la *Hemileia* del café a las Americas. *Turrialba* 2:47–50

75. Wellman, F. L. 1953. The Americas face up to the threat of coffee rust. *For. Agric.* 17(3):1–7

76. Wellman, F. L. 1955. Past and present investigation on the common coffee rust, and their importance for tropical America. (Mimeo.) 74 pp.

77. Wellman, F. L. 1961. *Coffee: Botany, Cultivation and Utilization.* London: Leonard Hill. 488 pp.

78. Zimmer, D. E. 1965. Rust infection and histological response of susceptible and resistant safflower. *Phytopathology* 55:296–301

VARIATION AND SPECIATION IN THE FUSARIA[1]

♦3612

T. A. Toussoun
Institute for Fungus Research, 2417 Franklin Street, San Francisco, California 94123

Paul E. Nelson
Fusarium Research Center, Department of Plant Pathology, The Pennsylvania State University, University Park, Pennsylvania 16802

In the beginning there was confusion and disorder in *Fusarium,* with more than 1000 species, varieties, and forms named on the basis of superficial observations, with little or no regard to culturing of these specimens. Much of this work was done before pleomorphism and variation in fungi were recognized. Systematists of that era generally did not describe the whole organism, and made no attempts to look for similarities or relationships between fungi. Rather it was a period best described as being an orgy of naming new fungi. The necessity for a precise and reliable system of classification became apparent when it was realized that fusaria caused serious diseases of many plants. Wollenweber took up this challenge, making a career of the study of *Fusarium,* culminating in Wollenweber & Reinking's *Die Fusarien* (28). The genus *Fusarium* was reduced to 143 species, varieties, and forms grouped in 16 sections. To illustrate the scope of this contribution, Wollenweber & Reinking (28) list 77 synonyms for *F. avenaceum* and 133 for *F. lateritium* and its perfect state.

Since then, and as a result of increased knowledge of variation in fungi, further advances in the taxonomy of *Fusarium* have been achieved. These have led to differing viewpoints. Today we have 9 species with no varieties or forms, according to Snyder & Hansen (19–21); 44 species and 7 varieties grouped in 12 sections according to Booth (4); about 50 to 60 species and varieties according to Gerlach (5); 26 species and 29 varieties in 9 sections according to Bilai (3); 33 species and

[1]Contribution Number 827, Fusarium Research Center, Department of Plant Pathology, Pennsylvania Agricultural Experiment Station.

14 varieties in 13 sections according to Joffe (11); or 9 species and 9 varieties according to Messiaen & Cassini (14). It may seem to the reader that progress since Wollenweber's time may not be the most appropriate word to use. Nevertheless, progress has been made, for behind the differing numbers of *Fusarium* species accepted by various authorities lies an increasing area of agreement as to what constitutes a "good" species. We think that this may lead to a unified taxonomic system in the near future, an achievement hoped for by many *Fusarium* research workers (10).

At present there are basically two or perhaps three schools of thought on classification in *Fusarium*. It is essential to remember that all modern systems of *Fusarium* classification are based on the work of Wollenweber & Reinking (28). We shall deal with these in chronological order, for it is important to understand the philosophies involved to obtain a proper perspective.

SNYDER & HANSEN'S SYSTEM

Snyder & Hansen recognized that the most pressing need in fungal classification was for a practical and dependable taxonomic system. This became their overriding goal when they found Wollenweber's system undependable for identification of *Fusarium* species they encountered as plant pathologists. They realized that Wollenweber & Reinking's (28) separation of species and varieties were too narrowly drawn because they were based on small differences in length and width of conidia, septation, position of chlamydospores, pigmentation of colonies, and other cultural characteristics. The uncertainties of the Wollenweber & Reinking system were compounded when Snyder & Hansen began to analyze, by Hansen & Smith's (8) single-spore method, the variability of the clones they had assembled. They found that the progeny of a single conidium gave rise to individuals which fell into different varieties, species, and even into different subsections established by Wollenweber & Reinking (28). When Snyder & Hansen expanded their studies of variation to a worldwide collection, the weakness of the Wollenweber system became more evident, and they proposed what has since become known as the nine-species system (19–21).

The species proposed were formed to reflect the morphologic features of whole sections or several sections of Wollenweber & Reinking (28). The separation of species was based on the shape of the macroconidia, the sole criterion ascertained by Snyder & Hansen to be wholly reliable, as well as such secondary characteristics as presence or absence of chlamydospores or microconidia and the shape of certain microconidia. All other morphologic criteria were deemed unreliable. While this proposal seemed drastic to many, it is worth noting that Wollenweber (26) considered conidium shape as the most important character used to divide the genus into sections, although this was not emphasized in *Die Fusarien* (28).

Snyder & Hansen thought that fungus species should be based solely on morphologic characteristics, and that the only subspecific categories necessary were those of forma specialis, race, and clone. Of these, only the clone could be based on morphology; forma specialis and race were to be based on behavioral characteristics (pathogenicity).

Snyder & Hansen dealt with perfect states of *Fusarium* in a characteristic manner. They felt that the International Code of Botanical Nomenclature had strong mycological traditions that bound mycology to a past when variation in fungi was unrecognized. An example of this traditional approach is that the Code (1966) avoided reference to form genus in fungi and applied it only to fossils (9). The result of the rules of Nomenclature is frequently four different names for a fungus. Snyder & Hansen suggested that the specific epithet remain the same for both the perfect and imperfect state, thus, *Nectria episphaeria* where the perfect state is used and *Fusarium episphaeria* where the imperfect state is described.

The philosophy of the Snyder & Hansen species concept has been previously summarized (23) but it is probably worthwhile to repeat the salient points. 1. It should be possible to identify any fungus to species level irrespective of whether it is alive or dead, if all required morphological characters are present. 2. Species are based on similarities among different individuals that make up the species. If species are based on differences, then every individual becomes a species. 3. Fungi are variable and species should be determined by an experimental analysis of variability. 4. Only morphologic characters should be used to delineate species and all higher categories. 5. Personal judgment on the part of the systematist should be replaced by the experimental method wherever possible; the systematist should use all knowledge at his command, and should know in its entirety the fungus he is describing. 6. A satisfactory taxonomic system should be readily usable by any biologist.

Cultivars

The great reduction of species proposed by Snyder & Hansen eliminated the convenience of naming certain fungi which previously had been known as species, particularly in the large group which Snyder & Hansen placed in their species, *F. roseum.* This objection was met by proposing nonbotanical, horticultural variety names or cultivars (22). Snyder et al (22) stated: "The cultivar provides a means of informally naming plants. It has nothing to do with taxonomy or classification, and, therefore, is entirely independent of the botanical variety which implies relationship and position in a scheme of plant classification. These two systems of naming serve different purposes and may supplement one another, but neither take the place of the other." It may also be added that the same distinction holds between cultivar and forma specialis which, like the botanical variety, has taxonomic standing. Snyder et al (22) further stated: "It seems to us that the designation of the transitory or minor variations in conidial morphology or appearances, which are too unstable to find a place in botanical classification are best labelled . . . as cultivars." Many earlier species such as *F. redolens, F. coeruleum, F. culmorum,* and varieties such as *F. moniliforme* var. *subglutinans* were given cultivar names which were written *F. oxysporum* 'Redolens', *F. solani* 'Coeruleum', etc. In addition, two cultivars were proposed which grouped all the small-spored and all the large-spored forms of the section Eupionnotes. Snyder et al (22) suggested that the description for Redolens and Coeruleum be the same as those given by Gordon (6), but did not specify descriptions for the other nine cultivars they proposed. Because the cultivar was to be an informal device, no formal guidelines for their delineation were laid down, nor was a definitive listing made. No doubt the authors felt that additional cultivars and unofficial descriptions would eventually be proposed as workers felt the need for the separation of groups of clones according to specialized requirements.

No further changes were made in the nine-species system by Snyder and his co-workers, but a paper summarizing their species concept was published in 1965 (23). It is also worth noting that no key was ever presented by Snyder & Hansen. Keys for their system were published by Messiaen (13) and Matuo (12), and a pictorial guide to the nine species was made by Toussoun & Nelson (25).

MESSIAEN & CASSINI'S SYSTEM

Messiaen & Cassini (14) followed the Snyder & Hansen system with some modification. They stated that while a *Fusarium* specialist might hesitate in choosing between the Snyder & Hansen and the Wollenweber systems, the practical worker, the person desiring quick, easy identification, would prefer the former system, once again pointing out the twin virtues of practicality and reliability that this taxonomic system offers.

The major modification made by Messiaen & Cassini was to substitute botanical varieties in place of cultivars. The majority of the cultivars thus became varieties, with the exception of Redolens in *F. oxysporum* and Coeruleum in *F. solani,* which were rejected because of the presence of numerous intergradations between the typical cultivar and the species. The *F. roseum* cultivars Scirpi and Acuminatum were grouped in the variety *gibbosum,* and a new variety, *arthrosporioides* was named, grouping together clones having microconidia and generally fusiform conidia. Snyder et al (22) did not commit themselves on the characteristics that delineate their cultivars, merely noting that they "are recognized chiefly on appearances." Messiaen & Cassini on the other hand included their varieties in the key to the species because they are part of formal classification, a situation that Snyder & Hansen undoubtedly sought to avoid with their "informal" cultivars. As a result, in addition to the shape of macroconidia, presence and absence of chlamydospores, Messiaen & Cassini also used size and width of macroconidia (large, small, thin), abundance of chlamydospores, growth rate, and colony color to delineate varieties. These are characteristics that Snyder & Hansen found to be unreliable in their experimental taxonomic studies, and they were particularly critical of pigmentation as a taxonomic criterion. This may be the price that has to be paid to prepare a key, as opposed to an informal system with no descriptions.

RAILLO & BILAI'S SYSTEMS

At about the time when Snyder & Hansen were working on their revision of *Fusarium,* Raillo (16) in the USSR undertook a critical study of Wollenweber's *Fusarium* taxonomy. While accepting in general Wollenweber's system, she felt that his species concept was deficient, particularly because he failed to determine the chief characters that delineate species and failed to base his taxonomy on them. She did this by analyzing the variability of the morphological characters within species by means of cultures from single spores. She concluded that the most fundamental characters were the curvature of the conidia and the shape of the top or apical cell of the macroconidia. She considered as unusable for species and varietal delineation such characters as length and width of conidia, septation of conidia, presence or absence of sclerotia, mode of sporulation, pigmentation, and other cultural characters. These conclusions, remarkably similar to those of Snyder & Hansen, led her nevertheless to a system more complicated than Wollenweber's, in which she recognized 55 species, 10 subspecies, 55 varieties, and 61 forms grouped in 17 sections and 12 subsections (17).

This line of investigation was continued by Bilai (2) who experimentally studied the variability of individual isolates. She also investigated the effects of moisture, temperature, and nutrition, as well as aging and germination of conidia, in single-spore isolates. She concluded "that the amplitude of variability is considerably wider than that accepted in the diagnoses of many species and often includes the features of the whole section" (3). This type of observation, again similar to that of Snyder & Hansen, led her to merge the section Arthrosporiella into Roseum principally because of the wide variability in spore shape observed in *F. avenaceum.* Likewise, the section Ventricosum was merged into Martiella as Snyder & Hansen proposed. Bilai went further, however, and merged the section Liseola into Elegans. She also went further in merging the section Gibbosum into Discolor together with *F. lateritium* of the section Lateritium and, even more surprisingly, *F. gigas* (which has the largest macroconidia ever observed and is in the section Macroconia), maintaining that in these, curvature of conidia, shape of the upper cells of conidia, and presence of chlamydospores is too variable and cannot be used for differentiating sections. At the species level, however, Bilai was more conservative than Snyder & Hansen and recognized 26 species and 29 varieties.

BOOTH'S SYSTEM

Booth's classification is based on that of Gordon (6, 7), who in essence used Wollenweber & Reinking's (28) taxonomic system modified to accept some of Snyder & Hansen's concepts (19–21). Booth (4) stated in his introductory chapters that spore morphology is the major character in the identification of *Fusarium* species but that the morphology of the sporogenous cell is of more fundamental importance. In fact, the emphasis on conidiophore morphology and mode of production of conidia, in line with current thinking of mycological systematists, is Booth's most conspicuous contribution to *Fusarium* taxonomy. As a consequence, Booth modified the sections Sporotrichiella, Arthrosporiella, and Gibbosum, and discarded the section Roseum. Briefly, Sporotrichiella was split on the basis of the sporogenous cell, and the two species with polyblastic conidiogenous cells were transferred to Arthrosporiella in which this type of spore formation is the unifying character. *Fusarium avenaceum,* formerly in Wollenweber & Reinking's section Roseum, was transferred to Arthrosporiella for the same reason. The remaining species in Roseum were transferred to Gibbosum in which the simple phialide is the unifying character. Morphology of microconidiophores was also used to distinguish readily *F. moniliforme* var. *subglutinans* from *F. oxysporum*; this obviates the sometimes lengthy wait for chlamydospore formation, which used to be the most reliable criterion. The morphology of the microconidiophore can also be used to separate *F. oxysporum* from *F. solani.*

In his general key to the imperfect states, Booth used morphology of sporogenous cells, conidiophores, and conidia, size of conidia, pigmentation, presence or absence and position of chlamydospores, growth rate, and substrate (insects, other fungi) to separate sections and species.

Booth's philosophy of speciation may perhaps be best seen in the preambles to his 12 sections. Section Arachnites is formed of species having closely related perithecial states (*Micronectriella*), with the exception of *F. dimerum* which has no perfect state. Phialide proliferation, though not unique, is unusual in *F. tabacinum.* This is a species that Booth no doubt had in mind when he stated at the outset of his monograph: "Even in species with only phialidic conidiogenous cells the structure is often characteristic of the species."

The section Martiella, which includes Wollenweber & Reinking's section Ventricosum (in agreement with Snyder & Hansen), contains four species and one variety. *Fusarium solani* is the characteristic species, and Booth accepted its limits as being almost as wide as those of Snyder & Hansen; it contains the var. *coeruleum* and a number of formae speciales. *Fusarium illudens* was separated from *F. solani* on the basis of the former's larger ascospores, although perithecia and ascospore shape are similar in both species. *Fusarium ventricosum* was separated on shape of macroconidia and by ascospores that are rough walled rather than striate. The perfect states are in *Nectria; Hypomyces* was rejected. Finally, *F. tumidum* is somewhat of an interloper, since it used to be in one of the subsections of Discolor (27) which also contained *F. graminearum, F. culmorum,* and *F. sambucinum,* which Booth still includes in Discolor.

Section Episphaeria contains seven species and two varieties separated mainly on the basis of ascospore size and perithecium morphology, and on size and septation of macroconidia. Phialide morphology is also described and may have added weight in speciation, as for example in *F. merismoides.* Booth formed the section Coccophilum primarily on the basis of perithecial characteristics. Section Spicarioides has only one species, *F. decemcellulare;* this agrees with Hansen & Snyder, but the latter chose the specific epithet from the perfect state, *Calonectria rigidiuscula,* in order to avoid having four names for one fungus.

Booth placed six species, one variety, and several formae speciales in the section Lateritium, the basic character of which is shape of macroconidia. Presence or absence of microconidia split the species, which ultimately were differentiated on the basis of size of ascospores and macroconidia. *Fusarium stilboides* was separated from *F. lateritium,* which it closely resembles, because they did not cross. Section Liseola contains one species and one variety differentiated on phialide morphology. Section Elegans contains one species, one variety, and numerous formae speciales. In these two sections Booth and Snyder & Hansen are in considerable agreement.

Section Discolor contains eight species and one variety divided by shape of macroconidia and further divided according to size of macroconidia. We have already mentioned how Booth reworked sections Sporotrichiella and Arthrosporiella on morphology of the sporogenous cell. The last section we deal with is

Gibbosum. Booth included four species, which he stated were separated principally by colony pigmentation and the fact that one, *F. arthrosporioides,* has no chlamydospores. He also stated that there was a difference in size of ascospores of *F. equiseti* and *F. acuminatum,* and we may assume that this was the motivation for separating these two species. The other two species have no known perfect states.

Although it is not possible in such a cursory overview to give full credit to Booth's work, we hope that the reader has an idea of his philosophy in *Fusarium* taxonomy. This is anchored in the precepts of formal mycological thinking and is centered on the perfect state. He made this interesting comment (4): "Perithecial states are for the most part soundly based on a type specimen and these names may be used for species when the perithecial state is known. However, so far as the plant pathologist is concerned or anyone else who has to work with *Fusarium* isolates the perithecial states are seldom seen and therefore do not help us greatly with routine identification." The foundation of Booth's speciation was thus the perfect state, and some of his species are based on ascospore size, wall sculpturing, or incompatibility in crossing. Booth's statement above also shows concern for the practical aspects of *Fusarium* taxonomy. He made full use of asexual spore morphology and, to his credit, pointed out the value of conidiophore morphology. Of course, species separated primarily on one aspect of their perfect state may not necessarily show great differences in their asexual state. For this and other reasons Booth frequently used size of macroconidia and pigmentation of cultures in his keys, two characteristics held in great disregard by Snyder & Hansen. This may be an unavoidable consequence when such a large group of similar fungi are classified according to classical methods.

We have seen how Messiaen & Cassini (14), while adhering to the Snyder & Hansen concept, needed to use pigmentation and spore size in their key. It is also probably not obvious from our cursory examination how "wide" some of Booth's species are, for example *F. culmorum, F. avenaceum,* and *F. solani.* His *F. oxysporum* equates almost exactly with Snyder & Hansen's ideas of this species, as does his appraisal of *F. moniliforme.* There is no doubt that Booth took variation into account in delimiting his species, and these are nowhere as narrowly defined as Wollenweber & Reinking's (28). After all, Booth reduced the 143 species, varieties, and forms to 44 species and 7 varieties. He could only have done this by enlarging his species concept if he was to cover the same ground.

OUR VIEW OF EXISTING SYSTEMS

Snyder & Hansen's plea for experimental analysis of variability is heeded more and more in fungal taxonomy. There is more to variation, however, than noting the type of variants that appear in the progeny of a few cultures. Snyder & Hansen stressed the need for accumulating individuals from as many areas of the world as possible. In so doing, natural variation is examined and intermediate strains discovered that bridge forms which otherwise could have been considered, when viewed in isolation, as distinct entities. It was precisely this type of intergradation that prompted Snyder & Hansen to unite the 51 species, forms, and varieties in Wollenweber & Reinking's sections Roseum, Arthrosporiella, Gibbosum, and Discolor into one species, *F. roseum.* Booth recognized 15 species and 2 varieties.

Obviously the difference lies in the species limits set by the investigator; for Snyder & Hansen, basing the limits solely on shape of the macroconidia, it is very broad; for Booth, using shape and size of macroconidia, pigmentation, conidiophore morphology, and the perfect state, it is narrower. Which is more correct? In our view it would depend a great deal on the stability of the morphologic characters used in delimiting species. Snyder & Hansen maintained that spore shape is the major reliable criterion. Booth maintains that the morphology of the sporogenous cell is taxonomically more fundamental than spore shape, but he did not indicate how variable this is. A study of the variability of this factor is certainly in order. Indeed, in certain cases these seem to be evanescent structures. For example, the first conidia in *F. avenaceum* and *F. semitectum* are formed from polyblastic cells, but the latter ones form from phialides. In *F. fusarioides* phialospores are produced at the base of polyblastic conidiogenous cells after these have ceased to function. Both phialospores and blastospores are produced in *F. sporotrichioides* and *F. camptoceras.* This is not just a simple difference of morphology (such as size), but entails a basic difference in spore ontogeny.

If an organism can bridge the gap between blastospore formation and phialospore formation, how much meaning should we place on size and other morphologic characteristics of conidiophores in species delimitation? It is crucial that variation in conidiophore morphology be studied. As we have already noted, Booth relied strongly on the perfect state. For example, *F. illudens* was separated from *F. solani* primarily because its *Nectria* perfect state ascospores were larger (20–25 X 8–13μ as compared to 11–18 X 4–7μ). Size in asexual structures is variable and not dependable, as anyone can attest who has studied spore sizes in *Fusarium.* In Graminearum, for example, mutants are readily formed having macroconidia twice as long as those from wild-type cultures. Is size or other morphological criteria in perfect states that much more dependable because of a meiotic division? We need not belabor the point. Suffice to say that Booth recognizes finer limits to his species than do Snyder & Hansen.

It is interesting to speculate what Snyder & Hansen's response to Booth's classification would have been. They might have incorporated many of Booth's species as cultivars. They might also have done nothing, for one of the tragedies is that Snyder & Hansen never completed their system. They never published a key, and there are very few drawings, photographs, or descriptions published by these authors to illustrate their system. Indeed, their system never went far beyond the stage of a proposal which could benefit from added published experimental evidence, particularly on species other than *F. oxysporum* and *F. solani.* It is probably not fortuitous that the revisions of these two species have been accepted by most *Fusarium* workers. One interesting aspect of their system was their refusal to admit any subspecific division, aside from the clone based on morphology, allowing only forma specialis and race based on pathogenicity. This was to forestall any fragmentation of their system, any dilution of their species concept, or other compromise.

Workers such as Messiaen & Cassini (14) who erected botanical varieties are, strictly speaking, no more in agreement with Snyder & Hansen than is Booth. This stand of Snyder & Hansen is a fundamental one and entails no small consequences, for it causes problems in identification of the multiplicity of forms that exist. For example, Booth (4) cautioned: "Records of *F. tricinctum* are to a large extent unreliable due to misidentification or over simplification of the species concept." Nowhere is this problem more acute than in Snyder & Hansen's concept of '*F. roseum*'. Indeed, this precipitated their proposal of the cultivar concept, but the idea was never pursued to its conclusion and the cultivars, with the exception of Redolens and Coeruleum were never properly identified or described. The cultivars proposed were a miscellany; some represented former species and varieties and others represented groups of species. Cultivars were added de facto as the need arose, and some could almost be classified as clones (15). As Messiaen & Cassini (14) pointed out, cultivars were sometimes used to identify the extreme forms in a Snyder & Hansen species, and at other times represented the major groupings within a species.

Nevertheless, in spite of the stopgap arrangement in which it was left, the cultivar concept is an elegant solution to the dilemma posed by Snyder & Hansen's nine species. It allowed room for expansion without compromising their principles, so that fusaria could be identified with a precision fitting the occasion. Booth's (4) comment that the use of cultivars introduced complex stages of nomenclature without simplification of identification is not entirely correct, for it added a needed flexibility to the Snyder & Hansen system. It served also to bring out relationships. *Fusarium roseum* 'Graminearum' is shorthand that places this cultivar in the species *F. roseum* and thus one closely allied, for example, to Culmorum. It does better than the notation *F. graminearum,* which gives no indication of its placement in the section Discolor. It also allows a certain leeway in the diagnosis of intermediate forms which can vary, depending on the judgment of the investigator, without causing too great a disruption, because the species, owing to the latitude allowed therein, would still be correct. Finally it served as a bridge between the Snyder & Hansen system and that of others. The point has been raised that cultivars cannot be employed in fungi, as this implies that they have been selected by man. Today such cultivars exist, but may not be so identified, in the various fungi selected by man for the improved yields of their industrially useful byproducts. Need the line be drawn between this and the selection pressure that crops, selected by man, undoubtedly exert on fungi such as *Fusarium?* It has also been stated that cultivars have known genetic connections and that their relationship can be traced. This is true enough for diploid higher plants but is not always attainable in the haploid Fungi Imperfecti. In our view the advantages outweigh the disadvantages and the rather petty criticisms.

What is at fault here is the incompleteness of the concept as proposed originally by Snyder et al (22) and its informal nature, which left too much to the discretion of each investigator with nothing but vague guidelines. An informal designation is not the best simply because it is informal. Accurate descriptions are needed and guidelines should be laid so that cultivars are not proliferated endlessly. Only by putting the finishing touches to the cultivar concept, can the Snyder & Hansen system compete on an equal basis with other taxonomic systems for attention and use by workers everywhere.

The use of the same specific epithet for all stages of a fungus species, as suggested by Snyder & Hansen, would reduce confusion in fungus taxonomy. However, such usage would require that the Code be changed accordingly, and we doubt that this will happen. We agree with Booth (4) when he states that the perfect states are so rarely encountered as to be of little value in the identification of *Fusarium* species. Hopefully then, emphasis on the perfect state will have commensurate weight in the delimitation of species.

The use of forma specialis as advocated by Snyder & Hansen has been accepted for *F. oxysporum* and *F. solani* by Booth (4) and Messiaen & Cassini (14). The Armstrongs have worked principally with the formae speciales and races in *F. oxysporum* and have published a complete listing of these (1). We only wish to echo the caution of Messiaen & Cassini (14) that experiments on seedlings are not necessarily applicable to mature plants. Snyder & Hansen (21) also proposed the f. sp. *cerealis* in *F. roseum* to indicate pathogenic forms of this species which attack cereals. They apparently thought that a host specificity similar to that found in *F. oxysporum* existed in *F. roseum,* and suggested that new formae speciales be named as they were discovered. However, Tammen (24) showed that isolates of *F. roseum* from wheat were pathogenic on carnation, and isolates from carnation were pathogenic on wheat. He suggested that the f. sp. *cerealis* be retained to separate the nonpathogenic members from those that are pathogenic, even though the line between saprogenicity and pathogenicity may not always be clear. Schneider (18) studied isolates of *F. avenaceum (F. roseum)* from cereals, *Lupinus* sp., and carnation for pathogenicity on a variety of host plants, and found no indication of host specialization. She concluded that f. sp. *cerealis* had no place in the taxonomic system for *Fusarium.* We agree with Schneider (18) that f. sp. *cerealis* should not be used. Continued use of this designation only adds further to the existing confusion about *F. roseum.*

The reader is probably now aware that from our viewpoint, none of the existing systems have been so perfected as to be clearly superior to the others. A good deal of refining, rounding out, and finishing remain to be done. Recent ideas and concepts such as that of the sporogenous cell have to be evaluated experimentally and incorporated. Only then can the various systems be presented and evaluated on the basis of practicality and dependability by the world at large. Let not the reader despair, however, and think that the present situation is a hopeless morass of conflicting systems. A glance at Table 1 will show, at least for the economically important fusaria, that a consensus of the groups that should be recognized are agreed upon by most workers. The disagreement centers mostly on the rank that should be accorded to these entities, and is therefore not drastic. It is only fundamental to the extent, as we have tried to show, that it reflects the basic philosophies involved, and these in turn determine the usefulness of the system to the everyday needs of all those interested in *Fusarium.*

Table 1 A comparison of the economically important species, varieties, and cultivars recognized by Snyder & Hansen, Messiaen & Cassini, and Booth in their taxonomic systems for the genus *Fusarium*

Snyder & Hansen	Messiaen & Cassini	Booth
oxysporum	*oxysporum*	*oxysporum*
'Redolens'		var. *redolens*
solani	*solani*	*solani*
'Coeruleum'		var. *coeruleum*
'Argillaceum'		*illudens*
		ventricosum
		tumidum
moniliforme	*moniliforme*	*moniliforme*
'Subglutinans'	var. *subglutinans*	var. *subglutinans*
rigidiuscula	*rigidiusculum*	*decemcellulare*
lateritium	*lateritium*	*lateritium*
		var. *buxi*
		udum
		xylarioides
		stilboides
roseum	*roseum*	
'Sambucinum'	var. *sambucinum*	*sambucinum*
		var. *coeruleum*
'Culmorum'	var. *culmorum*	*culmorum*
'Graminearum'	var. *graminearum*	*graminearum*
'Avenaceum'	var. *avenaceum*	*avenaceum*
'Concolor'	var. *arthrosporioides*	*arthrosporioides*
'Gibbosum'	var. *gibbosum*	*concolor*
'Equiseti'		*equiseti*
'Acuminatum'		*acuminatum*
'Reticulatum'		*camptoceras*
'Heterosporium' (sic)		*semitectum*
		var. *majus*
		trichothecioides
		buharicum
		heterosporum
		flocciferum
		sulphureum
tricinctum	*tricinctum*	*tricinctum*
		poae
		sporotrichioides
		fusarioides
nivale	*nivale*	*nivale*
		stoveri
		tabacinum
		dimerum

Literature Cited

1. Armstrong, G. M., Armstrong, J. K. 1968. Formae speciales and races of *Fusarium oxysporum* causing a tracheomycosis in the syndrome of disease. *Phytopathology* 58:1242–46
2. Bilai, V. I. 1955. *The Fusaria (Biology and Systematics)* Kiev: Akad. Nauk. Ukr. SSR. 320 pp.
3. Bilai, V. I. 1970. Experimental morphogenesis in the fungi of the genus *Fusarium* and their taxonomy. *Ann. Acad. Sci. Fenn. A, IV Biologica* 168:7–18
4. Booth, C. 1971. *The Genus Fusarium.* Kew, England: Commonwealth Mycol. Inst. 237 pp.
5. Gerlach, W. 1970. Suggestions to an acceptable modern *Fusarium* system. *Ann. Acad. Sci. Fenn. A, IV Biologica* 168:37–49
6. Gordon, W. L. 1952. The occurrence of *Fusarium* species in Canada II. Prevalence and taxonomy of *Fusarium* species in cereal seed. *Can. J. Bot.* 30:209–51
7. Gordon, W. L. 1960. The taxonomy and habitats of *Fusarium* species from tropical and temperate regions. *Can. J. Bot.* 38:643–58
8. Hansen, H. N., Smith, R. E. 1932. The mechanism of variation in imperfect fungi: *Botrytis cinerea. Phytopathology* 22:953–64
9. Hennebert, G. L. 1971. Pleomorphism in Fungi Imperfecti. In *Taxonomy of Fungi Imperfecti,* ed. B. Kendrick, 202–23. Toronto: Univ. Toronto Press. 309 pp.
10. Jamalainen, E. A. 1970. Report of the meeting of the *Fusarium* discussion group in Finland. *Ann. Acad. Sci. Fenn. A, IV Biologica* 168:3–6
11. Joffe, A. Z. 1974. A modern system of *Fusarium* taxonomy. *Mycopathol. Mycol. Appl.* 53:201–28
12. Matuo, T. 1972. Taxonomic studies of phytopathogenic fusaria in Japan. *Rev. Plant Prot. Res.* 5:34–45
13. Messiaen, C. M. 1959. La systematique du genre *Fusarium* selon Snyder et Hansen. *Rev. Pathol. Vég. Entomol. Agric. Fr.* 38:253–66
14. Messiaen, C. M., Cassini, R. 1968. Recherches sur les fusarioses IV. La systématique des *Fusarium. Ann. Épiphyt.* 19:387–454

15. Nash, S. M., Snyder, W. C. 1965. Quantitative and qualitative comparisons of *Fusarium* populations in cultivated fields and noncultivated parent soils. *Can. J. Bot.* 43:939–45
16. Raillo, A. 1935. Diagnostic estimation of morphological and cultural characters in the genus *Fusarium. Bull. Plant Prot. II, Leningrad (Phytopathol.)* 7:1–100
17. Raillo, A. 1950. Griby roda Fuzarium. State Publ. Moskva: Gos. izd-vo selkhoz. lit-ry. 415 pp.
18. Schneider, R. 1958. Untersuchungen über Variation und Pathogenität von *Fusarium avenaceum* (Fr.) Sacc. *Phytopahtol. Z.* 32:129–48
19. Snyder, W. C., Hansen, H. N. 1940. The species concept in *Fusarium. Am. J. Bot.* 27:64–67
20. Snyder, W. C., Hansen, H. N. 1941. The species concept in *Fusarium* with reference to section Martiella. *Am. J. Bot.* 28:738–42
21. Snyder, W. C., Hansen, H. N. 1945. The species concept in *Fusarium* with reference to Discolor and other sections. *Am. J. Bot.* 32:657–66
22. Snyder, W. C., Hansen, H. N., Oswald, J. W. 1957. Cultivars of the fungus, *Fusarium. J. Madras Univ. B* 27:185–92
23. Snyder, W. C., Toussoun, T. A. 1965. Current status of taxonomy in *Fusarium* species and their perfect stages. *Phytopathology* 55:833–37
24. Tammen, J. 1958. Pathogenicity of *Fusarium roseum* to carnation and to wheat. *Phytopathology* 48:423–26
25. Toussoun, T. A., Nelson, P. E. 1968. *A Pictorial Guide to the Identification of Fusarium Species according to the taxonomic system of Snyder and Hansen.* Univ. Park, Pa: Pa. State Univ. Press. 51 pp.
26. Wollenweber, H. W. 1913. Studies on the *Fusarium* problem. *Phytopathology* 3:24–50
27. Wollenweber, H. W. 1943. Fusarium-Monographie. II. Fungi parasitici et saprophytici. *Zentralbl. Bakteriol. Parasitenkd. Infektionskr. 2* 106:104–35, 171–202
28. Wollenweber, H. W., Reinking, O. A. 1935. *Die Fusarien, ihre Beschreibung, Schadwirkung und Bekämpfung.* Berlin: Paul Parey. 355 pp.

THE PRESENT STATUS OF *FUSARIUM* TAXONOMY

❖3613

C. Booth
Commonwealth Mycological Institute, Ferry Lane, Kew, Surrey, England

Taxonomy is the science of systematic classification of living or fossilized organisms. Classification is the orderly arrangement of the categories into which related groups (taxa) may be divided or amalgamated. The names of these groups are governed by the International Code of Botanical Nomenclature.

This contribution is not concerned with the higher divisions of classification. All genera of fungi, unless they are monotypic, have frayed edges; that is, they merge into related genera so that it is often a matter of opinion in which genus an intermediate species should be placed. In this respect, the limits of *Fusarium* are more homogenous than many other common genera.

Any problems that exist in *Fusarium* taxonomy apply at species or subspecies level and may be considered under two quite distinct headings: (*a*) the problem relating to the correct application of the name, and (*b*) the problems of assigning a collection or isolate to its correct taxon, be it species, variety, or forma specialis.

CORRECT APPLICATION OF THE NAME

The first problem which most plant pathologists blatantly ignore, although those dealing with fusaria have more justification than most for doing so, is the typification of the species.

The Code of Botanical Nomenclature in its jargon tells us: "The application of names of taxa of the rank of family or below is determined by means of *nomenclatural types.* A nomenclatural type (typus) is that constituent element of a taxon to which the name of the taxon is permanently attached whether as a correct name or as a synonym." In other words the collection on which the name was based is the type, more specifically the Holotype. If in fact the exact specimen was never indicated, then we can select a Lectotype from the author's original material if more

83

than one collection was cited. If this has all been lost or destroyed, a requirement often very difficult to prove, then we can select a Neotype or a new type on which to base the name. With these kinds of provisions the "founding fathers" who wrote the Code obviously thought they were making allowances for all eventualities, but they had little appreciation of the problems posed by *Fusarium*.

The disease-causing capability of a *Fusarium* species is expressed by the hypha and its ability to penetrate the tissue of its host. In this form, at the present state of knowledge, the species is unidentifiable. In active growth periods species are usually spread by means of asexually produced conidia, and their macroconidia are the characteristic spore form of the genus. These may be produced from an effuse mold or, as more frequently happens in nature, as conidial pustules borne on a well-developed stroma (sporodochium). This stromatic form of fructification was Link's (12) original basis for the genus which he described as follows: "Stroma subglobosum, Sporidia fusiformia non septata instrata." Later Fries (9) validated the genus, in terms of the Code, by placing it, because of the sporodochia, with *Aegerita, Atractium,* and *Tubercularia* in the Tuberculariaceae. The most important member of the genus, *F. oxysporum,* was described by Schlechtendal (17) as follows: "Stroma convexum erumpens varium roseum superficie inaequali rugulosa, sporidiis parvis curvatis untrinque acutissimis. In tuberibus semiputridis *Solani tuberosi.*" Corda (8) described *F. equiseti* in somewhat more detail as: "minutum rotundatum carnosum carneo-roseum, dien pulverulentum, stromate convexo pallido, sporis fusiformibus curvatis acutis flaccidis, 5–6 septatis, cellulis vacuis longit spor. 0.001460 p.p.p. (approx. 40 μ)." These descriptions may comply with the Code of Nomenclature insofar as they validate the species, but they are totally inadequate for the identification of *Fusarium* species.

As we must have some comparative basis for our identification, the only procedure available to us is to go back to the original (type) material and draw up a modern usable description. My experience with such material has shown it to be quite inadequate, even if it can be found. Enquiries concerning Schlechtendal's material in Berlin led to my being informed that it had been destroyed during the last war. I have since learned that part of his herbarium is in Halle, Germany, so it is possible that type material of *Fusarium oxysporum* still exists. Even so the question remains: if it was found, would it be of any value in confirming our present-day usage of this name. The loan of Corda's type material of *Fusarium equiseti* from the Prague Herbarium showed it to be without spores or any other fructifications. This illustrates what so often happens to the type material of microfungi after it has been in use for many years; various investigators gradually remove all the material. Presumably in this case Corda's figure of *F. equiseti* (8, Table ix, Figure 32) is taken as the type, a perfectly legitimate assumption under certain terms of the Code.

The purpose of these examples, which could be expanded to cover almost all the established *Fusarium* names, is to demonstrate that both the original descriptions and the so-called type material are both inadequate as a basis for *Fusarium* nomenclature. The basis of the type concept is that it should provide a stable unit (specimen) on which the name is based. At the most these *Fusarium* types can only be said, in some cases, to possibly be the same species as is represented by the present-day usage of the name; in most cases they cannot even do that. The importance of this is in the basic instability it creates in *Fusarium* nomenclature. Our aims are, or should be, stability of nomenclature, and yet there is no foundation for it. Unless we agree to conserve these epithets of *Fusarium* species in common usage and insist that the rules of nomenclature be changed to allow lyophilized or liquid-nitrogen-stored living cultures to be accepted as replacements for the classical types we cannot prevent a wholesale change of names that would be legitimate under the Code, and which would undoubtedly result if a widespread investigation of existing type specimens was undertaken.

To fully understand the present situation we have to look briefly at the history of the genus. It was, as stated, established by Link (12), and throughout the 19th century the taxonomy of the genus, as would be expected, was in the hands of the contemporary mycologists. Even so, by 1849 when Fries listed *Fusarium* species, only 9 were included in his 2 sections. Although no general work on the genus was published between 1849 and Saccardo's publication of the *Sylloge Fungorum IV* in 1886, the numbers of species described had grown to 178, with numerous varieties. All were described from various plants, most of which showed some decay. The early mycological concept that microfungi were host specific meant that every new host record warranted the description of a new species, and this together with the concept, which still exists in some quarters, that the easiest way to establish a new record is to describe it as a new species has led to the introduction of over 1000 *Fusarium* names.

Serious work on the genus and the meaningful use of names began with the investigations by Dutch and German pathologists [Martius (13), Harting (11), Schacht (16), Reinke & Berthold (14)] into the effect of *Fusarium* species causing disease of potatoes. Initially it was thought that the presence of *Fusarium* reduced the resistance of potato tubers, and this allowed other organisms to enter. Even in this early work it gradually became apparent that *Fusarium* species could also be responsible for the actual disease.

It was, however, the paper by Smith (20) on wilt diseases of cotton, watermelon, and cowpea which opened up the real significance of the *Fusarium* problem to plant pathologists, although Smith was wrong in his assumption that a genetic connection existed between *F. vasinfectum* (*F. oxysporum*) and the ascomycete *Neocosmospora vasinfecta*. The fact that he clearly demonstrated that fusaria caused watermelon wilt following inoculation of the plant with a culture of the fungus stimulated tremendous interest and almost all subsequent work on the genus for many years was carried out by individuals trained primarily as plant pathologists. The result was that knowledge of the genus progressed by the use of cultures usually obtained from diseased material and tested for pathogenicity by inoculation experiments. From this work approximately 60 *Fusarium* names of some meaning came into general usage. Many of the hundreds of others were never accepted, in fact never used, following publication and were finally reduced to synonymy (5, 33). The way different authorities have used these names is summarized in the contribution in this volume by Toussoun & Nelson.

THE PROBLEM OF ASSIGNING ISOLATES TO THEIR CORRECT TAXA

The division of the genus into sections was first proposed by Wollenweber (27) and later extended by him and co-workers (30, 33). These sections have been one of the most stabilizing influences in the genus. We are here concerned with the difficulties that workers often experience in placing isolations in the correct section and, moreover, in the correct species within the section. One may be excused for assuming that 70 years of intensive work on the genus by a large number of workers in many countries would have been sufficient to solve any problems of either identification or pathology that may exist. That problems, particularly of identification, still exist may be explained in part by the variability of *Fusarium* species themselves under relatively minor environmental changes and in part by the fact that many plant pathologists never really look at a fungus.

The variability of *Fusarium* isolates is more an expression of methods of isolation and subsequent growth and, in particular, the length of time they have been growing on artificial media, than a real problem of identification. Brown from 1924 to 1926 published a series of papers (6) which demonstrated how fusaria could adapt to a wide range of substrates, but in doing so undergo morphological changes. He demonstrated how the carbon/nitrogen ratio of the media affected spore length and septation, and how sensitive the pigmentation is to the pH of the media. These or similar experiments have since been presented perennially by candidates for higher degrees; however, the only lesson they teach us is that if we desire to make comparative identifications then we must standardize the media and conditions of growth. What is also equally important is that, until the range of variability of a particular species is understood, only single-spore isolates should be used. It can be categorically stated that, if fusaria are isolated as single spores from fructifications on plant tissue or grown out of plant tissue on neutral media without antibiotics, and single-spore cultures taken as soon as spores appear, the range of single-spore isolates will show no more variation than would be expected of any other microfungus.

If we look at problems posed by the more economically important sections of the genus we must first look at the section Elegans, basically the wilt-causing fusaria.

Elegans

The most important member of this section, in fact to most workers the only member, is *Fusarium oxysporum*. Considerable controversy has arisen among some workers over the separation of this species from the somewhat related *Fusarium solani* in the Martiella section. I regard any similarity between the two as an example of parallel evolution rather than evidence of a close relationship. They can easily be separated after a few days growth of fresh isolates by examination of the microconidiophores. In *F. oxysporum* these are short, stubby phialides borne laterally on the hyphae or at the apices of short lateral branches. They are much more beautiful and elaborate in *F. solani*, with well-developed conidiophores that branch extensively, and with long delicate phialides often bearing a well-marked apical collarette (Figure 1*a*, *b*).

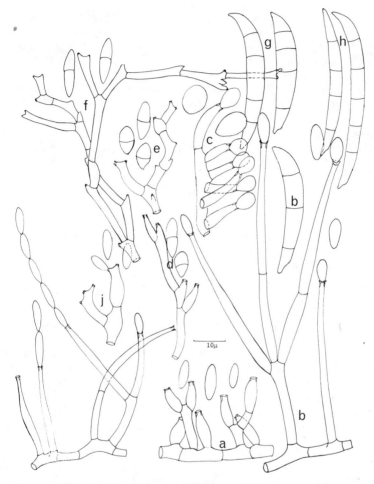

Figure 1   (*a-f*) Microconidia and microconidiophores of (*a*) *Fusarium oxysporum*, (*b*) *F. solani*, (*c*) *F. poae*, (*d*) *F. tricinctum*, (*e*) *F. sporotrichioides*, (*f*) *F. fusarioides*. Macroconidia of (*g*) *F. culmorum*, (*h*) *F. graminearum*. (*i,j*) Microconidia and microconidiophores of (*i*) *F. moniliforme* and (*j*) *F. moniliforme* var. *subglutinans*.

Species separation in the section Elegans is interesting because it is quite easy to separate the isolates into six or seven distinct cultural forms. What has been clearly demonstrated (2) is that these cultural forms bear little relationship to the pathogenic potential of the particular isolate. The section was based on *F. oxysporum* Schlechtendal, and important independent plant pathological work gradually threw up other *Fusarium* names which were finally incorporated into the section. Atkinson (3), investigating the disease of cotton in Alabama, proposed *F. vasinfectum*, and this name was taken up by Smith for the strain causing melon wilt. Smith (20) also proposed *Nectriella tracheiphila* (*Fusarium tracheiphilum*) (Sm.) Wr. (1913) for a probable saprophyte of cowpea. Cooke & Massee investigating the disease of *Narcissus* bulbs proposed *F. bulbigenum* (7), and Saccardo (15) split *F. lycopersici* from *F. aurantiacum* as the cause of disease in tomatoes. In creating the names of these early members of the Elegans section no reference was made to or comparison with the basic species, *F. oxysporum*. All were described without any critical study of the morphology of the species.

The first of the species to be placed in the section, and to be described after critical examination, was *F. orthoceras* Appel & Wollenweber (1). This species was beautifully illustrated, showing all the important diagnostic characters, and is probably the first species in the section for which the name can be authenticated. Somewhat undue emphasis was given to cultural differences between the various strains, and Wollenweber and his co-workers separated them as distinct species on presence or absence of sporodochia or pionnotes, odor, pigmentation on steamed rice, and minor difference in conidial size. Wollenweber (27) said that morphology should be the basis of pathology. On the basis of pathology these cultural differences have no relevance. This is not to say that genes for pathogenicity may only be carried by a characteristic strain, but that the relationship between a cultural form and pathogenicity cannot be relied upon to remain constant.

In dealing with this kind of situation one can only follow Snyder & Hansen's (21) proposal of using the name *F. oxysporum* for all Elegans type isolates and maintaining pathogenic strains as formae speciales. The one exception may be *F. oxysporum* var. *redolens* (Wr.) Gordon which has both microconidia and macroconidia somewhat intermediate between *F. oxysporum* and *F. solani*. The pathogenic potential of isolates of this variety do appear, however, to follow that of *F. oxysporum*, and the method of conidial production is also similar to this species. Gerlach & Pag (10) found isolates that had a distinct pathogenicity pattern, and they in fact restored the variety to species level, *F. redolens* Wr. This distinction has not been discernible in work in this country.

Martiella

This section was established by Wollenweber (27) for species with inequilateral conidia, the widest dimension being in the upper half of the spore (Figure 1b); chlamydospores are usually present. It was named after Martius who first described *Fusarium* (*Fusisporium*) *solani*. Wollenweber included two other species, *F. coeruleum* and *F. martii;* the latter was later (33) reduced to a variety of *F. solani*. Throughout Wollenweber's publications the number of species and varieties placed by him in Martiella grew to 19, but these were finally reduced by him (32) to 14.

In Snyder & Hansen's opinion (22) this system was unworkable. They reduced all species and varieties in both the Martiella and Ventricosum sections to synonyms of *F. solani*.

This is obviously the central species in the section and the most important to plant pathologists. The two basic species in Wollenweber's various systems (27, 33) are *Fusarium solani* and *F. javanicum*. The former has macroconidia 5.2–5.5 μ wide, whereas in the latter they are 4.6–4.8 μ wide. This character may appear to give a valid separation into two species, especially as those isolates with the narrower spores often have a brownish pigment. However, *F. solani* has a known perithecial state which is *Nectria haematococca* [*Hypomyces solani* is untenable (4)]. It occurs in nature, as do many other Hypocreales, as both homo- and heterothallic strains. Mating experiments carried out by the late W. L. Gordon, and repeated at the Commonwealth Mycological Institute (CMI), showed that these two apparently different isolates were merely strains of the same species.

In Europe we have tended to maintain *Fusarium solani* var. *coeruleum* for the strains that cause powdery rot, sometimes called dry or white rot, of potatoes. This is a strain with a strong violet-blue pigmentation, but there is no great justification on morphological grounds for maintaining it as a variety. There are three other members of this section, *Fusarium illudens, F. ventricosum,* and *F. tumidum,* which are clearly separated either by their distinct conidia, distinct perithecial states, or both. They appear to have little significance in causing plant disease.

Sporotrichiella

Workers who have examined a wide range of isolates in the section Sporotrichiella have generally agreed that four species belong to it (18). These species may be important either as plant pathogens or as toxin-producing molds.

Two of these, *Fusarium poae* and *F. tricinctum,* produce their markedly different microconidia from simple phialides (Figure 1*c, d*). The other two species, *F. sporotrichioides* and *F. fusarioides* (*F. chlamydosporum*), produce their microconidia from polyblastic conidiogenous cells (Figure 1 *e, f*) and the microconidia of these two species also differ markedly in their morphology. This character of the conidiogenous cells, which in many other sections of the hyphomycetes would be taken as justification for a new genus, was used by Booth (5) as the reason for transferring these two species names into the section Arthrosporiella where other species of fusaria have the same character. Irrespective of whether one follows this reasoning or not, no one who has looked at individual isolates of these four species could possibly believe they belonged to one species, namely *Fusarium tricinctum* sensu Snyder & Hansen (23).

Discolor

Members of the Discolor section are often referred to as the cereal fusaria. They do not form microconidia except under certain cultural conditions, and are generally distinguished on the morphology of the macroconidia which are comparatively thick walled, distinctly septate, fusiform to falcate with a beaked or fusoid apical cell; chlamydospores are usually present and may form either from the hyphae or from the cells of the macroconidia. The species of the section most frequently encountered by pathologists are *F. culmorum* (Figure 1*g*) (probably the most stable of all *Fusarium* species), *F. graminearum* (Figure 1*b*), *F. sambucinum, F. sambucinum* var. *coeruleum, F. heterosporum,* and *F. sulphureum.* Four of these five species have known *Gibberella* perithecial states, the exception being *F. culmorum.* These perithecial or perfect states are *Gibberella zeae* (*F. graminearum*), *G. gordonii*

(*F. heterosporum*), *G. cyanogena* (*F. sulphureum*), and *G. pulicaris* (*F. sambucinum*). They are clearly separated on ascospore size, but one should be careful to take measurements from asci containing eight spores. *Gibberella* species, in common with many other members of the Hypocreales, often have asci with abortive spores, and one may get three, five, or seven spored asci; the spores in these asci tend to be somewhat larger than normal.

On the evidence of the perithecial states alone there is no justification for lumping all these into one species, as did Snyder & Hansen (23). In fact they reduced all members of the sections Discolor, Roseum, Arthrosporiella, and Gibbosum [as used by Wollenweber & Reinking (33)] to synonyms of *Fusarium roseum*. The last four of these sections also include both important and minor pathogens such as *F. avenaceum* (*Gibberella avenacea*), *F. equiseti* (*G. intricans*), *F. semitectum,* and *F. acuminatum* (*G. acuminatum*).

Later Snyder et al (24) suggested a cultivar concept to bridge the differences between those workers who preferred to keep morphologically distinct forms as separate species and those who preferred to combine them in the *F. roseum* concept. Thus Graminearum type isolates would be referred to as *Fusarium roseum* 'Graminearum'. If pathogenicity to cereals was indicated then it should be referred to as *Fusarium roseum* f. sp. *cerealis* 'Graminearum'.

This rather cumbersome four-stage nomenclature has the added misleading implication that f. sp. *cerealis* is specific to cereals, which is not necessarily correct. The system has a further practical disadvantage, as can be seen from the literature, in that many workers tend to use *F. roseum* only, which has little value to any subsequent workers wishing to repeat an investigation.

Liseola

Some of the smaller but pathologically important sections of the genus have never created major problems in identification. *Fusarium moniliforme* was the original member of the section Liseola. The long, delicate phialides of the microconidiophores produce microconidia in chains (Figure 1*i*); the beige to purple pigmentation together with a complete lack of chlamydospores are definitive characters of this important species. A variety, *F. moniliforme* var. *subglutinans,* is maintained because it shows a further variation of conidial production within the fusaria. Here the microconidia are produced from polyphialides and do not form chains (Figure 1*j*).

Fusarium moniliforme was described by Sheldon (19) in 1904 following an investigation of an outbreak of ergotism-like disease in domestic animals fed with moldy corn in Nebraska. No ergot was found, but a clear relationship between the diseased animals and moldy corn was established. In view of the recent interest in toxin production by this species, which causes estrogenism in swine and leucoencephalomalacia in horses and other Equidiae, this early investigation is interesting and should be noted by exponents of the theory that plant pathology literature is out of date after ten years. *Fusarium moniliforme* also produces "Bakanae" disease of rice, which led to the discovery of the growth-promoting substance gibberellic acid, named from its perithecial state *Gibberella fujikuroi.* It should be noted that many tropical strains of *F. moniliforme* form macroconidia very sparsely, and identifications have to be based on the microconidial state alone. In fact, all isolates of *Cephalosporium sacchari* received for confirmation at the Commonwealth Mycological Institute have proved to be microconidial strains of *F. moniliforme*.

Lateritium

The section Lateritium was proposed by Wollenweber (28, 29) for species with narrow conidia having an acute inequilateral apex (beaked), blue sclerotia, rare chlamydospores, and cultures usually carmine. This description was later somewhat modified and major emphasis was placed on shape of the macroconidia (30). The central species of the section is *Fusarium lateritium* (*Gibberella baccata*) and, according to the Snyder & Hansen (23) system, the only one. *Fusarium udum* placed in this section by Booth (5) is a somewhat doubtful member, and it might be better placed in the Elegans group as Subramanian suggested (25, 26). *Fusarium stilboides* (*Gibberella stilboides*), the cause of scaly-bark disease of coffee in East Africa, and *F. xylarioides* (*G. xylarioides*), which causes tracheomycosis of coffee in West Africa, are both distinct species.

Spicarioides and Arachnites

The other two economically important species are *Fusarium decemcellulare* and *F. nivale*. *F. decemcellulare* (*Calonectria rigidiuscula*) is the cause of green point gall and dieback of cocoa. (*Fusarium rigidiuscula* (Brick) Snyder & Hansen (23) is a taxonomically invalid name.) It belongs to the section Spicarioides proposed by Wollenweber et al (30) at the *Fusarium* conference held in Madison, Wisconsin in 1924. *F. decemcellulare* is the only species ever to be placed in the section, which is therefore the most stable and uniform in the genus.

Fusarium nivale belongs to the Arachnites section, together with some other minor pathogens such as *F. tabacinum* and *F. stoveri*. *F. nivale* (*Micronectriella nivalis*) is of major importance as a cereal disease, especially of winter-sown wheat in colder regions. The perithecial state, which belongs to the Hypocreales, is not congeneric with *Calonectria*, and one cannot accept Muller & Arx's transfer into the Sphaeriales by following their transfer to *Griphosphaeria nivalis*. Therefore Hohnel's genus *Micronectriella* had to be taken up for the perithecial state of this species.

SUMMARY AND DISCUSSION

Fusarium taxonomy is in the first instance the systematic identification of the individual species, but without an adequate basic reference point (the type of the species) there is no firm starting point on which comparative identification can be based. In this genus, a pustule of dead macroconidia on plant material or soil (which may represent the type specimen of a *Fusarium* name) is quite inadequate for identification of the species. Thus, there is an urgent need to standardize the use of names by establishing new starting-point (type) material. Such reference material would only be useful in this genus if it was maintained as lyophilized living cultures.

Such stable and durable reference cultures should be available in all the major mycological culture collections.

Reference cultures of the species described in *The Genus Fusarium* (5) are deposited at the American Type Culture Collection (ATCC), Centraalbureau voor Schimmelcultures (CBS) and CMI culture collections, but such cultures or any replacements would need to be examined by members of the International Fusarium Committee before agreement on their validity could be achieved. It would then be necessary to seek amendments to the International Code of Botanical Nomenclature to allow living cultures to be used as Neotypes. The alternative, which is now a real possibility, is the wholesale change of *Fusarium* names at present in general usage.

The second point in the taxonomy of fusaria is the necessity to stabilize the status of formae speciales; these are the physiological strains, morphologically and culturally indistinguishable from saprophytic strains of the same species, which show different physiological properties in their ability to parasitize specific hosts. The initial belief that these strains were specific to one host led to specific host names such as f. sp. *callistephii, apii, mori,* and about 60 others being applied. The situation is not so clear-cut, and the true position has been dealt with by Armstrong & Armstong (2). It is sufficient to state that formae speciales are not necessarily specific to one host and there may be more than one pathogenic race in a forma specialis (i.e. ff. sp. *lycopersici* and *melonis* have 3 races, f. sp. *pisi* 5, and f. sp. *vasinfectum* 5). Neither formae speciales nor pathogenic races are covered by any Code of Nomenclature governing usage or application. The potential confusion that this situation may cause is even more apparent when it is appreciated that the identification of the race or strain is as important as the identification of the species. To avoid confusion about origin the author citation should, for the present at least, always follow the name. Armstrong & Armstrong (2) stated, "If formae speciales are eventually covered by the Botanical Code, we suggest that to establish a new forma specialis, a more detailed account of the pathogenicity of the organism and the symptomatology of the disease be required than is now available for many of those so far described." The deposition of a lyophilized culture in a national collection of all new formae speciales or races would provide an additional and probably more useful starting point for the name.

Biochemical aids to identification, particularly serology, have demonstrated equal differences between strains as between species. From general observation one would suggest that such investigations should be pursued at subspecies rather than species level. Certainly any technique that would distinguish different formae speciales or races in the laboratory without the necessity for inoculation experiments would be of immense value to plant pathologists.

Literature Cited

1. Appel, O., Wollenweber, H. W. 1910. Grundlagen einer Monographie der Gattung *Fusarium* (Link). *Arb. Kais. Biol. Anst. Land-Forstwirtsch.* 8:1–217
2. Armstrong, G. M., Armstrong, J. K. 1968. Formae speciales and races of *Fusarium oxysporum* causing a tracheomycosis in the syndrome of disease. *Phytopathology* 58:1242–46
3. Atkinson, G. F. 1892. *Fusarium vasinfectum.* Some diseases of cotton. *Ala. Agric. Exp. Stn. Bull.* 41:19–29
4. Booth, C. 1960. Studies of Pyrenomycetes: V. Nomenclature of some fusaria in relation to their nectrioid perithecial states. *Mycol. Pap.* 74:2
5. Booth, C. 1971. *The Genus Fusarium.* Kew, England: Commonwealth Mycol. Inst. 237 pp.
6. Brown, W., Horne, A. S. 1926. Studies in the genus *Fusarium.* III. An analysis of factors which determine certain microscopic features of *Fusarium* strains. *Ann. Bot.* 40:203–21
7. Cooke, M. C. 1887. New British fungi. *Grevillea* 16:49
8. Corda, A. C. J. 1838. *Icon. Fung.* 2:3
9. Fries, E. M. 1821. *Syst. Mycol.* 1:XLI (Intro.)
10. Gerlach, W., Pag, H. 1961. *Fusarium redolens* Wr., seine phytopathologische Bedeutung und eine *Dianthus*-Arten gefässparasitäre Form (*F. redolens* Wr. f. *dianthi* Gerlach). *Phytopathol. Z.* 42:349–61
11. Harting, P. 1846. Recherches sur la nature et les causes de la maladie des pommes de terre en 1845. *Nieuwe Verh. eerste Kl. Kond. Ned. Inst. Wetensch.* 12:203–97
12. Link, H. F. 1809. Observationes in Ordines plantarum naturales. *Mag. Ges. Naturforsch. Freunde Berlin* 3:10
13. Martius, C. F. P. 1842. *Die Kartoffel-Epidemie der letzten Jahre, oder die Stockfäule und Räude der Kartoffeln, geschildert und in ihren ursächlichen Verhaltnissen erörtert.* München: Akad. Deut. Wiss. 70 pp.
14. Reinke, J., Berthold, G. 1879. Die Zersetzung der Kartoffel durch Pilze. *Untersuchungen aus dem Botanischen Laboratorium der Universität Göttingen* I:1–100
15. Saccardo, P. A. 1886. *Syll. Fung.* 4:705
16. Schacht, H. 1854. *Bericht an das Konigl. Landes-Ökonomie-Kollegium über die Kartoffelpflanze und deren Krankheiten.* Berlin: G. Rosselmann. 29 pp.

17. von Schlechtendal, D. F. L. 1824. *Flora Berolinensis* 2 (Crypt.):139
18. Seemuller, E. 1968. Untersuchungen über die morphologische und biologische Differenzierung in der *Fusarium*-Sektion Sporotrichiella. *Mitt. Biol. Bundesanst. Land-Forstwirtsch. Berlin Dahlem* 127:1–93
19. Sheldon, J. L. 1904. A corn mold (*Fusarium moniliforme* n. sp.). *Rept. Nebr. Agric. Exp. Stn.* 17:23–32
20. Smith, E. F. 1899. Wilt disease of cotton, watermelon, and cowpea. *U.S. Dep. Agric. Div. Veg. Physiol. Pathol. Bull.* 17:10
21. Snyder, W. C., Hansen, H. N. 1940. The species concept in *Fusarium. Am. J. Bot.* 27:64–67
22. Snyder, W. C., Hansen, H. N. 1941. The species concept in *Fusarium* with reference to section Martiella. *Am. J. Bot.* 28:738–42
23. Snyder, W. C., Hansen, H. N. 1945. The species concept in *Fusarium* with reference to Discolor and other sections. *Am. J. Bot.* 32:657–66
24. Snyder, W. C., Hansen, H. N., Oswald, J. W. 1957. Cultivars of the fungus, *Fusarium. J. Madras Univ. B* 27:185–92
25. Subramanian, C. V. 1954. Studies on south Indian fusaria. III. Fusaria isolated from some crop plants. *J. Madras Univ. B* 24:21–46
26. Subramanian, C. V. 1971. *Hyphomycetes. An account of Indian species, except Cercosporae.* New Delhi: Indian Counc. Agric. Res. 930 pp.
27. Wollenweber, H. W. 1913. Studies on the *Fusarium* problem. *Phytopathology* 3:24–50
28. Wollenweber, H. W. 1917. Fusaria autographice delineata. *Ann. Mycol.* 15:1–56
29. Wollenweber, H. W. 1917 (1918). Conspectus analyticus Fusariorum. *Ber. Dtsch. Bot. Ges.* 35:732–42
30. Wollenweber, H. W. et al 1925. Fundamentals for taxonomic studies of *Fusarium. J. Agric. Res.* 30:833–43
31. Wollenweber, H. W. 1931. Fusarium-Monographie. Fungi parasitici et saprophytici. *Z. Parasitenkd.* 3:269–516
32. Wollenweber, H. W. 1943. Fusarium-Monographie. II. Fungi parasitici et saprophytici. *Zentralbl. Bakteriol. Parasitenkd. Infektionskr.*, 2, 106:104–35
33. Wollenweber, H. W., Reinking, O. A. 1935. *Die Fusarien, ihre Beschreibung, Schadwirkung und Bekämpfung.* Berlin: Paul Parey. 355 pp.

REFLECTIONS ON THE WILT FUSARIA

❖3614

G. M. Armstrong and Joanne K. Armstrong
Department of Plant Pathology, Georgia Agricultural Experiment Station,
Experiment, Georgia 30212

INTRODUCTION

Involvement of the senior author with wilt fusaria began when, as a boy, he helped harvest watermelons on his uncle's farm in South Carolina. The work of Atkinson on cotton wilt (19), Smith on cotton, cowpea, and watermelon wilt (26), and Orton on breeding these crops for wilt resistance (24) was then just beginning to reach the farmer. Later as a student at college, in the laboratory he saw for the first time the hyphae and spores of the *Fusarium* that caused these devastating wilt diseases. Shortly thereafter he inspected Orton's wilt plots in South Carolina where, among other crops, Conqueror, the first wilt-resistant watermelon [*Citrullus lunatus* (Thunb.) Mansf.], resulting from a cross with citron [*Citrullus lunatus* var. *citroides* (Bailey) Mansf.], was produced. Although not a delicacy, the melon was at least edible.

The first official *Fusarium* project of the senior author was with cotton wilt at the South Carolina Experiment Station. A cooperative project was also initiated with the crop-oriented Cotton Office of the US Department of Agriculture. Cotton and the cotton-wilt fungus were therefore always given the major role in the early work; every *Fusarium* isolate from any crop was used on cotton, and every isolate from cotton on other crops. A wide host range was found for the cotton-wilt *Fusarium*. If it lacked host specificity, did other wilt fusaria behave likewise? A project to study all *Fusarium* wilts was eventually undertaken and, after the junior author joined the staff, we started collecting isolates from all available regions of the world and using them in cross-inoculation experiments.

Some findings about wilt fusaria were purely accidental. The attention of the senior author was called to the presence of diseased plants of the weed, *Cassia tora* L., in a cotton-wilt plot. *Fusarium oxysporum* was obtained from these plants and later identified as *F. oxysporum* f. sp. *vasinfectum* (6). Thus, this was not a new *Fusarium* but a new host for f. sp. *vasinfectum*.

95

A pathologist on the staff, in making a survey of diseases of ornamentals, purchased seeds of various plants, among them Mexican sunflower [*Tithonia rotundifolia* (Mill.) Blake], and planted them in our cotton-wilt plot. Near the end of the season an inspection of the field revealed a plant of Mexican sunflower with what appeared to be a *Fusarium* wilt. Subsequent extensive investigations showed that the fungus isolated from the plant was f. sp. *apii* (10), the celery-wilt organism, most likely brought to the cotton field by infested seed. Without this observation, Mexican sunflower probably would never have been added to the plants used in the cross inoculations.

METHODS FOR TESTING

In a study of the pathogenic potentialities and host relations of wilt fusaria, the methods used are important considerations in comparing the results of workers. A lack of agreement in results is interpreted by us as due to the use of isolates of reduced virulence, inadequate or excessive inoculum concentrations, inoculations in the early seedling stage which may lead to damping-off and an inaccurate evaluation of mature plant reaction, or an unfavorable temperature for the development of disease. We have found that it is essential to use (*a*) virulent monospore isolates, (*b*) an adequate inoculum concentration, (*c*) definitive cultivars (cvs) of the proper age, (*d*) a suitable medium for plant growth, and (*e*) temperatures of 27–28° C, which are favorable for disease development.

Several methods for growing and inoculating the plants have been used during our investigations. Most soils available to us were toxic to the plants after sterilization. The addition of inoculum in a wheat-oats mixture to one soil that was not toxic caused the plants to damp-off when enough was added to give a satisfactory inoculum concentration. A solution-culture method (1) was used for awhile because a large number of plants could be grown in a given area, but this method was abandoned because greater care was necessary to prevent contaminations. A sand-culture technique (2) was found to be a very satisfactory method and has been used regularly.

NONSUSCEPTIBLE HOSTS AS CARRIERS OF WILT FUSARIA

Armstrong et al (18) reported that the cotton-wilt *Fusarium* could be obtained from a fairly high percentage of inoculated cotton plants of a resistant variety that showed no external symptoms of wilt or discoloration of the internal tissues. When inoculations of plants of various species were made, as for example sweet potato inoculated with the cotton-wilt fungus, fusaria were likewise obtained from a fairly high percentage of the stalks, although either no or slight external or internal symptoms of wilt were apparent. The sweet potato-wilt fungus also was obtained from naturally infected cotton and the weed, Mexican clover (*Richardia scabra* L.) (2), thus showing for the first time that wilt fusaria may invade plants without causing external symptoms of disease. Involved in these phenomena are an exchange of pathogens between crop- and wild-plant populations, symptomless carriers, and survival of pathogens. These are discussed in the review by Dinoor (20).

VARIATIONS IN CULTURAL AND MORPHOLOGICAL CHARACTERS OF THE WILT FUSARIA IN THE SECTION ELEGANS (*FUSARIUM OXYSPORUM*)

One of the first problems encountered was the identification of a *Fusarium* on potato-dextrose agar (PDA) after isolation from the plant. We were bewildered by the diversity of organisms obtained from plating root sections of wilt-diseased cotton (*Gossypium hirsutum* L.) from the field, but, by using the first few millimeters of the stem above the soil line, fewer kinds of fusaria were observed, but not all belonged in the section Elegans. It was soon seen that all isolates from wilting cotton plants in the pathogenicity tests could not be identified as *Fusarium vasinfectum* Atk. of Wollenweber & Reinking (30). Some of the variations were described (18). Later, 120 pathogenic isolates from sweet potato [*Ipomoea batatas* (L.) Lam.] and tobacco (*Nicotiana tabacum* L.) provided abundant material for a more extensive study of variability (5). When these were grown on PDA, potato plugs, and rice, the cultural characters of practically every species and form in the section Elegans could be found among them. Monospore lines were grown on a carefully prepared batch of PDA in carefully filled and uniform test tubes under uniform light and temperature conditions. One hundred 3-septate macroconidia from some of the tubes were measured and were found to vary in size from one side of the slant to the other and thus could be identified as spores of two different *Fusarium* species (29). Cultural and morphological variation among the wilt fusaria is now generally accepted as a banal truism.

Borderline forms in the sections make identification difficult. The difficulties of deciding whether *Fusarium udum* var. *crotalariae* Padwick belonged in the section Lateritium or the section Elegans have been presented (15, 25). When fresh isolates from pine canker (28) and wilted *Crotalaria* (*C. spectabilis* Roth) were compared, the former, but not the latter, was considered to be a typical Lateritium. The final determinants in placing the *Crotalaria* isolate in Lateritium rather than in Elegans was the hooked macrospore produced in the "hochkultur" of Wollenweber (30) and the greenish color produced in the fungal nutrient solution, which has never been observed in our experiments with Elegans fusaria.

What appear to be borderline isolates in sections Martiella and Elegans are frequently encountered. Matuo (23) gives for *F. oxysporum*, "microconidia formed on short conidiophore (with no septum)" and for *F. solani*, "microconidia formed on long conidiophore (with septum)," the conidiophore being the only distinct difference between them except for an ascigerous stage in *F. solani.* We also have used the long conidiophore as a distinguishing feature of *F. solani.* Overlapping of symptom expression in hosts attacked by these species frequently occurs, but we think that *F. oxysporum* and *F. solani* usually are sufficiently distinct morphologically and as pathogens to be retained as separate species.

HOST SPECIFICITY OF THE WILT FUSARIA IN THE SECTION ELEGANS (*FUSARIUM OXYSPORUM*)

The concept of limited host specificity of the 25 formae speciales listed by Snyder & Hansen in the revision of the section Elegans was supported by the evidence at that time (27). Our cross inoculations beginning about 1930 indicated that the host range of the cotton-wilt *Fusarium* was surprisingly broad, encompassing plants in the families Malvaceae, Solanaceae, and Leguminosae. In addition to cotton, f. sp. *vasinfectum* was isolated from wilting plants in the field, namely, alfalfa (*Medicago sativa* L.), *Cassia*, soybean [*Glycine max* (L.) Merr.], and tobacco. If isolates from these plants had been used to inoculate only each of them, respectively, ff. sp. *medicaginis, cassiae, glycines* or *tracheiphilum* race 1, and *nicotianae* would have been the proper designations, but we have shown that only f. sp. *vasinfectum* was involved. Isolates from tobacco designated as f. sp. *nicotianae* have not been shown to be specific in pathogenicity on tobacco, and, for the reason given below, it seemed logical to delete this f. sp. from our list (13). A fairly large collection of fusarium-wilt isolates from upland cotton, sweet potato, and tobacco from several states of the USA, were either f. sp. *batatas* race 1 or 2 or f. sp. *vasinfectum* race 1 or 2. All were pathogenic on either Burley 5 or Gold Dollar tobacco, but none were specific in pathogenicity on tobacco only. The American Type Culture Collection (ATCC) culture No. 10913 of *F. oxysporum* f. sp. *nicotianae* caused wilt of tobacco and sweet potato and, therefore, was f. sp. *batatas*.

Abundant evidence for a lack of host specificity in numerous other forms has accumulated during the intervening years. On the other hand, some forms have shown host specificity when tested on some or all of the plants in approximately 50 different genera, species, or cvs that have been useful in defining ff. sp. and races. The ff. sp. in our experiments that have caused wilt of only a single host as judged by external symptoms are *betae, cyclaminis, fragariae, glycines, lycopersici, medicaginis, passiflorae, perniciosum, ricini, sesami,* and *voandzeae*. Four of these caused considerable internal discoloration in *Voandzeia subterranea* (L.) Thou. (Armstrong & Armstrong, in press).

An interesting aspect of host specificity was noted in inoculations of cotton and *Cassia tora*. We were puzzled by variations and low percentages of wilt of *Cassia* caused by the *Fusarium* isolated from a mildly wilted *Cassia* plant growing in a cotton-wilt plot. The assumption was that this seed lot was from an apparently highly resistant *C. tora* (17). After an extensive survey, another isolate from a severely wilted plant was found to be highly virulent on plants from the same lot of seed. The first isolate proved to be f. sp. *vasinfectum* race 1, and the second isolate a new form later called f. sp. *cassiae* (11). Generally, f. sp. *cassiae* was weakly virulent or avirulent on upland cotton, with usually less than 50% of the plants showing mild external symptoms of wilt, and f. sp. *vasinfectum* was likewise mildly virulent or avirulent on *C. tora,* but each was highly virulent on its principal host (6, 11). The unusual reciprocal relationship in pathogenicity between these forms has not been encountered with other wilt fusaria.

A concept of primary and secondary hosts has evolved after investigating forms of *F. oxysporum* that are not specific in pathogenicity on a single host. Differences in two or more hosts in the degrees of virulence of the pathogen and the comparative rates of development, and types of symptoms of disease are the chief criteria that have been used to establish this concept (11). If the virulence of a f. sp. for one host remains fairly stable for a time, but an appreciable reduction or loss of pathogenicity occurs for another, the first is considered the primary host and the other the secondary host (12, 16). However, no support for this concept is furnished with some hosts, e.g. alfalfa on which the syndromes of disease with ff. sp. *cassiae, vasinfectum* races 1 and 2, and *medicaginis* are similar in all respects (4, 6, 7).

VARIATIONS IN VIRULENCE OF PATHOGENS

Some isolates of f. sp. *vasinfectum* cultured on PDA for a long time were shown to be less virulent than those recently isolated, and a single passage through the host did not significantly modify their relative pathogenicity (18). We have not performed many passages of an isolate through a host in contiguous repetitions but have used various ff. sp. numerous times without noticing increased virulence. The ever present problem of complete loss of pathogenicity of some isolates has been stressed, and for this reason cultures should be tested soon after isolation so that their full potentialities can be determined (5, 6). Loss of pathogenicity for one host but not for another also has been noted (3, 12, 16). Partial loss of virulence of isolates of f. sp. *batatas,* as well as their cultural characters and host relations, have been discussed in detail (5). Other workers also have noted a loss of virulence of this f. sp. (22). The difference in degrees of virulence between isolates may not be evident if tested only on very susceptible cvs, e.g. f. sp. *batatas* on the very susceptible sweet potato cv, Porto Rico, but a series of sweet potato cvs with varying degrees of resistance soon reveal these differences (5).

FORMAE SPECIALES AND RACES

In the revision of the section Elegans, formae speciales were established on the basis of limited or highly selective pathogenicity (27). By 1940, our investigations had revealed a multiplicity of hosts for some forms, suggesting some modification of the system of classification based on limited host specificity. The main features of the revised system are generally accepted, but a redefinition of numerous ff. sp. is being made due to their wide pathogenic capabilities and the indications that common genes for pathogenicity must exist among forms such as *apii, cassiae,* and *vasinfectum* (Table 1) and others. Many of the ff. sp. are poorly defined, and, if they are eventually covered by the International Code of Botanical Nomenclature, a more detailed account of the pathogenicity of a f. sp., including the names of cvs of the differential hosts, and the symptomology of the disease should be required to establish a new one. This may be a formidable task (8), because the name given a *Fusarium* causing a wilt disease of a new host may be uncertain until (*a*) inocula-

Table 1 Host relationships of *Fusarium oxysporum* ff. sp. *apii, cassiae,* and *vasinfectum*

Hosts of f. sp. *apii* primary or secondary	Forma specialis when host is primary	Forma specialis when host is secondary
Celery (*Apium graveolens* var. *dulce*)	*apii*	*cassiae*
Mexican sunflower (*Tithonia rotundifolia*)	*apii*	*vasinfectum* race 3
Eggplant (*Solanum melongena* var. *esculentum*)	*melongenae*	*apii*
Garden pea (*Pisum sativum*)	*pisi* races 1, 2, 4 to 11	*apii*
Asparagus (*Asparagus officinalis* var. *altilis*)	*asparagi*	*apii*
Cotton (*Gossypium arboreum*) 'Rozi'	*vasinfectum* races 3, 4	*apii*
Cotton (*G. barbadense*) 'Sakel'	*vasinfectum* races 1, 2, 3	*apii*
Hosts of f. sp. *cassiae* primary or secondary		
Cassia tora	*cassiae*	*vasinfectum* races 1, 2
Alfalfa (*Medicago sativa*)	*medicaginis*	*vasinfectum* races 1, 2
Physalis alkekengi		*vasinfectum* races 1, 2
P. alkekengi		*cassiae*
Alfalfa (*M. sativa*)		*cassiae*
Celery (*A. graveolens* var. *dulce*)	*apii*	*cassiae*
Chrysanthemum morifolium 'Encore'	*tracheiphilum* race 1	*cassiae*
Cotton (*G. hirsutum*)	*vasinfectum* races 1, 2	*cassiae*
Lupine (*Lupinus* sp.)	*lupini* races 1, 2, 3	*cassiae*
Tobacco (*Nicotiana tabacum*) 'Burley'	*batatas* races 1, 2	*cassiae*
P. alkekengi		*batatas* race 2
Hosts of f. sp. *vasinfectum* primary or secondary		
Cotton (*G. arboreum*) 'Rozi'	*vasinfectum* races 3, 4	*apii*
Cotton (*G. barbadense*) 'Sakel'	*vasinfectum* races 1, 2, 3	*apii*
Cotton (*G. hirsutum*)	*vasinfectum* races 1, 2	*cassiae*
Alfalfa (*M. sativa*)	*medicaginis*	*cassiae*
Alfalfa (*M. sativa*)	*medicaginis*	*vasinfectum* races 1, 2
C. tora	*cassiae*	*vasinfectum* races 1, 2
P. alkekengi		*vasinfectum* races 1, 2
Lupine (*Lupinus* sp.)	*lupini* races 1, 2, 3	*vasinfectum* races 1, 2
Mexican sunflower (*T. rotundifolia*)	*apii*	*vasinfectum* race 3
Tobacco (*N. tabacum*) 'Burley'	*batatas* races 1, 2	*vasinfectum* races 1, 2
Tobacco (*N. tabacum*) flue-cured	*batatas* race 2	*vasinfectum* race 2
Soybean (*Glycine max*)	*glycines*	*vasinfectum* race 2
Soybean (*G. max*)	*tracheiphilum* race 1	*vasinfectum* race 2
Lupine (*Lupinus* sp.)	*lupini* races 1, 2, 3	*vasinfectum* race 2
Lupine (*Lupinus* sp.)	*lupini* races 1, 2, 3	*tracheiphilum* race 1

tions of the new host with virulent isolates of all ff. sp. and races of *F. oxysporum* and (*b*) inoculations of the differential hosts of each f. sp. with the supposedly new *Fusarium* are made.

Pathogenic races were recognized in only one form in 1940, f. sp. *pisi* races 1 and 2, with cultivars of *Pisum sativum* L. as differentials. In the early stages of our investigations, we recognized races 1 and 2 of f. sp. *vasinfectum* after the inoculation of plants in widely separated families (16). However, eight cvs of *G. hirsutum* with very different genetic backgrounds surprisingly did not reveal races among the many isolates of f. sp. *vasinfectum* from upland cotton (6). Races 3 and 4 later were evident after inoculation of cvs in four species of *Gossypium* with cotton wilt isolates from Egypt, India, and USA (6). These isolates could not be separated as races on the basis of the reactions of all cvs in a species, as had been claimed (21), due to the specificity of cultivar reaction. If seven cvs from four species of *Gossypium* were inoculated, only two of the races 1, 3, and 4 could be identified, but by using only two cultivars, one in *G. arboreum* and one in *G. barbadense,* the three races were evident (6). As stated previously, other differentials were needed to separate races 1 and 2 (16). The necessity for giving the names of cvs in race delineation is obvious, but cvs with the same name but with different genes for resistance have made the separation of races difficult with cowpea [*Vigna unguiculata* (L.) Walp.] (3) and garden pea (*Pisum sativum* L.) (14).

We think that external symptoms of wilt are the best criteria for evaluating races. Although vascular discoloration has been used by others to measure degrees of virulence or severity of disease, we do not depend upon this symptom. For example, plants of *C. tora* inoculated with a wilt *Fusarium* weakly virulent on soybean showed no external symptoms of disease and were similar in appearance to the uninoculated plants. A *Fusarium* was isolated from the basal section of most of the 17 plants which were decidedly blackened internally to the tops of the plants. A similar reaction has been noted occasionally in other plants inoculated with wilt fusaria, thus indicating the uncertainty of using only vascular discoloration as a measurement of disease.

The number of races of a wilt *Fusarium* that may be established depends upon the criteria for separation of races and the diligence of the worker. If stable pure lines of the fungus could be tested under uniform environmental conditions on genetically pure lines of a host with cvs distinctly different in wilt resistance, an accurate delineation of pathogenic races would not be difficult. However, cultures kept for long periods in the laboratory, and even some freshly collected isolates, may show considerable variation in virulence. Slight to complete loss of pathogenicity by wilt fusaria when grown on common media is apparently not appreciated by some workers. Such cultures used in inoculations of hosts with varying degrees of wilt resistance may result in as many so-called races as there are cultures. However, the sharp difference in reaction of cvs of host plants to some clones of a f. sp. indicate that these clones must have distinctly different genotypes for pathogenicity and should be recognized as races. Good examples of this are the races of ff. sp. *pisi, lycopersici,* and *tracheiphilum.* Furthermore, differences in pathogenicity of clones of some forms are shown by using not only plants of different cvs of a species but plants of different species, genera, or families as differential hosts (e.g. ff. sp. *batatas, vasinfectum,* and *tracheiphilum*).

Other authentic races or subraces are to be expected when experiments include a wider selection of genetically pure differential hosts and relatively stable fungus cultures obtained through fresh collections from nature or by some method of preservation whereby pathogenicity is retained. Many workers concerned with the genetics of host and pathogen have emphasized the desirability of using isolines or near isolines of host, pathogen, or both for comparative studies of host-pathogen interaction in disease-susceptible and disease-resistant cvs. Where knowledge of the host genotypes is available, the selection of different types would be advantageous in identifying a large number of races of the fungus. With an organism such as *F. oxysporum,* the use of a large number of preserved cultures that retain selective pathogenicity and a high degree of virulence also would be desirable.

The problem of separating races among the pathogenic ff. sp. of *F. oxysporum* is often complex, as is evident from the preceding pages. With the advent of wilt in the wilt-resistant cvs of pea, it was logical to designate the new causal agent as race 2 of f. sp. *pisi.* With the discovery that the celery-wilt fungus, *F. oxysporum* f. sp. *apii,* caused wilt of pea, it might have been reduced in rank to a race of f. sp. *pisi.* We decided otherwise because, in our experiments, f. sp. *pisi* attacks only pea, but f. sp. *apii* has a wide host range in more than one family of plants (Table 1). The concept of primary and secondary hosts was evolved to cover this and similar cases.

With common hosts for the ff. sp. attacking plants in one family, as in the Cruciferae, it seemed logical to consider the ff. sp. *raphani* (race 2) and *mathioli* (races 3 and 4) as races of f. sp. *conglutinans* (race 1) because the former are not so specific in pathogenicity as was supposed. Mustard [*Brassica juncea* (L.) Coss var. *crispifolia* Bailey], a cultivated vegetable, and several cultivated ornamentals are common hosts for all of them (9).

Literature Cited

1. Armstrong, G. M. 1941. A solution-culture infection method used in the study of fusarium wilts. *Phytopathology* 31: 549–53
2. Armstrong, G. M., Armstrong, J. K. 1948. Nonsusceptible hosts as carriers of wilt fusaria. *Phytopathology* 38: 808–26
3. Armstrong, G. M., Armstrong, J. K. 1950. Biological races of the *Fusarium* causing wilt of cowpeas and soybeans. *Phytopathology* 40:181–93
4. Armstrong, G. M., Armstrong, J. K. 1954. Alfalfa—a common host for the wilt fusaria from alfalfa, cotton, and *Cassia. Plant Dis. Reptr.* 38:221–22
5. Armstrong, G. M., Armstrong, J. K. 1958. The fusarium wilt complex as related to the sweet potato. *Plant Dis. Reptr.* 42:1319–29

6. Armstrong, G. M., Armstrong, J. K. 1960. American, Egyptian, and Indian cotton-wilt fusaria: their pathogenicity and relationship to other wilt fusaria. *US Dep. Agric. Tech. Bull.* 1219. 19 pp.
7. Armstrong, G. M., Armstrong, J. K. 1965. Further studies on the pathogenicity of three forms of *Fusarium oxysporum* causing wilt of alfalfa. *Plant Dis. Reptr.* 49:412–16
8. Armstrong, G. M., Armstrong, J. K. 1965. Wilt of chrysanthemum caused by race 1 of the cowpea *Fusarium. Plant Dis. Reptr.* 49:673–76
9. Armstrong, G. M., Armstrong, J. K. 1966. Races of *Fusarium oxysporum* f. *conglutinans;* race 4, new race; and a new host for race 1, *Lychnis chalcedonica. Phytopathology* 56:525–30
10. Armstrong, G. M., Armstrong, J. K. 1966. Wilt of Mexican sunflower caused

by the celery-wilt *Fusarium. Plant Dis. Reptr.* 50:391–93

11. Armstrong, G. M., Armstrong, J. K. 1966. *Fusarium oxysporum* f. *cassiae* form. nov., causal agent of wilt of *Cassia tora* and other plants. *Phytopathology* 56:699–701

12. Armstrong, G. M., Armstrong, J. K. 1967. The celery-wilt *Fusarium* causes wilt of garden pea. *Plant Dis. Reptr.* 51:888–92

13. Armstrong, G. M., Armstrong, J. K. 1968. Formae speciales and races of *Fusarium oxysporum* causing a tracheomycosis in the syndrome of disease. *Phytopathology* 58:1242–46

14. Armstrong, G. M., Armstrong, J. K. 1974. Races of *Fusarium oxysporum* f. sp. *pisi*, causal agents of wilt of pea. *Phytopathology* 64:849–57

15. Armstrong, J. K., Armstrong, G. M. 1951. Physiological races of the crotalaria wilt *Fusarium. Phytopathology* 41:714–21

16. Armstrong, J. K., Armstrong, G. M. 1958. A race of the cotton wilt *Fusarium* causing wilt of Yelredo soybean and flue-cured tobacco. *Plant Dis. Reptr.* 42:147–51

17. Armstrong, G. M., Hawkins, B. S., Bennett, C. C. 1942. Cross inoculations with isolates of fusaria from cotton, tobacco, and certain other plants subject to wilt. *Phytopathology* 32:685–98

18. Armstrong, G. M., MacLachlan, J. D., Weindling, R. 1940. Variation in pathogenicity and cultural characteristics of the cotton-wilt organism, *Fusarium vasinfectum. Phytopathology* 30:515–20

19. Atkinson, G. F. 1892. Some diseases of cotton. *Ala. Agric. Exp. Stn. Bull.* 41. 65 pp.

20. Dinoor, A. 1974. Role of wild and cultivated plants in the epidemiology of plant diseases in Israel. *Ann. Rev. Phytopathol.* 12:413–36

21. Fahmy, T. 1928. The fusarium disease of cotton (wilt) and its control. *Egypt Min. Agric. Tech. Sci. Serv. Bull.* 74. 106 pp.

22. Harter, L. L., Weimer, J. L. 1929. A monographic study of sweet-potato diseases and their control. *US Dep. Agric. Tech. Bull.* 99. 118 pp.

23. Matuo, T. 1972. Taxonomic studies of phytopathogenic fusaria in Japan. *Rev. Plant Prot. Res.* 5:34–45

24. Orton, W. A. 1909. The development of farm crops resistant to disease. *US Dep. Agric. Yearb.* 1908:453–64

25. Padwick, G. W. 1940. The genus *Fusarium* V. *Fusarium udum* Butler, *F. vasinfectum* Atk. and *F. lateritium* Nees var. *uncinatum* Wr. *Ind. J. Agric. Sci.* 10:863–78

26. Smith, E. F. 1899. Wilt disease of cotton, watermelon, and cowpea (*Neocosmospora* nov. gen.). *US Dep. Agric. Div. Veg. Physiol. Pathol. Bull.* 17. 73 pp.

27. Snyder, W. C., Hansen, H. N. 1940. The species concept in *Fusarium. Am. J. Bot.* 27:64–67

28. Snyder, W. C., Toole, E. R., Hepting, G. H. 1949. Fusaria associated with mimosa wilt, sumac wilt, and pine pitch canker. *J. Agric. Res.* 78:365–82

29. Wollenweber, H. W. 1914. Identification of species of *Fusarium* occurring on the sweet potato, *Ipomoea batatas. J. Agric. Res.* 2:251–85

30. Wollenweber, H. W., Reinking, O. A. 1935. *Die Fusarien, ihre Beschreibung, Schadwirkung und Bekämpfung.* Berlin: Paul Parey. 355 pp.

THE USE OF PROTOPLASTS IN   ❖3615
PLANT VIROLOGY[1]

Itaru Takebe
Institute for Plant Virus Research, Chiba, Japan

INTRODUCTION

Understanding virus infection in higher plants depends heavily upon knowledge of infection in the individual cells that constitute the whole plant body. Unfortunately, however, experimental materials suitable for studying infection at the cellular level have not been available. Materials such as plants, organs, or tissues are inadequate for this purpose, because the number of cells initially infected by inoculation is extremely small and because the stage of infection in individual cells becomes random as infection spreads.

The concept of one-step virus growth was introduced into virology when Ellis & Delbrück (24) devised a procedure to simultaneously inoculate large numbers of *Escherichia coli* cells with a bacteriophage and to eliminate the chance of secondary infection. With the cells inoculated in this way, it is possible readily to follow one complete replication cycle of a virus. Such one-step growth experiments are one of the methodological foundations of modern bacterial and animal virology.

In order to realize one-step growth of plant viruses, two major requirements must be fulfilled in an experimental material: (*a*) a substantial proportion of cells present must be infected simultaneously and, (*b*) the material must consist of separated single cells to minimize the possibility of infection being spread. Both of the requirements are not fulfilled by plant cells grown either as callus or in suspension cultures (41). The leaf systems developed by Nilsson-Tillgren et al (59) and by Dawson & Schlegel (21) satisfy the first but not the second requirement, whereas the separated cells from infected leaves (1, 8, 38, 75, 94, 100) meet the second but not the first. Since plant viruses are unable to penetrate rigid plant cell walls, one way to achieve a high frequency of cell infection is to remove the walls, that is, to use cell protoplasts.

[1]Abbreviations for virus names: AMV (alfalfa mosaic virus), BMV (brome mosaic virus), CCMV (cowpea chlorotic mottle virus), CGMMV (cucumber green mottle mosaic virus), CMV (cucumber mosaic virus), CPMV (cowpea mosaic virus), PEMV (pea enation mosaic virus), PVX (potato virus X), PVY (potato virus Y), TMV (tobacco mosaic virus), TRV (tobacco rattle virus).

The pioneering work of Cocking (15) stimulated rapid progress in the enzymatic isolation of plant protoplasts (18). Inoculation of protoplasts with TMV was first tried by Cocking (16) and was then shown by Aoki & Takebe (2), Cocking & Pojnar (19), and Takebe & Otsuki (92) to give rise to infection in substantial numbers of protoplasts. Subsequent studies by Takebe & co-workers (66, 95) showed that active one-step growth of TMV was achieved in the inoculated protoplasts. Since then, protoplasts were successfully inoculated with several other viruses, and information pertaining to the process of infection in individual cells is beginning to accumulate. Indeed, protoplasts are becoming one of the standard laboratory materials in plant virus research.

This article reviews the current status of the use of protoplasts in plant virology and briefly discusses their potentiality. Protoplasts of lower plants are outside the scope of this review. Two recent reviews by Zaitlin & Beachy (101, 102) may also be of interest.

ISOLATION AND CULTURE OF PROTOPLASTS FOR USE IN PLANT VIROLOGY

Isolation of Protoplasts

Protoplasts for inoculation with plant viruses have been obtained mostly from the mesophyll tissue of tobacco. Besides being susceptible to a wide variety of viruses, leaves of this plant contain large numbers of parenchymatous mesophyll cells. For rapid and efficient isolation of protoplasts using enzymes, it is usually necessary to expose the mesophyll tissue by peeling the lower epidermis (94). Tobacco leaves allow relatively easy removal of the entire lower epidermis.

The methods currently used for isolating tobacco mesophyll protoplasts are based on the procedure of Takebe et al (94) by which large numbers of active mesophyll cells and protoplasts were obtained from tobacco leaves for the first time. The procedure consists of two steps: dissociation of the exposed mesophyll into single cells using pectinase (polygalacturonase), and digestion of walls of the isolated mesophyll cells by cellulase. A population of cells consisting mainly of palisade cells can be obtained by the first step and is used as the source of protoplasts in the second step. The current version of the procedure (67) permits isolation of 10^7 palisade protoplasts from 1 g (fresh weight) of mature tobacco leaves in about 2 hr.

Several modifications of the original procedure have been devised. Among these is the one-step method in which the exposed mesophyll is treated with a mixture of pectinase and cellulase to release protoplasts directly (42, 61, 71). The one-step method is simpler than the two-step procedure but yields more heterogeneous preparations, because it does not permit the separation of spongy- and palisade-parenchyma cells. In addition, considerable numbers of subprotoplasts, small spherical bodies containing only a portion of the cell content, are always produced by the one-step method. In a modified two-step method devised by Watts et al (98), the exposed mesophyll is briefly treated with pectinase to soften but not to dissociate the tissue, which is then digested with cellulase. This procedure was found useful for isolating tomato mesophyll protoplasts (F. Motoyoshi, personal communication). Other modifications include use of enzymes (82) or abrasives (7) to remove or to injure the epidermis and use of salts to replace non-ionic osmoticum (49).

The yield of active protoplasts is influenced by the age and the physiological state of plants as well as by the leaf position (94). Seasonal variations in yield are experienced in some laboratories (42, E. C. Cocking, personal communication). Influence of these factors was studied in some detail by Watts et al (98). The optimal condition of plant growth may differ with the method used for isolation of protoplasts (86).

Although protoplasts can be obtained from the leaves of many other species of plants besides tobacco (61), only cowpea, barley, *Petunia,* and tomato have been used as the source of mesophyll protoplasts for virus inoculation. Cowpea (33) and barley have an advantage in that the leaves suitable for protoplast isolation can be obtained by growing plants for a short period. It is likely that mesophyll protoplasts from other species, for example *Brassica* (61), will be useful for virus studies. Preliminary studies show that protoplasts infectible with TMV and CMV can be obtained from suspension-cultured *Vinca rosea* cells (I. Takebe & Y. Otsuki, unpublished).

Culture of Protoplasts

In most cases, protoplasts inoculated with viruses are cultured in the liquid medium of Aoki & Takebe (2) with or without minor modification. This medium originates from the medium of Takebe et al (94) and contains several inorganic salts, an auxin and a cytokinin, two antibiotics, but no metabolizable carbon source. Protoplasts support active virus replication in this medium but do not regenerate walls nor do they divide at least for several days; the cells survive as protoplasts. No DNA synthesis takes place (3) whereas RNA and protein are actively synthesized (77). Another medium that supported wall regeneration as well as sustained cell division and growth (58) did not give a better virus yield (Takebe & Y. Sugimura, unpublished). To what extent the nutrients in Aoki & Takebe's medium contribute to virus synthesis has not been studied systematically. The plant growth substances added to this medium are probably not necessary for virus replication (55, 60) but are beneficial to survival of protoplasts. Requirements for growth substances may differ according to the varieties of tobacco used (98). Two antibiotics, cephaloridine and rimocidin, are added to prevent growth of contaminating bacteria and fungi, respectively. Cephaloridine may be replaced by chloramphenicol or by aureomycin (81); they also do not affect virus replication (8, 63, 67, 92). Motoyoshi et al (55) studied the effects of various antibiotics on virus yield in protoplasts and suggest a combination of gentamycin, carbenicillin, and mycostatin as an alternative to cephaloridine and rimocidin.

The yield of virus in mesophyll cells or protoplasts is decreased when they are cultured in the dark (33, 42, 94). Addition of sucrose increases the yield of virus in the dark (B. Kassanis, personal communication) but not in the light (55).

INOCULATION OF PROTOPLASTS

Inoculation with Virus Particles

In 1969, scientists from two laboratories reported that TMV infection had been established in protoplasts from tomato fruits (19) and tobacco leaves (92) without mechanical injury to the cells. The procedure developed by Takebe & Otsuki (92) was used to infect large numbers of protoplasts using a low concentration of virus. By 1974, this procedure had been used to inoculate protoplasts from 6 species of plants with 10 different viruses (Table 1).

A major factor responsible for the high frequency of infection by the procedure of Takebe & Otsuki is the addition of a high molecular weight polycation to the inoculum virus solution. Poly-L-ornithine, a linear polymer of a basic amino acid, is most widely used for this purpose. Purified TMV dissolved in citrate buffer is treated for 10 min with poly-L-ornithine before being added to the suspension of protoplasts. During this time, poly-L-ornithine is believed to bind to the negatively charged TMV particles to form a virus-poly-L-ornithine complex (96). Subsequent incubation of protoplasts with a virus-poly-L-ornithine mixture for 10 min results in adsorption and penetration of the virus into the protoplasts (66). Unadsorbed virus is then removed by washing with mannitol solution, and the inoculated protoplasts are incubated to permit virus replication.

The current version (66) of the procedure of Takebe & Otsuki consistently gives TMV infection in about 90% of the protoplasts in suspension. An operational modification called the "direct method" was used by Motoyoshi et al (55) to inoculate tobacco mesophyll protoplasts with CCMV. In this method, pelleted protoplasts are suspended directly in the virus-poly-L-ornithine solution instead of being suspended first in mannitol solution and then exposed to virus. The direct method is reported to be effective not only with freshly isolated protoplasts but also with protoplasts stored or cultured for some time (55).

Another advantage of protoplasts is the ease with which the number of cells infected can be determined with a virus antibody labeled with a fluorescent dye (62). Like animal cells, protoplasts will adhere to glass slides and can be processed and stained without tedious embedding and sectioning (62). Because the entire staining procedure requires only a few hours, it is possible to know the amount of infection on the day after inoculation. Although fluorescein isothiocyanate (FITC) is the standard dye to label an antibody, lissamine rhodamine B also may be used (13). The combined use of FITC- and rhodamine-labeled antibodies should provide an opportunity to differentially stain two different viruses within the same protoplast, because the two dyes have entirely different emission spectra.

The efficiency of infection in protoplasts is defined as the ratio of the number of virus particles in the inoculum to the number of cells infected. It can be determined readily by the fluorescent-antibody technique. Because the frequency of virus infection is a function of the logarithm of virus concentration (Figure 1), inoculation at lower concentrations results in higher efficiency. Thus the efficiency of TMV infection in protoplasts is about 80,000:1 at the inoculum concentration of 1 μg/ml and about 2,500:1 at 0.01 μg/ml. The efficiency of TMV infection in leaf tissues has not

Table 1   Viruses used to inoculate higher plant protoplasts

Tobacco mesophyll protoplasts

Virus	Cultivar of tobacco	Percentage of protoplasts infected	Virus yield (particles/cell $\times$ 10^{-6})	References
TMV	Bright yellow	30	1.1	92
	Xanthi	90	1(3)	3, 60, 66, 78, 79
	Xanthi nc	37(34), 80(20)	1(34), 5.8(20)	20, 34, 60
	Samsun	84(60)		60, 68
	Samsun NN	89		60
	White burley	40(50), 80(42)	9.3(50), 4(42)	42, 50
TMV-RNA	Bright yellow	7	0.55	2
	Samsun	90	2.3	81
CMV	Xanthi	90		63
PVX	Burley 21	16		83
	Xanthi	70		67
CCMV	White burley	65	10	50
CCMV-RNA	White burley	31	15	50
PEMV	White burley	90	0.05	53
PEMV-RNA	White burley	3		53
BMV (V5)	White burley	77	8	52
AMV	White burley	35	18	54
TRV	Xanthi	98		44, 45
CPMV	Samsun NN	80	4	33
CGMMV	Xanthi	70		Y. Sugimura, personal communication

Protoplasts from plants other than tobacco

Virus	Plant & tissue	Percentage of protoplasts infected	Virus yield (Particles/cell $\times$ 10^{-6})	References
TMV	Tomato locule	40	0.26	19
	Tomato mesophyll	50		F. Motoyoshi, personal communication
	Petunia mesophyll	47		35; T. Hibi, personal communication
	Cowpea mesophyll	57		T. Hibi, personal communication
	Vinca suspension culture			I. Takebe & Y. Otsuki, unpublished
CMV	Cowpea mesophyll	95		T. Hibi, personal communication
	Vinca suspension culture	48		I. Takebe & Y. Otsuki, unpublished
CPMV	Cowpea mesophyll	96(33)	1(33), 0.5(7)	7, 33
AMV	Cowpea mesophyll	6		T. Hibi, personal communication
BMV	Barley mesophyll	30		I. Furusawa, personal communication

been determined accurately but appears to be of the order of 1,000,000:1 (87). Protoplasts are thus definitely superior to leaf tissues with respect to the efficiency of infection. Efficiency of infection by CMV and PVX is less than that by TMV (Figure 1). CMV has a particle weight only one seventh as large as that of TMV or PVX. Its efficiency is thus comparable to that of PVX—roughly 10,000,000:1.

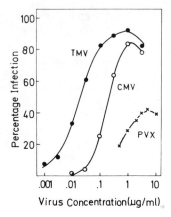

Figure 1   Effect of inoculum concentration on the frequency of infection of tobacco mesophyll protoplasts by TMV, CMV, and PVX.

The lower efficiency of CMV may be due at least partly to its multicomponent nature (30, 69). CCMV, another multicomponent virus (5), shows a higher efficiency of infection (400,000:1) in protoplasts than in tobacco or *Chenopodium* plants (50). The efficiency of infection in a CPMV-cowpea protoplast system is reported to be 100,000:1 (33).

Apart from the efficiency of infection, the average number of virus particles actually adsorbing to protoplast could be estimated in some cases. Takebe et al (95) estimated by infectivity assay that about 10^3 TMV particles are associated with each protoplast immediately after inoculation. Motoyoshi et al (51) used ^{14}C-labeled virus to determine the number of CCMV particles adsorbed per protoplast and report values ranging from 380 to 11,000 depending on the virus concentration used. About 10^3 CPMV particles adsorb to each cowpea protoplast (33).

Factors Influencing Infection of Protoplasts

VIRUS INOCULUM   The frequency of infection of protoplasts depends on the concentration of virus used for inoculation. With TMV, CMV, and PVX, the number of cells that become infected is roughly proportional to the logarithm of virus concentration within a certain range (Figure 1). A much steeper dose-infection relationship is reported with PEMV (53) which requires two component particles for infection. A peculiar feature of the dose-infection relationship in protoplasts is that the frequency of infection decreases when the concentration of virus is too high (Figure 1). This is probably because the number of virus particles fully complexed by poly-L-ornithine decreases at exceedingly high virus concentrations. The amount of infection also depends on the specific infectivity of the virus preparation used to inoculate protoplasts. Purified preparations of plant viruses often contain large numbers of noninfectious particles (25); also, protoplasts appear to be more exacting with respect to the specific infectivity of the virus inoculum than systemically infectible leaves. CPMV preparations lose infectivity toward cowpea protoplasts much more rapidly than toward cowpea plants (A. van Kammen, personal communication).

POLY-L-ORNITHINE   The presence of poly-L-ornithine is an absolute requirement for high frequency of infection of protoplasts by TMV (92), CMV (63), PVX (67), CCMV (50), and AMV (54). All of these viruses have an acidic isoelectric point and are, therefore, more or less negatively charged at inoculation pHs. One possible function of poly-L-ornithine is to neutralize or even reverse the surface charge of virus particles. This would facilitate their adsorption onto the protoplast surfaces which also may be negatively charged (29). The finding that virus particles must be incubated with poly-L-ornithine for some time before addition to the protoplasts (55, 67, 96) indicates that time must be provided for binding of the polycation to the virus particles. Poly-L-ornithine is not essential for infection of protoplasts by PEMV (53) and BMV (52), viruses charged positively at inoculation pHs. Although poly-L-ornithine may act primarily at the stage of virus adsorption, it may also stimulate virus entry into protoplasts (96). While poly-L-ornithine is not essential for the infection of protoplasts by PEMV and BMV, it increases the efficiency of infection by these viruses (52, 53). Basic polyamino acids, including poly-L-ornithine, are known to enhance the uptake of protein (76) and starch granules (43) by animal cells.

The complexity of the action of poly-L-ornithine in infection of protoplasts is illustrated by the recent findings of Hibi and co-workers. They showed that infection of cowpea mesophyll protoplasts by TMV is highly dependent on poly-L-ornithine (T. Hibi, personal communication), whereas infection of the same protoplasts by CPMV, another negatively charged virus, is not (33). CPMV requires poly-L-ornithine, on the other hand, to infect tobacco mesophyll protoplasts (33). More work is needed to understand fully the mechanism of action of poly-L-ornithine.

The effect of poly-L-ornithine apparently depends on its structure as a polycation. Other polycations such as poly-L-lysine, poly-L-histidine, and poly-L-arginine show similar but less pronounced effects (55, 95). Diethyl amino ethyl (DEAE) dextran is deleterious to protoplasts and enhances TMV infection only to a small extent (Takebe, unpublished, 55). Molecular size influences the effect of poly-L-ornithine to enhance the infection of protoplasts. One preparation with a molecular weight of 0.9×10^5 was significantly less effective than the commonly used preparation with a molecular weight of 1.3×10^5 (96). Another preparation with a molecular weight of 1.6×10^4 had little activity (Takebe, unpublished). Commercial preparations of poly-L-ornithine are not homogeneous in molecular size (degree of polymerization) or electrophoretic behavior in polyacrylamide gels (Takebe & Oishi, unpublished). It is possible that these or other differences in poly-L-ornithine preparations from different manufacturers may cause differences in their capacity to enhance infection of protoplasts.

pH AND BUFFER   The infection of protoplasts by many viruses is enhanced when the pH of the inoculation medium is lowered. The optimum for inoculation is usually about pH 5.0, because protoplasts become more or less unstable at pHs below this value. Infection by PVX (67) and TRV (44) exhibits a quite different response to pH of the inoculation medium; infection by these viruses is relatively insensitive to pH, and PVX shows a broad optimum around pH 5.8 (67). The reason for these differences from other viruses is not known.

Inoculation of protoplasts is usually performed in citrate buffer (92). Kubo et al (44) report that greater efficiency of infection of tobacco mesophyll protoplasts by TRV can be attained with phosphate than with citrate buffer. A similar effect of phosphate was also found for infection by TMV (Y. Otsuki & I. Takebe, unpublished).

OTHER FACTORS   Protoplasts to be inoculated are usually pelleted by low speed centrifugation and then resuspended in fresh mannitol solution immediately before they are mixed with virus solution (66). Protoplasts become less susceptible to infection when they are left in suspension before being exposed to a virus. It is possible that this is due to excretion of some unknown substance that is inhibitory to infection. If the mannitol solution in which protoplasts were previously pelleted is used to resuspend protoplasts, only a low infection efficiency can be attained even when protoplasts are resuspended immediately before inoculation (Y. Otsuki, unpublished).

The infection of protoplasts may be influenced by the age of the plant and the position of the leaves from which they are isolated (55). Kubo et al (44) state that this influence can be minimized by using phosphate buffer for the inoculation of protoplasts.

Inoculation with Viral RNA

Inoculation of protoplasts with free viral RNA was first accomplished by Aoki & Takebe (2) with a procedure different from that for inoculation with complete virus particles. The efficiency of infection by RNA was very low, a high concentration of TMV-RNA (1 mg/ml) being required to infect less than 10% of protoplasts (2). The low efficiency is due to inactivation of inoculum RNA by cellular enzymes. Much lower concentrations of CCMV-RNA were found to be sufficient to infect tobacco mesophyll protoplasts (50). Low efficiency has also been reported with PEMV-RNA (53).

A breakthrough was reported very recently by Sarkar et al (81) for the inoculation of protoplasts with viral RNA. They found that the efficiency of infection of protoplasts by TMV-RNA is strikingly improved when inoculation is performed in weakly alkaline medium of high ionic strength. Nearly all protoplasts are reported to be infected with TMV-RNA at 4–20 μg/ml, giving an efficiency of 100,000:1 to 1,000,000:1. The conditions used are believed to effectively reduce the activity of RNA-degrading enzymes. The finding by Sarkar et al should promote many types of experiments using protoplasts which heretofore have not been feasible because of the low efficiency of infection by viral RNA. The method might prove to be useful also for inoculating protoplasts with viroids.

VIRUS REPLICATION IN PROTOPLASTS

Virus replication in protoplasts inoculated in vitro can reasonably be assumed to be of the one-step type for the following reasons. First, infection is established by inoculation in a short time in the majority of cells present. Because protoplasts represent a homogeneous population of mesophyll cells, virus replication in this material should be synchronous within the range of variation in metabolic activity among individual cells. Second, the possibility of secondary infection by progeny virus is negligible. Progeny virus is not released into the medium except from a small number of protoplasts that might be damaged during culture. The cultural conditions, especially the absence of poly-L-ornithine, preclude infection during post-inoculation incubation.

Growth Curve

A plot of virus growth in protoplasts represents the course of virus replication in individual infected cells. Figure 2 illustrates the typical growth curve for TMV in tobacco mesophyll protoplasts. The infectivity apparent at zero time is due to inoculum virus adsorbed to protoplasts, and its decrease is assumed to reflect uncoating of the virus particles (92). Replication of virus is evident at 6 hr post infection (p.i.), and proceeds exponentially until 12 hr p.i. The rate of virus replication then decreases gradually but continues until about 72 hr p.i. An essentially similar curve was obtained by counting TMV particles in thin sections of protoplasts (34). Growth curves also have been determined for PVX in tobacco protoplasts (67)

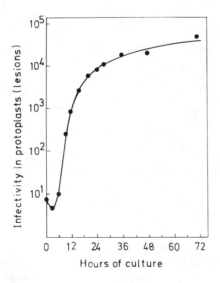

Figure 2 Growth curve of TMV in synchronously infected tobacco mesophyll protoplasts.

and for CPMV in cowpea protoplasts (33). The course of replication of other viruses in protoplasts has not been followed in detail because of limitations in the sensitivity of assay methods.

Yield of Virus

Virus yield, in terms of the number of progeny particles produced per infected protoplast, has been determined mostly by comparing the infectivity extractable from protoplasts with that of purified virus (2, 33, 50, 92). An assumption underlying this determination is that virus in a protoplast extract has a specific infectivity comparable to that of purified virus. This assumption appears to be valid at least for TMV, because virus yield estimated in this way agrees well with the values obtained by chemical determination (3). Yields ranging from 0.5 to 9.3 X 10^6 particles per cell have been reported for TMV in tobacco mesophyll protoplasts (Table 1). Comparable data for leaf tissues are 0.5–2.5 X 10^6 (32, 72, 99). It is thus clear that virus yield in protoplasts is of the same order of magnitude as that in leaf tissues. Similar yields have been reported for other viruses (Table 1), except for PEMV which is produced in much smaller amount (53).

Number of Replication Sites

Inoculation of protoplasts usually results in adsorption of large numbers of virus particles per protoplast. Although the exact number of particles that initiate replication in each protoplast is not known, some evidence indicates that it is more than one. Sakai & Takebe (79) irradiated tobacco mesophyll protoplasts with ultraviolet light immediately after inoculation with TMV and determined the number of protoplasts that became infected. Infection in protoplasts exhibited a pattern of response to ultraviolet inactivation which is expected with multiple targets, suggesting that more than one virus particle initiates replication in each protoplast. The hypothesis that a tobacco cell has multiple sites for TMV replication is also supported by the results of double infection experiments with protoplasts. Otsuki & Takebe (65) demonstrated that a large proportion of tobacco mesophyll protoplasts became doubly infected when inoculated with a common and a tomato strain of TMV. Double infection by the two strains could take place only if there were at least two replication sites in each protoplast.

A theory has been developed on the basis of experiments with leaf tissues that only one TMV particle can participate in the infection of a leaf cell (87). The results obtained with protoplasts are clearly not consistent with this theory, and suggest that the issue of exclusion (73) needs to be reexamined.

Virus Replication in Protoplasts of Necrotic Hosts

Replication of TMV in protoplasts from necrotic and systemic varieties of tobacco was compared by Otsuki et al (60). No difference was found in the rate and extent of TMV replication between protoplasts from Samsun NN and Samsun, showing that the cells of necrotic varieties have a capacity to produce as much virus as the cells of systemic varieties. This finding indicates that the smaller yield of virus in leaf cells of necrotic varieties (99) is the result of death of infected cells, which prevents continued virus synthesis. It was also found in this work that active virus replication does not cause cell death in protoplasts from necrotic varieties (60). This lack of necrotic reaction in protoplasts was not due to the absence of walls, and thus was interpreted to indicate the importance of cell-cell interaction in the necrotic reaction. These interactions are prevented in protoplasts by separation of the cells (60).

Synthesis of Viral RNA and Proteins

Protoplasts offer some technical advantages for studying RNA metabolism; RNA can be extracted without mechanically grinding the cells to yield clean preparations free from contamination by wall polysaccharides (77). Synthesis of TMV-RNA during the one-step virus growth in protoplasts was studied by Aoki & Takebe (3) by analyzing ^{32}P-labeled RNA from protoplasts using polyacrylamide gel electrophoresis. Actinomycin D was used to suppress host RNA synthesis (77). Synthesis of TMV-RNA was detectable at 4 hr p.i. by which time as much as 10^3 viral RNA molecules per protoplast were already synthesized. Viral RNA increased exponentially until 8 hr p.i., but the rate became linear at 10 hr p.i. and remained so thereafter. Formation of complete virus particles followed viral RNA synthesis with a delay of about 4 hr, so that a considerable amount of free viral RNA accumulated during the initial period of infection. Viral RNA synthesized in later stages was immediately encapsidated.

Double-stranded replicative forms of TMV-RNA were synthesized in protoplasts in much larger amounts than in leaf tissues (3). The time course of production of these forms was consistent with the hypothesis that they are intermediates of viral RNA replication. More direct evidence for the involvement of the replicative forms in TMV-RNA replication could not be obtained, however, because a large precursor pool in tobacco mesophyll protoplasts interfered with pulse-chase experiments. The "low molecular weight" component of TMV-RNA reported to occur in infected leaf tissues (38, 88) was not detected in protoplasts, suggesting that this component does not play an active role in virus replication.

Sakai & Takebe (78, 79) studied the synthesis of virus-specific proteins in TMV-inoculated protoplasts. In addition to viral coat protein, two TMV-specific proteins with high molecular weights, similar to those previously detected in infected leaf tissues (104), were identified in infected protoplasts. The molecular weight of one of these proteins (1.4×10^5) approximates that of solubilized TMV-RNA replicase (103). Coat protein was synthesized in far greater amount than the other proteins. A distinct difference was found in the time course of synthesis between these proteins; synthesis of the high-molecular-weight protein paralleled that of viral RNA, whereas coat protein synthesis lagged by several hours and was closely followed by the formation of progeny virus particles. Sakai & Takebe (79) concluded from these results that coat and the high-molecular-weight proteins are separately translated from TMV-RNA. Essentially similar results were recently reported also by Paterson & Knight (68). Takebe et al (89) discuss some of the possible mechanisms regulating replication and translation of TMV-RNA in protoplasts.

Bancroft et al (6) studied the synthesis of viral RNAs in tobacco mesophyll protoplasts inoculated with CCMV and BMV (V5) both of which have a divided genome (5, 46). Newly synthesized viral RNA was detectable in protoplasts at 7 hr p.i. RNA 3 appeared to be made first and was synthesized in greater amounts than the other components. RNase-resistant forms were found for RNAs 1, 2, and 3 but not for RNA 4, suggesting that the first three components replicate separately whereas RNA 4 is made probably on the minus strand of RNA 3. As in the case of TMV-infected protoplasts (3), the double-stranded forms were present in larger amounts than in leaf tissues.

Synthesis of virus-related proteins in CCMV-infected protoplasts also has been reported by Bancroft et al (6). Besides viral coat protein, a high-molecular-weight protein was detected in protoplasts infected for 24 hr, but was not considered to be involved in virus replication. More recently, an additional protein with a molecular weight of 35,000 was demonstrated in CCMV-infected protoplasts (F. Sakai, personal communication). The size of this protein is very close to that of BMV-RNA replicase (31). The early appearance of this protein is also suggestive of the possibility that it is the replicase of CCMV-RNA.

CYTOLOGICAL OBSERVATIONS

Entry of Virus

The process of virus entry into protoplasts has been studied by examining the sections of protoplasts fixed immediately after inoculation. Cocking found in one of his earliest studies (16) that one end of the TMV particles becomes attached to the invaginating plasmalemma of tomato fruit protoplasts. Virus particles were also found in cytoplasmic vesicles, and Cocking interpreted these observations to indicate that the virus enters protoplasts by a pinocytosis-like process (17). Observations indicating pinocytic uptake of a virus were subsequently made with tobacco mesophyll protoplasts inoculated with TMV (34, 66), CMV (37), PVX (96), and cauliflower mosaic virus (96). In all of these investigations, sections of protoplasts showed virus particles adsorbing to the plasmalemma, the plasmalemma invaginating at the site of virus attachment, and the cytoplasmic vesicles containing virus particles. Interestingly enough, viruses with linear morphology always adsorb at one end (16, 34, 66, 96), but it is not known whether one or both ends have the capacity to adsorb. On the basis of these observations, Takebe et al (96) postulated that the adsorption of a virus, probably in the form of virus/poly-L-ornithine complex, induces endocytic activity in the plasmalemma of protoplasts. There seems to be little doubt that virus enters protoplasts by an endocytosis-like process. It should be pointed out, however, that there is still no direct proof that the virus particles taken up by this process do initiate infection in protoplasts. Because large numbers of particles usually adsorb to protoplasts, the possibility cannot be excluded that some particles enter protoplasts via other routes and that infection is caused by these particles.

Pinocytic uptake was suggested as a mechanism of virus entry in the mechanical inoculation of leaf cells (56). The current state of knowledge is too incomplete to draw any conclusion about the similarity or dissimilarity of the process of virus entry in protoplasts and in cells in tissues. Addition of poly-L-ornithine does not increase the number of lesions formed on assay plants (I. Takebe, unpublished). On the other hand, poly-L-ornithine is reported to enhance the attachment and subsequent uncoating of TMV, when it is applied to tobacco leaves or is added to TMV inoculum (85).

A different idea for the process of virus entry into protoplasts was derived by Burgess et al (9, 10) from their electron microscopic observations. They observed various types of membrane lesions in inoculated protoplasts and state that virus binds preferentially to the sites of membrane damage. They assumed that poly-L-ornithine damaged the plasmalemma of protoplasts and that the virus directly entered the protoplasts as the damage was repaired. In direct contradiction to this hypothesis, however, Takebe et al (96) found no electron-microscopically detectable damage in the plasmalemma when protoplasts are treated with poly-L-ornithine alone. Burgess et al also finds it difficult to explain their own results for the infection by PEMV, a virus that does not require poly-L-ornithine (53). No lesion of the type they implicate as the port of virus entry is found in the protoplasts inoculated with PEMV, yet a large number of protoplasts are infected (10). For PEMV infection, these workers resort to the "presence of yet undemonstrated low level of intrinsic damage in freshly prepared protoplasts" (10). Cytoplasmic vesicles containing virus were noticed also by Burgess et al with TMV, CCMV, and PEMV (9, 10), but they presume that the virus is later eliminated into medium.

Motoyoshi et al (55) present evidence that might argue for the nonphysiological nature of virus entry process into protoplasts. Among these is the observation that infection occurs at a low temperature and in the presence of azide, conditions that should suppress cellular energy metabolism. Their experiments do not exclude the possibility, however, that infection occurred after the protoplasts were returned to their normal environment.

Progeny Virus

Electron microscopic detection of progeny virus particles in early periods of infection is much easier with protoplasts than with leaf tissues, because infection occurs synchronously in large numbers of the protoplasts. The earliest time at which progeny TMV particles are detected in tobacco mesophyll protoplasts is 6 hr p.i. (66), the time at which infectivity in protoplasts starts to rise (Figure 2). The course of production of progeny TMV particles followed by electron microscopy largely confirmed the results obtained by the infectivity assay and by staining with the fluorescent antibody (34, 66). Examination of large numbers of protoplasts sampled at a given time after inoculation showed that there is relatively little variation in the amount of progeny virus among individual protoplasts, indicating that infection proceeds synchronously (66).

In general, the distribution of progeny virus in protoplasts does not differ from that in the cells of leaf tissues—TMV (34, 66) and PVX (67) form large aggregates, whereas CCMV (50) and CPMV (33) are more or less randomly scattered in the cytoplasm. The distribution of CMV in protoplasts is somewhat unusual, however.

As has been reported with leaf materials (23, 36), this virus accumulates not only in the cytoplasm but also in the nucleus (37). In contrast to the random distribution in the cytoplasm of leaf materials (36), however, CMV in protoplasts forms aggregates which are very frequently associated with the plasmalemma and tonoplast (37). The same features appear when protoplasts are prepared from leaf tissues in which CMV has accumulated (37). Occurrence of PEMV in the nucleus of tobacco mesophyll protoplasts was demonstrated by staining with a fluorescent antibody (53) and by electron microscopy (12).

Effects of Infection on Cellular Fine Structure

An early paper by Cocking & Pojnar (19) describes extensive degradation of the cytoplasm in TMV-infected tomato fruit protoplasts. But this has not been reported in more recent investigations. For example, Honda et al (37) showed that the cytoplasm, chloroplasts, and mitochondria of tobacco mesophyll protoplasts are quite normal 48 hr after inoculation with CMV, although the nucleolus showed some unusual morphology. This is in marked contrast to disorders in the cellular ultrastructure which accompany infection of leaf tissues by CMV (36).

Some of the ultrastructural changes characteristic of virus infection are reproduced in protoplasts. The laminate inclusion body characteristic to PVX infection (84) is formed also in tobacco mesophyll protoplasts (67, 83), and the peculiar "cytopathic structure" (22) appears in CPMV-infected cowpea protoplasts (33). The high frequency and synchronous nature of infection in protoplasts enable one to follow the genesis and decay of these structures and to correlate them with virus replication. The observation that the PVX inclusion body appears after large amounts of virus have accumulated led Shalla & Petersen (83) and Otsuki et al (67) to conclude that this structure is not involved in virus replication. A similar time-course study yielded evidence that the virus-like rods formed in the chloroplasts of TMV U5-infected tobacco protoplasts are a product of encapsidation of chloroplast RNA with the viral coat protein (T. A. Shalla, personal communication). Ultrastructural changes accompanying the infection of tobacco mesophyll protoplasts by CCMV, BMV, and PEMV are reported by Burgess et al (11, 12). Proliferation and modification of endoplasmic reticulum and of nuclear membrane are characteristic of infection by these viruses. Vesicles containing fibrillar material are formed before progeny virus is produced. With PEMV-infected protoplasts, the fibrillar material was shown to contain RNA (12).

USEFULNESS OF PROTOPLASTS FOR PLANT VIRUS RESEARCH

Nature of Protoplasts

Protoplasts are an unusual form of plant cells which do not occur in natural environments. They have to be maintained in a hypertonic medium which usually differs nutritionally from the milieu in which plant cells normally exist. Although isolated protoplasts appear to be largely normal with respect to ultrastructure (97) and metabolism (77), there are some indications that the metabolism in protoplasts is under the influence of the hypertonic condition (40, 47, 97). In particular, the observation by Lázár et al (47) that osmotic shock triggers an increase in RNase level in tobacco leaf tissues and protoplasts might have serious consequences for the use of protoplasts in plant virus research. However, the vigorous virus replication in tobacco mesophyll protoplasts is a fact well established with many viruses (Table 1). It is probable, therefore, that the RNase induced by the hypertonic condition is confined in lysosome-like structures (48) and consequently does not interfere with synthesis of viral RNA.

The most important advantage of protoplasts for plant virus research is that they permit one-step virus growth experiments. It is only with an experiment of this type that the consecutive stages of virus replication in individual cells can be identified, followed, and analyzed. Without one-step growth experiments it is impossible to study the kinetics of the processes involved in virus replication, to pinpoint the stage which is affected by a given treatment, or to correlate a host response to a particular stage of replication. In addition, protoplasts permit one to attack these problems under simple, well-defined, and uniform conditions, because they are separated single cells, and are free from possible influence by other tissues. Another great advantage of protoplasts comes from their ability to form a uniform suspension in a liquid medium. Aliquots equivalent to each other qualitatively or quantitatively can be readily obtained by simply pipetting a suspension of known cell density. Furthermore, isolation of virus-related molecules or structures is easy because of the absence of rigid walls. All of these features are absent in the conventional tissue materials and may be exploited to approach a number of problems in plant virology.

Potential Uses of Protoplasts in Plant Virus Research

Some of the findings described earlier in this paper illustrate the usefulness of protoplasts for studying the basic mechanisms of plant virus infection and replication. In particular, studies of virus replication at the molecular level are very difficult without using protoplasts. As discussed below, protoplasts may be used also to investigate many other aspects of plant virus infection.

Infection of a plant with two viruses results in various types of interaction between viruses, such as cross protection, interference, and enhancement of replication. Mechanisms underlying these phenomena are poorly understood, and it is not clear in most cases whether the interaction occurs within doubly infected cells or between cells separately infected with one or the other of the viruses. Protoplasts may be used to obtain better understanding of these phenomena. Goodman & Ross (27) isolated protoplasts from tobacco leaves doubly infected by PVX and PVY (or by PVX and TMV) and demonstrated by fluorescent-antibody staining that double infection as well as the enhancement of PVX synthesis (74) takes place within individual cells. These results formed the basis of the conclusion reached in a later work (28) that heterologous encapsidation does not occur between these viruses. Protoplasts from healthy leaves should provide an excellent system to study interaction between plant viruses, because they can be doubly infected under a defined regime with both nonrelated (64) and related (65, 93) viruses. For example, Otsuki & Takebe (65) showed that the phenomenon of cross protection can be simulated using tobacco mesophyll protoplasts doubly infected with a common and a tomato strain of TMV. Replication of one strain in doubly infected protoplasts is supressed when the other strain predominates in the mixed inoculum or when protoplasts are infected previously with the other strain.

The high frequency of double infection obtained when protoplasts are inoculated with strains of TMV (65) should facilitate a detailed analysis of phenotypic mixing (4, 80). A recent investigation with protoplasts doubly infected with two strains of TMV (Y. Otsuki, unpublished) indicates that coat protein subunits from the both strains occur together in individual progeny particles. Preliminary evidence also indicates that RNA of a temperature-sensitive coat-protein mutant of CCMV is encapsidated with wild-type protein, when tobacco protoplasts are doubly infected at a nonpermissive temperature with the mutant and the wild-type viruses (6).

Protoplasts are probably useful also for the identification of genes in the plant viral genome. It would be interesting to see whether the different classes of temperature-sensitive mutants of TMV (39) can complement each other in doubly infected protoplasts. For viruses with a divided genome, analyses of the outcome of infection with incomplete sets of components should yield information about the genes located in individual RNA segments. As expected, infection of tobacco protoplasts with long particles of TRV results in the formation of infectious RNA but not of complete particles (45). Analysis of the proteins synthesized in this system should contribute to the identification and characterization of genes present in the long component.

Both viral and host genomes must be involved in the determination of the host range of a plant virus, but how it is determined is completely unknown. Protoplasts offer a highly sensitive material for determining whether or not the cells of a given plant can support replication of a given virus. If there are two related viruses one of which can infect the protoplast whereas the other cannot, these viruses provide an opportunity to pinpoint the viral genome determining the host range. The wild-type and V5 variant of BMV in tobacco protoplasts is an example (52). It also would be of great interest to determine if the ability to infect tobacco plants can be conferred on the wild type BMV by substituting its particular component with the counterpart from V5.

The influence of plant genes on the host range of a given virus may be studied by comparing the infection of protoplasts from susceptible and resistant varieties. A detailed analysis of the events following the inoculation of resistant protoplasts could uncover the stage at which infection by the virus is prevented. Tomato appears to be a good plant for this purpose, because it has genetically characterized varieties that are immune to TMV infection (70).

Protoplasts also are likely to open up new possibilities to study the mechanism of pathological response of plants to virus infection. Otsuki et al (60) showed that the N gene of tobacco determining the necrotic response to TMV is not phenotypically expressed in protoplasts. A search for a factor causing the in vitro expression of the N gene should be rewarding, and once such a factor is found, protoplasts should offer an excel.ent system to investigate the biochemical basis of the necrotic reaction. L. Hirth et a (personal communication) recently found that phenylalanine ammonia-lyase, a key enzyme in the biosynthesis of phenolic compounds, disappears when TMV-infected Samsun NN leaves are dissociated into single cells or protoplasts. It is possible that experiments along this line will yield a clearer picture of the role of phenolic compounds in the necrotic reaction.

Quite different lines of research may be envisioned using protoplasts for more practical aspects of plant virology. It has been shown that isolated protoplasts regenerate cell walls in vitro (57) and subsequently develop into callus-like colonies from which whole plants can be eventually regenerated (58, 90, 91). It may therefore be possible to utilize protoplasts for eliminating virus from naturally infected plants. Protoplasts have an advantage over meristem culture in that it is possible to obtain a vast number of clonal individuals (58). A more radical way to improve crop plants would be breeding for resistance to virus infection. Protoplasts may open up entirely new avenues in plant breeding (26), for example, by permitting mutagenesis in haploid protoplasts (91) and by selecting resistant clones while they grow on agar plates as callus colonies. A prerequisite to such an attempt will be a selection method which permits only resistant clones to grow on agar plates. Other possibilities to increase genetic variability in plants include somatic hybridization through protoplast fusion, gene transfer, and transplantation of organelles (14). Although these methods for modifying plant genomes by nonsexual means are still in a rudimentary stage, some of them may prove to be useful for introducing a resistance gene into existing plants.

CONCLUDING REMARKS

It is reasonable to postulate that virus-protoplast systems will be extended to new combinations of plant and virus. We can also expect technical improvements and refinements in the use of plant protoplasts as our knowldege about the system enlarges. These improvements in turn should make it possible to use protoplasts to solve still unforeseen challenges in plant virology.

ACKNOWLEDGMENTS

For providing unpublished materials I am grateful to Drs. J. B. Bancroft, G. Bruening, J. Burgess, A. C. Cassels, I. Furusawa, B. D. Harrison, T. Hibi, L. Hirth, B. Kassanis, S. Kubo, G. Melchers, F. Motoyoshi, Y. Otsuki, R. Paterson, T. A. Shalla, J. F. Shepard, Y. Sugimura, A. van Kammen, and J. W. Watts.

Literature Cited

1. Albouy, J., Kusiak, C., Lapierre, H., Laquerrière, F., Maury, Y. 1970. Isolement et propriétés de cellules de mésophylle de chou de chine infectées par le virus de la mosaique jaune du navet. *Ann. Phytophathol.* 2:607–16
2. Aoki, S., Takebe, I. 1969. Infection of tobacco mesophyll protoplasts by tobacco mosaic virus ribonucleic acid. *Virology* 39:439–48
3. Aoki, S., Takebe, I. 1975. Replication of tobacco mosaic virus RNA in tobacco mesophyll protoplasts inoculated in vitro. *Virology* 65:In press
4. Atabekov, J. G., Schaskolskaya, N.D., Atabekova, T. I., Sacharovskaya, G. A. 1970. Reproduction of temperature-sensitive strains of TMV under restrictive conditions in the presence of temperature-resistant helper strain. *Virology* 41:397–407
5. Bancroft, J. B. 1971. The significance of the multicomponent nature of cowpea chlorotic mottle virus RNA. *Virology* 45:830–34
6. Bancroft, J. B., Motoyoshi, F., Watts, J. W., Dawson, J. R. O. 1975. Cowpea chlorotic mottle and brome mosaic viruses in tobacco protoplasts. In *Modification of the Information Content of Plant Cells*, ed. R. Markham, D. R. Davis, D. A. Hopwood, R. W. Horne. Amsterdam: North-Holland. In press
7. Beier, H., Bruening, G. 1975. The use of an abrasive in the isolation of cowpea leaf protoplasts which support the multiplication of cowpea mosaic virus. *Virology* 64:272–76
8. Bosch, F. X., Jockusch, H. 1972. Temperature-sensitive mutants of TMV: Behavior of a non-coat protein mutant in isolated tobacco cells. *Mol. Gen. Genet.* 116:95–98
9. Burgess, J., Motoyoshi, F., Fleming, E. N. 1973. Effect of poly-L-ornithine on isolated tobacco mesophyll protoplasts: Evidence against stimulated pinocytosis. *Planta* 111:199–208
10. Burgess, J., Motoyoshi, F., Fleming, E. N. 1973. The mechanism of infection of plant protoplasts by viruses. *Planta* 112:323–32
11. Burgess, J., Motoyoshi, F., Fleming, E. N. 1974. Structural changes accompanying infection of tobacco protoplasts with two spherical viruses. *Planta* 117:133–44
12. Burgess, J., Motoyoshi, F., Fleming, E. N. 1974. Structural and autoradiographic observations of the infection of tobacco protoplasts with pea enation mosaic virus. *Planta* 119:247–56
13. Cassells, A. C., Gatenby, A. A. 1975. The use of lissamine rhodamine B conjugated antibody for the detection of tobacco mosaic virus antigen in tomato mesophyll protoplasts. *J. Gen. Virol.* In press
14. Chaleff, R. S., Carlson, P. S. 1974. Somatic cell genetics of higher plants. *Ann. Rev. Genet.* 8:267–78
15. Cocking, E. C. 1960. A method for the isolation of plant protoplasts and vacuoles. *Nature* 187:927–29
16. Cocking, E. C. 1966. An electron microscopic study of the initial stages of infection of isolated tomato fruit protoplasts by tobacco mosaic virus. *Planta* 68:206–14
17. Cocking, E. C. 1970. Virus uptake, cell wall regeneration, and virus multiplication in isolated plant protoplasts. *Int. Rev. Cytol.* 28:89–124
18. Cocking, E. C. 1972. Plant cell protoplasts—isolation and development. *Ann. Rev. Plant Physiol.* 23:29–50
19. Cocking, E. C., Pojnar, E. 1969. An electron microscopic study of the infection of isolated tomato fruit protoplasts by tobacco mosaic virus. *J. Gen. Virol.* 4:305–12
20. Coutts, R. H. A., Cocking, E. C. 1972. Infection of tobacco mesophyll protoplasts with tobacco mosaic virus. *J. Gen. Virol.* 17:289–94
21. Dawson, W. O., Schlegel, D. E. 1973. Differential temperature treatment of plants greatly enhances multiplication rates. *Virology* 53:476–78
22. De Zoeten, G. A., Assink, A. M., Van Kammen, A. 1974. Association of cowpea mosaic virus-induced double-stranded RNA with a cytopathological structure in infected cells. *Virology* 59:341–55
23. Doi, Y., Yora, K. 1972. Comparative observation of cell nuclei infected with plant viruses. *Ann. Phytopathol. Soc. Jpn.* 38:216 (Abstr., in Japanese)
24. Ellis, E. L., Delbrück, M. 1939. The growth of bacteriophage. *J. Gen. Physiol.* 22:365–84
25. Furumoto, W. A., Wildman, S. G. 1963. The specific infectivity of tobacco mosaic virus. *Virology* 20:53–61
26. Gamborg, O. L. 1975. Plant protoplasts in genetic modification and production of intergeneric hybrids. See Ref. 6

27. Goodman, R. M., Ross, A. F. 1974. Enhancement of potato virus X synthesis in doubly infected tobacco occurs in doubly infected cells. *Virology* 58:16–24

28. Goodman, R. M., Ross, A. F. 1974. Independent assembly of virions in tobacco doubly infected by potato virus X and potato virus Y or tobacco mosaic virus. *Virology* 59:314–18

29. Grout, B. W. W., Willison, J. H. M., Cocking, E. C. 1972. Interaction at the surface of plant cell protoplasts; an electrophoretic and freeze-etch study. In *The role of membrane structure in biological energy transduction. J. Bioenerg.* (special book issue)

30. Habili, N., Francki, R. I. B. 1974. Comparative studies on tomato aspermy and cucumber mosaic virus. I. Physical and chemical properties. *Virology* 57:392–401

31. Hariharasubramanian, V., Hadidi, A., Singer, B., Fraenkel-Conrat, H. 1973. Possible identification of a protein in brome mosaic virus infected barley as a component of viral RNA polymerase. *Virology* 54:190–98

32. Harrison, B. D. 1955. *Studies on the multiplication of plant viruses in inoculated leaves.* PhD thesis. Univ. London, London

33. Hibi, T., Rozeman, G., Van Kammen, A. 1975. Infection of cowpea mesophyll protoplasts with cowpea mosaic virus. *Virology.* In press

34. Hibi, T., Yora, K. 1972. Electron microscopy of tobacco mosaic virus infection in tobacco mesophyll protoplasts. *Ann. Phytopathol. Soc. Jpn.* 38:350–56

35. Hibi, T., Yora, K., Asuyama, H. 1968. TMV infection of *Petunia* protoplasts. *Ann. Phytopathol. Soc. Jpn.* 34:375 (Abstr., in Japanese)

36. Honda, Y., Matsui, C. 1974. Electron microscopy of cucumber mosaic virus-infected tobacco leaves showing mosaic symptoms. *Phytopathology* 64:534–39

37. Honda, Y., Matsui, C., Otsuki, Y., Takebe, I. 1974. Ultrastructure of tobacco mesophyll protoplasts inoculated with cucumber mosaic virus. *Phytopathology* 64:30–34

38. Jackson, A. O., Zaitlin, M., Siegel, A., Francki, R. I. B. 1972. Replication of tobacco mosaic virus III. Viral RNA metabolism in separated leaf cells. *Virology* 48:655–65

39. Jockusch, H. 1968. Two mutants of tobacco mosaic virus temperature-sensitive in two different functions. *Virology* 35:94–101

40. Jones, R. L., Armstrong, J. E., Taiz, L. 1973. Barley aleurone protoplasts: Some observations on their structure and physiology. In *Protoplastes et Fusion de Cellules Somatiques Végétales,* ed. J. Tempé, 109–17. Paris: Inst. Nat. Rech. Agron. 549 pp.

41. Kassanis, B. 1967. Plant tissue culture. In *Methods in Virology,* ed. K. Maramorosch, H. Koprowski, 1:537–66. New York: Academic. 640 pp.

42. Kassanis, B., White, R. F. 1974. A simplified method of obtaining tobacco protoplasts for infection with tobacco mosaic virus. *J. Gen. Virol.* 24:447–52

43. Katchalski, E., Sela, M., Silman, H. I., Berger, A. 1964. Polyamino acids as protein models. In *The Proteins,* ed. H. Neurath, 2:405–602. New York: Academic. 2nd ed. 4 vols. 840 pp.

44. Kubo, S., Harrison, B. D., Robinson, D. J. 1974. Effect of phosphate on the infection of tobacco protoplasts by tobacco rattle virus. *Intervirology* 3:382–87

45. Kubo, S., Harrison, B. D., Robinson, D. J., Mayo, M. A. 1975. Tobacco rattle virus in tobacco mesophyll protoplasts: infection and virus multiplication. *J. Gen. Virol.* In press

46. Lane, L. C., Kaesberg, P. 1971. Multiple genetic components in bromegrass mosaic virus. *Nature* 232:40–43

47. Lázár, G., Borbély, G., Udvardy, J., Premecz, G., Farkas, G. L. 1973. Osmotic shock triggers an increase in ribonuclease level in protoplasts isolated from tobacco leaves. *Plant Sci. Lett.* 1:53–57

48. Matile, P. 1969. Plant lysosomes in biology. In *Front. Biol.* 14:406–30

49. Meyer, Y. 1974. Isolation and culture of tobacco mesophyll protoplasts using a saline medium. *Protoplasma* 81:363–72

50. Motoyoshi, F., Bancroft, J. B., Watts, J. W., Burgess, J. 1973. The infection of tobacco protoplasts with cowpea chlorotic mottle virus and its RNA. *J. Gen. Virol.* 20:177–93

51. Motoyoshi, F., Bancroft, J. B., Watts, J. W. 1973. A direct estimate of the number of cowpea chlorotic mottle virus particles absorbed by tobacco protoplasts that become infected. *J. Gen. Virol.* 21:159–61

52. Motoyoshi, F., Bancroft, J. B., Watts, J. W. 1974. The infection of tobacco protoplasts with a variant of brome mosaic virus. *J. Gen. Virol.* 25:31–36

53. Motoyoshi, F., Hull, R. 1974. The infection of tobacco protoplasts with pea

enation mosaic virus. *J. Gen. Virol.* 24:89–99
54. Motoyoshi, F., Hull, R., Flack, I. H. 1975. Infection of tobacco mesophyll protoplasts by alfalfa mosaic virus. *J. Gen. Virol.* 27:In press
55. Motoyoshi, F., Watts, J. W., Bancroft, J. B. 1975. Factors influencing the infection of tobacco protoplasts by cowpea chlorotic mottle virus. *J. Gen. Virol.* 25:245–56
56. Mundry, K. W. 1963. Plant virus-host cell relations. *Ann. Rev. Phytopathol.* 1:173–96
57. Nagata, T., Takebe, I. 1970. Cell wall regeneration and cell division in isolated tobacco mesophyll protoplasts. *Planta* 92:301–8
58. Nagata, T., Takebe, I. 1971. Plating of isolated tobacco mesophyll protoplasts on agar medium. *Planta* 99:12–20
59. Nilsson-Tillgren, T., Kolehmainen-Sevéus, L., von Wettstein, D. 1969. Studies on the biosynthesis of TMV. I. A system approaching a synchronized virus synthesis in a tobacco leaf. *Mol. Gen. Genet.* 104:124–41
60. Otsuki, Y., Shimomura, T., Takebe, I. 1972. Tobacco mosaic virus multiplication and expression of the N gene in necrotic responding tobacco varieties. *Virology* 50:45–50
61. Otsuki, Y., Takebe, I. 1969. Isolation of intact mesophyll cells and their protoplasts from higher plants. *Plant Cell Physiol.* 10:917–21
62. Otsuki, Y., Takebe, I. 1969. Fluorescent antibody staining of tobacco mosaic virus antigen in tobacco mesophyll protoplasts. *Virology* 38:497–99
63. Otsuki, Y., Takebe, I. 1973. Infection of tobacco mesophyll protoplasts by cucumber mosaic virus. *Virology* 52:433–38
64. Otsuki, Y., Takebe, I. 1973. Multiple infection of tobacco protoplasts by TMV, CMV and PVX. *Ann. Phytopathol. Soc. Jpn.* 39:224 (Abstr., in Japanese)
65. Otsuki, Y., Takebe, I. 1975. Double infection of isolated tobacco leaf protoplasts with two strains of tobacco mosaic virus. In *Induction Mechanism of Biochemical and Cytological Response in Diseased Plants,* ed. K. Tomiyama. Tokyo: Tokyo Univ. Press. In press
66. Otsuki, Y., Takebe, I., Honda, Y., Matsui, C. 1972. Ultrastructure of infection of tobacco mesophyll protoplasts by tobacco mosaic virus. *Virology* 49:188–94

67. Otsuki, Y., Takebe, I., Honda, Y., Kajita, S., Matsui, C. 1974. Infection of tobacco mesophyll protoplasts by potato virus X. *J. Gen. Virol.* 22:375–85
68. Paterson, R., Knight, C. A. 1975. Protein synthesis in tobacco protoplasts infected with tobacco mosaic virus. *Virology* 64:10–22
69. Peden, K. W. C., Symons, R. H. 1973. Cucumber mosaic virus contains a functionally divided genome. *Virology* 53:487–92
70. Pelham, J. 1966. Resistance in tomato to tobacco mosaic virus. *Euphytica* 15:258–67
71. Power, J. B., Cocking, E. C. 1970. Isolation of leaf protoplasts: Macromolecule uptake and growth substance response. *J. Exp. Bot.* 21:64–70
72. Rappaport, I., Wildman, S. G. 1957. A kinetic study of local lesion growth on *Nicotiana glutinosa* resulting from tobacco mosaic virus infection. *Virology* 4:265–74
73. Rappaport, I., Wu, J. H. 1962. Release of inhibited virus infection following irradiation with ultraviolet light. *Virology* 17:411–19
74. Rochow, W. F., Ross, A. F. 1955. Virus multiplication in plants doubly infected by potato viruses X and Y. *Virology* 1:10–27
75. Roméro, J. 1973. Synthèse d'ARN dans des cellules et des protoplastes isolés de feuilles de *Vicia faba* infectées par le virus de la marbrure de la fève. *Ann. Phytopathol.* 5:225–32
76. Ryser, H. J. P. 1968. Uptake of protein by mammalian cells: An underdeveloped area. *Science* 159:390–96
77. Sakai, F., Takebe, I. 1970. RNA and protein synthesis in protoplasts isolated from tobacco leaves. *Biochim. Biophys. Acta* 224:531–40
78. Sakai, F., Takebe, I. 1972. A non-coat protein synthesized in tobacco mesophyll protoplasts infected by tobacco mosaic virus. *Mol. Gen. Genet.* 118:93–96
79. Sakai, F., Takebe, I. 1974. Protein synthesis in tobacco mesophyll protoplasts induced by tobacco mosaic virus infection. *Virology* 62:426–33
80. Sarkar, S. 1969. Evidence of phenotypic mixing between two strains of tobacco mosaic virus. *Mol. Gen. Genet.* 105:87–90
81. Sarkar, S., Upadhya, M. D., Melchers, G. 1974. A highly efficient method of inoculation of tobacco mesophyll protoplasts with ribonucleic acid of tobacco

mosaic virus. *Mol. Gen. Genet* 135:1–9
82. Schilde-Rentschler, L. 1973. A simpler method for the preparation of plant protoplasts. *Z. Naturforsch.* 27b:208–9
83. Shalla, T. A., Petersen, L. J. 1973. Infection of isolated plant protoplasts with potato virus X. *Phytopathology* 63:1125–30
84. Shalla, T. A., Shepard, J. F. 1972. The structure and antigenic analysis of amorphous inclusion bodies induced by potato virus X. *Virology* 49:654–67
85. Shaw, J. G. 1972. Effect of poly-L-ornithine on the attachment of tobacco mosaic virus to tobacco leaves and on the uncoating of viral RNA. *Virology* 48:380–85
86. Shepard, J. F., Totten, R. E. 1975. Isolation and regeneration of tobacco mesophyll cell protoplasts under low osmotic conditions. *Plant Physiol.* In press
87. Siegel, A., Zaitlin, M. 1964. Infection process in plant virus diseases. *Ann. Rev. Phytopathol.* 2:179–202
88. Siegel, A., Zaitlin, M., Duda, C. T. 1973. Replication of tobacco mosaic virus IV. Further characterization of viral related RNAs. *Virology* 53:75–83
89. Takebe, I., Aoki, S., Sakai, F. 1975. Replication and expression of tobacco mosaic virus genome in isolated tobacco leaf protoplasts. See Ref. 6
90. Takebe, I., Labib, G., Melchers, G. 1971. Regeneration of whole plants from isolated mesophyll protoplasts of tobacco. *Naturwissenschaften* 58:318–20
91. Takebe, I., Nagata, T. 1973. Culture of isolated tobacco mesophyll protoplasts. See Ref. 40, 175–87
92. Takebe, I., Otsuki, Y. 1969. Infection of tobacco mesophyll protoplasts by tobacco mosaic virus. *Proc. Nat. Acad. Sci. USA* 64:843–48
93. Takebe, I., Otsuki, Y. 1974. *Multiple Infection of Tobacco Mesophyll Protoplasts by Plant Viruses.* Presented at 3rd Int. Congr. Plant Tissue & Cell Culture, Leicester, England
94. Takebe, I., Otsuki, Y., Aoki, S. 1968. Isolation of tobacco mesophyll cells in

intact and active state. *Plant Cell Physiol.* 9:115–24
95. Takebe, I., Otsuki, Y., Aoki, S. 1971. Infection of isolated tobacco mesophyll protoplasts by tobacco mosaic virus. In *Les Cultures de Tissus de Plantes,* ed. L. Hirth, 503-11. Paris: Centre Nat. Rech. Sci. 511 pp.
96. Takebe, I., Otsuki, Y., Honda, Y., Matsui, C. 1974. Penetration of plant viruses into isolated tobacco leaf protoplasts. *Proc. 1st Intersect. Congr. Int. Assoc. Microbiol. Soc.,* ed. T. Hasegawa, 3:55–64. Tokyo: Jpn. Assoc. Microbiol. 5 vols. 662 pp.
97. Takebe, I., Otsuki, Y., Honda, Y., Nishio, T., Matsui, C. 1973. Fine structure of isolated mesophyll protoplasts of tobacco. *Planta* 113:21–27
98. Watts, J. W., Motoyoshi, F., King, J. 1974. Problems associated with the production of stable protoplasts of cells of tobacco mesophyll. *Ann. Bot.* 38:667–71
99. Weintraub, M., Ragetli, H. W. J. 1964. Studies on the metabolism of leaves with localized virus infections. Particulate fractions and substrates in TMV-infected *Nicotiana glutinosa. Can. J. Bot.* 42:533–40
100. Zaitlin, M. 1959. Isolation of tobacco leaf cells capable of supporting virus multiplication. *Nature* 184:1002–3
101. Zaitlin, M., Beachy, R. N. 1974. The use of protoplasts and separated cells in plant virus research. *Adv. Virus Res.* 19:1–35
102. Zaitlin, M., Beachy, R. N. 1974. Protoplasts and separated cells: Some new vistas for plant virology. In *Tissue Culture and Plant Science 1974,* ed. H. E. Street, P. J. King, 265–85. London & New York: Academic. 502 pp.
103. Zaitlin, M., Duda, C. T., Petti, M. A. 1973. Replication of tobacco mosaic virus V. Properties of the bound and solubilized replicase. *Virology* 53:300–11
104. Zaitlin, M., Hariharasubramanian, V. 1972. A gel electrophoretic analysis of proteins from plants infected with tobacco mosaic and potato spindle tuber viruses. *Virology* 47:296–305

HOST SPECIFICITY
OF PLANT VIRUSES

❖3616

J. G. Atabekov
Laboratory of Bioorganic Chemistry and Department of Virology, Moscow State University and Academy of Agricultural Sciences, Moscow, USSR

INTRODUCTION

What factors determine the host specificity of viruses? This is not a simple question and a simple answer is not expected. More often than not we do not know which features of the host and the virus are responsible for the tendency of a plant cell to resist or be susceptible to infection. For this reason this paper is a review of comparatively few experimental data; to a much larger extent it is an attempt to discuss several quite simple suppositions and speculations concerning the host specificity of viruses.

It will not be possible to understand the reasons behind success or failure of infection in one or another plant cell, until we can evaluate the contribution of the viral and host cell genomes to the process of virus replication. It is reasonable to assume that the genome of a plant virus lacks certain "host range genes," which are specifically designed for host range control. Rather, the host range is controlled by the combined activities of various genes, and failure of infection in a given cell type may be due to the inability of quite different genes (or maybe of a single gene) essential for virus replication. Conceivably, the mechanism (or rather mechanisms) controlling host range of a virus may operate at any phase of its replication; that is, if any single function (or a number of functions) not be performed in the infected plant, it would serve as a restrictive mechanism of host range control. Thus, the solution to the problem of which factors control host-range, could be tentatively reduced to the elucidation of the functions performed by a virus, on the one hand, and by a host cell, on the other, in the course of infection.

How many of these functions are there and which are they? Unfortunately, we are unable to give even an approximate number of virus- and cell-specific products involved in virus replication. There is no evidence for the number or functional significance of the genes (except for two or three of them) even in the genome of tobacco mosaic virus (TMV), the classical object of plant virology. This lack of experimental evidence makes it enormously difficult to answer our question, but at the same time, it renders the task somewhat simpler, for it justifies discussing certain evidence available from other RNA-containing viruses (bacteriophages and animal viruses).

The processes involved in the expression of the genetic information of RNA-containing viruses may be tentatively divided into two main stages: (*a*) the first stage includes adsorption of the virus, its penetration into the cell, uncoating its RNA, and possibly other functions yet unknown. (As discussed below, it is possible that penetration by plant viruses precedes adsorption); (*b*) later stages include translation[1] and replication of viral RNA, and maturation of the virus particles.

In this paper we discuss the problem of host range control in terms of the possibility of expression of the genetic information of a plant virus in different types of cells. To state it in more narrow terms, we are not concerned with the techniques of virus transmission, or increasing infection effectiveness, or with the effects of age, or of environmental and certain physiological conditions as factors affecting susceptibility of a plant to virus infection, and, therefore, acting on the host-range of a plant virus. By a "susceptible plant" (host plant) we mean a plant in whose cells the virus may replicate yielding progeny, irrespective of the type of host reaction to the virus (e.g. local lesions, systemic infection). Therefore, neither resistance connected with hypersensitivity, nor resistance to virus increase and spread in the infected plant are considered to be genuine resistance. We maintain increase of virus in inoculated cells as the only reliable criterion of susceptibility.

EXPERIMENTAL HOST RANGES OF PLANT VIRUSES

Host specificity is one of the major characteristics of a plant virus. Naturally, the literature on virus diseases contains ample data on susceptibility or resistance of various plant species to different viruses. These data are often contradictory, which is inevitable, for plant susceptibility to a virus depends upon many factors, such as the conditions of growth and of inoculation, the concentration of inoculum, and the peculiarities of a virus strain. Exact evaluation of the host range of a plant virus sometimes meets with obstacles already discussed in the literature (38, 56). Thorough, comparative studies of experimental host ranges show that different plant species differ greatly in the number of viruses to which they are susceptible (37, 38, 68). For example, *Chenopodium amaranticolor* Coste and Reyn, *Comphrena globosa* L., and *Tetragonia expansa* Murr. are susceptible to many more viruses than are some other species tested. On the other hand, *Portulaca oleracea* L. and *Lythrum salicaria* L. are resistant to many viruses (37).

Different plant viruses vary greatly in the range of plant species they are able to infect. For example, cucumber mosaic virus (CMV) has a very wide host range and multiplies in more than 40 dicotyledon and monocotyledon families (68). Tomato

[1]The present survey deals with the host range problem as regards the viruses that contain messenger RNA within the virion. By analogy with rabdoviruses and diplornaviruses of animals, it seems highly likely that certain plant viruses may contain virion RNA incapable of acting directly as a messenger.

spotted wilt virus (TSWV) also has a very wide host range, including dicotyledons and monocotyledons. TSWV infects about 170 species from 34 families (17). Alfalfa mosaic virus (AMV) can infect more than 300 species in about 50 families (40). On the other hand there are viruses of narrow host range: cucumber viruses 3 and 4 (CV3 and CV4) are mainly restricted to Cucurbitaceae (43, 68). American wheat striate mosaic virus is restricted to a few species of the Gramineae family (84). Thus, the plant species susceptible to certain viruses are distributed between taxonomically distant groups, whereas species susceptible to other viruses (e.g. CV3) may all belong to the same family.

It is well known that unrelated viruses may have overlapping host ranges, and related viruses (or strains) can differ in the plant species they infect. For example, the host ranges of TMV and tobacco etch virus (TEV) overlap, and though plants susceptible to TMV but not to TEV are known, very few plants susceptible to TEV and resistant to TMV are known. On the other hand, CV3 is antigenically related and structurally identical with TMV, but the only host common to both viruses is cucumber. The similarities or differences in host range need not correspond to similarity or difference among the structural, antigenic, and other properties of virions of the virus. Attempts to establish correlations between host range of the virus and its properties seem as much doomed to failure as, for example, the attempts to relate different types of viruses to different types of symptoms caused by them in different plant species.

Bald & Tinsley (10–12) examined the experimental host ranges of a number of mechanically transmitted viruses, and showed some taxonomic groups of plant species to be significantly different from others in their susceptibility to viruses. It was suggested by these authors that "the most susceptible groups of plants were relatively advanced in the phylogenic, and, therefore, probably in evolutionary, sense" and that "the more primitive groups of families were generally less suscepti-ble to the viruses examined."

Bald & Tinsley (11) postulated 1. that the genome of plant species susceptible to a given virus possesses a set of "factors for susceptibility," 2. that the genome of a virus capable of infecting this type of plant has a set of "infectivity factors," and 3. that the two sets must be complementary if the infection is to be a success. The postulate that the infectivity factors of the virus genome should match complemen-tary factors for susceptibility of the host genome sounds very attractive, though a little formal. Conceivably, in general idea, it is consistent with the supposition discussed above, that each of the essential functions coded by the virus genome can be performed only in a certain type of cell.

The idea of complementary factors may embrace concrete molecular-biological phenomena and may be illustrated by concrete examples. 1. It is well known that numerous animal viruses and bacteriophages are in need of special cell receptors (susceptibility factors), on which a virus can be adsorbed by means of special virus receptors (or infectivity factors). But both types of receptors must match each other. 2. Active RNA replicase of bacteriophage $Q\beta$ is an aggregate made up by four polypeptide chains, only one of which is virus-coded, the rest being cell-coded. In this system the virus-specific component of replicase seems to play the role of the infectivity factor matching the cell-specific components of replicase (susceptibility factors). The scope of such examples may be broadened. As a matter of fact, any one of the following mechanisms, considered as a possible host range controlling device, is based on the principle of functional match between virus- and cell-coded functions essential for virus replication.

FIRST STAGES OF VIRUS INFECTION AND THE HOST RANGE PROBLEM

The genetic information of a virus remains unexpressed as long as the virus genome is protected from the contact with an active cell's RNA-translating and replicating machinery. In other words, at least, partial uncoating of the virus RNA is an essential step in virus replication.

Quite obviously, the peculiarities of the initial phases of infection are largely determined by the properties of the "virus host" system. Hopefully, despite the diversity of modes of virus penetration into the cell, the initial phases of infection with different viruses have some basic features in common.

Nonhost animal and bacterial cells often are not susceptible to a certain virus because they fail to adsorb it and to allow its penetration. Cell receptors play an important role in the determination of the cell susceptibility to animal viruses and phages. "Cell receptors" include host cell structures showing affinity for virus particles and the ability to interact with the latter providing specific attachment of virions to the infected cell.

The concept of recognition of cell receptors by virus upon infection probably can be applied to plant viruses; yet, despite certain progress in the investigation of the initial phases of the interaction between plant viruses and their host cells, evidence for the chemistry of cell receptors or even their location in a cell is still lacking.

When trying to elucidate the role of the initial phases of infection in the plant host range control, we need answers to several questions: 1. Does a plant cell that is susceptible to a certain virus have specific sites (cell receptors) that can be recognized by the coat protein of this virus? 2. If the answer is yes, then does a cell susceptible to a different virus have a set of different receptors specific for definite viruses? 3. May a lack or artificial block of such receptors be a factor rendering the cell virus-resistant? In other words, is the possession of appropriate cellular receptors an essential condition for cell susceptibility to a plant virus? 4. Where are these receptors located in a plant cell? 5. Are the receptors specific for various viruses located in different cellular organelles? 6. Must a virus be adsorbed on the receptors before deproteinization of the virus genome? 7. Is the mechanism of deproteinization of viral RNA specific for a given virus (or a group of viruses), or is the cell capable of uncoating any virus admitted to it?

Outline of the Initial Phases of Infection

Penetration of virus particles into host cells presumably precedes their deproteinization. Although most plant viruses are incapable of infecting noninjured plant cells

(56, 61), some plant viruses can infect apparently nonwounded mesophyll cells by moving into them from the xylem (73). It is important that not every plant virus is capable of infecting noninjured cells after transport in the xylem. For example, TMV does not cause infections in noninjured leaves of bean, whereas tobacco necrosis virus (TNV) and southern bean mosaic virus (SBMV) do. It is not known how SBMV and TNV enter nondamaged protoplasts of some hosts so that infections result (maybe, by pinocytosis). These observations suggest that some mechanism controlling the host range of a plant virus may be operative at the initial phases of infection.

The effectiveness of plant inoculation with viruses is relatively low for several reasons, including imperfect inoculation techniques and low specific infectivity of plant viruses (56, 61, 93). It is pertinent to ask whether available evidence for the host range of plant viruses has been influenced by the above two circumstances and whether the experimental host ranges of plant viruses would broaden, if conditions ensuring more effective introduction of viral particles into plant cells were provided.

Recently, several methods of isolation of plant protoplasts have been developed. Takebe and co-workers (1, 64, 87) have succeeded in infecting with different viruses protoplast suspensions isolated from their respective host plants.

The virus particles probably enter cultured protoplasts by pinocytosis (18, 19, 65). Poly-L-ornithine increases the initial interaction of virus and host cell by enhancing pinocytic activity of the protoplasts. No pinocytic uptake of virus occurred when the protoplasts were inoculated without poly-L-ornithine (65). Poly-L-ornithine stimulated infection not only of protoplasts but also of tobacco leaves with TMV (82). However, the specific infectivity of TMV was not very high even when tobacco protoplasts were inoculated using poly-L-ornithine. As calculated by Kimura & Black (47) from the data of Takebe & Otsuki (87), about 22,000 to 210,000 TMV particles are required to infect one protoplast. Poly-L-ornithine increases the initial interaction of TMV with the host cell; there are, however, no reports on the possibility of widening host range by the use of this polycation. Parental virions attached to the protoplast and uptaken by a pinocytic vesicle were not detectable by electron microscopy (18, 65). Possibly the pinocytic vesicle serves as a device for the transport of the intact virus to the site of its further replication. For example, it can be transported to specific cellular receptors on which the virus is adsorbed and deproteinized. These receptors may be located in the endoplasmic reticulum, chloroplast, mitochondria, or nuclei. Alternatively, the pinocytic vesicle cannot serve as a virus-transporting system but can ensure deproteinization of the virus genome, after which the free viral RNA can reach the further virus replication site within the cell. This suggests that the specific cell receptors to which the virus is initially attached are localized on the cytoplasmic membrane, but not within the cell.

The most thorough investigation into the fate of TMV particles entering the host cell was undertaken by Shaw (79–81, 83), who showed virus deproteinization to be a two-step process. The author claimed that the first stage of the uncoating of TMV RNA was controlled by some physical forces at certain sites within the infected cell and that this mechanism did not operate specifically in different host cells. The deproteinization occurred in virus-resistant plants (corn, hollyhock) as well as in *Nicotiana tabacum* (80).

The next step of TMV deproteinization is inhibited by cycloheximide and low temperature. Possibly it is mediated by virus-induced and/or virus-coded enzymes. The second stage of deproteinization also seems to be a nonhost-specific process, in that it can be performed in different plant species.

The Virus Coat Protein and Host Range of Plant Viruses

This section deals with diverse and sometimes contradictory information on the role of the coat protein in controlling the host range of plant viruses.

INTERFERENCE BETWEEN VIRUSES   The ability of one virus strain to protect against infection by a second related strain has been recognized for almost half a century (58). This phenomenon has been designated by different terms (interference, cross protection, mutual antagonism, cross immunization). The subject of interference of plant viruses has been extensively reviewed by several authors (16, 46, 68, 76; see also 27, 28). In this paper, the cases of interference between plant viruses (or between the virus and viral coat protein) during the first stages of infection (virus attachment interference) are considered. In a typical case the number of local lesions decreases upon mixed inoculation of a local lesion-producing strain and a mottle-forming strain or with a viral coat protein preparation (see below).

As a rule, interference between related viruses is observed; however, cases of antagonism between biologically distant and serologically unrelated viruses also are known. On the other hand, antigenically related viruses do not necessarily interfere with one another (26). Unfortunately, we still do not know for sure which factors lead to interference between viruses, or whether there are different types of interference resulting from different causes. The phenomenon of interference was discussed in a remarkable review by Kassanis (46). This author was inclined to believe that the plant cell contained several infectible sites and that their number could vary from cell to cell. This hypothesis assumes that the plant cell possesses several independent sites, each potentially capable of interacting with a virus.

The cell acquires resistance, that is, it becomes a nonhost for a certain virus when all receptors are occupied by the virus antagonist. We have no knowledge of the nature of the structures referred to as "cellular receptors" or "infectible sites." However, the phenomenon of interference suggests that the interaction between the structural components of the virus capsid (virus receptors) and the cellular receptors is the necessary stage in the replication of a plant virus, that is, it may be the factor controlling the host range of the virus at early stages of the virus-cell interaction.

It is important that the preparations of the coat protein isolated from TMV are capable of interfering with the intact virus (14, 39, 77). Presumably, the virus coat protein preserves (at least, partially) the ability to recognize the cell receptors specific for the intact virus: as the saturation of cell receptors by the interfering protein occurs, particles of the virus cannot participate in the infection.

Experiments on the interference between the virus protein and the virus can provide some indirect evidence concerning the role of the adsorption mechanism (i.e. attachment of the virus particle to the cell receptors) as one of the possible mechanisms controlling virus host range specificity. Therefore, the specificity of interference between plant viruses and homologous viral protein could be tentatively equated with the specificity of interaction between virus particles and receptors.

The phenomenon of competition among related viruses (or between a virus and its coat protein preparation) for the cell receptors (virus-attachment interference) may offer an experimental approach to the evaluation of the receptors in the cell. Logically, only a specific inhibitory effect, produced either by the same but noninfective virus (or homologous viral protein) or by related strains (or their proteins), may be interpreted as reflecting the virus-receptor interaction.

In the experiments of Novikov & Atabekov (62) and Atabekov et al (5), *Chenopodium amaranticolor* was used as host because it is susceptible to a great number of different viruses and can be regarded as a "universal host." So it seemed reasonable to suggest that the cells of *C. amaranticolor* may possess different types of cellular receptors specific for different viruses, though it could hardly be expected that any of a hundred viruses capable of infecting *C. amaranticolor* would find in this plant cell its own receptors, available only for this virus and not available for others.

Our data (62) showed that there is specific interference between the virus [TMV or Barley stripe mosaic virus (BSMV)] and its homologous coat proteins in *C. amaranticolor*. Viral protein had an inhibitory effect against the homologous, but not against the heterologous, virus. The specificity of the interference suggested specific competition between virus and the virus protein for the cell receptors. These results suggest that there are sets of different receptors in *C. amaranticolor,* some of them specific for a certain virus but not for other viruses. For example, receptors specific for BSMV do not adsorb TMV and CV4; receptors specific for TMV do not adsorb BSMV and CV4.

The data on the effect of TMV protein on TMV RNA infectivity are rather contradictory (39, 77). Our results are in agreement with the data of Wu & Rappaport (92), who showed that there was no interference between the RNAs of two TMV strains and intact viruses in *Nicotiana glutinosa.*

The cells of *C. amaranticolor,* which became resistant to the intact virus as a result of interference, remained, however, susceptible to infective viral RNA (62). These data suggest that (a) the barrier of receptors is really operative in the plant cell under the conditions of interference, and (b) virus particles that are not adsorbed on the receptors (when the latter are occupied) are also unable to release their RNA; thus the virus must be adsorbed before uncoating takes place. If so, the structural protein of plant viruses may be expected to play a specific role in host range control.

One more question arises from the comparison of the above data with the results of Troutman & Fulton (89) and Thomas & Fulton (88), who demonstrated a correlation between the number of ectodesmata of the leaf and susceptibility of the plant to the virus. The resistance studied by Thomas & Fulton (88) was nonspecific and equally effective against different unrelated viruses. On the other hand, the process of adsorption of virus particles on receptors should be highly specific. Thus there may be different types of resistance—the efficiency of infection may depend directly on the number of ectodesmata which may function as "gates" of infection. On the other hand, the efficiency and, in some cases, the possibility of infection might depend on the presence in the plant cell of receptors specific for a given virus (62).

STRAIN SPECIFICITY AND HOST RANGES OF PLANT VIRUSES Different strains of a virus may show differences in the host range. The TMV strain U5 is not able to infect tomato under greenhouse conditions (9). Bald et al (9) recently reported that a strain almost indistinguishable from U5 could be isolated under certain conditions of cultivation and several transfers in *Nicotiana glauca* from U1 strain. The newly isolated strain (M5) differed by 15 amino acid units from common TMV coat protein and corresponded with natural U5 according to the host range. The mechanism of U1-U5 transition is obscure; however, it should involve some changes in the gene(s) responsible for host range control.

Some well-known strains of TMV are differentiated by host specificity on certain *Lycopersicon* hosts (20, 59). Different tomato strains of TMV isolated from different countries were shown to possess similar chemical properties (the presence of methionine and C-terminal serine). On the other hand, the protein subunit of common TMV strains contains C-terminal threonine and no methionine (91). However, in this case the differences in the amino acid content of the coat protein are not the cause of the host specificity of TMV strains. It was shown that the host range of TMV, reassembled from protein and RNA from two strains differing in their ability to infect tomato, did not depend upon the properties of the coat protein, being exclusively dependent upon the type of RNA used for the reassembly (20, 91).

Different species of insect vector vary in their ability to transmit plant viruses. The reasons for this are obscure. We shall not discuss this problem as a whole, but dwell only on an aspect which was raised by studies of strains of the barley yellow dwarf virus (BYDV). BYDV is a polyhedral circulative plant virus that is usually transmitted by aphids.

Some isolates of BYDV have been studied by Rochow (75). One isolate was transmitted specifically by *Rhopalosiphum padi* (RPV), another isolate was transmitted nonspecifically by both *R. padi* and *Macrosiphum avenae* (PAV), and a third isolate was transmitted specifically by *M. avenae* (MAV). MAV and PAV are serologically related but not identical viruses. On the other hand, RPV was antigenically distinct from MAV and PAV.

It turned out that *R. padi*, which would not transmit MAV from MAV-infected plants, was capable of transmitting it together with RPV from plants previously infected by both RPV and MAV. The mechanism of the so-called "dependent transmission of viruses" by aphids was studied by Rochow, who showed that the vector specificity of virus transmission in this system was determined by properties of the capsid protein of the particular BYDV strains used. *R. padi*, which does not regularly transmit MAV from singly infected cells acquires the ability to transmit this isolate from plants doubly infected by MAV and RPV. No transmission of MAV occurred, however, when *R. padi* aphids were injected with or fed through membranes containing inocula made by mixing concentrated preparations of RPV and MAV. Both viruses replicate together in the same mixedly infected plant with the resulting effect of dependent transmission. Direct experiments (74, 75) have shown that the ability of MAV to be transmitted by *R. padi* arises because in the mixedly infected plants some MAV nucleic acid becomes coated with RPV capsid protein. Thus, in this particular virus-host combination the coat protein of the helper virus plays a critical role in the determination of the biological specificity of the virus.

HOST RANGE OF FREE VIRAL RNA AND OF HYBRID VIRUSES   Infection of certain cells that lack virus-specific cellular receptors with free nucleic acid is a common experimental procedure with animal viruses and bacteriophages (23, 35, 85). There are a few reports of attempts to extend the host range of TMV by using free RNA for inoculation of nonhost cells (8, 41, 86).

Occasionally, the protein coat of a virus may restrict the infectivity of the virus for plants that would normally serve as hosts for this virus (e.g. see 4, 7, 15, 29). Despite this, there are no convincing reports in the literature on the possibility of extending the host range of a given virus by infecting plants immune to the intact virus with free RNA of the same virus.

Animal cells resistant to a certain virus sometimes can be infected by so-called "mixed" virions in which the genome of a virus which is unable to infect the cells is enclosed in the capsid of another virus capable of infecting them (23). "Mixed" particles can be also reconstituted in vitro using RNA and protein from foreign viruses.

Hiebert et al (34) reported that protein from a spherical plant virus normally unable to infect a certain host does not restrict the infectivity of foreign viral RNA enclosed in the mixed spherical virus obtained in vitro.

The contribution of the structural protein to the control of the host specificity of plant viruses was studied by the reassembly in vitro of hybrid viruses from the protein of TMV, CV4 or BSMV, and RNA from different viruses (6).

The results are summarized in Figure 1.[2] The general scheme of the experiment is seen in Figure 1*A*. Here the viruses designated as RNA PROT and rna prot are infective for plant species P' and P'', respectively. A plant designated as P''' is a common host susceptible to both viruses. Using different combinations of protein and RNA from the two viruses, one can reconstitute the hybrid viruses of different composition and study their host range (Figure 1*A*). With the help of this basic scheme several questions could be answered: (*a*) Will the RNA from virus RNA PROT preserve the ability to use plants P' as hosts if this RNA is coated with the protein from virus rna prot? (*b*) Will this RNA acquire the ability to infect plants P'' (nonhosts for RNA PROT) when coated with protein prot from the virus capable of infecting P''?

[2]The common strain of TMV was not infective for *Hordeum vulgare* var. Viner used in these experiments.

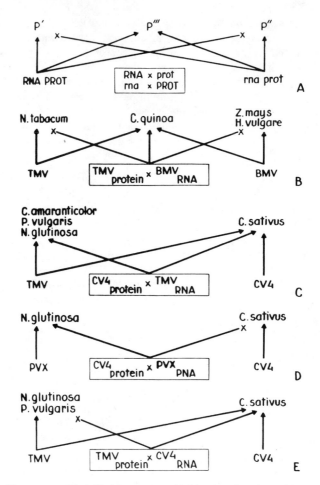

Figure 1   Host ranges of hybrid viruses reassembled in vitro from heterologous viral RNA and protein (3). The symbols ⟶ and —× stand for the infectivity and noninfectivity of the virus to the appropriate host (see text).

It was shown that the hybrid virus consisting of TMV protein and Brome Mosaic Virus (BMV) RNA was not able to infect *Zea mays* and *Hordeum vulgare* plants susceptible to BMV (Figure 1*B*). On the other hand, free RNA isolated from hybrid virus was infective to barley plants (5).

The $TMV_{protein}$ × BMV_{RNA} hybrid was absolutely unable to infect *Nicotiana glutinosa* and *N. tabacum,* but this hybrid virus was infective to *Chenopodium guinoa* which is a common host for TMV and BMV.

A probable explanation for the noninfectivity of the BMV_{RNA} × $TMV_{protein}$ hybrid in *Z. mays* and *H. vulgare* is that at early stages of infection some barriers render TMV protein in barley and corn cells nonfunctional. These plants are probably unable to remove the TMV protein from the hybrid particles, that is, in this case the foreign coat protein plays a restrictive role, making BMV RNA inert in the host cells. One more possibility should not be ignored. It cannot be ruled out that the BMV_{RNA} X $TMV_{protein}$ hybrid particles normally enter the cell and uncoat TMV particles upon their adsorption on cell receptors specific for TMV, but not for BMV. In this case, the subsequent blocking of the infection process in barley and corn cells could be due to lack of transport of BMV RNA to the proper replication site.

It appeared that the heterologous coat protein was unable to broaden the host range of a plant virus RNA which was a part of the mixed virion. For example, CV4 protein failed to render potato virus X (PVX) RNA infective to *Cucumis sativus* (nonhost for PVX), and TMV protein failed to make RNA CV4 infective to *N. glutinosa* or *Phaseolus vulgaris* (Figure 1D, E). This observation is in agreement with the data of Kado & Knight (43), who showed that the $TMV_{protein}$ X $CV4_{RNA}$ hybrid virus had the same host specificity as CV4.

On the other hand, it is important that the coat protein of some plant viruses normally unable to attack a host, does not necessarily restrict the host range of heterologous viral RNA enclosed in the hybrid particles. For example, the hybrid containing TMV RNA or PVX RNA coated with CV4 protein maintains its ability to infect some plants resistant to CV4 (Figure 1C, D). These data suggest that the mechanism controlling the host range does not always operate at the early stages of plant virus replication.

These results seem to be in contradiction with our earlier suggestion that the early stages of infection are specific in plant virus-host interaction. However, this contradiction can be explained, at least in the case considered below.

DO C. AMARANTICOLOR CELLS CONTAIN RECEPTORS FOR A FOREIGN VIRUS?   *C. amaranticolor* is resistant to CV4. It was found (62) that neither CV4 nor CV4 protein interfered with TMV infectivity in *C. amaranticolor* plants. This observation shows that TMV and CV4 do not compete for the same cell receptors, but it does not mean that *C. amaranticolor* cells do not contain receptors specific for CV4. This problem may be approached by the use of the hybrid ($CV4_{protein}$ X TMV_{RNA}) virus in the interference experiments.

It was shown that infectivity of the $CV4_{protein}$ X TMV_{RNA} hybrid in *C. amaranticolor* could be specifically inhibited by CV4 protein but not by TMV protein (5). Thus, *C. amaranticolor* cells may contain receptors specific for CV4, interacting with CV4 but not with TMV protein.

It is possible that the set of receptors in a given plant cell sometimes includes receptors for a certain virus(es) unable to infect this cell. If so, it would be natural to ask how many types of receptors the plant cell possesses. A great many different receptors ought to render this phase of replication ineffective as a host range controlling mechanism.

Attachment and uncoating are the necessary but not always sufficient conditions for virus replication. In particular cases the cell is abortively infected even after successful adsorption, penetration, and uncoating (23, 35).

The plant cell may have multiple cellular barriers controlling the host range of plant viruses at different stages of infection. For example, TMV protein is probably nonfunctional in barley cells, because it renders noninfective the hybrid virus BMV $_{RNA}$ X TMV$_{protein}$. On the other hand, free TMV RNA is also nonfunctional in barley, and therefore there is probably a double-blocking system of TMV infection in barley cells. Rnas from different plant viruses were completely inert in plants not susceptible to the virus itself; that is, the host range of these viruses coincides with that of their free RNA (5, 43). These observations are indicative of the existence of one (or many) more intracellular barrier(s) operating at the later stages of virus replication. Such barriers may constitute mechanisms controlling translation or replication of the viral RNA responsible for the specific defense of the plant cell against foreign genetic material.

HOST SPECIFICITY AND VIRAL RNA REPLICATION

Any intact RNA-containing virus is capable of coding the virus-specific enzyme(s), RNA-dependent RNA-polymerase(s). Virus-specific RNA polymerase was isolated from plants infected with turnip yellow mosaic virus (TYMV) (2), CMV (34, 57), tobacco ringspot virus (66), BMV (32, 33), and TMV (94). Solubilized BMV-specific RNA polymerase was isolated when the enzyme was stimulated specifically by homologous viral RNA (32). It could be suggested from the data of Hadidi & Fraenkel-Conrat (32) that the active enzyme consisted of different polypeptides. However, only two virus-specific polypeptides were found in BMV-infected cells, one of which corresponded with the coat protein (33). Presumably, plant virus RNA polymerase may consist of several peptide chains, only one (or only some) of which is virus-coded, while the rest of the subunits of active enzyme are host specific.

Active Qβ phage-specific replicase is known to consist of four polypeptides, only one of which is virus-coded (44, 45, 50). Moreover, the ability of phage replicase to recognize viral RNA upon its replication depends on the presence of two additional host-specific protein factors (24, 78). Thus, at least five host-specific proteins are necessary for the replication of bacteriophage RNA.

A similar phenomenon may also occur in plant viruses, that is, a plant virus is in need of some components of the normal host cell for the replication of viral RNA. The properties of cell components involved in the activities of the RNA-polymerase system of different viruses might differ in different cells. Imagine a cell-virus combination in which the cell lacks appropriate protein(s) and its susceptibility or resistance depends on its ability to contribute to the replication of the viral RNA. Such a possibility holds up even though the cell may be capable of correct translation of the viral RNA, producing a complete set of virus-specific proteins.

HOST SPECIFICITY AND VIRAL MESSAGE TRANSLATION

No direct evidence is yet available to show that the resistance of a plant to a virus can be due to the impossibility of translation of viral RNA in the cells of certain plant species. Nevertheless, some indirect evidence encourages one to imagine situations in which blocking of translation may be part of a host range controlling mechanism.

Selective Translation of Viral Messengers

Conceivably the protein-synthesizing machinery of the cell may recognize certain types of mRNAs, translating them selectively and discriminating them from other mRNAs. The selective translation of mRNA, or even of a certain cistron within one mRNA molecule rests on the ability of ribosomes to interact with a specific site at which translation begins. The choice of the starting point of mRNA translation may depend upon (a) the ribosome itself (e.g. on the structural proteins of ribosomal subunits) (5, 52, 53); (b) the ribosome plus factors governing initiation of translation and/or so-called "interference factors" (70) controlling initiation of mRNA translation; and (c) the peculiarities of the secondary and/or tertiary structures of the mRNA molecule (54).

It is quite likely that cells of different species might differ in their ability to initiate the translation of some viral RNA genetic message and to translate them in a different way or at different rates. The presence or absence of adequate translation initiation and/or interference factors in a given type of cell may be responsible for the possibility of mRNA translation of certain viruses, that is, these factors serve as a device controlling the host range of the virus.

Another factor restricting the expression of the viral RNA-genome in the nonhost cell may be the absence of affinity between the ribosome of the cell and the viral mRNA to which this cell is resistant. Two examples have been recently described for the RNA-bacteriophage-bacterial cell system (52, 53).

It must be noted, however, that several studies (on the in vitro translation of viral mRNAs in heterologous cell-free systems from nonhost cells and in vivo translation in amphibian oocytes) argue against the possibility of host-specific control at the translational level (13, 22, 49, 51, 60, 71, 72, 90).

These data support the idea that viral messengers can be read by heterologous systems without addition of homologous components from the host cell. However, we would not dare to claim that the translation machinery of the cell is nonspecific for viral mRNA of widely different origins. It would also be premature to claim this because, in the majority of investigations that demonstrated lack of specificity upon translation of heterologous messengers, there was no evidence that the whole molecule of heterologous mRNA was read absolutely correctly, leading to the production of a full set of virus-specific proteins indistinguishable from naturally occurring ones. Moreover, even if such evidence is presented, a claim like this could be made only after it is shown that the sequence and rate of synthesis of separate polypeptide chains coded by viral mRNA are identical in vivo in the host cell and in vitro in the heterologous cell-free system. In this way the protein synthesizing machinery could operate as a barrier to infection of nonhost cells by discriminating between the viral mRNAs.

Fidelity of Translation

Although it is widely accepted that the genetic code is universal for different groups of organisms, available experimental evidence suggests that the set of tRNAs and aminoacyl-tRNA synthetases sometimes vary with species. So it seems plausible that different cells may be somewhat different in their translational language.

It is well known that fidelity of reading of mRNA in different species of bacteria depends on the presence of nonsense and missense suppressors and that the mutations affecting the ribosomal proteins can influence the fidelity of translation (30, 31). The phenomenon of informational suppression of missence mutations has not been demonstrated in plant viruses so far.

An elegant approach to the study of fidelity of TMV RNA translation in vivo by different plant species was that of Okada et al (63). They demonstrated that the amino acid sequences in TMV coat protein were identical when TMV was grown in tobacco (Solanaceae) and garden zinnia (Compositae) plants. These data suggest that different cells translate the same viral RNA genetic message identically, that is, the translational languages of tobacco and garden zinnia are similar. Many more plants will need to be used in similar investigations before this idea can be extended beyond these two plant species.

The rate of protein synthesis can be controlled by modulation of specific tRNAs and synthetases. Also, variations in the specificity of reading of the viral message in different plant species may be a factor controlling the host range of plant viruses. So, for example, small differences in the translational language of host and nonhost plant species might give rise to nonfunctional virus-specific protein(s) in nonhost cells which might possibly account for their resistance to a virus. However, we must be aware that this is still only a fascinating hypothesis.

Mode of Translation of Viral Message

While considering the process of translation of polycistronic mRNA, two alternative mechanisms may be proposed: 1. Several ribosomes become attached to the mRNA molecule, simultaneously inducing, more or less independently, the translation of different cistrons, as in the case of RNA-containing bacteriophages. 2. Initiation and termination of translation occur as single events at the beginning (the 5'-end) and the end (the 3'-end) of the mRNA molecule, that is, the translation goes continuously by the type of the mRNA translation of picornaviruses. (36, 42). As a result of such translation, a giant polypeptide corresponding to all (or almost all) genetic information of the virus, is produced. Later this polypeptide is specifically cleaved into individual active virus-specific proteins. Unfortunately, up to now we have no exact knowledge of the mode of translation of any of the plant virus mRNAs; however, there is some evidence that the precursor polyprotein formation is possible upon translation of plant virus mRNA (72).

We may speculate that cleavage of polyprotein is effected by proteolytic enzyme(s) which are cell-specific and capable of correctly cleaving only the polyproteins coded by certain viruses. In the latter case, resistance to the virus should be due to the inability of the plant cell to cleave the virus-specific precursor polyprotein into active virus-specific proteins.

Intracellular Location of Viral Message Translation

Different viruses might be markedly different in their demands on subcellular sites of replication. Different strains of the same virus may differ in the intracellular localization of their biosynthesis and maturation. If these suppositions prove valid, the possibility exists that the host range of a virus could be dependent on mechanism(s) controlling the transport of the virus (or viral RNA) to its infection site.

Curious results have been obtained recently by Dingjan-Versteegh and Van Vloten-Doting (21) regarding the intracellular localization of virus-specific synthesis. According to their data, the coat protein of the cycloheximide-resistant strain of alfalfa mosaic virus (AMV) added to the mixture of three AMV-RNA species (M–, B–, Tb) enables the parental virus RNA to be translated in the chloroplasts of the infected cells. It was suggested, therefore, that the coat protein played some part in the intracellular localization of viral RNA translation. These observations prompt the speculation that the ability or inability of certain plant virus RNAs to be translated in appropriate cell organelles of a certain plant species may depend on the properties of the structural or some other virus-specific protein controlling the fidelity of the intracellular transport of the viral RNA. Presumably, this stage of virus replication may be controlled by both the virus and cell genomes. Unfortunately, the intracellular location of plant virus replication remains obscure although considerable efforts have been made to gain insight into the problem.

INHIBITION OF THE HOST CELL SYNTHESES BY THE VIRUS

We may suppose, largely by analogy with the action of many animal viruses, that one of the functions performed by plant viruses in the infected cell is inhibition of the host cell RNA and protein synthesis, or in other words, inhibition of the cell genome. Unfortunately, the evidence for such activity in plant viruses is too scarce and fragmentary. It is known that TMV severely inhibits cellular RNA synthesis and causes about 50% breakdown of chloroplast ribosomal RNA in the infected leaf. Several TMV strains also inhibit cytoplasmic ribosomal RNA synthesis (25); both RNA and protein synthesis in chloroplasts of TMV-infected plants are severely depressed. It was reported (69) that degradation products of host RNA were utilized for TMV-RNA biosynthesis. Kiraly et al (48) suggested that there was direct correlation between the rate of protein and nucleic acid synthesis and plant resistance to TMV. It is not clear why the inhibition of the host macromolecular synthesis is advantageous for susceptibility to the virus. Success or failure of infection is likely to depend on whether the virus succeeds in gaining control over the macromolecular synthesis in the infected cell or not. Here the following questions are to be answered: 1. Does the plant virus genome contain a special gene(s) whose products are responsible for virus pathogenicity and the ability to inhibit host synthesis? 2. Do such situations occur so that the possibility of virus reproduction depends on its ability to depress the host-synthesis, that is, is this function essential for virus replication? 3. Are there any effective mechanisms of cell protection against inhibition of macromolecular synthesis, which can resist the virus infection?

INHIBITION OF VIRUS REPRODUCTION BY THE HOST CELL

Several authors recently reported (55) that the formation of local lesions after virus infection made other tissues (not involved in the necrosis) resistant to subsequent infection by different plant viruses. This phenomenon is known as acquired resistance. It can be suggested that the antiviral state of the cell (acquired resistance) is due to some antiviral cell-specific substance(s) spreading beyond the virus-induced local lesions. Relatively little is known about the mechanisms underlying acquired resistance. Yet it is quite likely that this phenomenon is similar in some respects to the antiviral state caused by interferon in the animal cell. The latter renders animal cells incapable of supporting the multiplication of viruses by blocking the expression of the virus genome. The antiviral activity of interferon induced by some RNA viruses is selectively directed against the translation of viral and other mRNAs foreign for the cell.

We may speculate that success or failure of infection (i.e. the host range of the virus) in different "virus-cell" combinations could be determined by the ability of the virus to induce and the ability of the cell to acquire, the antiviral state.

CONCLUDING REMARKS

The problem of host range specificity of plant viruses has been treated in terms of the possibility of expression of the plant virus genetic information in different types of cells. It has been assumed that different functions coded by the viral genome, if not performed in certain plant cells, may play the role of the factor controlling the host range of the virus. Unfortunately, we cannot identify, even approximately, the virus- and cell-specific products involved in the replication of a plant virus. The present work is an attempt to consider the possible role of some virus-coded and a few cell-coded functions and their contribution in the host range control of plant viruses.

ACKNOWLEDGMENTS

I wish to thank Mrs. S. Glozman for her assistance in translation of the text.

Literature Cited

1. Aoki, S., Takebe, I. 1969. Infection of tobacco mesophyll protoplasts by tobacco mosaic virus ribonucleic acid. *Virology* 39:439–48
2. Astier-Manifacier, S., Cornuet, P. 1965. Isolation of turnip yellow mosaic virus RNA-replicase and assymetrical synthesis of polynucleotides identical to TYMV-RNA. *Biochem. Biophys. Res. Commun.* 18:283–87
3. Atabekov, J. 1971. Some properties of the coat protein of plant viruses, including the function of host-range control. *Acta Phytopathol. Acad. Sci. Hung.* 6:57–60
4. Atabekov, J. G., Novikov, V. K. 1966. Some properties of the nucleoprotein of barley mosaic virus and of its structural components. *Biochemistry USSR* 31: 157–66

5. Atabekov, J. G., Novikov, V. K., Vishnichenko, V. K., Javakhia, V. G. 1970. A study of the mechanisms controlling the host range of plant viruses. II. The host range of hybrid viruses reconstituted in vitro and of free viral RNA. *Virology* 41:108–15
6. Atabekov, J. G., Novikov, V. K., Vishnichenko, V. K., Kaftanova, A. S. 1970. Some properties of hybrid viruses reassembled in vitro. *Virology* 41:519–32
7. Atabekov, J. G., Novikov, V. K. 1971. Barley stripe mosaic virus. *Commonw. Mycol. Inst./Assoc. Appl. Biol. Descriptions of Plant Viruses* N68
8. Atherton, J. G. 1968. Formation of tobacco mosaic virus in an animal cell culture. *Arch. Gesamte Virusforsch.* 24: 406–18

9. Bald, J. C., Gumpf, D. J., Heick, J. 1974. Transition from common tobacco mosaic virus to the *Nicotiana glauca* form. *Virology* 59:467–76
10. Bald, J. G., Tinsley, T. W. 1967. A quasi-genetic model for plant virus host ranges. I. Group reactions within taxonomic boundaries. *Virology* 31:616–24
11. Bald, J. G., Tinsley, T. W. 1967. A quasi-genetic model for plant virus host ranges. II. Differentiation between host ranges. *Virology* 32:321–27
12. Bald, J. G., Tinsley, T. W. 1967. A quasi-genetic model for plant virus host ranges. III. Congruence and relatedness. *Virology* 32:328–36
13. Ball, L. A., Minson, A. C., Shih, D. S. 1973. Synthesis of plant virus coat proteins in an animal cell-free system. *Nature New Biol.* 246:206–8
14. Bawden, F. C., Pirie, N. W. 1957. The activity of fragmented and reassembled tobacco mosaic virus. *J. Gen. Microbiol.* 17:80–95
15. Bawden, F. C. 1961. The susceptibility of *Rhoeo discolor* to infection by tobacco mosaic virus. *J. Biol. Chem.* 236:2760–61
16. Bennett, C. W. 1953. Interactions between viruses and virus strains. *Adv. Virus Res.* 1:39–67
17. Best, R. J. 1968. Tomato spotted wilt virus. *Adv. Virus Res.* 13:65–146
18. Cocking, E. C. 1966. An electron microscopic study of the initial stages of infection of isolated tomato fruit protoplasts by tobacco mosaic virus. *Planta* 68:206–14
19. Cocking, E. C., Pojnar, E. 1969. An electron microscopic study of the infection of isolated tomato fruit protoplasts by tobacco mosaic virus. *J. Gen. Virol.* 4:305–12
20. Dahl, D., Knight, C. A. 1963. Some nitrous acid-induced mutants of tomato atypical mosaic virus. *Virology* 21:580–86
21. Dingjan-Versteegh, A., Van Vloten-Doting, L. 1974. The effect of inhibitors of protein biosynthesis on the infectivity of alfalfa mosaic virus: role of coat protein. *Virology* 58:136–40
22. Efron, D., Marcus, A. 1973. Translation of TMV-RNA in a cell-free wheat embryo system. *Virology* 53:343–48
23. Fenner, F. 1968. *The Biology of Animal Viruses.* Vol. 1. *Molecular and Cellular Biology.* New York: Academic. 474 pp.
24. Franze de Fernandez, M. T., Hayward, W. S., August, J. T. 1972. Bacterial proteins required for replication of phage Qβ ribonucleic acid. *J. Biol. Chem.* 247:824–31
25. Fraser, R. S. S. 1969. Effects of two TMV strains on the synthesis and stability of chloroplast ribosomal RNA in tobacco leaves. *Mol. Gen. Genet.* 106:73–79
26. Fulton, R. W. 1950. Cross protection tests with cucumber viruses 3 and 4 and tobacco mosaic virus. *Phytopathology* 40:219–20
27. Furumoto, W. A., Mickey, R. 1970. Mathematical analyses of the interference phenomenon of tobacco mosaic virus: theoretical considerations. *Virology* 40:316–21
28. Furumoto, W. A., Mickey, K. 1970. Mathematical analyses of the interference phenomenon of tobacco mosaic virus: experimental tests. *Virology* 40:322–28
29. Gordon, M. P., Smith, C. 1961. The infection of *Rhoeo discolor* by tobacco mosaic virus ribonucleic acid. *J. Biol. Chem.* 236:2762–63
30. Gorini, L. 1970. Informational suppression. *Ann. Rev. Genet.* 4:107–34
31. Gorini, L. 1974. Streptomycin and misreading of the genetic code. In *The Ribosome*, ed. M. Nomura, A. Tissiers, P. Lengyel. Cold Spring Harbor, NY. 930 pp.
32. Hadidi, A., Fraenkel-Conrat, H. 1973. Characterization and specificity of soluble RNA polymerase of brome mosaic virus. *Virology* 52:363–72
33. Hariharasubramanian V., Hadidi, A., Singer, B., Fraenkel-Conrat, H. 1973. Possible identification of a protein in brome mosaic virus infected barley as a component of viral RNA polymerase. *Virology* 54:190–98
34. Hiebert, E., Bancroft, J. B., Bracker, C. E. 1968. The assembly in vitro of some small spherical viruses, hybrid viruses and other nucleoproteins. *Virology* 34:492–508
35. Holland, J. J. 1964. Enterovirus entrance into specific host cells and subsequent alteration of cell protein and nucleic acid synthesis. *Bacteriol. Rev.* 28:3–13
36. Holland, J. J., Kiehn, E. D. 1968. Specific cleavage of viral proteins as steps in the synthesis and maturation of enteroviruses. *Proc. Nat. Acad. Sci. USA* 60:1015–22
37. Hollings, M. 1959. Host range studies with fifty-two plant viruses. *Ann. Appl. Biol.* 47:98–108

38. Holmes, F. O. 1946. A comparison of the experimental host ranges of tobacco-etch and tobacco-mosaic viruses. *Phytopathology* 36:643–59
39. Holoubek, V. 1964. Effect of tobacco mosaic virus protein on tobacco mosaic virus infectivity. *Nature London* 203:499–501
40. Hull, R., 1969. Alfalfa mosaic virus. *Adv. Virus Res.* 15:365–433
41. Humm, D. G., Humm, J. H. 1968. Identification of tobacco mosaic coat protein after the injection of tobacco mosaic virus RNA into the common slime mold. *Fed. Proc.* 27:772
42. Jacobson, M. F., Baltimore, D. 1968. Polypeptide cleavages in the formation of poliovirus proteins. *Proc. Nat. Acad. Sci. USA* 61:77–84
43. Kado, C. I., Knight, C. A. 1970. Host specificity of plant viruses. I. Cucumber virus 4. *Virology* 40:997–1007
44. Kamen, R. 1970. Characterization of the subunits of Qβ replicase. *Nature London* 228:527–33
45. Kamen, R., Kondo, M., Römer, W., Weissmann, C. 1972. Reconstitution of Qβ replicase lacking subunit α with protein-synthesis-interference factor i. *Eur. J. Biochem.* 31:44–51
46. Kassanis, B. 1963. Interaction of viruses in plants. *Adv. Virus Res.* 10:219–55
47. Kimura, I., Black, L. M. 1972. The cell-infecting unit of wound tumor virus. *Virology* 49:549–61
48. Király, Z., Hammady, M. El, Pozsár, B. I. 1968. Susceptibility to tobacco mosaic virus in relation to RNA and protein synthesis in tobacco and bean plants. *Phytopathol. Z.* 63:47–63
49. Klein, W. H., Nolan, C., Lazar, J. M., Clark, J. M. 1972. Translation of satellite tobacco necrosis virus ribonucleic acid. I. Characterization of *in vitro* procaryotic and eucaryotic translation products. *Biochemistry* 11:2009–14
50. Kondo, M., Gallerani, R., Weissmann, C. 1970. Subunit structure of Qβ replicase. *Nature London* 228:525–27
51. Laskey, R. A., Gurdon, J. B., Crawford, L. V. 1972. Translation of encephalomyocarditis viral RNA in oocytes of *Xenopus laevis. Proc. Nat. Acad. Sci. USA* 69:3665–69
52. Leffler, S., Szer, W. 1973. Messenger selection by bacterial ribosomes. *Proc. Nat. Acad. Sci. USA* 70:2364–68
53. Lodish, H. F. 1970. Specificity in bacterial protein synthesis: role of initiation factors and ribosomal subunits. *Nature London* 226:705–7
54. Lodish, H. F. 1970. Secondary structure of bacteriophage f2 ribonucleic acid and the initiation of *in vitro* protein biosynthesis. *J. Mol. Biol.* 50:689–702
55. Loebenstein, G. 1972. Localization and induced resistance in virus-infected plants. *Ann. Rev. Phytopathol.* 10:177–206
56. Matthews, R. E. F. 1970. *Plant Virology.* New York: Academic. 778 pp.
57. May, J. T., Gilliland, J. M., Symons, R. H. 1969. Plant virus-induced RNA polymerase: Properties of the enzyme partly purified from cucumber cotyledons infected with cucumber mosaic virus. *Virology* 39:54–65
58. McKinney, H. H. 1929. Mosaic diseases in the Canary Islands, West Africa, and Gibraltar. *J. Agric. Res.* 39:557–78
59. McRitchie, J. J., Alexander, L. J. 1963. Host-specific Lycopersicon strains of tobacco mosaic virus. *Phytopathology* 53:354–98
60. Morrison, T. G., Lodish, H. F. 1973. Translation of bacteriophage Qβ RNA by cytoplasmic extracts of mammalian cells. *Proc. Nat. Acad. Sci. USA* 70:315–19
61. Mundry, K. W. 1963. Plant virus-host cell relations. *Ann. Rev. Phytopathol.* 1:173–96
62. Novikov, V. K., Atabekov, J. G. 1970. A study of the mechanisms controlling the host range of plant viruses. I. Virus-specific receptors of *Chenopodium amaranticolor. Virology* 41:101–7
63. Okada, Y., Nozu, Y., Ohno, T. 1969. Demonstration of the universality of the genetic code in vivo by comparison of the coat proteins synthesized in different plants by tobacco mosaic virus RNA. *Proc. Nat. Acad. Sci. USA* 63:1189–95
64. Otsuki, J., Takebe, I., 1972. Infection of tobacco protoplasts by PVX. *Ann. Phytopathol. Soc. Jpn.* 38:220 (Abstr., in Japanese)
65. Otsuki, J., Takebe, I., Honda, Y., Matsui, C. 1972. Ultrastructure of infection of tobacco mesophyll protoplasts by tobacco mosaic virus. *Virology* 49:188–94
66. Peden, K. W. C., May, J. T., Symons, R. H. 1972. A comparison of two plant virus-induced RNA polymerases. *Virology* 47:498–501
67. Price, W. C. 1940. Acquired immunity from plant virus diseases. *Quart. Rev. Biol.* 15:338–61

68. Price, W. C. 1940. Comparative host ranges of six plant viruses. *Am. J. Bot.* 27:530–41
69. Reddi, K. K. 1972. Tobacco mosaic virus with emphasis on the events within the host cell following infection. *Adv. Virus Res.* 17:51–94
70. Revel, M. et al 1973. IF-3-interference factors: protein factors in *Echerichia coli* controlling initiation of mRNA translation. *Biochimie* 55:41–51
71. Roberts, B. E., Mathews, M. B., Bruton, C. J. 1973. Tobacco mosaic virus RNA directs the synthesis of a coat protein peptide in a cell-free system from wheat. *J. Mol. Biol.* 80:733–42
72. Roberts, B. E., Paterson, B. M., Sperling, R. 1974. The cell-free synthesis and assembly of viral specific polypeptides into TMV particles. *Virology* 59:307–13
73. Roberts, D. A., Price, W. C. 1967. Infection of apparently uninjured leaves of bean by the viruses of tobacco necrosis and southern bean mosaic. *Virology* 33:542–45
74. Rochow, W. F. 1970. Barley yellow dwarf virus: phenotypic mixing and vector specificity. *Science* 167:875–78
75. Rochow, W. F. 1972. The role of mixed infections in transmission of plant viruses by aphids. *Ann. Rev. Phytopathol.* 10:101–24
76. Ross, A. F. 1959. The interaction of viruses in the host. *Plant Pathology—Problems and Progress 1908–1958,* ed. C. S. Holton, 511–20. Madison, Wis.: Univ. Wisconsin Press
77. Santilli, V., Piacitelli, J., Wu, J. H. 1961. The effect of tobacco mosaic virus protein on virus incubation period and infectivity. *Virology* 14:109–23
78. Shapiro, L., Franze de Fernandez, M. T., August, J. T. 1968. Resolution of two factors required in the Qβ-RNA polymerase reaction. *Nature London* 220:478–80
79. Shaw, J. G. 1967. *In vivo* removal of protein from tobacco mosaic virus after inoculation of tobacco leaves. *Virology* 31:665–75
80. Shaw, J. G. 1969. In vivo removal of protein from tobacco mosaic virus after inoculation of tobacco leaves. II. Some characteristics of the reaction. *Virology* 37:109–16
81. Shaw, J. G. 1970. Uncoating of tobacco mosaic ribonucleic acid after inocula-

tion of tobacco leaves. *Virology* 42:41–48
82. Shaw, J. G. 1972. Effect of poly-L-ornithine on the attachment of tobacco mosaic virus in tobacco leaves and on the uncoating of viral RVA. *Virology* 48:380–85
83. Shaw, J. G. 1973. In vivo removal of protein from tobacco mosaic virus after inoculation of tobacco leaves. III. Studies on the location on virus particles for the initial removal of protein. *Virology* 53:337–42
84. Sinha, R. C., Behki, R. M. 1972. American wheat striate mosaic virus. *C.M.I. / A.A.B. Descriptions of plant viruses N99*
85. Stent, G. S. 1963. *Molecular Biology of Bacterial Viruses.* San Francisco: Freeman. 474 pp.
86. Süss, R., Sander, E., Röttger, B., Senger, H. 1965. Versuche zur vermehrung infektiöser Ribonukleinzäure aus Tabakmosaikvirus in *Chlorella. Biochim. Biophys. Acta* 95:388–97
87. Takebe, I., Ōtsuki, Y. 1969. Infection of tobacco mesophyll protoplasts by tobacco mosaic virus. *Proc. Nat. Acad. Sci. USA* 64:843–48
88. Thomas, P. E., Fulton, R. W. 1968. Correlation of ectodesmata number with nonspecific resistance to initial virus infection. *Virology* 34:459–69
89. Troutman, J. L., Fulton, R. W. 1958. Resistance in tobacco to cucumber mosaic virus. *Virology* 6:303–16
90. Twardzik, D., Simonds, J., Oskarsson, M., Portugal, F. 1973. Translation of AKR-murine leukemia viral RNA in an *E. coli* cell-free system. *Biochem. Biophys. Res. Commun.* 52:1108–14
91. Wang, A. L., Knight, C. A. 1967. Analysis of protein components of tomato strains of tobacco mosaic virus. *Virology* 31:101–16
92. Wu, J. H., Rappaport, I. 1962. Competition between infectious tobacco mosaic virus nucleic acid and intact virus on *Nicotiana glutinosa. Nature London* 193:908
93. Yarwood, C. E. 1964. Assay of infectivity. *Techniques in Experimental Virology,* ed. R. J. C. Harris, 80–107. New York: Academic. 450 pp.
94. Zaitlin, M., Duda, C. T., Petti, M. A. 1973. Replication of tobacco mosaic virus. V. Properties of the bound and solubilized replicase. *Virology* 53:300–11

APPRESSORIA ❖3617

R. W. Emmett
Victorian Plant Research Institute, Department of Agriculture, Mildura 3500, Australia

D. G. Parbery
School of Agriculture and Forestry, University of Melbourne, Parkville 3052, Australia

The initiation, formation, and action of the appressorium are integral parts of the infection process of many parasitic fungi. In some species, formation of appressoria may be obligatory for infection, while in others it may be optional or unnecessary. An understanding of appressorium formation can benefit several areas of plant pathology. For example, the influence of phyllosphere microorganisms on the infection process (89) and the effect on infection of the genetic control of chemical and physical properties of the host are of considerable importance (35).

Traditionally, only those structures produced on germtubes are known as appressoria, while adhesion and penetration structures produced on hyphae have been described as "hyphopodia," or by other names. In this review the restrictive morphological concept has been abandoned in favor of a broader concept of structures whose basic function is entry into a host. Some fungi, however, gain direct entry into a host without the aid of any special structure, and some fungi appear to require appressoria only under certain conditions.

There are two major ecological groups of plant parasitic fungi that need to be considered: the epiphytic parasites, such as members of the Erysiphales and Meliolales which, with the exception of haustoria, mainly produce their mycelium on the host surface; and the many endophytic parasites, such as the Polystigmataceae and Uredinales, which apart from germtubes and appressoria produce their mycelium inside the host. Differences in these groups are important when evaluating the roles of appressoria (85).

In addition, spore populations of the same species may be composed of different genotypes that require different environmental conditions for germination (76). Further, the environmental conditions under which spores develop can determine their requirements for germination. It is possible, therefore, that not all spores in a population will be homogeneous in conditions required for appressorium formation.

THE CONCEPT OF THE APPRESSORIUM

For over 120 years, in studies on infection processes of parasitic fungi, various concepts of the term "appressorium" have arisen. While there has been general agreement over the functional meaning of the term, opinion has varied over which structures constitute an appressorium. Thus, the concept has become unwieldy and requires clarification.

The term "appressorium" was introduced in 1883 by Frank (40) for "spore-like organs" formed on germtubes of *Colletotrichum lindemuthianum, Polystigma rubrum,* and *Fusicladium tremulae.* He believed appressoria were adhesive discs attaching a parasitic fungus to its host during the early stages of infection. Hasselbring (46) confirmed Frank's belief by experimenting with *Gloeosporium fructigenum.* He refuted the interpretation of many early authors that appressoria were also spore-like in function. The conflict of ideas about the nature of appressoria has already been reviewed (37).

The morphological concept of the appressorium has become progressively more loosely defined. Frank (40) saw them as spore-like. As more fungi were studied, it appeared that appressoria formed on hyphae as well as germtubes and could take different forms. Some, such as in *Colletotrichum* species, were morphologically well defined, whereas others were not. It was difficult, therefore, to define appressoria morphologically, and this led to many inaccurate or inadequate definitions. Dictionaries and textbooks refer to them as flattened, thickened, or swollen hyphal branches. Ainsworth defined them as "swellings on germtubes, or hyphae." These definitions are inadequate on morphological grounds alone, and most are qualified by saying they give rise to infection hyphae or haustoria (3). Difficulties also arise in distinguishing them on morphological criteria. Simple appressoria range from well-defined, dark, thick-walled structures delimited by septa to barely swollen, hyaline hyphal apices. When no septum delimits the appressorium, distinction between germtube or hypha and appressorium becomes arbitrary. Then, morphological delimitation of the appressorium becomes vague, and precise definition impossible.

This problem could be resolved in either of two ways. The scope of the concept could be reduced to include only morphologically distinct entities separated from the germtube or hypha by a septum. This would then include only structures produced by fungi such as species of *Colletotrichum.* The term would be restricted to an easily identifiable structural form with the capacity to adhere firmly to the host surface. Alternatively, the scope of the concept could be expanded to incorporate all structures adhering to host surfaces to achieve penetration, regardless of morphology. The only criteria delimiting an appressorium would then be its capacity to adhere to a host surface, and ability to germinate and penetrate the host. Thus the terms, appressorium and "infection structure," would become synonymous, and artificial restrictions imposed by morphology would be bypassed. It is in this latter sense that the term appressorium is used in this review.

MORPHOLOGY OF APPRESSORIA

Some fungi produce appressoria at different stages of growth. Some produce more than one kind. The dematiaceous epiphytic parasites mostly produce appressoria on germtubes identical with those produced later on hyphae. Other fungi such as *Gaeumannomyces graminis* may do likewise (117). In the Erysiphaceae most species produce appressoria different in form on hyphae from those on germtubes, each type being morphologically constant for a species (21). Appressoria produced by both ascospores and conidia of species of *Glomerella* (J. M. Lenné & D. G. Parbery, unpublished), *Dimerosporium,* and Meliolales (85) are identical. This may not be true for all fungi.

Numerous reports on species such as *Sclerotinia libertiana* and *Rhizoctonia solani* demonstrate great variation in morphology in compound appressoria. This can be related to differences in isolates (54), or hosts (128), or in environment (14).

Parbery & Emmett (85) have surveyed the variety of appressoria formed by various ecological and phylogenetic groups of fungi. The different types can be divided into two basically different groups: simple appressoria formed from a single, modified cell, and compound appressoria made up of many cells. The first comprises the larger group and contains many different structures. Fewer fungi produce compound structures that are likely to be more variable within one species.

Simple appressoria occur terminally on germtubes but may be terminal, lateral, or intercalary when produced by mycelial hyphae. Appressoria produced on germtubes are epiphytic, and three types can be distinguished. The first, "protoappressoria," are little more than slightly swollen, hyaline germtube apices adhering to the host and giving rise to an infection peg. They are rarely delimited by a septum and occur in *Venturia inaequalis* (43), *Pythium* spp. (2, 71), and various other fungi (85). The second group, hyaline appressoria, occur in the Erysiphales, Uredinales, and other species (85). They become swollen and are usually delimited from the germ-tube by a septum. Appressorium morphology is relatively uniform for all rusts, but in powdery mildews, while being constant within species, it is often sufficiently variable between species to be taxonomically useful (21, 130). The third type, the dark appressoria, may be sessile or subtended by a germtube (83) from which they are delimited by one or more septa (109). They are usually thick walled, relative to the spore or germtube producing them. They are produced by most epiphytic parasites except powdery mildews (85) as well as many endophytic species in genera such as *Polystigma, Glomerella (Colletotrichum), Phyllachora,* and *Guignardia (Phyllosticta).* Many authors recognize their taxonomic value (83, 85, 109, 110, 114).

Appressoria are also produced on epiphytic and endotrophic hyphae. Those on hyphae of epiphytic fungi may serve a dual role, for as well as aiding host penetration, they persist as the connective between infection thread and epiphytic mycelium through which nutrient absorbed by the haustorium passes to the mycelium (85). Such structures may be terminal, lateral, or intercalary.

Epiphytic appressoria are hyaline or dark. Hyaline terminal ones occur in the Erysiphaceae (21) and species of *Alternaria* (129), *Botrytis* (68), *Rhizoctonia* (28), and *Ustilago* (118). Dark ones have been found in *Colletotrichum* spp. (110, 116). Lateral appressoria (21, 128), best known as "hyphopodia" of dematiaceous epiphytic parasites (85) are sometimes called "stigmopodia" (3). They are a specialized group of appressoria whose constant morphology is valuable taxonomically (21, 85, 114). Intercalary appressoria have been called "les ampoules perforatrices" (64), "node cells," and "stigmatocysts" (3). They are common among epiphytes, function as hyphopodia, and are also useful taxonomically (85).

Endotrophic appressoria produce infection pegs that allow intracellular hyphae to pass from cell to cell. They are best known in species of *Phyllachora* (83), Ustilaginales (9), *Pseudopeziza* (W. C. Morgan & D. G. Parbery, unpublished), *Gibberella* (87), *Deightoniella* (72), and *Mycosphaerella* (29). In wood invaded by blue stain and softrotting fungi they have been called "transpressoria" (98).

Compound appressoria vary from a few cells to multicelled structures called "infection cushions." They are divided into two groups according to their manner of development. Some originate from a single hypha, others from several. They form more commonly on hyphae than on germtubes.

Lobate appressoria are the simplest type consisting of a single lobate cell that produces an infection peg from near the extremity of each lobe, e.g. *Pythium irregulare* (2). Two-celled to multicelled types produce germpores in most cells, e.g. some species of *Colletotrichum* (109), *Phyllosticta,* and Meliolales (85). Further complexity is seen in multicellular, lobate types occurring in species of *Botrytis* on glass (68), and *Rhizoctonia* (54) and *Sclerotinia* (14, 91) on host surfaces. Under certain conditions the latter type may develop into infection cushions (54).

Compound appressoria built up from several hyphae become "infection cushions" or "infection plaques." They may produce many infection threads (54, 64). The hyphae frequently become dark and thick walled (84). Infection cushions are produced by species of *Rhizoctonia* (54, 78, 112), *Sclerotinia* (14, 18, 59), and *Sclerotium* (17). Infection plaques, composed of numerous intercalary appressoria, occur in *Leptosphaeria herpotichoides* and *Gaeumannomyces graminis* (64). Most compound appressoria are epiphytic, but endophytic structures sometimes form to break barriers within hosts (49, 59).

VARIATION IN APPRESSORIUM MORPHOLOGY

Morphological variation is induced by several sets of factors. Physical factors may include temperature, water availability, light intensity, and hardness of surface. Chemical factors include the complex of chemicals acting at the plant surface. Genetic control of appressorium formation is strongly indicated by the constant phenotypes and patterns of development of appressoria of given species (21, 85, 109, 110, 114). Time is also important. It is suspected that some structures described as protoappressoria may have been incorrectly reported through premature observations or from germination occurring in an environment unsuitable for complete formation. Appressorium formation takes 6 to 12 hr in most species germinating under favorable conditions. It is important to elucidate the reasons for variation in a particular case, and to avoid the assumption that the same factors induce variation in all cases. In the following section morphological variation is discussed further.

PATTERNS OF DEVELOPMENT

Appressoria usually form on only one primary germtube per cell and often per spore. This occurs in species of *Phyllachora*, (83), powdery mildews (89), and *Colletotrichum* (52), but some Meliolales produce two structures. Sometimes, in rusts, more than one germtube is produced, and in species of *Cochliobolus* and *Pyrenophora* a single germtube branches. Appressorium formation, however, may be negatively correlated with degree of branching (16).

Studies of developmental sequences define several stages in appressorium development (35, 83, 89, 93, 103). Three stages are recognized in the primary development of powdery mildews (75, 89): appressorium initiation, appressorium maturation, and infection peg production or penetration. The third stage is sometimes called appressorium germination. Similar stages are recognized in other fungi. Maturation of appressoria is considered a critical process during infection by some fungi (35, 80) but not in all (61).

MORPHOLOGICAL STUDIES

Initiation

This is generally the cessation of germtube or hyphal elongation and the swelling of the hyphal apex. Dey (24) described the formation of appressorium initials (protoappressoria) in *C. lindemuthianum* once the germtube apex contacted the host surface. A similar pattern of development has since been described for many fungi (5, 6, 67, 73, 86, 101, 103). Sutton (110) describes this development as blastic. Many authors seem to consider that appressoria develop apically as additions to germtubes rather like spores developing terminally on sporophores. It is doubtful whether this ever happens. In *Phyllachora* species once the germtube stops elongating, the apex begins to swell and this swelling proceeds back along the germtube towards the spore (83). Thus, part or all of the germtube is converted into an appressorium. The diameter of the germtube across the septum is often greater than at that point prior to appressorium formation. This pattern occurs among other fungi including species of *Colletotrichum* and powdery mildews (D. G. Parbery, unpublished).

Maturation

Whether development proceeds beyond the protoappressorium stage or not depends on either or both genetic or environmental factors. If it proceeds, the protoappressorium continues to expand, the swollen apex may be cut off from the germtube or hypha by a septum, and in some cases the swelling begins to darken. Masri & Ellingboe (70) regard elongation and bending of the appressorium and its being cut off by a septum as indicating maturity in the powdery mildews. Such criteria of maturity are not universal. Sometimes septa are formed well before the full expansion of appressoria (110). Similarly, bending does not occur in all species. It is common in the Erysiphaceae and some species of *Phyllachora* but is rare among species of *Colletotrichum* and other fungi studied. In *Phyllachora* spp. pigmentation can begin in the protoappressorial stage and be well advanced before the septum is formed (83). In some *Colletotrichum* spp., pigmentation follows septum formation (109). Wall thickening often accompanies pigmentation.

At appressorium maturity the parent spore often appears empty and sometimes collapses. This has been noted in species of *Phyllachora* (D. G. Parbery, unpublished), *Colletotrichum* (J. M. Lenné, unpublished), *Phytophthora* (90), and other fungi (73). In some rusts a septum forms between appressorium and spore before the appressorium contents pass into the substomatal vesicle, sometimes regarded as an extension of the appressorium (25). In powdery mildews the parent spore and primary appressorium are incorporated into the developing mycelium (128).

Endotrophic appressoria are structurally simple, being similar to protoappressoria except that the point of contact with the cell wall is more or less at right angles to the direction of growth of the hypha. They produce infection pegs in the same way as other appressoria (56, 84).

The development of compound appressoria from single cells is not well documented. Germtubes from ascospores of *Sclerotinia libertiana* branch dichotomously on contact with a surface (14). Each branch produces a hand-like structure on which each digital apex swells and adheres to the surface. The circular area of each of these appendages is the only part of the fungus attached to the surface (14). *Sclerotinia sclerotiorum* behaves similarly (91) as do the lobate appressoria of *Rhizoctonia solani* (28) except that with the latter, the final structure is different in appearance from those of *Sclerotinia*. In *Pythium irregulare* each lobe appears as a separate, simple appressorium on one short, multibranched hypha (2).

Two patterns of development occur in *Colletotrichum dematium* and *C. trichellum* (109). Sometimes a large, simple appressorium lays down one or more septa to produce several cells each with a germ pore. Alternatively the hypha branches at the base of a simple appressorium, grows a short distance around it, and produces a second appressorium in juxtaposition to it. Sometimes an appressorium from another hypha forms in contact with such a group. It is not known whether species of *Colletotrichum* on a host surface behave in this way.

Detailed studies of the development of cushion-like, compound appressoria have been made with *Rhizoctonia solani* (20, 27, 28, 126). Dodman & Flentje (27) found different patterns of development according to differences in isolate and host. The pattern of development on bean and radish hypocotyls was typical of the species. The first stage was cessation of hyphal elongation and formation of many side branches close to each other, which later curled back on themselves and branched repeatedly. In contrast to the usual type, vegetative side branches forming appressoria were short, swollen, irregular, and often bent or curled.

The extent to which other fungi produce infection cushions is doubtful. Mangin's illustrations suggest that *Leptosphaeria herpotrichoides* does (64). Originally, infection-cushion production as a prelude to infection was claimed for *Helicobasidium purpureum*, but it was found that these so-called infection cushions developed in the periderm of the host (50) after infection had occurred from solitary superficial hyphae, raising doubts about earlier observations.

CYTOLOGICAL STUDIES

Nuclear Behavior

This has been studied in relatively few species. Williams (125), in reviewing infection structure development in *Puccinia graminis* states, "cytological studies of germtube development show that differentiation of an infection structure is accomplished by a co-ordinated sequence of morphological and cytological events." In *Erysiphe graminis* (111), *E. polygoni* (89), and *Colletotrichum lagenarium* (4) the nucleus divides either immediately before or after the initiation of a germtube. One daughter nucleus remains in the conidium; the other migrates into the germtube and then into the appressorium. This is seen as the reason only one germtube produces an appressorium in *E. graminis* (89) and may explain a similar observation in *E. polygoni* (89). In fungi that produce dikaryotic spores, appressoria develop only on germtubes containing a dikaryon, e.g. *P. graminis* (125), and are suspected in *Ustilago scitaminea* (J. M. Waller, personal communication). In all cases a further nuclear division occurs in the appressorium prior to penetration (63, 103).

Nucleic Acid and Protein Synthesis

Dunkle et al (31) concluded that in *P. graminis* the activity of template RNA and synthesis of new kinds of protein occur when germinating uredospores form infection structures. Similar conclusions have been drawn for other rusts (93, 125) and *C. lagenarium* (74).

THE ROLE OF THE APPRESSORIUM

The main role of appressoria is the direct penetration of a host. To achieve this, appressoria must attach firmly to the host surface and produce infection pegs that pierce the cuticle and usually the epidermal cell wall. However, appressoria produced by germinating uredospores of rusts and *Mycosphaerella citri* (123) penetrate through stomata. Few recent authors disagree that this is the primary role of the appressorium, but some suggest appressoria may play a secondary, if not auxiliary role (85).

Adhesion

The ability of appressoria to adhere fast to glass or leaf surfaces when washed under running water is well documented (68, 84, 102, 104). Many authors attributed this to the production of a mucilaginous sheath (12–14, 24, 34, 61, 67, 96, 101, 103), but in many cases the evidence for the production of a sheath was poor and other means of adherence were postulated. These include intermolecular forces between appressorium and host (11), appressorium shape (101), and appressorial weight to surface area ratio (112).

Recent studies using modern techniques support the hypothesis that appressoria and attendant structures are fixed to surfaces by binding substances (4, 34, 61). Blackman & Welsford (12) first proposed the hypothesis and regarded the binder as mucilage. Mucilage production in some fungi may be influenced by the host (39). It has been suggested more recently that in *Puccinia coronata* the binder incorporates fungal secretions plus dissolved host wax (82). This could also explain recent observations on powdery mildews (107).

Penetration

The most commonly held view supported by seemingly strong evidence is that penetration is mechanical (59, 67, 89, 103). This does not mean, however, that penetration is not aided by chemical dissolution of cuticle or cell wall in some cases (89, 103). Nor does it mean that penetration is not wholly or predominantly enzymic in at least some species.

The classical experiment by Brown & Harvey (15) that showed that *B. cinerea* could exert enough force to penetrate gold leaf encouraged the view that penetration is mechanical. However, McKeen (61) seriously questioned the reliability of Brown & Harvey's experiment. He concluded that passage of *Botrytis cinerea* through cuticle of *Vicia faba* is not accomplished by either chemical or mechanical means alone, but by pressure and a limited excretion of cutinase for a short time. McKeen's conclusion summarized a growing body of evidence. Such direct evidence has not been available previously through lack of sophisticated techniques such as microchemical methods, electron microscopy, and scanning electron microscopy. The mechanics of penetration were described by Marks et al (67) who conceded that penetration could be aided by chemical softening of cuticle in advance of penetration. This hypothesis has long been advocated (23, 61), but Marks et al found no evidence to support it. They supported an equally long line of advocates for the mechanical theory (12, 15, 24, 46, 64).

The discovery of the capacity of some pathogens to produce cutinase (48, 58) provided a firmer basis for the chemical penetration theory. Observations that the penetration channel through cuticle can be of greater diameter than the penetration peg itself suggests chemical dissolution rather than mechanical breaching (7, 34). EM studies of cucumber and barley leaves inoculated with powdery mildew show entire areas of cuticle dissolved to accommodate the appressorium and a hole dissolved by the infection peg (99, 107). There now seems no doubt that chemical dissolution of cuticle and cell wall occur. It is likely that this is aided by mechanical means, or vice versa. It is sometimes claimed that cuticle is penetrated mechanically and the underlying cell wall enzymatically (89, 90). The opposite is also claimed (34, 107). This raises the question of whether internal appressoria penetrate from cell to cell mechanically, or by enzymes; the latter way is usually proposed (43, 126). It is unlikely that the pattern of penetration is the same in each host-parasite relationship.

Survival

Parbery & Emmett (85) support the view that many appressoria play a role in survival that is not secondary to that of an infection structure, but is an auxiliary capacity that increases the chance of successful infection. Appressoria are not long-term survival structures, but possess the capacity to endure conditions of light intensity, desiccation, or antagonism potentially lethal to unprotected germtubes or hyphae during penetration. In some species if adverse conditions halt germination of an appressorium, it can remain dormant until conditions again favor germination and infection (103).

Pigmented thick-walled appressoria are produced consistently by comparatively few fungi, especially the dematiaceous epiphytic parasites (85) and certain endophytic parasites producing hyaline one- to few-celled spores. It may be significant, therefore, that of claims for the capacity of appressoria to survive adverse conditions, all relate to species of *Colletotrichum,* and most claims that stress induces appressorium production relate also to colletotrichums or their near relatives, the phyllachoras.

Parbery & Emmett (85) proposed, first, that the evolution of dematiaceous mycelium and attendant structures in epiphytic parasitic fungi (except powdery mildews) affords protection against their exposed habitat. Such protection has not generally evolved in endophytic parasites except in the exposed germination-to-infection phase. Second, protection of the epiphytic phase of the life cycle of endophytes has arisen in various ways, two of which are reflected in morphology. Either, characters have evolved that provide spores with a high survival potential, or a similar potential has evolved in appressoria. Few endophytes produce both spores and appressoria with a high survival capacity while few have evolved neither one nor the other.

Site of Chemical Activity

Staub et al (107) demonstrate the capacity of appressoria and germtubes of *Erysiphe graminis* and *E. cichoracearum* to liberate cutinase from their ventral surfaces. Higgins (50) first suggested that appressoria of *Colletotrichum piperatum* may produce both cutinase and phytotoxins. Apparently infection pegs produced from appressoria secrete cutinase and other enzymes, although these enzymes could be released through the appressorial pore. It is possible that appressoria of other parasites such as *Drechslera* spp. also act as sites for toxin production and liberation. This is suggested by the presence of such toxins in spores of some species (79, 111, 121).

FACTORS INFLUENCING APPRESSORIUM FORMATION

The two initial stages, initiation and maturation of appressoria, may respond to different sets of factors. Providing this is appreciated, there is no need to deal with each stage separately.

A vast literature proposes many mechanisms for appressorium initiation. Within a species there is frequent controversy over which mechanism is important. However, it is often possible to reinterpret the findings claimed to support one or other mechanism. It is not possible here to deal with this literature in detail, but two points have emerged that warrant consideration. First, much work on appressorium formation starts with the assumption that appressoria form only in response to stimulation. Consequently, the fundamental problem is seen as deciding which specific stimuli are involved. Second, in a high proportion of species, opinion frequently differs as to which stimuli are important. Some controversial examples have been studied for many years without any satisfactory conclusion. Perhaps this would not have happened had answers been sought on which factors exert an influence on appressorium formation instead of what are the stimuli. It is suggested that in most cases an erroneous assumption has led to the wrong questions.

Early authors were mainly preoccupied with contact stimuli. This lasted until the second decade of this century when interest developed in the search for chemical stimuli. Gradually, the idea gained acceptance that combinations of chemical and physical stimuli may be involved, and more recently, the importance of genotype interacting with these combinations of stimuli has been emphasized (125).

It may be helpful to consider Gäumann's (43) view that appressorium formation is the endpoint of spore germination. Germination usually occurs not because of the presence of specific stimuli to germination, but because of a conducive set of external conditions at a time when the internal environment of the spore is itself favorable for germination (45, 108). We may then explore the possibility that few fungi require stimuli for appressorium formation, while accepting the probability that well-defined environmental conditions (a conducive environment) are required in most species. Blakeman (89) implies a similar idea when referring to factors affecting growth of pathogens on plant surfaces. He states, "Form of growth depends primarily upon species, but may be modified by environment." It is proposed that appressorium formation is primarily controlled by genotype whose expression may require a specific conducive environment.

Many authors agree that appressorium initiation occurs in response to a check in germtube elongation. It may be more accurate to think of it as "coincident" with a halt in germtube extension. This suggests that in rusts, cessation in germtube elongation is triggered by some environmental shock that need not be specific. This may apply to other fungi.

Sometimes the conditions required for the germination of spores is predetermined in part by the conditions under which they formed. It is possible that such preconditioning also influences appressorium formation in some species (102). The sensitivity of species producing appressoria to fluctuations in environmental conditions distinguishes two groups: sensitive species where there may be little tolerance of fluctuations in one or more components of the conducive environment and less sensitive species, which, although they are sensitive to major fluctuations, may be more tolerant of lesser variations than the more sensitive group. Many factors, inherent and external, have been identified as important in appressorium initiation and development.

Factors Inherent in Fungi

Studies with *Phyllachora* species strongly suggest that appressorium formation is the terminal act of germination (83), although adverse environmental conditions can either prevent or modify development. This is supported by other work (31, 32, 43). The importance of genotype on appressorium morphology and development has already been mentioned. Thus, it is considered that the primary factors leading to appressorium initiation and development are inherent.

The interaction of the genotype with the environment determines the metabolic climate within the fungus and this may determine whether or when appressoria will form. Environmental conditions during vegetative growth can influence appressorium formation in fungi (59, 119). Similarly, light intensities and temperatures prevailing during sporulation of *Colletotrichum falcatum* can influence appressorium formation (102). Metabolic changes in mature conidia of *C. falcatum* and *C. trichella* appear necessary before germination will result in appressorium formation (106, 109). Ungerminated uredospores or actively growing germtubes of some rusts contain substances that inhibit germtube elongation while promoting appressorium formation (6, 41). These substances are not particularly specific because extracts from one rust species can promote appressorium formation in some but not all species (6). There is controversy over the nature of these promoters (62).

Factors External to Fungi

THIGMOTROPISM   There have been numerous reports to support Busgen's (18) hypothesis that appressorium initiation results from the germtube or hyphal apex contacting a solid surface (24, 25). However, the general lack of attempts to produce appressoria without contact confounds evaluation of the need for this requirement. The production of appressoria by two species of *Phyllachora* in the absence of contact shows that in those species at least, contact was not required for initiation although it was necessary for normal maturation (83). Many believe that contact is only one among several requirements for appressorium formation (43, 59, 62, 113, 126).

The question here may not be, what is the effect of contact but rather what instigates apical contact? It is hard to believe that germtubes are not frequently in contact with the surface soon after germination and well before appressorium initiation. Staub et al support this view (107). The key to understanding the role of surface contact may lie in the discovery of why apical contact occurs at a given point of development. The mechanism of thigmotropism may involve factors inherent both in the host and fungus as well as external physical factors.

FACTORS INHERENT IN PLANTS   Claims for importance of any single factor as a stimulus to appressorium formation are rarely convincing. The separation of the action of proposed chemical stimuli from thigmotrophic effects or the more general nutrient environment of the plant surface is rarely satisfactory. Indeed, the plant surface environment can be influenced by a complex of chemicals such as aerosols released into the atmosphere by plants (94), various pollutants (47, 89), byproducts of the metabolism of phyllosphere microorganisms (89), chemical constituents of waxes, and other components of the cuticle and chemicals released from cells within the plant that find their way to the plant surface (89). Several authors in Preece & Dickinson (89) recognize this complexity. While selected topics are discussed it must be appreciated that, as presented in the following section, the individual factors probably are interrelated components of one conducive environment.

Surface phenomena   While there are many claims that fungi require a general, if not specific, plant surface to form appressoria (35, 127), it is disconcerting to find that others can obtain equally good results with the same species on a glass surface under closely defined environmental conditions (130). Despite the extensive literature on the importance of plant surface phenomena in appressorium formation, in most cases it is still difficult to evaluate it. Specificity may be unimportant. Appressoria often form equally well on susceptible and resistant hosts (7, 56, 89, 97, 120), on living and dead tissue (62), and on a wide range of hosts and nonhosts alike (107). Although surface specificity may be important in some cases (127), generally it does not appear to be so.

Hardness of surface may sometimes be critical. As surface hardness increases, germinating conidia of *Colletotrichum gloesporioides* produce increasing proportions of appressoria (113). It is possible that rigidity sometimes is more important than hardness. On the other hand, softness of surface may affect formation. Appressoria of *Ustilago scitaminea* form only on soft bud tissue of sugarcane (118), although such tissue may liberate greater amounts of exudates than other tissues or possess other characters. It is often difficult to disentangle the many factors that might be involved.

Various leaf surface sites have specific attraction for some fungi (126). The subcuticular cell junction is an example. A greater rate of exosmosis over cell junctions is thought to stimulate germination in some fungi. It may also influence appressorium formation. Some fungi grow along cleavage lines of epidermal cells especially on roots (126) and some produce appressorial initials (54). Also, there is evidence of site specificity in *Erysiphe polygoni* and *Peronospora parasitica,* which produce appressoria only along junction lines in certain hosts (88). *Colletotrichum lindemuthianum* produces appressoria more commonly over veins and cell junctions on the upper surfaces of bean leaves, but randomly on the lower surface (89). Stomata are favored sites for rusts (5) and some other fungi (122). It has been suggested that this may be no more than a response to indentation (90, 126), but its frequency in the rusts (5, 62) renders this unlikely. Bulliform cells of rice are favored sites for appressorium formation and infection by *Piricularia oryzae* (53) and *Drechslera oryzae* (81). The regularity with which some fungi select particular sites for appressorium production is evidence of the importance of specific microhabitats.

Seasonal changes possibly associated with maturation and senescence of tissues may also be important. For example, early in the season germtubes of *Alternaria longipes* grow longer before appressoria form than when they grow later in the season. In addition, the early season appressoria are produced only at selected sites, whereas later they occur at random (115). It seems unlikely that leaf surface alone causes a change in pattern and site of development in *A. longipes.* Loss of site selectivity may be associated with increased exosmosis accompanying aging of the host.

Cuticular waxes   Ether-soluble components and hydrocarbon fractions of snapdragon cuticle stimulate appressorium formation in several rusts. In *Alternaria porri* (4) benzene and benzene-ether extracts of onion cuticle increase appressorium production three- to fourfold. Waxes of onion cuticle have no effect on nonpathogens of onion (4). Even so, it appears that the effects of cuticle are fairly nonspecific; this is probable because cuticle composition is similar in most plants (69). However, cuticles of some plants do contain substances lacking in others (51, 89).

Maheshwari et al (62) claimed the chain length of the active hydrocarbon was approximately 30 CH_2 units which would apply to several hydrocarbons found in cuticular wax (89). Dickinson's (26) studies suggest that it may not be the chemical nature as much as the physical size of the molecules that is critical in thigmotropism.

The claim that the physical structure of wax is important in appressorium formation in *Erysiphe graminis* (26) is not substantiated (107). Nor does work with other fungi suggest it is likely (62). Some plant waxes contain substances that are fungistatic (89) or fungitoxic (69). Where such substances impair germtube elongation rather than prevent spore germination, it is possible that they enhance appressorium formation. It is noteworthy that resistant hosts sometimes induce more appressoria than susceptible hosts (13).

Endogenous chemicals   In addition to cuticular chemicals and chemicals exogenous to the plant, those released from within the plant may also influence appressorium formation (89, 106). It may be difficult, therefore, to determine the source of a chemical influencing appressorium formation in a particular case. However, as differences have been noted between washed and unwashed leaves (89), there is some suggestion that surface chemicals influence appressorial behavior. The influence of chemicals at the root surface is well illustrated by studies on *Rhizoctonia solani.* Particular strains of *R. solani* form infection cushions only on roots of susceptible hosts; this is interpreted as specific stimulation (28, 54). Guttation fluid may influence appressorium formation in species of *Drechslera* (38) and *Mycosphaerella* (123). Blakeman (89) considered that the influence of chemicals on leaves may be determined by changes occurring in infection drops. This comment is relevant to all plant surfaces. Concentrations, as well as presence, of chemicals in infection drops or water films on plant surfaces may determine whether the effects are stimulatory or inhibitory. In attempting to elucidate the role of single factors, it is necessary to supplement studies at the plant surface with studies in vitro.

In view of current theories on the mechanism of hyphal growth (8), it is possible that certain metallic ions or their associated charges could influence appressorium formation. For example, there are claims that zinc is specifically required for appressorium development in *Puccinia coronata* (100) and *P. graminis* (16, 22). Others (62, 86) have been unable to substantiate this view. Yet others (working with the same species although probably different races) have obtained appressoria in the absence of zinc (42, 77, 124). Consequently, it seems unlikely that zinc per se is important.

Nutrient status   It has already been noted that changes in season and age of a host can directly influence the availability of chemicals at plant surfaces, and the character of plant waxes can indirectly influence their susceptibility. Plant waxes containing hydrophilic compounds allow a greater exudation of material from within the leaf to the leaf surface, than waxes containing hydrophobic compounds (89). The need for nutrients at the plant surface for full development of appressoria in *R. solani* and *E. cruciferarum* has already been noted. Chemicals exogenous to the fungus may also promote appressorium development in *S. sclerotiorum* (91), *Colletotrichum piperatum* (89), *Pythium irregulare* (2), and other fungi. It is possible that complete expansion of appressoria in some *Phyllachora* spp. (83) requires exogenous nutrients.

On the other hand, in the presence of high nutrient levels, germtubes of some fungi continue to elongate and branch, instead of remaining short and producing appressoria. This has been noted in *C. musae, C. gloeosporioides* (46, 103, 113), *C. dematium* (30), *C. graminicola* (103), and *S. libertiana* (23). It is not clear why wound exudates sometimes inhibit appressorium formation. Failure of *Glomerella cingulata* to produce appressoria on wounded leaves of *Ginkgo biloba* has been attributed to excess nutrient (1). It is possible, however, that the sap contained an inhibitor. Gilliver (44) found wound sap of 23% of 1915 plant species inhibitory to germination of *Venturia inaequalis.*

Inhibition of appressorium formation by nutrients need not always be due to specific inhibitors or inhibitory concentrations. The appressorium is a structure enabling a fungus to establish a nutritional relationship with its host so that in some fungi appressorium initiation is possibly a response to exhaustion of endogenous energy reserves. Starvation has been proposed as initiating appressorium formation in *Glomerella* sp. (33), *Colletotrichum dematium* (32), and *C. gloeosporioides* (10). The behavior of *C. lindemuthianum* (89) could also be explained on this basis.

Phyllosphere microorganisms   Although the surface microflora is external to the plant, the composition of the phyllosphere is strongly influenced by factors inherent in the plant, modified by external environment. The succession of wet and dry periods (89), seasonal changes (115), and ontogenetic changes in the host (89) combine to cause continuous fluctuations in the nature of the phyllosphere, thereby providing a succession of specialized habitats. Behavior of pathogens is affected by phyllosphere populations (89). Antagonistic populations of *Bacillus* sp. on citrus leaves promote appressorium formation in *C. gloeosporioides* (57).

FACTORS EXTERNAL TO PLANTS   The influence of any factor external to the plant may be difficult to evaluate because its indirect influence on the host is often difficult to separate from its direct effect on the fungus (89).

Exogenous chemicals   These can arrive at the plant surface from various sources. Monoterpenes and other volatiles liberated from one plant may be deposited on the surface of others (94) where they are used by phyllosphere microorganisms (95). Such chemicals can influence phyllosphere ecology and are potentially important in any microbiological activity, including appressorium formation. The presence of pollen sometimes promotes infection (89). Pollutants (47, 89), fungicides (97), and herbicides (104) can influence pathogens directly or indirectly through effects on the phyllosphere (19). Some fungicides arrest appressorium formation (97).

Water relations   Powdery mildews require relatively dry surfaces in a saturated atmosphere to germinate and form appressoria abundantly (130). Many other fungi require free water for germination, but little is known of requirements for appressorium formation. Drying out of infection drops, however, promotes appressorium formation in species of *Phyllachora* (83) and *Mycosphaerella* (123). Surface tension phenomena are important in *Colletotrichum trifolii* (74).

Temperature   In most cases the optimum temperature for appressorium formation coincides with that for spore germination (52, 74, 92, 103). Several effects of temperature have been reported. Pregermination treatments of spores at extreme temperatures can delay or prevent appressorium initiation in *Colletotrichum* spp. (52, 74, 92, 103). Appressoria formed at extreme temperatures may either fail to germinate or may remain dormant until the return of near optimal temperatures (103). In some fungi there is an inverse relationship between the rates of germtube growth and appressorium formation with changes in temperature, no appressoria developing at extreme temperatures (66, 83, 130). Low extremes tend to delay rather than prevent formation (66). In some species, e.g. *Puccinia graminis,* an even temperature regime may be unfavorable for appressorium formation, whereas an increase in temperature during germination can trigger their initiation (52, 58). Temperatures during spore formation (65), differences in race of fungus (105, 124), or ecotype may also modify response to temperature.

Light   In addition to the indirect effects on a host (89) light can directly influence appressorium formation. While light favors protoappressorium formation in *Clasterosporium carpophilum* (96), powdery mildews prefer dark (130). However, alternating dark and light may promote appressorium formation in some powdery mildews (128). Although light is not required for appressorium initiation in *E. graminis,* it is necessary for maturation (70, 75). Increasing light intensity depresses appressorium formation in *Puccinia helianthi,* but this is overcome by alternating periods of light and dark (107).

Light and temperature   Most claims for requirements for specific light-temperature regimes for appressorium formation apply to rusts. Several authors suggest that a two-phase regime is necessary to promote appressorium formation. Phase one is usually a 2 hr period at about 20°C, while phase two involves a similar period at a temperature increased by approximately 5°C (36, 55, 62, 86, 100). There is a controversy, however, over the need for dark during phase one (36, 62, 86, 100) or not (55, 105) and about the concentration of light required during phase two (36,

62, 100). There is no evidence of a light requirement for substomatal vesicle formation (36, 62), considered by some as an extension of appressorium development (25).

Other considerations   In particular species there are probably many factors making up the "conducive environment" for appressorium formation. Some may play a more important role than others, while some may be stimulatory. By showing that alteration of one of these factors can depress or stimulate appressorium production some authors believe they have found "the" controlling factor. However, different workers can find other "controlling" factors. In *Puccinia graminis*, temperature is certainly important and light may be, at least in some ecotypes or genotypes. The presence or absence of certain ions or charges may play a role, and so may various other factors. Increase of the CO_2 level from ambient to 5% for 10 min can bring about appressorium formation in *P. graminis* (60) as readily as a rise in temperature or exposure to uredospore extract (6). At this point, therefore, it seems relevant to ask whether any of these factors is important because of what it is or because of what it does. Is it really some kind of "shock reaction" triggering a "survival response" that really initiates appressorium development? How often would appressoria simply develop as a matter of ontogenetic course if left to their own genetic devices in a conducive environment?

CONCLUSIONS

While some authors regard appressorium formation as an independent, specialized process in fungi, it is considered here to be part of the sequence of expression of the genotype during the final phase of germination. The association of morphologically constant forms of appressoria with species is evidence of genetic control over morphology. The regular pattern of development of appressoria in so many fungi is evidence of genetic control over development of appressoria. In particular groups of fungi, however, the expression of the basic genotype may require fairly well-defined environmental conditions possibly including contact with a surface.

   Species of *Colletotrichum* often respond to exogenous nutrients by continuing germtube elongation and delaying appressorium formation. Nutrient stress and external stresses frequently favor appressorium formation. Usually temperatures between 20 and 25°C are also favorable. Nevertheless, for appressorium formation in powdery mildews, it is difficult to make a case supporting the need for requirements other than those proposed by Zaracovitis (130). Possibly, however, the more environmental conditions depart from the optimal conducive environment defined by Zaracovitis, the more exaggerated the influence of marginally important factors becomes. This might explain some of the less substantiated claims. It may also be true for other fungi. Some rusts apparently need a two-phase environment involving a rise in temperature from phase one to phase two, possibly accompanied by other changes. However, the apparent conflict in this section of the literature could be resolved by showing that an "environmental shock" initiates appressorium formation in rusts rather than specific changes in environment.

Fungi that form infection cushions generally do so in response to increases in nutrient availability at the host surface. Such fungi infect roots, or crowns. It appears from this review that appressorium formation is an inherent mechanism required by many fungi for penetration of host tissue. In some species this is only slightly influenced by environment, while in others there is a need for a fairly well-defined conducive environment. It is suggested that the evolution of appressoria in some fungi has provided protection for a stage in their life cycle otherwise vulnerable to environmental changes that could prevent the establishment of a nutritional relationship with a host.

ACKNOWLEDGMENTS

The authors wish to thank Professor N. H. White and Dr. J. W. Meagher for reading the manuscript and making many valuable suggestions.

Literature Cited

1. Adams, P. B., Sproston, T., Tietz, H., Major, R. T. 1962. Studies on the disease resistance of *Ginkgo biloba*. *Phytopathology* 52:233–36
2. Agnihotri, V. P. 1969. Production and germination of appressoria in *Pythium irregulare*. *Mycologia* 61:967–80
3. Ainsworth, G. C. 1971. Dictionary of the fungi. Kew, England: Commonw. Mycol. Inst. 663 pp.
4. Akai, S., Fukutomi, M., Ishida, N., Kunoh, H. 1967. An anatomical approach to the mechanism of fungal infections. In *The Dynamic Role of Molecular Constituents in Plant Parasite Interaction,* ed. C. J. Mirocha, I. Uritani, 1–20. St Paul, Minn.: Am. Phytopathol. Soc. 372 pp.
5. Allen, R. F. 1923. A cytological study of infection of Baart and Kanred wheats by *Puccinia graminis tritici. J. Agric. Res.* 23:131–51
6. Allen, P. J. 1957. Properties of a volatile fraction from uredospores of *Puccinia graminis* var. *tritici* affecting their germination and development. I. Biological activity. *Plant Physiol.* 32:385–89
7. Anderson, J. L., Walker, J. C. 1962. Histology of watermelon anthracnose. *Phytopathology* 52:650–53
8. Bartnicki-Garcia, S. 1973. Fundamental aspects of hyphal morphogenesis. *Symp. Soc. Gen. Microbiol.* XXIII: 245–67
9. Batts, C. C. V. 1955. Observation on the infection of wheat by loose smut (*Ustilago tritici*). *Trans. Br. Mycol. Soc.* 38:465–75
10. Biraghi, A. 1934. On the biological significance of alleged appressoria of the genus *Gloeosporium. Boll. Stn. Patol. Veg. Rome* XIV, 2:202–10
11. Blackman, V. H. 1924. Physiological aspects of parasitism. *Br. Assoc. Adv. Sci. Rept.* 92:233–46
12. Blackman, V. H., Welsford, E. J. 1916. Studies in the physiology of parasitism. II. Infection by *Botrytis cinerea. Ann. Bot. London* 30:389–98
13. Blasquez, C. H., Owen, J. H. 1963. Histological studies of *Dothidella ulei* on susceptible and resistant *Hevea* clones. *Phytopathology* 53:58–65
14. Boyle, C. 1921. Studies in the physiology of parasitism. *Ann. Bot. London* 35:337–47
15. Brown, W., Harvey, C. C. 1927. Studies in the physiology of parasitism. X. On the entrance of parasitic fungi into the host plant. *Ann. Bot. London* 41:643–62
16. Broyles, J. W. 1955. *Comparative studies of races and biotypes of Puccinia graminis with special reference to morphology of urediospore germination, chemical composition and factors affecting survival.* PhD thesis. Univ. of Minnesota. Minneapolis. 229 pp.
17. Buller, A. H. 1958. The gemmifers of *Sclerotium coffeicola*. In *Researches on Fungi* VI: 448–50. New York: Hafner. 513 pp.
18. Büsgen, M. 1893. Uber einige Eigerschaften der Keimlinge parasitischer Pilze. *Bot. Ztg. Leipzig* 51:53–72
19. Carter, M. V. 1971. Biological control of *Eutypa armeniacae. Aust. J. Exp. Agric. Anim. Husb.* 11:687–92
20. Christou, T. 1962. Penetration and host-parasite relationships of *Rhizoc-*

tonia solani in the bean plant. *Phytopathology* 52:381–89

21. Clare, B. G. 1964. Erysiphaceae of South Eastern Queensland. *Univ. Queensl. Pap. Dep. Bot.* IV(10):111–44

22. Couey, H. M., Smith, F. G. 1961. Effect of cations on germination and germtube development of *Puccinia coronata* uredospores. *Plant Physiol.* 36:14–19

23. De Bary, A. 1886. Uber einige *Sclerotinien* und Sclerotienkrankheiten. *Bot. Ztg. Leipzig* 44:377–474

24. Dey, P. K. 1919. Studies in the physiology of parasitism. V. Infection by *Colletotrichum lindemuthianum. Ann. Bot. London* 33:305–12

25. Dickinson, S. 1955. Studies in the physiology of obligate parasitism. V. Further differences between the uredospore germtubes and leaf hyphae of *Puccinia triticina. Ann. Bot. N.S.* 19:161–71

26. Dickinson, S. 1970. Studies in the physiology of obligate parasitism. VII. The effect of a curved thigmotropic stimulus. *Phytopathol. Z.* 69:115–24

27. Dodman, R. L., Flentje, N. T. 1970. The mechanism and physiology of plant penetration by *R. solani.* In *Rhizoctonia solani—Biology and Pathology,* ed. J. R. Parmeter Jr., 172–88. Berkeley: Univ. Calif. Press. 225 pp.

28. Dodman, R. L., Barker, K. R., Walker, J. C. 1968. A detailed study of the different modes of penetration by *Rhizoctonia solani. Phytopathology* 58: 1271–76

29. Dring, D. M. 1961. Studies on *Mycosphaerella brassicicola. Trans. Br. Mycol. Soc.* 44:253–64

30. Duke, M. M. 1928. The genera *Vermicularia* and *Colletotrichum. Trans. Br. Mycol. Soc.* 13:156–84

31. Dunkle, L. D., Maheshwari, R., Allen, P. J. 1969. Infection structures from rust urediospores: Effect of RNA and protein synthesis inhibitors. *Science* 163:481–82

32. Dunkle, L. D., Wergin, W. P., Allen, P. J. 1970. Nucleoli in differentiated germ tubes of wheat stem rust uredospores. *Can. J. Bot.* 48:1693–95

33. Edgerton, C. W. 1908. The physiology and development of some anthracnoses. *Bot. Gaz.* 45:384

34. Edwards, H. H., Allen, P. J. 1970. A fine-structure study of the primary infection process during infection of barley by *Erysiphe graminis* f. sp. *hordei. Phytopathology* 60:1504–9

35. Ellingboe, A. H. 1972. Genetics and physiology of primary infection by

Erysiphe graminis. Phytopathology 62:401–6

36. Emge, R. G. 1958. The influence of light and temperature on the formation of infection type structures of *Puccinia graminis* var. *tritici* on artificial substrates. *Phytopathology* 48:649–52

37. Emmett, R. W. 1969. *Appressoria.* B. Agric. Sc. (Hons.) thesis. Univ. Melbourne, Parkville, Victoria. 180 pp.

38. Endo, R. M., Oertli, J. J. 1964. Stimulation of fungal infection of bentgrass. *Nature* 201:313

39. Flentje, N. T. 1957. Studies on *Pellicularia filamentosa* (Pat.) Rogers. III. Host penetration and resistance and strain specialization. *Trans. Br. Mycol. Soc.* 40:322–36

40. Frank, A. B. 1883. Uber einige neue und weniger bekannte Pflanzenkrankheiten. *Ber. Dtsch. Bot. Ges.* 1:I 29–34,II 58–63

41. French, R. C., Massey, L. M. Jr., Weintraub, R. L. 1957. Properties of a volatile fraction from uredospores of *Puccinia graminis* var. *tritici* affecting their germination and development. II. Some physical and chemical properties. *Plant Physiol.* 32:389–93

42. Fuchs, W. H., Gaertner, A. 1958. Untersuchungen zur keimungsphysiologie des Schwartzrostes *Puccinia graminis. Arch. Mikrobiol.* 28:303–9

43. Gäumann, E. A. 1950. *Principles of Plant Infection.* London: Crosby Lockwood. 543 pp.

44. Gilliver, K. 1947. The effect of plant extracts on the germination of the conidia of *Venturia inaequalis. Ann. Appl. Biol.* 34:136–43

45. Gottlieb, D. 1966. Biosynthetic processes in germinating spores. In *The Fungus Spore,* ed. M. F. Madelin. London: Butterworths. 338 pp.

46. Hasselbring, H. 1906. The appressoria of the anthracnoses. *Bot. Gaz.* 42: 135–60

47. Heagle, A. S. 1973. Interactions between air pollutants and plant parasites. *Ann. Rev. Phytopathol.* 11:365–88

48. Heinen, W. 1962. Uber den enzymatischen Cutin-Abbau. III. Die enzymatische Austüstung von *Penicillium spinulosum* zum Abbau der Cuticularbestandtede. *Arch. Mikrobiol.* 41: 268–81

49. Hering, T. F. 1962. Infection cushions of *Helicobasidium purpureum* Pat. *Trans. Br. Mycol. Soc.* 45:46–54

50. Higgins, B. B. 1926. Anthracnose of

pepper (*Capsicum annum* L.) *Phytopathology* 16:333–46
51. Horn, D. H. S., Kranz, Z. H., Lamberton, J. A. 1964. The composition of *Eucalyptus* and some other leaf waxes. *Aust. J. Chem.* 17:464–76
52. Ishida, N., Akai, S. 1969. Relation of temperature to germination of conidia and appressorium formation in *Colletotrichum lagenarium. Mycologia* 61: 382–86
53. Ito, S., Shimada, S. 1937. Studies on blast disease of rice plants with special reference to the infection process of the causal fungus and the varietal resistance of rice plants. *Contrib. Impr. Agric. Min. Agric. For. No.* 120:1–109
54. Kerr, A., Flentje, N. T. 1957. Host infection in *Pellicularia filamentosa* controlled by chemical stimuli. *Nature* 179:204–5
55. Kim, W. K., Rohringer, R. 1974. Ribonucleic acids of differentiating and non-differentiating uredosporelings of wheat stem rust. *Can. J. Bot.* 52:1304–17
56. Leach, J. G. 1923. The parasitism of *Colletotrichum lindemuthianum. Minn. Univ. Agric. Exp. Stn. Tech. Bull.* 14. 41 pp.
57. Lenné, J. M., Parbery, D. G. 1975. Phyllosphere antagonists and appressorium formation in *Colletotrichum gloeosporioides Trans. Br. Mycol. Soc.* In press
58. Linskens, H. F., Haage, P. 1963. Cutinase-Nachweis in Phytopathogenen Pilzen. *Phytopathol. Z.* 48:306–11
59. Lumsden, R. D., Dow, R. L. 1973. Histopathology of *Sclerotinia sclerotiorum* infection of bean. *Phytopathology* 63:708–15
60. Macko, V., Fuchs, E. 1970. Effect of carbon dioxide on uredospore germtubes of *Puccinia striiformis. Phytopathology* 60:1529–30
61. McKeen, W. E. 1974. Mode of penetration of epidermal cell walls of *Vicia faba* by *Botrytis cinerea. Phytopathology* 64:461–67
62. Maheshwari, R., Allen, P. J., Hildebrandt, A. C. 1967. Physical and chemical factors controlling the development of infection structures from urediospore germ-tubes of rust fungi. *Phytopathology* 57:855–62
63. Maheshwari, R., Hildebrandt, A. C., Allen, P. J. 1967. The cytology of infection structure development in urediospore germ tubes of *Uromyces phaseoli* var. *typica. Can. J. Bot.* 45:447–50

64. Mangin, M. L. 1899. Sur le pietin ou maladie du pied du blé. *Bull. Soc. Mycol.* 15:210–39
65. Manners, J. G. 1966. Assessment of germination. See Ref. 45, 165–73
66. Manners, J. G., Hossain, S. M. M. 1963. Effects of temperature and humidity on conidial germination in *Erysiphe graminis. Trans. Br. Mycol. Soc.* 46:225–34
67. Marks, G. C., Berbee, J. G., Riker, A. J. 1965. Direct penetration of leaves of *Poplus tremuloides* by *Colletotrichum gloeosporioides. Phytopathology* 55:408–12
68. Marshall-Ward, H. 1888. A lily disease. *Ann. Bot. London* 2:319
69. Martin, J. T. 1964. Role of cuticle in the defense against plant disease. *Ann. Rev. Phytopathol.* 2:81–100
70. Masri, S. S., Ellingboe, A. H. 1966. Germination of conidia and formation of appressoria and secondary hyphae in *Erysiphe graminis* f. sp. *tritici. Phytopathology* 56:304–8
71. Mellano, H. M., Munnecke, P. E., Endo, R. M. 1970. Relationship of seedling age to development of *Pythium ultimum* on roots of *Antirrhinum majus. Phytopathology* 60:935–42
72. Meredith, D. S. 1961. Fruit spot (speckle) of Jamaican bananas caused by *Deightoniella torulosa.* I. Symptoms of disease and studies on pathogenicity. *Trans. Br. Mycol. Soc.* 44:95–104
73. Meredith, D. S. 1964. Appressoria of *Gloeosporium musarum* on banana fruits. Detection of latent infection by direct microscopic examination of banana tissues. *Nature* 201:214–15
74. Miehle, B. R., Lukezic, F. L. 1972. Studies on conidial germination and appressorium formation by *Colletotrichum trifolii. Can. J. Microbiol.* 18: 1263–69
75. Nair, K. R. S. 1962. *Studies on the physiology of primary infection by Erysiphe graminis tritici, the cause of powdery mildew of wheat.* PhD thesis. Michigan State Univ., East Lansing. 79 pp.
76. Nair, K. R. S., Ellingboe, A. H. 1965. Germination of conidia of *Erysiphe graminis* f. sp. *tritici. Phytopathology* 55:365–68
77. Naito, N., Tani, T., Toide, H. 1964. Infection type structures produced on artificial media by uredospores of *Puccinia coronata. Tech. Bull. Fac. Agric. Kagawa Univ.* 16:44–49
78. Nakayama, T. A. 1940. A study of the infection of cotton seedlings by *Rhizoc-*

tonia solani. Ann. Phytopath. Soc. Jpn. 10:93–103

79. Nishimura, S., Scheffer, R. P. 1965. Interaction between *Helminthosporium victoriae* spores and oat tissue. *Phytopathology* 55:629–34

80. Ogle, H. J., Brown, J. F. 1971. Some factors affecting the relative ability of strains of *Puccinia graminis tritici* to survive when mixed. *Ann. Appl. Biol.* 67:157–68

81. Oku, H. 1962. Histochemical studies on the infection process of *Helminthosporium* leaf spot disease of rice plant with special reference to disease resistance. *Phytopathol. Z.* 44:39–56

82. Onoe, T., Tani, T., Naito, N. 1972. Scanning electron microscopy of crown rust appressorium produced on oat leaf surface. *Kagawa Daigaku Nogakubu Gakujutsu Hokoku* 24:42–47

83. Parbery, D. G. 1963. Studies on graminicolous species of *Phyllachora*. I. Ascospores—their liberation and germination. *Aust. J. Bot.* 11:117–30

84. Parbery, D. G. 1963. Studies of graminicolous species of *Phyllachora*. II. Invasion of the host and development of the fungus. *Aust. J. Bot.* 11:131–40

85. Parbery, D. G., Emmett, R. W. 1975. Hypothesis regarding appressoria, spores, survival and phylogeny in parasitic fungi. In press

86. Pavgi, M. S., Dickson, J. G. 1961. Influence of environmental factors on development of infection structures of *Puccinia sorghi. Phytopathology.* 51:224–26

87. Pearson, N. L. 1931. Parasitism of *Gibberella saubinetti* on corn seedlings. *J. Agric. Res.* 43:569–96

88. Preece, T. F., Barnes, G., Bayley, J. M. 1967. Junctions between epidermal cells as sites of appressorium formation by plant pathogenic fungi. *Plant Pathol.* 16:117–18

89. Preece, T. F., Dickinson, C. H. 1971. *Ecology of Leaf Surface Micro-Organisms.* Academic: London & New York. 640 pp.

90. Pristou, R. R., Gallegly, M. E. 1954. Leaf penetration by *Phytophthora infestans. Phytopathology* 44:81–86

91. Purdy, L. H. 1958. Some factors affecting penetration and infection by *Sclerotinia sclerotiorum. Phytopathology* 48:605–9

92. Rahe, J. E., Kuc, J. 1970. Metabolic nature of the infection limiting effect of heat on bean anthracnose. *Phytopathology* 60:1005–9

93. Ramakrishnan, L., Staples, R. C. 1970. Changes in ribonucleic acids during uredospore differentiation. *Phytopathology* 60:1087–91

94. Rasmussen, R. A., Went, F. W. 1965. Volatile organic material of plant origin in the atmosphere. *Proc. Nat. Acad. Sci. USA* 53:215—20

95. Rasmussen, R. A., Hutton, R. S., Garner, R. J. 1968. Factors in establishing microbial populations on biologically inert surfaces. In *Biodeterioration of Materials,* ed. A. H. Walters, J. J. Elphick. London: Elsevier

96. Samuel, G. 1927. On the shot-hole disease caused by *Clasterosporium carpophilum* and on the shot-hole effect. *Ann. Bot. London* 41:375–404

97. Schlueter, K., Weltzien, H. C. 1971. A contribution to the mode of action of systemic fungicides on *Erysiphe graminis. Meded. Fac. Landbouwwet. Rijksuniv. Gent.* 36:1159–64

98. Schmid, R., Liese, W. 1965. Elektronenmikroskopische Beobachtungen an Hyphen von Holzpilzen. Heft. 1, Holz und Organismen, *Int. Symp. Berlin-Dahlem* 1965:251–61

99. Schwinn, T. J., Dahmen, F. A. 1973. Beobachtungen zum Infektionsvorgung bei *Erysiphe graminis* D.C. *Phytopathol. Z.* 77:89–92

100. Sharp, E. L., Smith, F. G. 1953. Factors affecting germ tube development in *Puccinia coronata avenae. Phytopathology* 43:483 (Abstr.)

101. Simmonds, J. H. 1941. Latent infection in tropical fruits discussed in relation to the part played by species of *Gloeosporium* and *Colletotrichum. Proc. Roy. Soc. Queensl.* 52:92–120

102. Singh, P. 1973. Effect of light, temperature and substrate during spore formation on the germinability of conidia of *Colletotrichum falcatum. Physiol. Plant* 29:194–97

103. Skoropad, W. P. 1967. Effect of temperature on the ability of *Colletotrichum graminicola* to form appressoria and penetrate barley leaves. *Can. J. Plant Sci.* 47:431–34

104. Skoropad, W. P., 1972. The effect of some herbicides and fungicides on formation of appressoria in *Colletotrichum graminicola. Proc. Can. Phytopathol. Soc.* 39:41 (Abstr.)

105. Sood, P. N., Sackston, W. E. 1972. Studies on sunflower rust. XI. Effect of temperature and light on germination and infection of sunflowers by *Puccinia helianthi. Can. J. Bot.* 50:1879–86

106. Srinivasan, K. V., Narasimhan, R. 1971. The effect of certain phenolic and related compounds on spore germination and appressorium formation in *Colletotrichum falcatum. Proc. Indian Acad. Sci.* 73:81–91

107. Staub, T., Dahmen, H., Schwinn, F. J. 1974. Light- and scanning electron microscopy of cucumber and barley powdery mildew on host and non-host plants. *Phytopathology* 64:364–72

108. Sussman, A. S. 1966. Dormancy and spore germination. In *The Fungi II,* ed. G. C. Ainsworth, A. S. Sussman. New York & London: Academic. 805 pp.

109. Sutton, B. C. 1962. *Colletotrichum dematium* and *C. trichellum. Trans. Br. Mycol. Soc.* 45:222–32

110. Sutton, B. C. 1968. The appressoria of *Colletotrichum graminicola* and *C. falcatum. Can. J. Bot.* 46:873–76

111. Turner, M. T., Martinson, C. A. 1972. Susceptibility of corn lines to *Helminthosporium maydis* toxin. *Plant Dis. Reptr.* 56:29–32

112. Ullstrup, A. J. 1936. Leaf blight of China Aster caused by *Rhizoctonia solani. Phytopathology* 26:981–90

113. Van Burgh, P. 1950. Some factors affecting appressorium formation and penetrability of *Colletotrichum phomoides. Phytopathology* 40:29 (Abstr.)

114. Von Arx, J. A. 1968. *Pilzkunde.* Cramer: Lehre. 356 pp.

115. Von Ramm, C. 1962. Histological studies of infection by *Alternaria longipes* on tobacco. *Phytopathol. Z.* 45:391–98

116. Walker, J. C. 1921. Onion smudge. *J. Agric. Res.* 20:685–721

117. Walker, J. 1972. Type studies on *Gaeumannomyces graminis* and related fungi. *Trans. Br. Mycol. Soc.* 58:427–57

118. Waller, J. M. 1970. Sugarcane smut (*Ustilago scitaminea*) in Kenya. II. Infection and resistance. *Trans. Br. Mycol. Soc.* 54:405–14

119. Weinhold, A. R., Bowman, T., Dodman, R. L. 1969. Virulence of *Rhizoctonia solani* as affected by nutrition of the pathogen. *Phytopathology* 59: 1601–5

120. White, N. H., Baker, E. P. 1954. Host pathogen relations in powdery mildew of barley. I. Histology of tissue reactions. *Phytopathology* 44:654–62

121. White, G. A., Taniguchi, E. 1972. The mode of action of helminthosporal 11. Effect on the permeability of plant cell membranes. *Can. J. Bot.* 50:1415–20

122. Whiteside, J. O. 1973. Action of oil in the control of citrus greasy spot. *Phytopathology* 63:262–66

123. Whiteside, J. O. 1974. Environmental factors affecting infection of citrus leaves by *Mycosphaerella citri. Phytopathology* 64:115–20

124. Williams, P. G. 1971. A new perspective of the axenic culture of *Puccinia graminis* f. sp. *tritici* from uredospores. *Phytopathology* 61:994–1002

125. Williams, P. G., Scott, K. J., Kuhl, J. L. 1966. Vegetative growth of *Puccinia graminis* f. sp. *tritici* in vitro. *Phytopathology* 56:1418–19

126. Wood, R. K. S. 1967. *Physiological Plant Pathology.* Oxford & Edinburgh: Blackwell. 570 pp.

127. Yang, S. L., Ellingboe, A. H. 1972. Cuticle layer as a determining factor for the formation of mature appressoria of *Erysiphe graminis* on wheat and barley. *Phytopathology* 62:708–14

128. Yarwood, C. E. 1957. Powdery mildews. *Bot. Rev.* 23:235–301

129. Young, P. A. 1926. Penetration phenomena and facultative parasitism in *Alternaria diplodia. Bot. Gaz.* 81: 258–79

130. Zaracovitis, C. 1966. The germination in vitro of conidia of powdery mildew fungi. See Ref. 45, 273–86

SCLEROTIAL MORPHOGENESIS IN FUNGI

❖3618

I. Chet and Y. Henis
Department of Plant Pathology and Microbiology, The Hebrew University of Jerusalem,
Faculty of Agriculture, Rehovot, Israel

INTRODUCTION

In the last decade, the growing interest focused on the mechanism involved in cell differentiation has brought biologists to study similar basic problems in developmental biology with different living systems. The vast data from these studies have been summarized in a number of books and reviews (5, 6, 25, 29, 64, 68–70, 73, 86, 122, 123, 137, 140, 143, 166).

This review deals with recent developments in the study of fungal morphogenesis using sclerotium formation in slime molds and filamentous fungi as a model system. Of the many definitions used for unicellular and multicellular differentiation we chose the one referring to this process as the sum of a predetermined, programmed sequence of molecular, cytological, and morphological changes that take place during fungal vegetative growth, terminating in sclerotium production. Whereas sclerotium morphogenesis deals with the cytological and morphological changes, it is the result of the molecular changes which, in turn, are controlled by the cell genome and by the environment. It is the goal of developmental biologists to reveal the early events or the "morphogenetic triggers" that initiate the overall process of cell morphogenesis.

Sclerotia are asexual, multicellular, firm resting structures. In myxomycetes, such as *Physarum* sp., the sclerotium is a cluster of enwalled subunits named spherules (135), whereas sclerotia of filamentous fungi are composed of condensed vegetative hypha cells which become interwoven and aggregate together (101, 147). Fungal sclerotia are often irregular in shape, but the constituent cells tend to be more or less isodiametric, in contrast with the myxomycete plasmodia or elongated hyphal cells.

An attempt is made here to review the data so far known on sclerotium morphogenesis and to reveal common control mechanisms of the process.

MYXOMYCETES

The myxomycetes, also referred to as slime molds, are simple organisms whose vegetative stage is a large plasmodium of indefinite shape containing up to several million nuclei. The slime molds serve as an excellent model system for biologists and biochemists in the study of various aspects of cell metabolism, mitosis, plasmodial movement, and differentiation (5, 135). Morphogenesis and especially sclerotium formation were extensively studied in *Physarum polycephalum.*

Induction and Formation of Spherules and Sclerotia in Physarum polycephalum

When subjected to unfavorable conditions, the *Physarum* plasmodium forms dormant, resistant sclerotia which enable the organism to survive for years. While the sclerotia of the various myxomycetes differ somewhat in their appearance, all are ؛ nsters of hard-walled mono- or multinucleate units called "spherules" (93).

Sclerotization occurs when the plasmodium is cooled, slowly dried (69), or exposed to low temperature, low pH (pH 2), high osmotic pressure, and the presence of metal ions (93). The most common way of introducing transformation of plasmodia into sclerotia in the laboratory is by transferring a growing culture to a carbonless salt medium (52, 53).

The process of sclerotization in liquid media is commonly referred to as "spherulation" (or spherule formation). The spherules can be dried on a filter paper and stored at 4°C for many years. Under starvation conditions spherules are formed within 24–36 hr (62, 72).

Several light- and electron-microscopic studies have been done on mature spherules (73, 93, 132, 142). Goodman & Rusch (61) studied the ultrastructural changes occurring during spherulation. They found that the number of glycogen granules decreased during the early developmental period and a Golgi complex appeared. They confirmed that cytokinesis occurred by the fusion of vesicles after 24 hr of starvation and that protein synthesis was necessary for spherulation. After plasmodial cleavage, a cell wall was synthesized. Chemical characterization revealed that this hard wall contains galactosamine as the only sugar component and about 2% single glycoprotein. The phosphate content of isolated spherule walls was 9.8% (113). Chet & Rusch (40) induced spherule formation in *P. polycephalum* within 12–35 hr by transferring the culture to a synthetic medium containing 0.5 M mannitol or other polyols. In contrast with the clusters of spherules formed during starvation, the mature spherules obtained by this method appeared as a distinct morphological unit, probably as a result of the lower production of slime in this system. When transferred to a fresh synthetic medium, 90% of the spherules germinated within 24 hr. Similar spherules were also obtained in a salt solution (60).

Recently, the ultrastructure of starvation-induced sclerotia and mannitol-induced spherules were compared by scanning electron microscopy (SEM) (36, 96, 97). Each spherule in the berry-like sclerotium is not rounded and has lobe-like forms, whereas the separate spherules are round and significantly bigger. Laminar bead-like granules were observed inside and on spherules. Desiccation led to a separation of cytoplasm from walls and the appearance of "horny" spherules. A new technique of freeze-fracturing enabled observation of the ultrastructural organization by the SEM (100).

Biochemical Changes Occurring During Spherulation of Physarum polycephalum

During spherulation, carbohydrate metabolism declines (109) and a dramatic decrease in glycogen content is observed, 30% of the glycogen remaining after starvation at time of cleavage. Simultaneously, the number of glycogen granules decreases. The appearance of Golgi apparatus may be responsible for the high production of slime. The slime, a sulfated galactose polymer, cannot serve as a carbon source for the plasmodium (113). The slower rate of glycogen decrease, the uninhibited ^{14}C-glucose uptake, and its incorporation into glygocen during mannitol-induced spherulation (40), suggest that mannitol changes the ratio between glycogen synthesis and degradation but does not cause glucose deprivation. Incorporation of ^{14}C-mannitol into glycogen was very low as compared to ^{14}C-glucose, excluding the possibility that mannitol can replace glucose as a carbon source. However, the mode of spherule induction by mannitol is still unknown.

ENZYMES   The activity of various enzymes was studied during the first 24 hr of spherulation in starved cultures (90). A significant increase of about 8-fold was observed only with glutamate dehydrogenase (GLDA) and phosphodiesterase. Inhibition of protein synthesis by cycloheximide resulted in inhibition of plasmodial cleavage and wall formation (61), and in a decrease of enzymatic activity during the process (90). Using deuterated amino acids, Hüttermann et al (88) found a de novo synthesis of GLDA during spherulation. Two phosphodiesterase isoenzymes were detected during growth but only one was found during spherulation (85). Decrease in glycogen during spherulation could not be correlated with pyrophosphorylase activity, whereas synthesis of the slime during this process was not triggered by epimerase activity (89). Enzymatic composition of cells subjected to starvation-induced spherulation was somewhat different as compared with that of mannitol-induced spherulating cells. GLDA and phosphodiesterase activities in mannitol-induced spherulating cells increased much less than expected (87). The role of these enzymes in spherulation is not clear; however, GLDA was reversibly inhibited by glutaric acid, which also inhibited the whole process (90). Localization of nucleolar and chromatin residual acidic protein also changes during differentiation (102).

The data available so far suggest the existence of differential protein synthesis during spherulation. *P. polycephalum* is the first "model system" of differentiation where true differential protein synthesis has been demonstrated (86). These data corroborate the theory of differential gene activity during differentiation (119).

MACROMOLECULAR SYNTHESIS   Actinomycin D did not inhibit protein synthesis during the first 20 hr of starvation, but significantly inhibited amino acid incorporation shortly before spherule formation (139). An early inhibitory effect of actinomycin D on amino acid incorporation was observed in the case of mannitol-induced spherulation (42).

Spherules formed in the presence of actinomycin D were misshapen and sometimes incompletely cleaved from the plasmodium. Their germination potential was very low. It was suggested, therefore, that synthesis of new RNA was essential for the spherulation process (40). Nucleic acids and protein content significantly decreased when cultures of *P. polycephalum* were transferred to a mannitol-supplemented medium. When spherule formation was completed 46 hr after induction, RNA, protein, and DNA contents were 13.0, 23.0, and 52.5%, respectively, as compared with the values at the beginning of the differentiation process. Sauer et al (139) observed a similar decrease in the starvation-induced system. These authors also characterized RNA from growing and starving plasmodia using methylated albumin Kieselguhr (MAK) columns. They found that cytoplasmic RNA from starvation-induced spherulating cultures contained a large proportion of low molecular weight RNA, possibly as a result of a degrading activity of ribonucleases present in the plasmodia (21, 84). RNA-DNA hybridization and competition experiments with RNA extracted during the first 24 hr of starvation, did not reveal any significant changes in RNA composition (139). Chet & Rusch (41, 42) used this system to study the changes in RNA during spherulation induction by mannitol. The radioactivity profiles of short pulse-labeled spherulating cultures differed considerably from those produced in short labeling periods during growth. A 10 min ^{3}H-uridine labeling was carried out 2 hr after onset of induction. A relatively high specific RNA radioactivity was obtained in the light zone of the gradient, indicating an increase in the rate of RNA synthesis or in the synthesis of new RNA fraction(s). Evidence of synthesis of new classes of RNA with a concomitant repression of other classes during differentiation was shown by both liquid and filter DNA-RNA competition hybridizations (41). *P. polycephalum* was assumed to have ribosomal RNA classes of 26 S and 19 S (116). Chet & Rusch (41) characterized the hybridizable classes of RNA according to their mobility on acrylamide gels. The heavier classes of RNA (about 24 S) and some lighter classes from spherulating microplasmodia hybridized better than those from growing cultures. This confirms earlier data indicating the synthesis of mRNA during spherulation (40). "Pulse-Chase" experiments with ^{3}H-uridine showed a significant decrease in the specific activity of RNA during spherulation. This degradation increases the pool of ribonucleotides, which accounts for the initial low rate of ^{3}H-uridine incorporation into RNA at the beginning of the morphogenetic process. These results corroborate with findings that ribonuclease activity in *P. polycephalum* increases rapidly after the addition of mannitol for spherule induction (39). The use of the isoelectric focusing technique revealed that *P. polycephalum* contains 12 isozymes capable of degrading RNA. No new ribonucleases were detected in the cell-free extracts during spherulation, but the activity of several intracellular isoenzymes immediately increased after spherule induction. The dramatic increase in ribonuclease activity seems to depend on protein synthesis de novo and is not controlled by RNA synthesis.

Goodman et al (62) found a large difference in the amount and distribution of polyphosphates and other phosphorus-containing compounds during spherulation as compared with the growth period. It was assumed that polyphosphates were involved in energy relationship and nucleic acid synthesis. Polyphosphates may also serve as a storage product which becomes available during spherule germination. Further evidence for this suggestion was obtained by Sauer et al (138) who demonstrated a transfer of ^{32}P from polyphosphate to RNA when starving microplasmodia were returned to a growth medium.

Spherulation in *P. polycephalum* can be induced by many stimuli, and it is reasonable to assume that a plasmodium can be transformed to a resting body by any specific effect of heterogeneous stimuli that interferes with the growth cycle (137). It appears that this morphogenetic process is controlled by a rapid turnover of RNA and that ribonucleases are responsible for this degradation process. Consistent with the idea that one of the central mechanisms for controlling morphogenesis is at the level of transcription, different studies (29, 137) indicate that new RNA is synthesized during spherulation. Whether the control of RNA turnover during differentiation involves sequential synthesis of ribonucleases and other enzymes or preferential synthesis at specific cell sites is still an enigma.

FILAMENTOUS FUNGI

Fungal sclerotia are multicellular resting structures, resistant to adverse conditions, formed by some filamentous fungi. The term sclerotia includes various different bodies and is not entirely definite. Garrett (57), makes a fundamental distinction between sclerotia produced by air-borne fungal pathogens and those produced by root-infecting fungi. Sclerotia of the former group are variable in size and can attain maximal diameter of 2 cm. They usually produce fruiting bodies such as apothecia of *Sclerotinia* species or perithecia of *Claviceps* species. In contrast, the sclerotia of root-infecting fungi are smaller (1-2 mm in diameter), more uniform in size, and also more regular in shape, such as those of *Verticillium dahliae, Sclerotium rolfsii, S. cepivorum,* and *Phymatotrichum omnivorum.* The development of fungal sclerotia had been studied as early as the 19th century by Brefeld (22) and De Bary (55). Townsend & Willetts (147) investigated the development and structure of several types of sclerotia by light microscopy and found that a mature sclerotium of *S. rolfsii,* for example, is composed of four distinct cell layers: (*a*) a fairly thick skin, (*b*) a rind, made of broad and tangentially flattened cells, (*c*) a cortex of thin-walled cells with a densely stained cytoplasm, and (*d*) a medulla made of loosely arranged, ordinary filamentous hyphae. More details were revealed with transmission (37) and scanning electron microscopes (127). Similar studies were done on other fungi (165, 167). The types of sclerotial structures and their morphology were comprehensively reviewed by Butler (25), Coley-Smith & Cooke (45), and recently by Willetts (166) and, therefore, are not discussed here.

Factors Affecting Formation of Sclerotia

Three stages in sclerotia formation and development were distinguished by Townsend & Willetts (147): (*a*) initiation, the appearance of small distinct initials formed from the interwoven hyphae, (*b*) development, increase in size, (*c*) maturation, characterized by surface delimitation, internal consolidation, pigmentation (35), and often associated with droplets excretion (48, 49). There are some other patterns of development such as in *Claviceps purpurea* (55). In this fungus the sclerotium replaces preexisting mycelium in the host ovary. The hyphae become densely interwoven and closely septate with thick walls (25). The large sclerotia of some species of *Polyporus* are enclosed by a fungal rind but incorporate host roots or stones and soil (13).

There are some common factors affecting sclerotial formation that are mentioned here, but most attention is drawn to specific morphogenetic triggers that may contribute to a better understanding of the mechanisms involved in sclerotial formation.

LIGHT   Light may induce initiation and development of sclerotia in various fungi (26, 76, 130, 134, 145). Kaiser (95) found that in *Verticillium albo-atrum* blue light inhibited microsclerotia production while yellow-orange and red light promoted it. White fluorescent light inhibited melanin production in most isolates tested by Brandt (18). Near ultraviolet radiation (peak 3650 Å) completely inhibited both melanin and microsclerotial formation in one isolate but, on a yeast-extract–containing medium, septation and enlargement of cells were not inhibited (18).

Considerably greater numbers of *S. rolfsii* sclerotia are produced in cultures exposed to light (71, 112, 149). Trevethick & Cooke (149) found that in *S. rolfsii*, *S. delphinii*, and *Sclerotinia sclerotiorum* both number and size of sclerotia could be determined by length of photoperiods. Lately, Tan & Epton (144) found that sclerotial formation in *Botrytis cinerea* occurred in darkness, in yellow, red, and infrared light, and in cultures irradiated for less than 30 min with black light. Unfortunately, the role of light in sclerotial morphogenesis is not understood yet.

TEMPERATURE   In general, the optimum temperature range for mycelial growth is suitable for sclerotial formation, but not necessarily (74). Heale & Isaac (75) found that the optimal temperature for microsclerotia production in *V. albo-atrum* is 24°C. Temperature affects the shape of the microsclerotia (11). Wilhelm (164) found that at 10–22°C *V. albo-atrum* colonies are black and rich in microsclerotia, whereas at 25–31°C they appear white with few microsclerotia.

Sclerotial formation in *S. rolfsii* is maximal at or near the optimal temperature for mycelial growth (44, 63, 117, 118). Abeygunawardena & Wood (1) found the greatest weight of sclerotia on media incubated at 30°C. Formation of sclerotia was observed also at 10°C but maturation was very slow or nonexistent. There is some indication that sclerotial formation may be accelerated by a sudden and temporary rise in temperature (78). Apparently temperature has an important effect on sclerotial pigmentation. Light and yellow colors are more prevalent at 22°C (64).

pH   The most favorable pH for mycelial growth is usually the optimal for formation of sclerotia (134, 146); for example, the pH range for *Verticillium* growth is 4–10 and the optimal pH for production of microsclerotia is 5–6.

Chowdhury (43) obtained maximum sclerotial production in *S. rolfsii* on agar at pH 6.4. The rate of sclerotium formation was not affected when the fungus was grown on a medium with pH adjusted to 1.3 or 6.7 (3, 100).

OXYGEN AND CO₂   A low oxygen and/or a high CO_2 tension are probably the factors responsible for the inability of fungi to produce sclerotia in submerged cultures (Chet and Henis, unpublished; 11). Understanding the role of these ecologically important factors in sclerotial formation awaits future studies.

MECHANICAL FACTORS   Henis et al (81) noted that when vegetative mycelium, grown on an agar medium, was cut with a sterile cork borer, production of sclerotia along the cutting lines took place within 24 hr. On the other hand, on an intact mycelium, sclerotia produced were distributed sporadically and appeared after a much longer incubation period. Similar results were obtained with mycelia grown on a liquid medium. In flasks, where mycelium was once torn off by shaking, an immediate production of sclerotia took place at the torn edges. These results were confirmed by Wheeler & Waller (161) and Geeson (58). It was also noted (4, 81, 160) that sclerotial production usually occurred when the mycelium reached the Petri plate's walls and its linear growth was thus restricted. Mekhaimer (115) working with *S. delphinii,* found that sclerotia were never formed until the medium surface was completely covered by the fungus and mycelial growth was physically limited.

A similar phenomenon can be observed also when *S. sclerotiorum* is grown in Petri plates. Townsend (146) suggested that conditions unfavorable for vegetative growth lead to sclerotium formation in fungi. It is not clear, however, whether growth inhibition always results in sclerotial production in *Rhizoctonia solani* (80).

In *S. rolfsii,* inhibition of mycelial linear growth by itself is not a major single cause for sclerotium induction (34, 125).

STALING PRODUCTS, MICROBIAL EXCRETIONS, AND ANTIBIOTICS   Morphogenetic processes in fungi are often stimulated by microorganisms or by their products (12, 110). Bedi (9) found that the number of sclerotia formed by *S. sclerotiorum* significantly increased when staling products of other cultures or the same fungus were added to the growth medium. Sclerotia formed in the presence of these substances appeared earlier and were bigger. Mutants that normally did not form sclerotia produced them after the addition of staling substances to the growth medium. Culture filtrate of *S. rolfsii* increased the number of sclerotia, suggesting the presence of sclerotia-inducing substance(s) (107). Sclerotial production in *R. solani* Kuhn may be induced by unidentified, volatile substances produced by *S. griseus.* Henis & Inbar (80) found eight strains of *Bacillus subtilis* and one strain of *B. licheniformis* which inhibited mycelial growth of *R. solani* and induced sclerotial formation. The active principle produced by *B. subtilis* was heat-stable, alcohol- and ether-soluble, and mostly cell-bound. According to Sanford (136), sclerotium formation by *R. solani* in soil is correlated with the antagonistic activity of soil microflora. Chet (28) found that the antifungal antibiotics griseofulvin, trichomycin, and bacitracin induced sclerotial formation in *S. rolfsii*. All three affect either the cell wall or the cell membrane (121), thus suggesting a possible connection between formation of sclerotia and interference in cell wall synthesis.

NUTRITIONAL FACTORS   The effect of nutrients on fungal differentiation, especially on sporulation, is well known (24). Different nutritional conditions may be required for initiation, development, and maturation of sclerotia. Sclerotial initials may be formed on a relatively poor medium, but further development requires more nutrients than those essential for the growth of a vegetative mycelium (146). Unfavorable conditions may enhance, but are not essential for, sclerotial formation (135, 146). The nutritional factors and chemicals affecting this process can be classified as follows.

Plant tissues and extracts   Most of the sclerotia-forming fungi are facultative parasites and produce sclerotia on their host tissue during pathogenesis and on plant residues in soil. Little information is available concerning the specific effects of these substrates on sclerotial formation.

Boyle (15) observed that available food has considerable influence on which phase of growth predominates. He found that when oats of high test weight were used as a medium for *S. rolfsii*, sclerotium formation was scant, whereas using oats of low density induced development of sclerotia in great numbers.

Mycelial growth and production of sclerotia by *S. rolfsii* were studied on sterilized soil amended with wheat bran or on a defined agar medium (81). The highest numbers of sclerotia were produced under suboptimal condition for mycelial growth. These results corroborate earlier observations by Boyle (15), who noted an inverse relationship between mycelial growth and sclerotial production in the field. The production of fewer but bigger and heavier sclerotia on soil amended with large amounts of wheat bran would perhaps indicate that the "trigger" processes inducing sclerotium initiation are different from those involved in the further development and maturation of those structures. Addition of plant extracts to growth media, however, has been widely employed.

Higgins (83) found that the proportion of mycelia and sclerotia of *S. rolfsii* grown on bean and pepper pod decoctions, was similar to that observed on various fruits and vegetables in the field. Davey & Leach (54) observed that *S. rolfsii* sclerotia produced on sugar beet were characteristically larger than those produced in pure cultures. Tucker (150) reported that the largest *S. delphinii* sclerotia developed on an onion agar, and Bertus (10) found abnormally large *S. rolfsii* sclerotia up to 8 mm in diameter, on sweet potato agar.

Carbohydrates, nitrogen, and C/N ratio   The source and concentration of nutrients, particularly those supplying carbon and nitrogen, have been shown to affect sclerotial formation (28, 118, 134, 154, 160).

The number of *S. rolfsii* sclerotia produced on agar plates increased with a glucose concentration of up to 4% (81). However, if calculated on a mycelial weight basis, the numbers of sclerotia decreased at higher glucose concentrations.

Heale & Isaac (75) observed microsclerotial production in *Verticillium* grown on carbonless agar medium; the addition of carbohydrate, especially sucrose, increased their numbers.

Working on *S. sclerotiorum*, Bedi (8) found that maltose was the best supporter of both the number and the dry weight of sclerotia. Lactose and galactose were poor sources and mannitol, although very good for mycelial growth, totally inhibited sclerotial formation. Wang & LeTourneau (152) reported some different results, finding the highest sclerotial number with raffinose, sucrose, maltose, lactose, D-mannose, D-glucose, and D-fructose.

The addition of 0.5% lactose to a glucose-mineral medium partially inhibited mycelial linear growth and induced synchronous formation of sclerotia in concentric circles in *S. rolfsii*. The diameter and formation timing of the sclerotial circle were determined by the glucose concentration (126).

Sclerotium formation in *S. rolfsii* was inhibited by 2% ethanol (126, 159). On the other hand, ethanol is used as a sole carbon source in a selective medium employed for the isolation of *Verticillium* species from soil (120).

Both NH_4^+ and NO_3^- can be utilized by sclerotia-forming fungi for growth and sclerotial formation. No specific differential effect of inorganic nitrogen source on sclerotial formation has been reported.

The acidic amino acids were rapidly utilized in *S. rolfsii* (107). L-Glutamic acid and D-L- asparagine favored excellent growth and sclerotial formation in *S. sclerotiorum*, probably because of their role in the tricarboxylic acid cycle (153). This is of interest because *S. sclerotiorum* produces organic acids (151), and formation of sclerotia may be related to acid formation (152).

L-Threonine, at a concentration of $10^{-2}M$, induced abundant sclerotial formation in *S. rolfsii*, probably because of its role in the glyoxylic acid cycle (Okon, Chet & Henis, in press). On the other hand, sulfur-containing amino acids, especially cysteine, inhibited sclerotial formation in *S. rolfsii* (34) as well as in other fungi. These specific effects are discussed elsewhere in this review.

The dependence of sclerotial formation on carbon:nitrogen ratio in the growth medium has been shown with *B. cinerea* (128) and *V. albo-atrum* (171). Major differences in microsclerotial production in *Verticillium* occurred at ratios of 3.8 or lower when N was supplied as $NaNO_3$.

Minerals   The mineral constituents in culture medium of *S. sclerotiorum* do not operate to modify the dominant role of the carbohydrate components (8). The omission of Zn^{2+} from a glucose-salts medium decreased dry weight of *S. sclerotiorum* and prevented sclerotial production. One mg/liter of Zn^{2+} caused maximal sclerotium production, whereas higher amounts increased their total dry weight but in a smaller number of sclerotia (151). Chet & Henis (31) reported that in *S. rolfsii* chelating agents, such as Na_2EDTA, induced sclerotial formation in concentric circles around inoculum discs. However, the sclerotia were usually big, soft, and yellow in color. They also fused into each other and formed irregularly shaped masses. Upon the addition of Cu^{2+} ions, sclerotial formation and appearance resembled the untreated control. Other ions did not neutralize the effect of Na_2EDTA.

In the same fungus, low phosphate concentrations had a lesser effect on sclerotial formation than low K^+ levels. Without phosphate, considerable numbers of initials formed and the number of mature sclerotia was more than 10 times that on media lacking K^+ (160).

Vitamins   Kaiser (94) compared the vitamin requirements of 14 isolates of *V. albo-atrum* and found no absolute requirements. He found that biotin increased mycelial growth and was essential for microsclerotial development. On the other hand, Roth & Brandt (133) found that biotin increased mycelial growth in two isolates of *Verticillium* and that a mixture of biotin, thiamine, pyridoxine, and inositol had a more favorable effect than biotin alone. Thiamine and pyridoxine alone had no effect on linear growth. In one isolate, inositol alone severely suppressed growth, microsclerotial development, and melanin synthesis. Pyridoxine counteracted the inositol inhibition, and the mixture of pyridoxine and inositol synergistically stimulated microsclerotial development and melanin synthesis. The differences between those two reports may be due to the use of different fungus strains or techniques. It is interesting to note that just recently Zilberstein, Henis & Chet (unpublished) found that microsclerotia of *V. dahliae* which were formed in the presence of the B complex, germinated significantly better than those produced by cultures grown on vitamin-free media. *S. rolfsii* requires thiamine for growth and reproduction (4). Thus, increased mycelial growth and sclerotial formation in the vicinity of other organisms may be due, in part, to vitamins and other stimulatory substances produced by those organisms (83).

Sulfur-containing compounds   The effect of sulfur-containing amino acids on morphogenesis of microorganisms has been demonstrated by Nickerson & Falcone (124) and by Brachet (16).

Nickerson (122) and Nickerson & Bartnicki-Garcia (123) found that in budding yeast the rigidity of glucomannan protein in the cell wall is maintained by protein-S-S-bonds: budding occurs only at sites where these bonds are reduced. There are only a few reports on the presence of –SH groups in fungal walls, most of which are found near the hyphal tips (129, 172). Sclerotial formation can be inhibited in Houston black clay soil, which normally supports a good production of sclerotia in *Phymatotrichum omnivorum,* by autoclaving with dusting sulfur (56, 57). The inhibitory effect of sulfur on sclerotium production was also observed by Chet, Henis & Mitchell (34) in *S. rolfsii.* They found that glutathione and sulfur-containing amino acids, especially L-cysteine at 10^{-4}–10^{-5}M, inhibited sclerotial formation, without affecting mycelial weight and its linear growth. Higher concentrations inhibited both linear growth and sclerotial production. Other –SH-containing compounds, such as β-mercaptoethanol, 2,3-dimercaptoethanol, and thioglycollic acids, were much less active.

Iodoacetic acid, which is a known –SH antagonistic agent, induced sclerotium formation at 3×10^{-5} – 10^{-4}M, sclerotia being produced in circles around the inoculation point. The effect of L-cysteine was competitively antagonized by iodoacetic acid at a molar ratio of 30:1. L-Cysteine had a similar effect to that of –SH compounds, probably because of the reduction of its S–S bonds to –SH, since cysteine reductase was detected in the mycelium of *S. rolfsii* (31). Trevethick & Cooke (148) examined these effects in *S. rolfsii, S. delphinii,* and *S. sclerotiorum.* Their results showed that whereas sclerotial formation in the first two species was inhibited by –SH compounds, variable results were obtained for the last one. A similar inhibitory effect of sulfur-containing compounds on sclerotial formation was also found with dimethyl sulfoxide (77). Wong & Willetts (170) tested several isolates of *S. sclerotiorum, S. minor, S. rolfsii,* and *B. cinerea* and found that the acceptance of data obtained from only one isolate of a species could be misleading. Geeson (58) also found differences between isolates, and lately, Chet & Henis (33) reported that 10 isolates of *S. rolfsii* could be divided into two types (R and A) which differed in their response to factors affecting sclerotial formation, such as cysteine and iodoacetic acid.

INTERNAL MORPHOGENETIC FACTORS   Wheeler & Waller (161) suggested that the initiation of sclerotia in *S. rolfsii* is regulated by the accumulation, in the hyphae, of substances metabolized in a different way from that occurring in mycelial growth. Working on the same fungus, Henis & Chet (79) found that sclerotial initials reached the maturation phase when transferred onto a new growth medium only if taken from the original medium with their surrounding mycelium. They suggested that factors supplied by the mycelium are involved in the initiation, development, and maturation of the sclerotia. Phenylthiourea can stimulate these factors only at the initiation phase, whereas iodoacetic acid and Na_2EDTA are effective at both initiation and development. All these internal, naturally occurring, hypothetic factors are inhibited by L-cysteine. Goujon (65–67) and Geiger & Goujon (59) also concluded that a morphogenetic factor is produced in *S. rolfsii.* They found that one or several factors capable of initiating the formation of sclerotia migrate from old to young sclerotia through the hyphae. Goujon (67) obtained this "morphogenetic factor" in cell-free extracts and found that it was thermolabile, sensitive to polyphenols, and that its molecules were of a relatively high molecular weight. By using protein-synthesis inhibitors he delayed the appearance of the sclerotia and reduced their number in culture. The inhibitory action of the antibiotics apparently takes place at the level of the synthesis of one or several morphogenetically efficient compounds. According to Goujon (67), the morphogenetic factor must reach a sufficient concentration inside the hyphae of *S. rolfsii,* this level being reached, on the average, after four days of mycelial growth. Liu & Wu (108) suggested that *S. rolfsii* produces unidentified specific substances which trigger sclerotial formation.

Some homokaryotic isolates of *V. albo-atrum* vary in colony appearance. This variation is nonhereditary (17) and its cause appears to be a morphology-affecting, diffusible substance (DMF) (18, 20). At certain levels, DMF stimulates production of microsclerotia and melanin, while inhibiting hyphal elongation and sporulation. The hyphae swell, septation is increased, and the cell walls become thickened. These phenomena are part of the morphogenetic process of microsclerotial development. At higher concentrations the DMF can inhibit production of microsclerotia and melanin as well as conidial germination. It appears that near-ultraviolet radiation, which inhibits microsclerotial formation, acts by suppressing DMF synthesis. DMF is active in pH range of 4–10, seems to be nonvolatile, dialyzable, and water soluble, perhaps chemically related to phenol.

Morphological and Physiological Processes Occurring During Sclerotial Development and Maturation

The majority of sclerotia arise as discrete small initials among the hyphae, which grow and develop until the sclerotia reach maturation. Initiation, development, and maturation are complicated processes accompanied by both morphological and biochemical differentiation. The information known so far concerning the events occurring during these phases and their relationships is discussed.

HYPHAL BRANCHING   Goujon (67) studied extensively the morphological processes of formation of sclerotia in *S. rolfsii.* He noted that the mycelium of this fungus was composed mainly of leading hyphae with 240 μ long internodes containing 40 nuclei per cell, and lateral hyphae, with 10 nuclei per cell. Lateral branches first emerge from the leading hyphae, in acute angles; later, thinner branches emerge at right angles and give rise to sclerotia. Henis et al (82) studied the effect of various substances that influence sclerotial production on the branching process of the leading hyphae in two types of *S. rolfsii,* as related to sclerotial formation. Substances inhibiting sclerotial production also caused a reduction in lateral branching, whereas substances that induced sclerotial formation increased lateral branching at colony margins, long before the formation of sclerotial initials, indicating that in *S. rolfsii,* sclerotial morphogenesis and hyphal branching are closely related.

INTERWEAVING AND FUSION OF HYPHAE   The behavior of the hyphae of *S. rolfsii* during sclerotial production was described in detail by Townsend & Willetts (147). Sclerotial initials in *S. rolfsii* are formed from hyphal strands which consist of 3–12 hyphae lying parallel to each other. The initials may be produced from a single strand or where two strands cross. A spherical shape is formed, even though it consists merely of a network of hyphae that become increasingly more septate; no coalescence occurs at this stage. Differentiation inside the sclerotium proceeds when the sclerotium is larger and firmer. Higgins (83) suggested that the interweaving and coalescence of different hyphae to form a sclerotium is a sexual process, involving two different types of hyphae. Willetts (166) observed that the hyphal strands stick together after coming in contact. A possible explanation for this may be the presence of mucilaginous matrix in many sclerotia-forming fungi, or lytic and synthetic enzymes on the hypha surface (51, 162). Reports on *Dictyostelium discoideum* (141) show that adhesion of cells is specific and essential for morphogenesis. Buller (23) found fusions in sclerotia of *S. rolfsii* and suggested that this process favors transport of nutrients into the developing body. In anastomosis formation the possibility of chemotaxis among specific hyphae must be considered. Chemotaxis is a well-known phenomenon in both prokaryotes (2) and eukaryotes (14, 27).

TRANSLOCATION   Growth and developmental processes in fungi require an organized translocation of metabolites and macromaterials in the hyphae (172). Studies revealed that within certain sclerotia-forming fungi, translocation often results in the accumulation of substances in the developing sclerotia. Thus, Littlefield et al (105) showed that translocated ^{32}P accumulated in sclerotia of *R. solani*. Translocation was observed when sclerotial primordia were formed, but not in cultures that formed no sclerotia. Age, size, and location of the sclerotia had no consistent effect on degree of accumulation (104). Wilcoxson & Subbarayudu (163) found that ^{32}P was translocated and accumulated in sclerotia of *S. rolfsii* and to a significantly greater extent when the direction of translocation was from the old to the young region of the cultures than when in the reverse direction. Working on the same fungus, Chet et al (34) reported the accumulation of ^{14}C-labeled iodoacetic acid in the sclerotia. Goujon (67) described two types of hyphae in *S. rolfsii*—a translocatory and a lateral type. Cooke (49) also mentioned the importance of translocation during the development of sclerotia in *S. sclerotiorum*. Okon et al (125) found that synchronous formation of sclerotia occurred when *S. rolfsii* was grown on 0.5% (w/v) lactose-supplemented medium, whereas the addition of ethanol (2%, v/v) inhibited formation of sclerotia. In a later study (126) it was found that protein synthesis and translocation of L-^{14}C-leucine, D-^{14}C-glucose, and inorganic ^{32}P were enhanced in cultures grown with lactose. Ethanol inhibited translocation but not protein synthesis. Neither lactose nor ethanol affected the uptake of radioactive substances by *S. rolfsii*. These results support the suggestion that internal translocation is related to morphogenesis (157). It can be concluded, therefore, that translocation is essential for sclerotial formation.

SCLEROTIAL EXUDATION   Exudation is a common phenomenon during the early stages of sclerotium development (147). Remsberg (131) dried the exudates collected from sclerotia of *Typhula* and found crystalline residue. Cooke (48, 49) found that the droplets collected from the surface of sclerotia of *S. sclerotiorum* contained soluble carbohydrates, and suggested that these droplets were actively excreted during the rapid development of the sclerotia. Water loss was greatest when the initials began to increase in size, and this was also the period of maximum translocation. The excretion was correlated with a decrease in endogenous mannitol in both *S. sclerotiorum* and *S. trifoliorum,* and with a decrease in endogenous glucose and trehalose in *S. trifoliorum*. Jones (91) found salts, amino acids, proteins, and polyphenoloxidase activity in exudates of sclerotia of *S. sclerotiorum*. These findings were confirmed by Colotelo et al (46), who found that the liquid droplets are enclosed in small, thin-walled sacs. Chemical analyses of the exudate showed the presence of proteins, 10 fatty acids, cations, ammonia, and various enzymes such as polyphenoloxidases, peroxidase, catalase, β-glucosidase, cellulase, and polygalacturonase (47). These authors suggest that the exudation may be a phenomenon similar to the transformation of soils into gels by the process of syneresis. During cell wall formation and thickening, final polymerization of soluble compounds occurs, cell wall dehydration takes place, and the resulting liquid is then expelled. According to Cooke (48), during early development, when sclerotia are converting soluble carbohydrates to storage and structural compounds, the movement of materials into the tissues is greater than their conversion rate, and the tissues actively exude water, with excess carbon compounds to maintain an internal physiological equilibrium.

CARBOHYDRATE METABOLISM   When comparing the difference between chemical composition of hyphal and sclerotial walls of *S. rolfsii,* Chet & Henis (30) and Chet et al (35) found that the content of polysaccharides and hexosamines in sclerotial walls was much lower than that of the hyphal walls, although both contained glucose, mannose, and glucosamine. Batra et al (7) extracted glucose polymer from the mycelium of *S. rolfsii* and studied its biosynthesis. Jones et al (92) analyzed hyphal walls and sclerotial walls of several fungi. Sugar analysis of the hydrolysates revealed differences between sclerotia and hyphae in sugar composition.

Kitahara (98, 99) reported the presence of trehalose, mannitol, and β-glucan in sclerotia of *S. sclerotiorum.* Chet et al (35) also found β-1-3-glucan in sclerotia of *S. rolfsii.* LeTourneau (103, 154) reported that trehalose and mannitol were the two major alcohol-soluble carbohydrates and constituted 6–7% of the dry weight of *S. sclerotiorum* sclerotia. The major insoluble reserve substance is glycogen (99). Wang & LeTourneau (152) found mannitol biosynthesis in cell-free extracts of mycelia and sclerotia of *S. sclerotiorum.*

It has been suggested that mannitol and trehalose may serve as endogenous reserve substances in sclerotia of *S. sclerotiorum* (103). However, although trehalose is the major sugar found in these sclerotia, only few, if any, sclerotia were produced on a trehalose-salt medium, 65% of sclerotial carbon originating from glucose. This indicates a selective utilization of glucose for sclerotial formation. It was suggested, therefore, that external trehalose is directed to the mycelium and converted to other cell constituents (106).

Cooke (48, 50) studied the changes in soluble carbohydrates in sclerotia of *S. sclerotiorum* and *S. trifoliorum* during their development. He found that carbohydrate excretion is correlated with a decrease in trehalose and glucose, and suggested that the two tested fungi had a similar carbohydrate metabolism. *S. sclerotiorum* was also found to produce nonvolatile organic acids. The interrelationship between organic acid production, pH changes in the medium, and sclerotial formation is unknown yet (151).

PROTEIN SYNTHESIS AND ENZYME ACTIVITY   Cycloheximide applied to the colony margins of *S. rolfsii* inhibited protein synthesis and prevented sclerotial formation. On the other hand, lactose, which induced synchronous sclerotial formation, also enhanced preferential incorporation of ^{14}C-leucine into trichloroacetic acid-insoluble fraction, with a concomitant decrease in the amino acid pool, indicating an increase in protein synthesis before the appearance of sclerotial initials (126). Significant changes in total soluble proteins were also found during sclerotial formation as detected by gel isoelectric focusing (38). These experiments reveal that protein synthesis is essential for formation of sclerotia in *S. rolfsii.* Both structural proteins and enzymes are probably degraded and synthesized during differentiation. There are very few reports on the changes in structural proteins during sclerotial formation. Working on the cell wall of *S. rolfsii,* Chet et al (35) found that acid hydrolysates of sclerotial walls contained only 10 amino acids, whereas 13 were found in hyphal cell walls, L-arginine, L-serine, and L-proline being absent from sclerotial walls. Jones et al (92) compared sclerotial walls with hyphal walls in several plant pathogenic fungi, using X-ray diffraction and infrared absorption spectroscopy, but most of the results are related to components other than proteins.

Several workers studied the changes in enzymes during the morphogenetic process of sclerotial formation. The most studied ones include polyphenoloxidases, peroxidases, esterases, and some enzymes of carbohydrate metabolism.

Polyphenoloxidase and peroxidases   Fungal sclerotia are coated with a pigmented rind which, in most cases, contains melanin. Melanin is synthesized from phenols by polyphenoloxidase (PPO) and catecholperoxidase. In *Verticillium,* factors that inhibit or stimulate melanin synthesis usually affect microsclerotial development in the same way. One such factor is near-UV radiation that inhibits both melanin synthesis and microsclerotial development (18). The addition of melanin precursors such as catechol, reversed the near UV effect (19, 114). Brandt (19) suggested that light prevents the synthesis of melanin precursors rather than suppressing the melanin-initiating enzymes. Chet & Henis (31) found that Na_2EDTA caused the formation of white sclerotia. This inhibition of melanogenesis could be reversed by the addition of Cu^{2+}, suggesting that copper ions are essential for PPO activity in *S. rolfsii.* Addition of catechol to the growth medium induced melanin synthesis also in hyphal walls of *S. rolfsii,* thus increasing the resistance of the fungus to lytic enzymes (32). Isoenzyme patterns of PPO and peroxidase were studied in the normal mycelium of *S. rolfsii* and in three stages of sclerotial formation (38), using the gel isoelectric focusing technique. On the other hand, the appearance of new PPO isoenzymes during sclerotial formation, all capable of oxidizing 3,4-dihydrophenylalanine, catechol, and tyrosine, could be detected at early developmental stages. It seems, however, that their presence does not necessarily mean that melanogenesis will occur, unless their substrates reach a minimal level. Only one isoenzyme of peroxidase was found in gels of mature sclerotia but not in gels of other stages.

Esterases   Chet et al (38), who studied the patterns of esterase isoenzymes during sclerotial morphogenesis of *S. rolfsii,* found that gels of all developmental stages showed esterase activity, but marked differences were noticed in their isoenzyme patterns. Of the four isoenzymes appearing in the mycelium, only one was present throughout development and maturation initiation. The physiological significance of these changes is unknown yet, but the isoenzyme patterns can perhaps be used as indicators of differentiation before the morphological alterations.

Enzymes involved in carbohydrate metabolism   Cell-free extracts of mycelia and sclerotia of *S. sclerotiorum* contained D-mannitol-l-phosphate:NAD oxidoreductase and weak D-mannitol:NADP oxidoreductase activities. Accordingly, the major pathway for mannitol synthesis was determined (155). Trehalase degrades trehalose, which is a reserve carbohydrate, during growth and sclerotium formation of

S. sclerotiorum (154). This enzyme was partially inhibited by mannitol. However, mannitol did not inhibit trehalase of *Schizophyllum commune* (168). When *S. sclerotiorum* was grown on a liquid medium with an aldopentose as the sole carbon source, the corresponding pentitol as well as pentitol-oxidoreductase activity were found in both sclerotia and mycelia (156). Recently, Wong & Willetts (169) found that succinate dehydrogenase (SDH) and glucose-6-phosphate dehydrogenase (Glu-6-PDH) from *S. sclerotiorum* were moderately active in submerged mycelium while in nonsclerotial aerial mycelium, arylesterase and acid phosphatase were very active. In sclerotial initials, glyceraldehyde-3-phosphate dehydrogenase (Gly-3-PDH) and SDH were at their highest level of activity. Glu-6-PDH and phosphogluconate dehydrogenase (PGDH) were moderately active. In young compacting sclerotia, the activities of Glu-6-PDH and PGDH increased, while Gly-3-PDH and SDH showed lowered activities. Suppression of the glycolytic and the Krebs-cycle pathways and the stimulation of the pentose phosphate pathway seem important during the compaction and maturation of sclerotia.

CHANGES IN RNA   No information is available concerning the changes in ribonucleic acid during sclerotial formation in filamentous fungi. We have just started studying this subject in our laboratory and found a very active incorporation of ^{3}H-uridine into RNA of the developing sclerotia, indicating a synthesis of new RNA during the morphogenetic process.

Control Mechanisms

The control mechanisms regulating the processes of sclerotial initiation and development in fungi are unknown. Although in the slime mold *P. polycephalum* the need for transcription during differentiation was proved (41, 42), the trigger "factors" that initiate and regulate this process are still obscure.

Attempts have been made by many authors to reveal these early trigger events in filamentous fungi. Henis & Chet (79) suggested that some unknown factors are supplied to the developing sclerotia in *S. rolfsii* by the surrounding mycelium. The existence of morphogenetic factor(s) which trigger the initiation of sclerotia in *S. rolfsii* and microsclerotia in *V. albo-atrum* was suggested by Goujon (67) and Brandt (18, 19), respectively. Brandt (18) suggested that microsclerotial induction results from the inhibiting effect of the morphogenetic substance(s) on mycelial growth, thus switching the metabolism from hyphal extension to microsclerotial formation. According to Goujon (67) the function of this factor(s) is to inhibit growth of the leading hyphae and induce the development of lateral hyphae, mycelial aggregation, and initials formation.

Sclerotial formation in *S. rolfsii* was inhibited by sulfur-containing amino acids and induced by –SH antagonists (34). Induction was limited by hyphal tips in the periphery of the fungal colony, the sclerotia forming a circle around the inoculation point. Chet & Henis (31) found that Na_2EDTA induced formation of sclerotia in concentric circles around the inoculum, this effect being completely prevented only by Cu^{2+}. A similar phenomenon was observed with potassium iodate, indicating a possible role of oxidation processes. On the basis of this observation, they suggested

that iodoacetic acid, chelating agents, and potassium iodate induce sclerotial formation in *S. rolfsii* by modifying a sulfhydryl-containing copper-like protein entity
which acts as a repressor of sclerotial formation. From further studies of this group,
as well as of Wong & Willetts (169) and Trevethick & Cooke (148), it seems that
these compounds interfere nonspecifically with some biochemical pathways and
indirectly stimulate or inhibit sclerotial production.

The importance of internal translocation in fungal morphogenesis was emphasized by Zalokar (172) and Watkinson (157, 158). Okon et al (126) suggested that
the enhanced translocation, with the concomitant partial inhibition of mycelial
linear growth, could be the cause of the formation of a circle of sclerotia at the
colony margins of *S. rolfsii*. Translocation of various substances was significantly
enhanced in a lactose-supplemented medium. In contrast to lactose, ethanol drastically decreased translocation of the compounds tested with only a slight inhibition
of linear growth. In addition to its effect on translocation, lactose induced a preferential incorporation of ^{14}C-leucine into trichloroacetic acid-insoluble material at the
colony margins, with a concomitant decrease in the free amino acid pool, indicating
an increase in protein synthesis before the appearance of sclerotial initials at this site.
Application of cycloheximide to colony margins prevented sclerotial formation but
not the translocation of radioactive compounds, either in the presence or absence
of lactose. It was therefore concluded that lactose stimulates amino acid translocation and incorporation independently, both processes being involved in sclerotial
formation.

Conditions that favor evaporation also enhance translocation (166). No information is available concerning the relationship between evaporation and sclerotial
formation. The relationship between translocation and the degree of lateral branching at colony margins of *S. rolfsii* suggests that the translocation activity in *S. rolfsii*
depends on the number of lateral hyphal tips functioning as "active metabolic
pumps." A higher rate of translocation in *S. rolfsii* colonies and preferential uptake
of metabolites by sclerotia initials (126) and of ^{32}P by sclerotia (163) possibly reflect
extensive branching at these sites, and may be a secondary, rather than a primary,
factor in sclerotial formation.

Recent work by Henis et al (82) indicates that substances that preferentially
inhibit development of lateral branches, such as ethanol and acetate, also inhibit
sclerotial formation, whereas threonine and lactose, which increase the proportion
of lateral hyphae, also induce sclerotial production. Although branching seems to
be related to sclerotial formation in *S. rolfsii* the "efficiency" of the mycelium to
produce sclerotia may be different in various types (4, 33). However, no differences
in branching patterns at the colony margins could be observed between two types
of *S. rolfsii* which differ in their ability to produce sclerotia.

The relationship between branching and sclerotial formation may be considered
as a part of the basic problem of branching control in fungi. If apical dominance
by hormone-like substances is involved in repression of lateral branching (24), a
similar regulation mechanism may function in the control of sclerotial formation
(31). Hypothetical repressors of sclerotial formation may be produced by the growing hyphal tip and move backwards. On the other hand, it is possible that trophic

competition between leading and lateral hyphae also plays a role in sclerotial initiation (160). Wang & LeTourneau (154) found that amino acids that are closely related to the tricarboxylic acid cycle are most suitable for supporting growth and sclerotial formation in *S. sclerotiorum*.

Recently, Kritzman, Okon, Henis, and Chet (unpublished) found that L-alanine and L-threonine, which induced formation of sclerotia in *S. rolfsii*, also increased the activity of isocitrate lyase but did not change the activity of glucose-6-phosphate dehydrogenase and SDH. Sodium oxalate inhibited the stimulatory effect of L-threonine on sclerotial formation as well as the isocitrate lyase activity. Oxalate inhibition of this enzyme has already been noted by Maxwell & Bateman (111).

It seems possible that the glyoxylate cycle is involved in sclerotial formation, and favors also hyphal branching by increasing the efficiency of the citric acid cycle that supplies both energy and amino acid, and in the synthesis of purine and pyrimidine precursors which are required for the synthesis of macromolecules during sclerotial formation. It seems that the genetic information for sclerotium production is present in every fungal cell, its expression being regulated by mechanisms, yet unknown, on the transcription and/or on the translation level. Because of the genetic complexity of most sclerotia-forming fungi, future studies on the molecular basis of this control call for the use of genetically pure strains, obtained from monokaryotic cells.

Literature Cited

1. Abeygunawardena, D. V. W., Wood, R. K. S. 1957. Factors affecting the germination of sclerotia and mycelial growth of *Sclerotium rolfsii* Sacc. *Trans. Br. Mycol. Soc.* 40:221–31
2. Adler, J. 1966. Chemotaxis in bacteria. *Science* 153:708–16
3. Avizohar-Hershenzon, Z., Palti, J. 1958. Small scale soil fumigation trials for the control of *Sclerotium rolfsii*. *Ktavim Engl. Ed.* 9:147–51
4. Aycock, R. 1966. Stem rot and other diseases caused by *Sclerotium rolfsii*. *N.C. Agric. Exp. Stn. Tech. Bull. No. 174.* 202 pp.
5. Baldwin, H. H., Rusch, H. P. 1965. The chemistry of differentiation in lower organisms. *Ann. Rev. Biochem.* 34:565–94
6. Bartnicki-Garcia, S. 1973. Fundamental aspects of hyphal morphogenesis. In *Microbial Differentiation*, ed. J. M. Ashworth, J. E. Smith, 245–67. London: Cambridge Univ. Press
7. Batra, K. K., Nordin, J. H., Kirkwood, S. 1969. Biosynthesis of β-D-glucan of *Sclerotium rolfsii* Sacc., direction of chain propagation and the insertion of the branch residues. *Carbohydr. Res.* 9:221–29
8. Bedi, K. S. 1956. Studies on *Sclerotinia sclerotiorum* (Lib.) De Bary. Part I. Some chemical factors affecting the formation of sclerotia. *Proc. Nat. Acad. Sci. India Sect. B.* 26:112–30
9. Bedi, K. S. 1958. The role of stale products in the formation of sclerotia of *Sclerotinia sclerotiorum* (Lib.) De Bary. *Indian Phytopathol.* 2:29–36
10. Bertus, L. S. 1927. A sclerotial disease of groundnuts caused by *Sclerotium rolfsii* Sacc. *Ceylon Dep. Agric. Yearb.* 41–43
11. Binkerhoff, L. A. 1969. The influence of temperature, aeration and soil microflora on microsclerotial development of *Verticillium albo-atrum* in abscised cotton leaves. *Phytopathology* 59:805–8
12. Bitancourt, A. A. 1951. Stimulation of growth of *Phytophthora citrophthora* by gas produced by *Mucor spinosus*. *Science* 113:531
13. Bommer, C. 1896. Sclerotes et condons myceliens. *Mem. Acad. R. Sci. Belg.* 54:1–116
14. Bonner, J. T., Hall, E. M., Sachsenmaier, W., Walker, B. K. 1970. Evidence for a second chemotactic system in the cellular slime mold, *Dictyostelium discoideum*. *J. Bacteriol.* 102:682–87
15. Boyle, L. W. 1961. Symposium on *Sclerotium rolfsii*: the ecology of *Sclerotium rolfsii* with emphasis on the role of saprophyric media. *Phytopathology* 51:117–19

16. Brachet, J. 1964. The role of nucleic acids and sulphydryl groups in morphogenesis. *Adv. Morphog.* 3:247–300

17. Brandt, W. H. 1964. Morphogenesis in Verticillium. A self-induced, nonhereditary variation in colony form. *Am. J. Bot.* 51:820–24

18. Brandt, W. H. 1964. Morphogenesis in Verticillium: Effect of light and ultraviolet radiation on microsclerotia and melanin. *Can. J. Bot.* 42:1017–23

19. Brandt, W. H. 1965. Morphogenesis in *Verticillium*: reversal of the near-UV effect by catechol. *BioScience* 15: 669–70

20. Brandt, W. H., Reese, J. E. 1964. Morphogenesis in Verticillium: a self-produced, diffusible morphogenetic factor. *Am. J. Bot.* 51:922-27

21. Braun, R., Behrens, K. 1969. A ribonuclease from Physarum. Biochemical properties and synthesis in the mitotic cycle. *Biochim. Biophys. Acta* 182: 511–22

22. Brefeld, O. 1877. *Botanische Untersuchungen über Schimmelpilze.* Vol. III. Leipzig: Felix

23. Buller, A. H. R. 1933. *Researches in Fungi.* Vol. 5. New York: Longmans, Green. 416 pp.

24. Burnett, J. H. 1968. *Fundamentals of Mycology.* London: Arnold

25. Butler, G. M. 1966. Vegetative Structures. In *The Fungi,* ed. G. C. Ainsworth, S. Alfreds Sussman, 2:83–112. New York & London: Academic

26. Carlile, M. J. 1956. A study of the factors influencing non-genetic variation in a strain of *Fusarium oxysporum. J. Gen. Microbiol.* 14:643–54

27. Carlile, M. J. 1970. Nutrition and chemotaxis in the Myxomycete *Physarum polycephalum.* The effect of carbohydrates on the plasmodium. *J. Gen. Microbiol.* 63:221–26

28. Chet, I. 1967. *The structure and behaviour of the fungus Sclerotium rolfsii Sacc.* PhD thesis. The Hebrew University of Jerusalem. 121 pp.

29. Chet, I. 1973. Changes in ribonucleic acid during differentiation of *Physarum polycephalum. Ber. Dtsch. Bot. Ges.* 86:77–92

30. Chet, I., Henis, Y. 1967. X-ray analysis of hyphal and sclerotial walls of *Sclerotium rolfsii* Sacc. *Can. J. Microbiol.* 14:815–16

31. Chet, I., Henis, Y. 1968. The control mechanism of sclerotial formation in *Sclerotium rolfsii* Sacc. *J. Gen. Microbiol.* 54:231–36

32. Chet, I., Henis, Y. 1969. Effect of catechol and disodium EDTA on melanin content of hyphal and sclerotial walls of *Sclerotium rolfsii* Sacc. and the role of melanin in the susceptibility of these walls to β-(1–3)-glucanase and chitinase. *Soil Biol. Biochem.* 1:131–38

33. Chet, I., Henis, Y. 1972. The response of two types of *Sclerotium rolfsii* to factors affecting sclerotium formation. *J. Gen. Microbiol.* 73:483–86

34. Chet, I., Henis, Y., Mitchell, R. 1966. The morphogenetic effect of sulphur-containing amino acids, glutathione and iodoacetic acid on *Sclerotium rolfsii* Sacc. *J. Gen. Microbiol.* 45:541–46

35. Chet, I., Henis, Y., Mitchell, R. 1967. Chemical composition of hyphal and sclerotial walls of *Sclerotium rolfsii* Sacc. *Can. J. Microbiol.* 13:137–41

36. Chet, I., Kislev, N. 1974. Scanning electron microscopy of spherules of *Physarum polycephalum. Tissue Cell* 5: 545–51

37. Chet, I., Kislev, N., Henis, Y. 1969. Ultrastructure of sclerotia and hyphae of *Sclerotium rolfsii* Sacc. *J. Gen. Microbiol.* 57:143–47

38. Chet, I., Retig, N., Henis, Y. 1972. Changes in total soluble protein and in some enzymes during morphogenesis of *Sclerotium rolfsii* as detected by the gel isoelectric focusing technique. *J. Gen. Microbiol.* 72:451–56

39. Chet, I., Retig, N., Henis, Y. 1973. Changes in ribonucleases during differentiation (spherulation) of *Physarum polycephalum. Biochim. Biophys. Acta* 294:343–47

40. Chet, I., Rusch, H. P. 1969. Induction of spherule formation in *Physarum polycephalum. J. Bacteriol.* 100:673–78

41. Chet, I., Rusch, H. P. 1970. Differences between hybridizable RNA during growth and differentiation of *Physarum polycephalum. Biochim. Biophys. Acta* 213:478–83

42. Chet, I., Rusch, H. P. 1970. RNA differences between spherulating and growing microplasmodia of *Physarum polycephalum* as revealed by sedimentation patterns and DNA-RNA hybridization. *Biochim. Biophys. Acta* 209: 559–68

43. Chowdhury, S. 1946. Effect of hydrogen-ion concentration on the growth and parasitism of *Sclerotium rolfsii* Sacc. *Indian J. Agric. Sci.* 16:290–93

44. Chowdhury, S. 1948. Disease of Pan (Piper Beetle) in Sylhet, Assam. Part VIII. Effect of temperature on the de-

188   CHET & HENIS

velopment of sclerotial wilt of Pan. *Proc. Indian Acad. Sci.* Sect. B 28:240–46

45. Coley-Smith, J. R., Cooke, R. C. 1971. Survival and germination of fungal sclerotia. *Ann. Rev. Phytopathol.* 9:65–92

46. Colotelo, N., Sumner, J. L., Voegelin, W. S. 1971. Presence of sacs enveloping the liquid droplets on developing sclerotia of *Sclerotinia sclerotiorum* (Lib.) De Bary. *Can. J. Microbiol.* 17:300–1

47. Colotelo, N., Sumner, J. L., Voegelin, W. S. 1971. Chemical studies on the exudate and developing sclerotia of *Sclerotinia sclerotiorum* (Lib.) De Bary. *Can. J. Microbiol.* 17:1189–94

48. Cooke, R. C. 1969. Changes in soluble carbohydrates during sclerotium formation by *Sclerotinia sclerotiorum* and *S. trifoliorum*. *Trans. Br. Mycol. Soc.* 53:77–86

49. Cooke, R. C. 1970. Physiological aspects of sclerotium growth in *Sclerotinia sclerotiorum*. *Trans. Br. Mycol. Soc.* 54:364–65

50. Cooke, R. C. 1971. Physiology of sclerotia of *Sclerotinia sclerotiorum* during growth and maturation. *Trans. Br. Mycol. Soc.* 56:51–59

51. Corner, E. J. H. 1950. *A Monograph of Clavaria and Allied Genera.* Oxford Univ. Press: London & New York

52. Daniel, J. W., Rusch, H. P. 1961. The pure culture of *Physarum polycephalum* on a partially defined soluble medium. *J. Gen. Microbiol.* 25:47–59

53. Daniel, J. W., Babcock, K. L., Sievert, A. H., Rusch, H. P. 1963. Organic requirements and synthetic media for growth of the myxomycete *Physarum polycephalum. J. Bacteriol.* 86:324–31

54. Davey, A. E., Leach, L. D. 1941. Experiments with fungicides for use against *Sclerotium rolfsii* in soils. *Hilgardia* 13:523–47

55. De Bary, A. 1887. *Comparative Morphology and Biology of Fungi, Mycetozoa and Bacteria.* London & New York: Oxford Clarenden. 525 pp.

56. Dunlap, A. A. 1943. Inhibition of *Phymatotrichum* sclerotia formation by sulphur autoclaved with soil. *Phytopathology* 33:1205–8

57. Garrett, S. D. 1970. *Pathogenic Root Infecting Fungi.* London: Cambridge Univ. Press. 294 pp.

58. Geeson, J. D. 1972. *Factors affecting the production of sclerotia by plant pathogenic fungi.* PhD thesis. Cambridge Univ.

59. Geiger, J. P., Goujon, M. 1970. Mise en evidence dans les extracts de thalle d'un facteur morphogene responsable de l'apparition des sclerotes du *Corticium rolfsii* Sacc. *C. R. Acad. Sci. Paris Ser. D* 271:41–44

60. Goodman, E. M., Beck, T. 1974. Metabolism during differentiation in the slime mold *Physarum polycephalum. Can. J. Microbiol.* 20:107–11

61. Goodman, E. M., Rusch, H. P. 1970. Ultrastructural changes during spherule formation in *Physarum polycephalum. J. Ultrastruct. Res.* 30:172–83

62. Goodman, E. M., Sauer, H. W., Sauer, L., Rusch, H. P. 1969. Polyphosphate and other phorphorus compounds during growth and differentiation of *Physarum polycephalum. Can. J. Microbiol.* 15:1325–31

63. Goto, K. 1952. *Sclerotium rolfsii* Sacc. in perfect stage. VII. Germination of basidiospores. *Ann. Phytopathol. Soc. Jpn.* 16:132–36

64. Goto, K. 1952. *Sclerotium rolfsii* in perfect stage, especially in reference to genetics, morphology in primary state, relation of environments to sporulation and the significance of the stage for the fungus. *Tokai Kinki Nat. Agric. Exp. Stn. Spec. Bull. No. 1.* 82 pp.

65. Goujon, M. 1968. Morphogenese— Mise en evidence dans le mycelium du *Corticium rolfsii* (Sacc.). Curzi d'un facteur morphogenetique responsable de l'apparition des sclerotes. *C. R. Acad. Sci. Paris* 267:409–11

66. Goujon, M. 1969. Etude comparative du *C. rolfsii* (Sacc.) curzi et du *S. coffeicolum* Stahl. *Cah. ORSTOM Ser. Biol.* 7:69–87

67. Goujon, M. 1970. Mecanismes physiologiques de la formation des sclerotes chez le *Corticium rolfsii* (Sacc.) curzi. *Physiol. Veg.* 8:349–60

68. Grant, W. D. 1973. RNA synthesis during the cell cycle in *Physarum polycephalum.* In *The Cell Cycle in Development and Differentiation,* ed. M. Balls, F. S. Billett, 77–109. London: Cambridge Univ. Press

69. Gray, W. D., Alexopoulos, C. J. 1968. *Biology of Myxomycetes.* New York: Ronald

70. Gross, P. R. 1968. Biochemistry of differentiation. *Ann. Rev. Biochem.* 37:631–55

71. Gruelach, V. A., Mohr, H. C. 1947. Greenhouse experiments with the southern blight disease of peanuts. *Texas Agric. Exp. Stn. Rept. 1097*

72. Guttes, E., Guttes, S. 1963. Starvation and cell wall formation in the myxomycete *Physarum polycephalum. Ann. Bot. N.S.* 27:49–53

73. Guttes, E., Guttes, S., Rusch, H. P. 1961. Morphological observation on growth and differentiation of *Physarum polycephalum* growth in pure culture. *Dev. Biol.* 3:588–614

74. Hawker, L. E. 1957. *The Physiology of Reproduction in Fungi.* London & New York: Cambridge Univ. Press

75. Heale, J. B., Isaac, I. 1965. Environmental factors in the production of dark resting structures in *Verticillium alboatrum, V. dahliae* and *V. tricorpus. Trans. Br. Mycol. Soc.* 48:39–50

76. Heath, L. A. F., Eggins, H. O. W. 1965. Effect of light, temperature and nutrients on the production of conidia and sclerotia by forms of *Aspergillus japonicus. Experientia* 21:385–86

77. Melhuish, J. H. Jr., Bean, G. A. 1971. Effect of dimethyl sulfoxide on the sclerotia of *Sclerotium rolfsii. Can. J. Microbiol.* 17:429–31

78. Hemmi, T., Endo, S. 1931. Studies on sclerotium diseases of the rice plant. III. Some experiments on the sclerotial formation and the pathogenicity of certain fungi causing sclerotium disease of the rice plant. *Auf. Dem. Geb. Pflanzenkr. Kyoto* 1:111–25

79. Henis, Y., Chet, I. 1968. Developmental biology of sclerotia of *Sclerotium rolfsii. Can. J. Bot.* 46:947–48

80. Henis, Y., Inbar, M. 1968. Effect of *Bacillus subtilis* on growth and sclerotium formation by *Rhizoctonia solani. Phytopathology* 58:933–38

81. Henis, Y., Chet, I., Avizohar-Hershenzon, Z. 1965. Nutritional and mechanical factors involved in mycelial growth and production of sclerotia by *Sclerotium rolfsii* in artificial medium and amended soil. *Phytopathology* 55:87–91

82. Henis, Y., Okon, Y., Chet, I. 1973. The relationship between early hyphal branching and sclerotial formation in *Sclerotium rolfsii. J. Gen. Microbiol.* 79:147–50

83. Higgins, B. B. 1927. Physiology and parasitism of *Sclerotium rolfsii* (Sacc.). *Phytopathology* 17:417–48

84. Hiramaru, M., Uchida, T., Egami, E. 1969. Studies on ribonucleases from *Physarum polycephalum:* purification and characterization of substrate specification. *J. Biochem. Tokyo* 65:693–700

85. Hüttermann, A. 1972. Isoenzyme pattern and de novo synthesis of phosphodiesterase during differentiation (spherulation) in *Physarum polycephalum. Arch. Mikrobiol.* 83:155–64

86. Hüttermann, A. 1973. Biochemical events during spherules formation of *Physarum polycephalum. Ber. Dtsch. Bot. Ges.* 86:55–76

87. Hüttermann, A., Chet, I. 1971. Activity of some enzymes in *Physarum polycephalum.* III. During spherulation (differentiation) induced by mannitol. *Arch. Mikrobiol.* 78:189–92

88. Hüttermann, A., Elsevier, S. M., Eschrich, W. 1971. Evidence for the de novo synthesis of glutamate dehydrogenase during spherulation of *Physarum polycephalum. Arch. Mikrobiol.* 77:74–85

89. Hüttermann, A., Gebauer, M. 1973. Inorganic pyrophosphatase during differentiation (spherulation) of *Physarum polycephalum. Cytobiology* 316:1–5

90. Hüttermann, A., Porter, M. T., Rusch, H. P. 1970. Activity of some enzymes in *Physarum polycephalum.* II. During spherulation (differentiation). *Arch. Mikrobiol.* 74:283–91

91. Jones, D. 1970. Ultrastructure and composition of the cell walls of *Sclerotinia sclerotiorum. Trans. Br. Mycol. Soc.* 54:351–600

92. Jones, D., Farmer, V. C., Bacon, J. S., Wilson, M. J. 1972. Comparison of ultrastructure and chemical components of cell walls of certain plant pathogenic fungi. *Trans. Br. Mycol. Soc.* 59:11–23

93. Jump, J. A. 1954. Studies on sclerotization in *Physarum polycephalum. Am. J. Bot.* 41:561–67

94. Kaiser, W. J. 1964. Biotic requirement of the Verticillium wilt fungus. *Phytopathology* 54:481–87

95. Kaiser, W. J. 1964. Effect of light on growth and sporulation of the Verticillium wilt fungus. *Phytopathology* 54:765–70

96. Kislev, N., Chet, I. 1973. Scanning electron microscopy of sporulating cultures of the Myxomycete *Physarum polycephalum. Tissue Cell* 5:349–57

97. Kislev, N., Chet, I. 1974. Scanning electron microscopy of freeze fractured sclerotia of *Physarum polycephalum. Tissue Cell* 6:209–14

98. Kitahara, M. 1950. Chemical components of the sclerotia of *Sclerotinia libertiana.* III. Trehalose. *Gifu Coll. Agric.*

Res. Bull. 68:64–67 (*Chem. Abstr.* 46:5535 e)

99. Kitahara, M. 1950. IV. Carbohydrates in the extracts with several solvents. *Gifu Coll. Agric. Res. Bull.* 68:117–23 (*Chem. Abstr.* 46:5535 f)

100. Lavee, S. 1955. *Studies in the physiology and pathology of Sclerotium rolfsii Sacc. and apple resistance.* MS thesis. The Hebrew University of Jerusalem. (In Hebrew)

101. Lentz, P. L., McKay, H. H. 1970. Sclerotial formation in *Corticium olivascens* and other Hymenomycetes. *Mycopathol. Mycol. Appl.* 40:1–13

102. Les Stourgeon, W. M., Rusch, H. P. 1973. Localization of nucleolar and chromatin residual acidic protein changes during differentiation in *Physarum polycephalum. Arch. Biochem. Biophys.* 155:144–58

103. LeTourneau, D. 1966. Trehalose and acyclic polyols in sclerotia of *Sclerotinia sclerotiorum. Mycologia* 58:934–42

104. Littlefield, L. J. 1967. Phosphorus-32 accumulation in *Rhizoctonia solani* sclerotia. *Phytopathology* 57:153–55

105. Littlefield, L. J., Wilcoxson, R. D., Sudia, T. W. 1965. Translocation of phosphorus-32 in *Rhizoctonia solani. Phytopathology* 55:536–42

106. Liu, S. C., LeTourneau, D. 1970. Trehalose utilization and sclerotium formation in *Sclerotinia sclerotiorum. Phytopathology* 60:1535 (Abstr.)

107. Liu, T. M., Wu, L. C. 1971. The effect of amino acids on the growth and morphogenesis of *Sclerotium rolfsii* Sacc. *Plant Prot. Bull.* 13:87–96

108. Liu, T. M., Wu, L. C. 1971. Some observations on the sclerotial formation of *Sclerotium rolfsii* Sacc. *Mem. Coll. Agric. Nat. Taiwan Univ.* 12:59–66

109. Lynch, T. J., Henney, H. R. Jr. 1972. Carbohydrate metabolism during differentiation (sclerotization) of the myxomycete *Physarum flaveconum. Arch. Mikrobiol.* 90:189–98

110. Manning, W. J., Crossan, D. F. 1966. Effect of a particular soil bacterium on sporangial production in *Phytophthora cinnamoni* in liquid culture. *Phytopathology* 56:235–37

111. Maxwell, D. P., Bateman, D. F. 1968. Glucose catabolism in *Sclerotium rolfsii. Phytopathology* 58:1630–34

112. McClellan, W. D., Borthwick, H. A., Marshall, B. H. Jr. 1955. Some responses of fungi to light. *Phytopathology* 45:465 (Abstr.)

113. McCormick, J. J., Blomquist, J. C., Rusch, H. P. 1970. Isolation and characterization of a galactosamine wall from spores and spherules of *Physarum polycephalum. J. Bacteriol.* 104:1119

114. McMillan, P. C., Brandt, W. H. 1966. Possible role of a peroxidative system in melanin synthesis in *Verticillium. Antonie van Loeuwenhoek J. Microbiol. Serol.* 32:202–11

115. Mekhaimer, S. G. 1950. The determination and isolation of Delphinium species immune to *Sclerotium delphinii* Welch. *Proc. Egypt. Acad. Sci.* 6:37–44

116. Melera, P. W., Chet, I., Rusch, H. P. 1970. Electrophoretic characterization of ribosomal RNA from *Physarum polycephalum. Biochim. Biophys. Acta* 209:569–72

117. Milthorpe, F. L. 1941. Studies on *Corticium rolfsii* Sacc. Curzi (*Sclerotium rolfsii* Sacc.) I. Cultural characters and perfect stage. II. Mechanism of parasitism. *Proc. Linn. Soc. NSW* 66:65–75

118. Misra, A. P., Haque, S. Q. 1962. Factors affecting the growth and sclerotial production in *Sclerotium rolfsii* Sacc. causing storage rot of potato. *Proc. Indian Acad. Sci. Sect. A* 56:157–68

119. Morgan, T. H. 1934. *Embryology and Genetics.* New York: Columbia Univ. Press

120. Nadakarukaren, M. J., Horner, L. E. 1959. An alcohol agar medium selective for determining Verticillium microsclerotia in soil. *Phytopathology* 49:527–28

121. Newton, B. A. 1965. Mechanisms of antibiotic action. *Ann. Rev. Microbiol.* 19:209–40

122. Nickerson, W. J. 1963. Symposium on biochemical bases of morphogenesis in fungi. IV. Molecular bases of form in yeasts. *Bacteriol. Rev.* 27:305–24

123. Nickerson, W. J., Bartnicki-Garcia, S. 1964. Biochemical aspects of morphogenesis in algae and fungi. *Ann. Rev. Plant Physiol.* 15:327–44

124. Nickerson, W. J., Falcone, G. 1956. Identification of protein disulfide reductase as a cellular division enzyme in yeasts. *Science* 124:722–23

125. Okon, Y., Chet, I., Henis, Y. 1972. Lactose-induced synchronous sclerotium formation in *Sclerotium rolfsii* and its inhibition by ethanol. *J. Gen. Microbiol.* 71:465–70

126. Okon, Y., Chet, I., Henis, Y. 1973. Effects of lactose, ethanol and cycloheximide on the translocation pattern of radioactive compounds and on scler-

otium formation in *Sclerotium rolfsii. J. Gen. Microbiol.* 74:251–58

127. Okon, Y., Chet, I., Kislev, N., Henis, Y. 1974. Effect of lactose in soluble glucan production and on the ultrastructure of *Sclerotium rolfsii* Sacc. grown in submerged culture. *J. Gen. Microbiol.* 81:145–49

128. Peiris, J. W. L. 1947. *The Botrytis disease of Gladiolus, together with a physiological study of certain Botrytis species.* PhD thesis. Univ. of London

129. Pitt, D. 1969. Cytochemical observations on the localization of sulphydryl groups in budding yeast cells and in the phialides of *Penicillium notatum* Westling during conidiation. *J. Gen. Microbiol.* 59:257–62

130. Rai, J. N., Tewari, J. P., Sinha, A. K. 1967. Effect of environmental conditions on sclerotia and cleistothecia production in *Aspergillus. Mycopathol. Mycol. Appl.* 31:209–24

131. Remsberg, R. E. 1940. Studies in the genus Typhula. *Mycologia* 32:52–96

132. Rhea, R. P. 1966. Electron microscopic observations on the slime mold, *Physarum polycephalum* with special reference to fibrillar structures. *J. Ultrastruct. Res.* 15:349–79

133. Roth, J. N., Brandt, W. H. 1964. Some effects of vitamins on growth and differentiation in Verticillium. *Plant Dis. Reptr.* 48:649–52

134. Rudolph, E. D. 1962. The effect of some physiological and environmental factors on sclerotial Aspergilli. *Am. J. Bot.* 49:71–78

135. Rusch, H. P. 1969. Some biochemical events in the growth cycle of *Physarum polycephalum. Fed. Proc.* 28:1761–70

136. Sanford, G. B. 1956. Factors influencing formation of sclerotia by *Rhizoctonia solani. Phytopathology* 46:281–84

137. Sauer, H. W. 1973. Differentiation in *Physarum polycephalum.* See Ref. 6, 375–466

138. Sauer, H. W., Babcock, K. L., Rusch, H. P. 1969. Changes in RNA synthesis associated with differentiation (sporulation) in *Physarum polycephalum. Biochim. Biophys. Acta* 195:410–21

139. Sauer, H. W., Babcock, K. L., Rusch, H. P. 1970. Changes in nucleic acid and protein synthesis during starvation and spherule formation in *Physarum polycephalum. Wilhelm Roux Arch. Entwicklungsmech. Org.* 165:110–24

140. Smith, J. E., Galbraith, J. C. 1971. Biochemical and physiological aspects of

differentiation in the fungi. *Adv. Microbiol. Physiol.* 5:45–134

141. Sonneborn, D. R., Sussman, M., Levine, L. 1964. Serological analysis of cellular slime-mold development. I. Changes in antigenic activity during cell aggregation. *J. Bacteriol.* 87:1321–29

142. Stewart, P. A., Stewart, B. J. 1961. Membrane formation during sclerotization of *Physarum polycephalum* plasmodia. *Exp. Cell Res.* 23:471–78

143. Sussman, M. 1965. Development phenomena in microorganisms and in higher forms of life. *Ann. Rev. Microbiol.* 19:59–78

144. Tan, K. K., Epton, H. A. S. 1973. Effect of light on the growth and sporulation of *Botrytis cinerea. Trans. Br. Mycol. Soc.* 61:147–57

145. Tarurenko, E. 1954. The influence of light on the development of mould fungi. *Mikrobiologya* 23:29–33 (Abstr. in *Rev. Appl. Mycol.* 34:803)

146. Townsend, B. B. 1957. Nutritional factors influencing the production of sclerotia by certain fungi. *Ann. Bot. London NS* 21:153–66

147. Townsend, B. B., Willetts, H. J. 1954. The development of sclerotia of certain fungi. *Trans. Br. Mycol. Soc.* 37:213–21

148. Trevethick, J., Cooke, R. C. 1971. Effects of some metabolic inhibitors and sulphur-containing amino acids on sclerotium formation in *Sclerotium rolfsii, S. delphinii* and *Sclerotinia sclerotiorum. Trans. Br. Mycol. Soc.* 57:340–42

149. Trevethick, J., Cooke, R. C. 1973. Non nutritional factors influencing sclerotium formation in some *Sclerotinia* and *Sclerotium* species. *Trans. Br. Mycol. Soc.* 60:559–66

150. Tucker, C. M. 1934. Botany investigation with *Sclerotium delphinii. Mo. Agric. Exp. Stn. Bull.* 340:29–30

151. Vega, R. R., LeTourneau, D. 1970. The effect of zinc on the formation of sclerotia by *Sclerotinia sclerotiorum. Phytopathology* 60:1537 (Abstr.)

152. Wang, S. Y., LeTourneau, D. 1971. Carbon sources, growth, sclerotium formation and carbohydrate composition of *Sclerotinia sclerotiorum. Arch. Mikrobiol.* 80:219–33

153. Wang, S. Y., LeTourneau, D. 1972. Amino acids as nitrogen sources for growth and sclerotium formation in *Sclerotinia sclerotiorum. Trans. Br. Mycol. Soc.* 59:509–12

154. Wang, S. Y., LeTourneau, D. 1972. Trehalose from *Sclerotinia sclerotiorum. Arch. Mikrobiol.* 87:235–41

155. Wang, S. Y., LeTourneau, D. 1972. Mannitol biosynthesis in *Sclerotinia sclerotiorum. Arch. Mikrobiol.* 81:91–99
156. Wang, S. Y., LeTourneau, D. 1973. Pentitol oxidoreductase in *Sclerotinia sclerotiorum. Arch. Mikrobiol.* 93:87–90
157. Watkinson, S. C. 1971. The mechanism of mycelial strand induction in *Serpula lacrimans*, a possible effect on nutrient distribution. *New Phytol.* 70:1079–88
158. Watkinson, S. C. 1971. Phosphorus translocation in the stranded mycelium of *Serpula lacrimans. Trans. Br. Mycol. Soc.* 57:535–39
159. Wheeler, B. E. J. 1972. Effect of ethanol on production of sclerotia by *Sclerotium rolfsii. Trans Br. Mycol. Soc.* 59:453–61
160. Wheeler, B. E. J., Sharan, N. 1965. The production of sclerotia by *Sclerotium rolfsii.* I. Effect of varying the supply of nutrients in an agar medium. *Trans. Br. Mycol. Soc.* 48:291–301
161. Wheeler, B. E. J., Waller, J. M. 1965. The production of sclerotia by *Sclerotium rolfsii.* II. The relationship between mycelial growth and initiation of sclerotia. *Trans. Br. Mycol. Soc.* 48:303–14
162. Whetzel, H. H. 1945. A synopsis of the genera and species of the Sclerotiniaceae, a family of stromatic inoperculate discomycetes. *Mycologia* 37:648–714
163. Wilcoxson, R. D., Subbarayudu, S. 1968. Translocation and accumulation of phosphorus-32 in sclerotia of

Sclerotinium rolfsii. Can J. Bot. 46:85–88
164. Wilhelm, S. 1948. The effect of temperature on the taxonomic characters of *V. albo-atrum. Phytopathology* 38:919 (Abstr.)
165. Willetts, H. J. 1968. The development of stomata of *Sclerotinia fructicola* and related species. *Trans. Br. Mycol. Soc.* 51:625–32
166. Willetts, H. J. 1972. The morphogenesis and possible evolutionary origins of fungal sclerotia. *Biol. Rev.* 47:515–36
167. Willetts, H. J., Calonge, F. D. 1969. Spore development in the brown rot fungi (*Sclerotinia* spp.). *New Phytol.* 68:123–31
168. Williams, C. F., Niederpruem, D. J. 1968. Trehalase in *Schizophyllum commune. Arch. Mikrobiol.* 60:377–83
169. Wong, A. L., Willetts, H. J. 1973. Electrophoretic studies of soluble proteins and enzymes of Sclerotinia species. *Trans. Br. Mycol. Soc.* 61:167–78
170. Wong, A. L., Willetts, H. J. 1974. Polyacrylamide-gel electrophoresis of enzymes during morphogenesis of sclerotia of *Sclerotinia sclerotiorum. J. Gen. Microbiol.* 81:101–9
171. Wyllie, T. D., DeVay, J. E. 1970. Growth characteristics of several isolates of *Verticillium albo-atrum* and *Verticillium nigrescens* from cotton. *Phytopathology* 60:907–10
172. Zalokar, M. 1959. Growth and differentiation of *Neurospora* hyphae. *Am. J. Bot.* 45:602–10

PREDISPOSITION, STRESS, AND PLANT DISEASE

❖3619

Donald F. Schoeneweiss
Section of Botany and Plant Pathology, Illinois Natural History Survey,
Urbana, Illinois 61801

INTRODUCTION

Most species of higher plants are either immune or resistant to attack by the majority of microorganisms with which they come in contact (19, 31). Although penetration or invasion of plant surfaces is occasionally prevented by physical or chemical barriers, in most cases the organisms enter resistant and susceptible hosts with equal frequency (15, 19, 31, 39, 88, 89). Whether a disease condition develops depends upon the influence of environmental factors on the genetically controlled response of the host plant to the presence of the pathogen or its metabolites. The tendency of nongenetic factors, acting prior to infection, to affect the susceptibility of plants to disease is called predisposition (100).

The concept of predisposition was first introduced into the field of plant pathology by Sorauer (81), who clearly recognized the importance of environmental factors in relation to plant diseases. Hartig (38) also realized the significance of the environment and expanded the concept of predisposition to include many internal and external factors, such as proximity of inoculum to the host. Ward (90) was a strong proponent of predisposition, although he excluded inherited variation from his concept and was primarily concerned with the influence of environment. Gäumann (31) and others have also recognized the importance of predisposition and have defined the term in various ways. For additional historical perspective on concepts of predisposition, the reader is referred to reviews by Colhoun (20) and Yarwood (100).

Yarwood (100) used the term predisposition to denote "an internal degree of susceptibility resulting from external causes." This definition is somewhat analogous to Gäumann's "disease proneness" (31), and the "abnormal predisposition" of Sorauer (81) and Hartig (38). Yarwood's concept includes induced changes toward greater or lesser susceptibility. The present discussion is concerned with the influence of stresses as predisposing factors in plant disease; consequently the emphasis is placed on those factors that predispose plants toward greater susceptibility.

193

Stress may influence plant disease through effect on the pathogen, effect on host susceptibility, or effect on the host-pathogen interaction. Predisposition implies an effect on the host rather than on the pathogen, although effects of stresses on the host-pathogen interaction are often difficult to separate from effects on the host itself (100).

Levitt (54) defines biological stress as "any environmental factor capable of inducing a potentially injurious strain in living organisms." He also separates strain caused by stress into "plastic strain," which results in an irreversible physical or chemical change, and "elastic strain," in which physical or chemical changes are reversible by removal of the stress. Although these terms are somewhat ambiguous and are derived from mechanics (54), they serve a useful purpose for the present discussion because both types of strain commonly occur in plants and may have an effect on disease susceptibility.

According to Ward's concept of predisposition (90), any disturbance in the "normal" state may predispose plants to disease; therefore, any deviation from environmental conditions optimum for expression of disease resistance could be considered stress. In this discussion, only extensive deviations from normal such as drought, flooding, freezing, defoliation, and transplanting "shock" will be considered stresses. In some cases a plastic strain may be produced; more commonly, however, an elastic strain results and plants normally recover without ill effects in the absence of pathogens (6).

The most common effects of stress on infectious diseases are illustrated in Figure 1. As in all biological systems, living organisms seldom follow man-made rules and exceptions occur to the pathways illustrated. Invasion of the host plant may occur before or after exposure to stress but the end result is usually the same. Therefore, the placement of the term invasion in the illustration is merely for convenience. Strain resulting from stress is separated into plastic strain and elastic strain only in the case of nonaggressive facultative parasites, because removal of an elastic strain has the most pronounced effect on disease susceptibility where nonaggressive or "weak" parasites are involved; these parasites seldom cause disease in nonstressed hosts. The relative potential of facultative parasites to colonize host tissues is called aggressiveness, in preference to virulence or pathogenicity, as suggested by Gäumann (31). Stresses that cause stomatal closure or formation of thicker cuticle may prevent invasion by pathogens, particularly obligate parasites (31). In most cases, however, pathogens enter regardless of stresses; therefore, the influence of stresses on disease susceptibility is usually on disease development (22, 89). Although a large volume of work has been done on substances which are present in plants prior to infection and are toxic or inhibitory to pathogens, it has seldom been demonstrated that resistance operates through this simple mechanism (99). According to Allen (2), "It is not a condition of the plant which constitutes resistance, but a process of response."

Sorauer (81) long ago recognized that disease resistance in plants is only relative and that plants constantly face conditions of predisposition and immunity during their normal course of development. Yet, in spite of the fact that the predisposition concept is nearly a century old, it is surprising how little work has been accom-

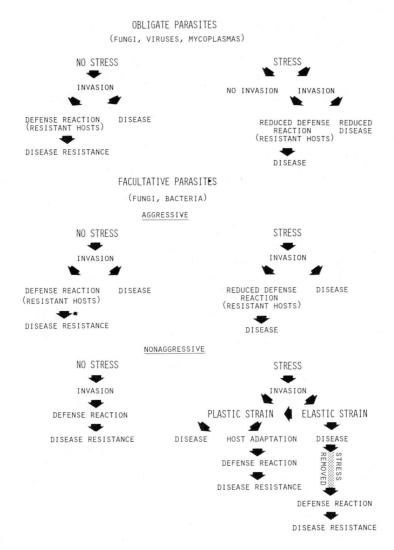

Figure 1   Common effects of stress on infectious diseases caused by obligate and facultative parasites.

plished on the influence of common stresses on disease susceptibility. Only in recent years, with new pesticide laws and increasing concern over man's impact on the environment, have pathologists begun to take a closer look at how and why stresses affect the susceptibility of plants to pathogens. Space limitations of this paper permit only a brief discussion of the accumulated knowledge on stress effects, and emphasis is placed on results of current research on individual stress factors. Host vigor is not treated separately, because all factors that increase the disease proneness of the host do so ultimately through alteration of host vigor or vitality (31).

WATER STRESS

Water Deficits and Drought

Variations in precipitation and the availability of moisture for plant growth occur from year to year throughout the world, except in areas where artificial irrigation is a regular practice. Major changes in climate over a period of years have been implicated as stress factors affecting the incidence and severity of many diseases. Ash dieback, maple decline, sweetgum blight, birch dieback, oak decline, dry face of slash pine, and pitch streak of slash pines have been associated with an extended period of below normal precipitation in the 1930s in the United States (44, 70). Although the relative importance of disease organisms in many of these diseases was not clearly stated, host predisposition to pathogens was one of the major factors involved in disease development. Leaphart & Stage (53) concluded that adverse growing conditions, particularly extended drought, during the period from 1916 to 1940 in the United States, was instrumental in the origin and severity of pole blight of western white pine. In this case, root-infecting organisms, combined with drought, reduced the root-regenerating capacity of the host and ultimately resulted in pole blight.

Short-term droughts of days or weeks during the growing season may also predispose plants to disease. Unfortunately, most reports on predisposition due to water stress are based on field observations and are not substantiated by experimental evidence (26). Perhaps the main reason for this lack of experimental data involves the difficulties and limitations of controlling the water potential of soil. Means to surmount these difficulties have been summarized by Couch et al (25) and by Cook & Papendick (22). The most common method of exposing plants to controlled water stress is by varying watering regimes (25). Couch & Bloom (24) and Moore et al (60) showed that water stress predisposed Kent bluegrass to *Sclerotinia homeocarpa* (24), and Highland bentgrass to *Pythium ultimum* (60). By exposing host plants to controlled water stress, increased susceptibility was reported in crab apples to *Physolospora obtusa* (52), in loblolly pines to *Fomes annosus* (87), in aspens to *Hypoxylon pruinatum* (4, 5), and in paper mulberry to *Fusarium solani* (77). Ghaffar & Erwin (32) reported that *Macrophomina phaseoli* produced root rot of cotton only on seedlings subjected to water stress prior to inoculation. Extending the period between waterings, however, merely subjects plants to wide variations in soil water potential over an extended period, and may have little relevance to conditions prevailing during drought periods in the field.

Bier (10–13), and later Bloomberg (14) and Landis & Hart (52), correlated critical levels of bark moisture content with increases in susceptibility of trees to several canker fungi. Filer (29), however, found no apparent correlation between bark moisture content and susceptibility of cottonwood stems to attack by *Cytospora,*

Phomopsis, and *Hypomyces.* Although bark moisture content may be a factor in predisposition of bark to colonization by certain pathogens, its relation to disease susceptibility in tissues other than bark has not been established.

Using a humidity cabinet to maintain relatively constant stem water potentials over a one-week incubation period in container-grown tree seedlings, Crist & Schoeneweiss (26) were able to associate stem water potentials, measured in a pressure bomb, with changes in susceptibility of European white birch stems to attack by *Botryosphaeria dothidea.* No cankers formed and stems were not colonized by the pathogen unless stem water potentials were lower (more negative) than −12 bars. When this procedure was modified to provide better control of stem water potentials during incubation (Schoeneweiss, submitted for publication), correlations were made between the water potential and extent of canker formation and/or stem colonization in sweetgum and red-osier dogwood seedlings inoculated with *B. dothidea.* The relation between stem water potential, canker formation, and colonization of wood and bark in these two species is presented in Figure 2. No cankers formed on plants exposed to water potentials less negative than −12 bars. The extent of canker formation and colonization of bark and wood increased with decreasing water potential. Because canker expansion and stem colonization ceased within one week after stem turgidity was restored by watering, water stress resulted in an elastic, reversible strain, at least from the standpoint of the effect on disease susceptibility. Increased disease susceptibility following exposure to water stress, with subsequent recovery of resistance after removal of the stress, was also reported for European white birch inoculated with *B. dothidea* (26).

The influence of length of exposure to various levels of water stress on disease susceptibility has been more difficult to determine, because it is difficult to wilt plants rapidly enough to eliminate the time factor involved. However, no increase in susceptibility of European white birch to *B. dothidea* occurred when stem water potentials were maintained at levels as low as −12 bars for 20 days (Crist and Schoeneweiss, unpublished data). This indicates that the level of water stress is a critical factor in predisposition. A delay in recovery of disease resistance following an increase in susceptibility due to induced water stress was reported by Edmunds (28). He found that *Macrophomina phaseoli* caused stalk rot in grain sorghum even though plant turgidity was restored by watering immediately prior to inoculation.

Although water stress apparently has a predisposing effect on host susceptibility to disease, the ability of certain pathogens to grow or even increase growth in vitro, on media with low water potential, is a complicating factor. Cook et al (23) reported that hyphal growth of *Fusarium roseum* 'Culmorum' was stimulated when the osmotic potential of the growth medium was decreased to −8 to −10 bars. Bagga & Smalley (4) reported increased colony diameter of *Hypoxylon pruinatum* on 8% agar as compared to 1% agar. Several other reports of increased growth of pathogens on media of low water potential have been cited (21, 22). Cook (21) suggested that the ability of certain pathogens to grow at low water potentials, which inhibited growth of antagonistic organisms, may be a factor in increased incidence of diseases under drought conditions. Cook & Papendick (22), however, state that the predis-

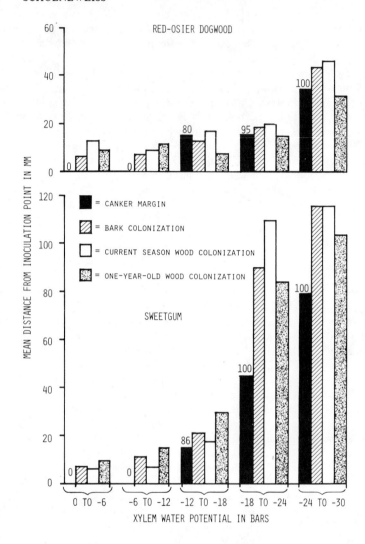

Figure 2 Effect of water stress on disease susceptibility in woody stems: relation between xylem water potentials during incubation (measured with a pressure bomb), canker size, and extent of colonization of bark and wood by *Botryosphaeria dothidea* in stems of two-year-old, container-grown seedlings of red-osier dogwood (*Cornus stolonifera*) and sweetgum (*Liquidambar styraciflua*). Plants were wilted to various levels, incubated for one week at relatively constant stem water potentials, then watered. Measurements of canker size and stem colonization were made one month after stem turgidity was restored by watering, and represent means of from five to ten replicate plants in each category of water potentials (D. F. Schoeneweiss, unpublished). Numbers indicate percentage of canker formation at inoculation points.

posing effect of stress generally is not on initial establishment of the pathogen in the host, but rather on development of established infections.

Excess Water and Flooding

In contrast to water deficits, few reports have appeared on the predisposing effects of excess water. Tinsley (86) reported that increasing the water supply of plants usually increased the amount of infection by viruses. Abundant water in such cases may act by causing structural differences in leaves, such as thinner cuticle layers and less regularly arranged palisade layers, that allow leaves to be more easily injured. Most root diseases are favored by wet soils (22), but the role of host predisposition in the majority of cases is not clear. High incidence and severity of littleleaf disease of shortleaf pine was reported on heavy, poorly drained clay soils in the southeastern United States (44, 102). According to Zak (102), any factor that reduced host vigor predisposed trees to attack by the feeder root pathogen, *Phytophthora cinnamomi*. The combined effects of feeder root necrosis, decreased host vigor in 20-year-old and older trees, and fewer mycorrhizae in poorly aerated soils ultimately produced the littleleaf symptoms.

Vascular diseases characteristically are more severe in wet than in dry soils, although symptom development resulting from vascular plugging may progress rapidly during hot, dry weather (21, 22). Cook (21) postulated that root infection by vascular pathogens may be more prevalent in drier soils, but that spread within the plant by means of spores is dependent upon the transpirational stream, hence the association between wet soils and wilt diseases.

Excess water and flooding, although not injurious directly, may produce a secondary, oxygen-deficient strain (54). This in turn, may restrict root growth, resulting in an accumulation of toxic metabolites which interfere with defense reactions of the host (85). Reduced soil aeration often inhibits water and nutrient uptake and may produce stresses similar to drought or physical root damage, resulting in wilting and chlorosis (54), and increased disease susceptibility (85). Host predisposition to stem-canker fungi and wood-boring insects was probably involved in dieback of mature white and bur oaks, following 3 consecutive years of abnormally heavy spring rains from 1971 to 1973 in the north-central United States (E. B. Himelick and D. F. Schoeneweiss, unpublished observations). Although Ward (90) long ago included flooding as a predisposing factor in plant disease, the literature is practically devoid of factual information on the effects of excess water on plant diseases.

TEMPERATURE STRESS

Nearly every paper written in plant pathology makes some reference to the influence of temperature on either the pathogen, the host, or disease development. In this paper, only the extremes of high and low temperature are considered as stress factors affecting host susceptibility to disease.

Low Temperature and Freezing

Levitt (54) separates low temperature stress into chilling stress and freezing stress. Chilling stress, which occurs above the freezing point, usually results in an elastic strain but may become a plastic strain, whereas freezing stress commonly results in a plastic strain. From the standpoint of disease susceptibility, however, strains from low temperature stress are often elastic, and increased disease susceptibility due to exposure to low temperatures may be reversible with time, provided the host tissues are not killed directly during exposure. Chilling stress is usually associated with exposure of subtropical or tropical plants (57), or plants in an active stage of growth (61) to unusually low temperatures above freezing. In contrast, freezing stress usually results from a rapid drop in temperature below freezing following a warm period in the spring or fall (26, 76, 95, 96), or following a period of mild weather at any time during the winter. Dormant woody plants are especially vulnerable to freezing stress if the chilling period required to break dormancy has been satisfied (66).

Most studies on the effects of low temperature stress in plants have been conducted in connection with plant hardiness investigations, and little is known about effects on disease susceptibility. Kable et al (49) reported increased incidence of *Cytospora* canker on sweet cherry following winter injury in New York State. Natural infections on freeze-injured stems resulted in colonization of injured tissues, followed by callus formation and healing. Helton (43) reported that stems of plum trees injured with dry-ice blocks were easily invaded by naturally disseminated *Cytospora* inoculum; symptoms typical of those observed in the field were produced. He concluded that natural winter injury from low temperatures is a primary predisposing factor (42). Brener et al (16) reported that *Sclerophoma pythiophila* caused a tip dieback of juniper in Wisconsin, which closely resembled the more commonly known *Phomopsis* blight. Apparently, winter injury or freezing damage on the previous season's growth either allows the fungus to enter or predisposes branches to colonization. They concluded that *S. pythiophila* is a weak (nonaggressive) facultative parasite that occasionally causes serious damage in conifer plantations under stress.

Crist & Schoeneweiss (26) reported increased susceptibility of European white birch stems to attack by *B. dothidea* following controlled freezing of dormant seedlings. The predisposing effect of freezing was most pronounced in xylem tissues, with little change in susceptibility of bark to colonization. By exposing dormant, container-grown seedlings of *Rhamnus frangula* 'Tallhedge' inoculated with *Tubercularia ulmea* to a rapid drop in ambient air temperature from above freezing to –10°C, –20°C, and –30°C, Schoeneweiss (76) found that disease susceptibility of stems increased with decreasing exposure temperature. In a similar test with sweetgum seedlings inoculated with *B. dothidea,* results were quite similar (Figure 3). Colonization of wood, particularly older wood, increased with decreasing exposure temperature. Bark colonization, however, was very limited when compared to plants exposed to water stress (Figure 2), and seldom extended to canker margins. These results are supported by the findings of Quamme et al (67), who reported that bark of apple stems is 20°C hardier than xylem or pith in winter and 12°C less hardy in spring. In most cases reported, if plants were not killed by freezing or girdled by pathogens, defense reactions eventually limited disease development and the strain resulting from freezing was elastic in relation to disease susceptibility.

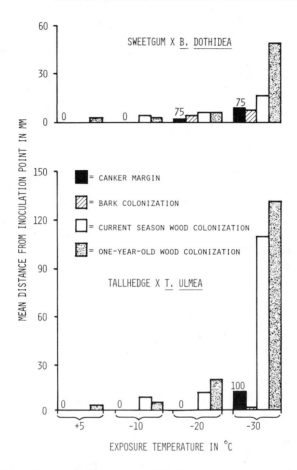

Figure 3   Effect of freezing on disease susceptibility in woody stems: relation between exposure temperature, canker size, and extent of colonization of bark and wood in stems of dormant
two-year-old, container-grown seedlings of sweetgum (*Liquidambar styraciflua*) inoculated
with *Botryosphaeria dothidea* (D. F. Schoeneweiss, unpublished), and Tallhedge (*Rhamnus
frangula* "Tallhedge") inoculated with *Tubercularia ulmea* (adapted from Schoeneweiss, 76).
Stems of test plants were inoculated and frozen from 5°C to the temperature indicated at a
rate of 4°C per hour. Measurements of canker size and stem colonization were made one
month after exposure and represent means of nine Tallhedge and twelve sweetgum plants at
each temperature. Numbers indicate percentage of canker formation at inoculation points.

High Temperatures

Chamberlain (17) and Chamberlain & Gerdemann (18) reported heat-induced susceptibility of soybeans to several pathogens and nonpathogens when hypocotyls
were heated to 47°C for 30 min prior to inoculation. They concluded that high

temperatures predisposed soybeans to fungal attack by suppressing or reducing production of phytoalexins. McGuire & Cooper (58) found that *Diplodia gossypina* caused collar rot of peanut if host plants were predisposed by heat prior to inoculation. Miller (59) enhanced susceptibility of apple stems to black rot canker by burning inoculation wounds with a Bunsen burner, although the relation of such treatment to field infections was not discussed. Yarwood & Hooker (101) increased susceptibility of corn to rust by immersing seedlings in a water bath at 55°C before inoculation. Wills & Moore (97) induced susceptibility in tobacco to *Phytophthora parasitica* var. *nicotiana* by heat-treating leaf strips or roots in a water bath for several minutes at 50°C before inoculating. Ikegami (46) increased susceptibility of beans to rust by a similar treatment. Predisposition by heat was reversible in beans (46) and soybeans (18), and plants regained normal disease resistance within three to five days after exposure to heat stress. Sommer (80) reported that the branch-wilt fungus, *Hendersonula toruloidea*, entered cracks caused by sun scald in walnut trees. After initial invasion, the fungus often remained saprophytic in dead tissue but became actively parasitic when infected branches were exposed to additional sun scald. He concluded that heat from sun scald predisposed walnut stems to attack by the pathogen.

DEFOLIATION STRESS

Loss of foliage during the growing season may predispose plants to attack by disease organisms. Siddiqui et al (79) found that frequent clipping of red clover predisposed plants to infection by species of *Fusarium* (which they considered weak pathogens), possibly by depleting food reserves in clover roots. Staley (82) and Stephens (83) reported increased mortality of oaks following defoliation by insects early in the growing season. Both authors concluded that defoliation depleted food reserves, causing loss of vigor and increased susceptibility to insects and other pathogens. Stephens & Hill (84) reported that repeated insect defoliation increased mortality of oaks in Connecticut forests, but that a single defoliation had little effect. Staley considered moisture stress a prime contributing factor in host predisposition (82), whereas drought had little effect on oak mortality in Connecticut (83, 84). In both cases, the authors considered disease organisms to be of secondary importance.

Defoliation has been shown to decrease the starch content and increase the amounts of reducing sugars in branches (34) and roots (63, 64, 91, 94) of trees. Wargo (91) reported increased growth of *Armillaria mellea* in media amended with extracts from sapwood of defoliated sugar maple trees and suggested that defoliation may predispose trees to attack by initiating chemical changes favorable for growth of the pathogen. Wargo & Houston (93) found that artificial defoliation in June or July resulted in higher mortality of sugar maple from *A. mellea* than defoliation in August, and that defoliation for two consecutive years gave a greater percentage of infection than a single defoliation. Rohrbach & Luepschen (69) reported an increase in certain sugars and sugar alcohols in bark of susceptible varieties of peach during fall and winter. Since these substances stimulate germination of *Cytospora leucostoma* spores, the authors suggested a relationship between carbohydrate content and disease susceptibility. Whether changes in sugar content resulting from defoliation predispose trees to disease has not been established.

Heichel et al (41) reported that regrowth of artificially defoliated oak and maple became heavily infected by anthracnose fungi. They theorized that trees weakened by repeated insect defoliation were predisposed to attack by fungi and wood-boring insects, which may contribute to mortality. Schoeneweiss (74) found that stems of cottonwood seedlings weakened by artificial defoliation became susceptible to attack by several genera of fungi; he stated that the appearance of blackstem symptoms indicated predisposition to fungal attack but was not characteristic of any particular pathogen.

Crist & Schoeneweiss (26) exposed European white birch stems, inoculated with *B. dothidea*, to controlled defoliation stress over a period of eight weeks. Cankers began forming after four weeks and canker incidence increased with length of exposure. On plants allowed to refoliate after four weeks of exposure, canker enlargement ceased, indicating a defense reaction. On those allowed to refoliate after five or six weeks, cankers continued to enlarge until stems were girdled and killed. Apparently, defoliation stress produced an initial elastic strain, which later became a plastic strain, with regard to disease susceptibility. Similar studies on defoliation stress were conducted with several other host-pathogen combinations (Schoeneweiss, unpublished data). A summary of results is presented in Figure 4. In these studies, inoculated plants were completely defoliated artificially at weekly intervals for eight weeks. Cankers were initiated after three weeks exposure in most cases, and canker incidence increased with length of exposure. Defoliation apparently predisposed European white birch, sweetgum, and European mountain ash, but not red-osier dogwood, to attack by *B. dothidea*. White birch stems were attacked by *Melanconium betulinum,* but only one small canker formed on one of twelve defoliation-stressed stems inoculated with *Cryptospora betulae. B. dothidea* and *M. betulinum* are commonly associated with stem cankers, whereas *C. betulae* is normally a saprophyte on dead stems.

An increase in sugars with a decrease in starch in the inner bark of loblolly pines following moisture stress was reported by Hodges & Lorio (45). Continuous flooding of roots had the same effect but required a longer period of exposure. If sugar content in stems and roots is involved in predisposition to pathogens, water stress, flooding, and defoliation may have similar effects on plant diseases, but proof of such a relationship awaits further research.

TRANSPLANTING STRESS

More than 70 years ago Ward (90) postulated the importance of transplanting stress in predisposing plants to disease; but little attention has been given this idea since then. This is partially understandable, because most plant pathologists work on diseases of annual crop plants that are normally grown from seed and of natural vegetation such as forest trees. The influence of transplanting on disease is of great importance in the nursery and landscape industries and in reforestation, where transplanting is an essential practice in the production and establishment of trees

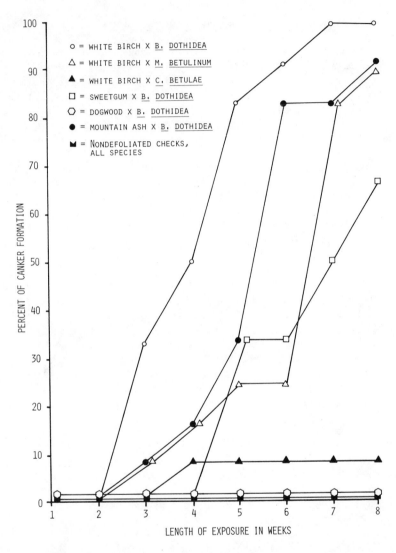

Figure 4 Effect of defoliation stress on disease susceptibility in woody stems: relation between length of exposure to defoliation stress and percentage of canker formation at inoculation points in two-year-old, container-grown seedlings of European white birch (*Betula alba*) inoculated with *Botryosphaeria dothidea, Melanconium betulinum,* and *Cryptospora betulae;* and sweetgum (*Liquidambar styraciflua*), red-osier dogwood (*Cornus stolonifera*), and European mountain ash (*Sorbus aucuparia*) inoculated with *B. dothidea.* Percentage of canker formation is based on twelve replicate plants in each host-pathogen combination (D. F. Schoeneweiss, unpublished). Plants were completely defoliated artificially at weekly intervals for seven weeks.

and shrubs. In highway, parkway, and urban plantings and in nursery fields and home plantings, the most severe economic losses caused by disease organisms often occur following transplanting (75).

Schoeneweiss (72) reported severe damage caused by *Fusicoccum* sp. (later reclassified as *B. dothidea,* 26) on recently transplanted European mountain ash. When established nursery trees were inoculated with the pathogen, no cankers developed until the trees were transplanted the following year; then nine of ten trees died from girdling cankers. In another study, Schoeneweiss (75) found that no disease developed when container-grown seedlings of European white birch and European mountain ash were inoculated with *B. dothidea,* removed from the containers in early spring, and repotted. However, when approximately 50% of the root system around the periphery of the soil ball was excised before repotting, simulating root pruning during transplanting, typical stem cankers appeared on all inoculated plants. Because all plants broke dormancy at the time of transplanting, and no visible symptoms of water stress appeared on new shoots, it is likely that transplanting did not result in plant water potentials low enough to account for the change in disease susceptibility. Schoeneweiss (73) also reported extensive damage caused by *Cytospora gleditschiae* on thornless honey locust trees weakened by transplanting.

The term "transplanting shock" is commonly used to describe stress associated with transplanting. Plants are undoubtedly exposed to severe stresses during and after transplanting, particularly if plants are not properly handled and maintained, but the nature of the stresses involved is largely unknown. Root and branch pruning during transplanting provide avenues of entrance for pathogens through fresh wounds, yet defense reactions of the hosts usually prevent the development of disease if plants are maintained in a vigorous condition after these operations. Root pruning also can decrease available food reserves. Much research is needed before the various stresses involved in transplanting can be elucidated fully.

NUTRIENT STRESS

More experiments have been conducted on the effects of mineral nutrients, particularly nitrogen, phosphorus, and potassium, on disease susceptibility than on any other potentially predisposing treatments (100). Most of these experiments, however, were not designed to separate predisposition from other effects (100). Examples of the effects of deficiencies and excesses of individual nutrients on plant diseases are given by Bollard & Matthews (15), Gäumann (31), and Yarwood (100). In general, most studies on nutrition have been conducted with obligate parasites and aggressive facultative parasites. Large amounts of nitrogen usually favor infection, excess potash reduces infection, and the effects of excess phosphorus are variable (31, 100). With certain viruses, addition of nitrogen and phosphorus increased the number of local lesions as long as the additions increased growth of the host; deficiencies or excesses of these elements reduced the number of lesions (9).

Lewis (55) proposed a balanced hypothesis of parasitism, in which host-parasite relations are governed by a combination of the biochemistry of the host and nutritional requirements of the parasite. Because nutrients present in the host in metabolic concentrations sometimes cause inhibition of the parasite, a certain nutrient balance may be a necessary prerequisite for infection. Garber (30) extended this hypothesis and suggested that only a combination of host nutritional environment adequate for the parasite plus an ineffective host inhibitory environment result in the establishment of a pathogenic relationship. Grainger (33) postulated that the disease proneness of a host could be measured by the ratio between the total carbohydrate pool available to the parasite and the dry weight of the shoot. Variation in precision of this ratio reflects varying needs for carbohydrates by different parasites.

In general, disease resistance has not been definitely associated with lack of host nutrients for parasites (36). The statement of Wingard (98) that the nutritional status of the host affects each disease differently appears to have some validity. Nutrient imbalance undoubtedly affects host vigor and may influence defense reactions (31), but the interactions among other environmental factors and nutrition have made valid interpretations of the predisposing effects of nutrient stresses nearly impossible (20).

OTHER STRESSES

Reduced light has been reported as a predisposing factor in many diseases (8, 15, 31, 68, 100). High light intensity and differences in light quality have also been implicated in disease (15, 31, 100). The quality and quantity of light available to host plants affects photosynthesis and consequently the amount of food reserves. The influences of light quality and quantity need to be considered in the design of greenhouse and growth chamber experiments and in investigating diseases of understory plants in natural ecosystems. Under normal agricultural field conditions, however, it is doubtful whether variations in the quality or quantity of light are often of sufficient magnitude to be considered as valid predisposing stress factors. Therefore, in the context of this review, light alone is not treated as a stress factor.

Toxic substances such as weed killers and other pesticides may cause undesirable side effects on crop plants, but little research has been reported on the predisposing effects of toxicants on infectious diseases. Atmospheric pollutants also affect plants and may occasionally predispose plants to pathogenic organisms. Much of the literature pertaining to the interaction between air pollutants and plant parasites has been summarized in a recent review by Heagle (40). In general, exposure to atmospheric pollutants often decreases parasitic disease damage. The main exceptions reported thus far are increased disease damage due to *Armillaria mellea* and several wood-rotting fungi on trees weakened by exposure to sulfur dioxide, increased infection by *Botrytis* on several hosts exposed to high levels of ozone, and increased damage by tobacco mosaic virus on pinto beans exposed to either ozone or fluoride (40). In most cases, however, the effects of air pollutants on host predisposition have not been distinguished from effects on the parasites themselves.

Wounding (31, 100), reduced host vigor due to attack by parasitic nematodes (56, 65), and many other factors may also predispose plants, directly or indirectly, to infectious diseases, but space limitations preclude such a comprehensive review.

DISCUSSION

Environmental conditions favorable for growth of higher plants usually favor infection by obligate parasites and resistance to facultative parasites (6). Therefore stress, by definition any factor capable of producing a potentially injurious strain (54), exerts the most pronounced effect in predisposing plants toward greater susceptibility to facultative parasites, particularly weak or nonaggressive parasites. Although extreme stresses may injure plants through direct irreversible plastic strains, in most cases plants tolerate or adapt to stresses without permanent injury in the absence of disease organisms (6).

Nonaggressive facultative parasites commonly enter host plants (15, 19, 31, 88, 89) and remain latent or nonpathogenic until the host is stressed (6). According to Chester (19), post-infectional mechanisms of immunity or resistance are present in most, if not all plants; immunity depends in large part on defense reactions resulting from stimulation of the host by the parasite. Much evidence has been presented to substantiate claims that facultative parasites cause hypersensitive reactions in weakened hosts that kill cells in advance of the parasite, which then grows saprophytically in dead tissue (15, 35, 36). On the other hand, many cases have been cited where resistance to facultative parasites is the result of a dynamic physiological response of living parenchyma, particularly where production of phytoalexins (6, 15, 27, 36, 51, 78), or accumulations of phenolics and other compounds toxic to pathogens (15, 36, 99), have been implicated in defense reactions. Although a substantial body of literature has accumulated on the role of phytoalexins in disease resistance (27, 50, 51), the influence of predisposing stress factors on phytoalexin synthesis and effectiveness has yet to be studied adequately (6, 37, 50, 88). Mechanical barriers such as cork layers, periderm, gums, lignin, and suberin (1, 3, 6, 7, 15, 31, 47), and abnormal xylem production (71) have also been reported as disease resistance mechanisms. Wargo (92) recently reported that chitinase and β,1-3 glucanase in the healthy inner bark and outer wood of several forest tree species could dissolve the cell walls of *Armillaria mellea*. He suggested that this mechanism could account for the resistance of nonstressed trees to invasion by *A. mellea* and other organisms. In general, however, much of the evidence is conflicting (15). Whether each case of disease resistance is unique, as suggested by Hart (37), or whether some common mechanism or mechanisms are involved in many defense reactions remains a controversial question.

Much research on disease resistance has been conducted on vigorous plants, with comparisons being made between genetically resistant and susceptible hosts. Because stresses often predispose normally resistant plants to attack by pathogens, it seems apparent that much basic information can be gained by working with stressed and nonstressed, genetically uniform hosts. Methods and techniques have been, and are being developed for exposing plants to common stresses under controlled conditions (26, 75, 76). Colhoun (20) cautioned that results obtained with environmental stresses under rigidly controlled conditions may have little relevance to those that occur under field conditions, and that more work is needed on interacting environmental factors that are cycled and varied. Although such criticism may be valid, controlling the many variables encountered in working with plants under stress presents a formidable task. The value of the evidence obtained in recent years by working with individual stress factors appears to me to warrant further effort in this direction.

Hartig (38) long ago cautioned that it is often a mistake to say whether a fungus is pathogenic on the basis of a few inoculations, without taking into account possible predisposing factors. Yet how often have we as pathologists rigidly followed Koch's postulates and discarded as "saprophytes" or "secondary parasites" those organisms that did not produce disease symptoms when inoculated into vigorous hosts? Much knowledge has been obtained on the responses of plants to stresses in the absence of pathogenic organisms (54, 62). Surely this knowledge can be useful in interpreting host resistance and response to parasites. Jones (48) made a strong plea over fifty years ago for cooperation between pathologists and physiologists in investigating the nature of disease resistance in plants. Hopefully, this plea is finally being heard.

Literature Cited

1. Akai, S. 1959. Histology of defense in plants. In *Plant Pathology,* ed. J. G. Horsfall, A. E. Dimond, 1:391-434. New York & London: Academic. 674 pp.
2. Allen, P. J. 1959. Physiology and biochemistry of defense. See Ref. 1, 435–67
3. Arnett, J. D., Wichter, W. 1974. Histochemical studies of yellow poplar infected with *Fusarium solani. Phytopathology* 64:414–18
4. Bagga, D. K., Smalley, E. B. 1967. Water stress in relation to initiation and development of *Hypoxylon* canker of aspen. *Phytopathology* 57:802 (Abstr.)
5. Bagga, D. K., Smalley, E. B. 1974. The development of *Hypoxylon* canker of *Populus tremuloides:* Role of interacting environmental factors. *Phytopathology* 64:658–62
6. Baker, K. F., Cook, R. J. 1973. *Biological Control of Plant Pathogens.* Freeman: San Francisco. 433 pp.
7. Barnett, H. C. 1959. Plant disease resistance. *Ann. Rev. Microbiol.* 13:191-209
8. Bawden, F. C., Roberts, F. M. 1948. Photosynthesis and the predisposition of plants to infection with certain viruses. *Ann. Appl. Biol.* 35:418–28
9. Bawden, F. C., Kassanis, B. 1950. Some effects of host nutrition on the susceptibility of plants to infection by certain viruses. *Ann. Appl. Biol.* 37:46–57
10. Bier, J. E. 1959. The relation of bark moisture to the development of canker diseases caused by native facultative

parasites. I. *Cryptodiaporthe* canker on willow. *Can. J. Bot.* 37:229–38
11. Bier, J. E. 1959. The relation of bark moisture to the development of canker diseases caused by native facultative parasites. II. *Fusarium* canker on black cottonwood. *Can. J. Bot.* 37:781–88
12. Bier, J. E. 1959. The relation of bark moisture to the development of canker diseases caused by native facultative parasites. III. *Cephalosporium* canker on western hemlock. *Can. J. Bot.* 37:1140–42
13. Bier, J. E. 1961. The relation of bark moisture to the development of canker diseases caused by native facultative parasites. IV. Pathogenicity studies of *Hypoxylon pruinatum* (Klotsch) Cke., and *Septoria musiva* Pk. on species of *Acer, Populus,* and *Salix. Can. J. Bot.* 39:1555–61
14. Bloomberg, W. J. 1962. *Cytospora* canker of poplars: Factors influencing development of the disease. *Can. J. Bot.* 40:1271–80
15. Bollard, E. G., Matthews, R. E. F. 1966. The physiology of parasitic disease. In *Plant Physiology,* ed. F. C. Steward, IVB: 417–550. New York & London: Academic. 599 pp.
16. Brener, W. D., Setliff, E. C., Norgren, R. L. 1974. *Sclerophoma pythiophila* associated with a tip dieback of juniper in Wisconsin. *Plant Dis. Reptr.* 58:653–57
17. Chamberlain, D. W. 1972. Heatinduced susceptibility to nonpathogens and cross-protection against *Phytoph-*

thora megasperma var. *sojae* in soybean. *Phytopathology* 62:645–46

18. Chamberlain, D. W., Gerdemann, J. W. 1966. Heat-induced susceptibility of soybeans to *Phytophthora megasperma* var. *sojae, Phytophthora cactorum,* and *Helminthosporium sativum. Phytopathology* 56:70–73

19. Chester, K. S. 1933. The problem of acquired physiological immunity in plants. *Q. Rev. Biol.* 8:129–54, 275–324

20. Colhoun, J. 1973. Effects of environmental factors on plant disease. *Ann. Rev. Phytopathol.* 11:343–64

21. Cook, R. J. 1973. Influence of low plant and soil water potentials on diseases caused by soilborne fungi. *Phytopathology* 63:451–58

22. Cook, R. J., Papendick, R. I. 1972. Influence of water potential of soils and plants on root diseases. *Ann. Rev. Phytopathol.* 10:349–74

23. Cook, R. J., Papendick, R. I., Griffen, D. M. 1972. Growth of two root rot fungi as affected by osmotic and matric water potentials. *Soil Sci. Soc. Am. Proc.* 36:78–82

24. Couch, H. B., Bloom, J. R. 1960. Influence of environment on diseases of turfgrasses. II. Effect of nutrition, pH, and soil moisture on *Sclerotinia* dollar spot. *Phytopathology* 50:761–63

25. Couch, H. B., Purdy, L. H., Henderson, D. W. 1967. Application of soil moisture principles to the study of plant disease. *Va. Polytech. Inst. Dept. Plant Pathol. Bull.* No. 4. 23 pp.

26. Crist, C. R., Schoeneweiss, D. F. 1975. The influence of controlled stresses on susceptibility of European white birch stems to attack by *Botryosphaeria dothidea. Phytopathology* 65:369–73

27. Cruickshank, I. A. M. 1963. Phytoalexins. *Ann. Rev. Phytopathol.* 1:351–74

28. Edmunds, L. K. 1964. Combined relation of plant maturity, temperature, and soil moisture to charcoal stalk rot development in grain sorghum. *Phytopathology* 54:514–17

29. Filer, T. H. Jr. 1967. Pathogenicity of *Cytospora, Phomopsis,* and *Hypomyces* on *Populus deltoides. Phytopathology* 57:978–80

30. Garber, E. D. 1956. A nutrition-inhibition hypothesis of pathogenicity. *Am. Nat.* 90:183–94

31. Gäumann, E. 1950. *Principles of Plant Infection.* Transl. W. B. Brierley. London: Crosby Lockwood. 543 pp.

32. Ghaffar, A., Erwin, D. C. 1969. Effect of soil water stress on root rot of cotton

caused by *Macrophomina phaseoli. Phytopathology* 59:795–97

33. Grainger, J. 1962. The host plant as a habitat for fungal and bacterial parasites. *Phytopathology* 52:140–50

34. Gross, H. L., Larsen, M. J. 1971. Nutrient content of artificially defoliated branches of *Betula papyrifera. Phytopathology* 61:631–35

35. Hadwiger, W. E., Schwochau, M. E. 1969. Host resistance responses—an induction hypothesis. *Phytopathology* 59:223–27

36. Hare, R. C. 1966. Physiology of resistance to fungal diseases in plants. *Bot. Rev.* 32:95–137

37. Hart, H. 1949. Nature and variability of disease resistance in plants. *Ann. Rev. Microbiol.* 3:289–316

38. Hartig, R. 1882. *Lehrbuch der Baumkrankheiten.* Berlin: Springer. 198 pp. (Engl. trans. W. Somerville, ed. H. M. Ward, 1894. London & New York: Macmillan. 331 pp.)

39. Hayward, A. C. 1974. Latent infections by bacteria. *Ann. Rev. Phytopathol.* 12:87–97

40. Heagle, A. S. 1973. Interactions between air pollutants and plant parasites. *Ann. Rev. Phytopathol.* 11:365–88

41. Heichel, G. H., Turner, N. C., Walton, G. S. 1972. Anthracnose causes dieback of regrowth on defoliated oak and maple. *Plant Dis. Reptr.* 56:1046–47

42. Helton, A. W. 1961. Low temperature injury as a contributing factor in *Cytospora* invasion of plum trees. *Plant Dis. Reptr.* 45:591–97

43. Helton, A. W. 1962. Effect of simulated freeze-cracking on invasion of dry-ice-injured stems of Stanley prune trees by naturally disseminated *Cytospora* inoculum. *Plant Dis. Reptr.* 46:45–47

44. Hepting, G. H. 1963. Climate and forest diseases. *Ann. Rev. Phytopathol.* 1:31–50

45. Hodges, J. D., Lorio, P. L. 1969. Carbohydrate and nitrogen fractions in the inner bark of loblolly pines under moisture stress. *Can. J. Bot.* 47:1651–57

46. Ikegami, H. 1968. Heat-induced susceptibility of beans to rust. *Phytopathology* 58:773–75

47. Jewell, F. F., Snow, G. A. 1972. Anatomical resistance to gall-rust infection in slash pine. *Plant Dis. Reptr.* 56:531–34

48. Jones, L. R. 1924. The relation of environment to disease in plants. *Am. J. Bot.* 11:601–9

49. Kable, P. F., Fliegel, P., Parker, K. G. 1967. *Cytospora* canker on sweet cherry in New York State: association with winter injury and pathogenicity to other species. *Plant Dis. Reptr.* 51:155–57

50. Kuć, J. 1965. Resistance of plants to infectious agents. *Ann. Rev. Microbiol.* 20:337–70

51. Kuć, J. 1972. Phytoalexins. *Ann. Rev. Phytopathol.* 10:207–32

52. Landis, W. R., Hart, J. H. 1967. Cankers of ornamental crabapples associated with *Physalospora obtusa* and other microorganisms. *Plant Dis. Reptr.* 51:230–34

53. Leaphart, C. D., Stage, A. R. 1971. Climate: a factor in the origin of the pole blight disease of *Pinus monticola* Dougl. *Ecology* 52:229–39

54. Levitt, J. 1972. *Responses of Plants to Environmental Stresses.* New York & London: Academic. 697 pp.

55. Lewis, R. W. 1953. An outline of the balanced hypothesis of parasitism. *Am. Nat.* 87:273–81

56. Lownsbery, B. F., English, H., Moody, E. H., Shick, F. J. 1973. *Criconemoides xenoplax* experimentally associated with a disease of peach. *Phytopathology* 63:994–97

57. Lyons, J. M. 1973. Chilling injury in plants. *Ann. Rev. Plant Physiol.* 24:445–66

58. McGuire, J. M., Cooper, W. E. 1965. Interaction of heat injury and *Diplodia gossypina* and other etiological aspects of collar rot of peanut. *Phytopathology* 55:231–36

59. Miller, P. W. 1973. Susceptibility of some apple cultivars to black rot canker. *Plant Dis. Reptr.* 57:676–77

60. Moore, L. D., Couch, H. B., Bloom, J. R. 1963. Influence of environment on diseases of turfgrasses. III. Effect of nutrition, pH, soil temperature, and soil moisture on *Pythium* blight of Highland bentgrass. *Phytopathology* 53:53–57

61. Olien, C. R. 1967. Freezing stresses and survival. *Ann. Rev. Plant Physiol.* 18:387–408

62. Parker, J. 1965. Physiological diseases of trees and shrubs. *Adv. Front. Plant Sci.* 12:97–248

63. Parker, J. 1970. Effects of defoliation and drought on root food reserves in sugar maple seedlings. *US Dep. Agric. For. Serv. Res. Pap.* NE-169. Upper Darby, Pa.: NE. For. Exp. Stn. 8 pp.

64. Parker, J., Houston, D. R. 1971. Effects of repeated defoliation on root and root

collar extractives of sugar maple trees. *For. Sci.* 17:91–95

65. Powell, N. T. 1963. The role of plant-parasitic nematodes in fungus diseases. *Phytopathology* 53:28–34

66. Prince, V. E. 1966. Winter injury to peach trees in Georgia. *Am. Soc. Hort. Sci. Proc.* 88:190–96

67. Quamme, H., Stushnoff, C., Weiser, C. J. 1972. The relationship of exotherms to cold injury in apple stem tissues. *J. Am. Soc. Hort. Sci.* 97:608–13

68. Read, D. J. 1968. Some aspects of the relationship between shade and fungal pathogenicity in an epidemic disease of pines. *New Phytol.* 67:39–48

69. Rohrbach, K. G., Luepschen, N. S. 1968. Seasonal changes in sugar alcohols and sugars in peach bark: a possible relationship to *Cytospora* canker susceptibility. *Am. Soc. Hort. Sci. Proc.* 93:135–40

70. Ross, E. W. 1966. Ash dieback: etiological and developmental studies. *NY State Coll. For. Tech. Publ.* No. 88. 80 pp.

71. Schoeneweiss, D. F. 1959. Xylem formation as a factor in oak wilt resistance. *Phytopathology* 49:335–37

72. Schoeneweiss, D. F. 1965. *Fusicoccum* canker of mountain ash in Illinois. *Plant Dis. Reptr.* 49:251–52

73. Schoeneweiss, D. F. 1966. *Cytospora* canker on thornless honey locust trees. *Plant Dis. Reptr.* 50:13–14

74. Schoeneweiss, D. F. 1967. Susceptibility of weakened cottonwood stems to fungi associated with blackstem. *Plant Dis. Reptr.* 51:933–35

75. Schoeneweiss, D. F. 1973. Progress and problems in stress-related pathology of landscape plants. *First Woody Ornamental Workshop Proc.,* University of Missouri, Columbia. 128 pp.

76. Schoeneweiss, D. F. 1974. *Tubercularia ulmea* canker of Tallhedge: influence of freezing stress on disease susceptibility. *Plant Dis. Reptr.* 58:937–40

77. Schreiber, L. P., Dochinger, L. S. 1967. *Fusarium* canker on paper mulberry (*Broussonetia papyrifera*). *Plant Dis. Reptr.* 51:531–32

78. Shain, L. 1967. Resistance of sapwood in stems of loblolly pine to infection by *Fomes annosus. Phytopathology* 57: 1034–45

79. Siddiqui, W. M., Halisky, P. M., Lund, S. 1968. Relationship of clipping frequency to root and crown deterioration in red clover. *Phytopathology* 58:486–88

80. Sommer, N. F. 1955. Sunburn predis-

poses walnut trees to branch wilt. *Phytopathology* 45:607–13

81. Sorauer, P. 1974. *Handbuch der Pflanzenkrankheiten.* Wiegandt, Berlin: Hempel und Parey. 406 pp.

82. Staley, J. M. 1965. Decline and mortality of red and scarlet oaks. *For. Sci.* 11:2–17

83. Stephens, G. R. 1971. The relation of insect defoliation to mortality in Connecticut forests. *Conn. Agric. Exp. Stn. Bull.* No. 723. 15 pp.

84. Stephens, G. R., Hill, D. E. 1971. Drainage, drought, defoliation, and death in unmanaged Connecticut forests. *Conn. Agric. Exp. Stn. Bull.* No. 718. 50 pp.

85. Stolzy, L. H., Letey, J., Klotz, L. J., Labanauskas, C. K. 1965. Water and aeration as factors in root decay of *Citrus sinensis. Phytopathology* 55:270–75

86. Tinsley, T. W. 1953. The effects of varying the water supply of plants on their susceptibility to infection with viruses. *Ann. Appl. Biol.* 40:750–60

87. Towers, B., Stambaugh, W. J. 1968. The influence of induced soil moisture stress upon *Fomes annosus* root rot of loblolly pine. *Phytopathology* 58:269–72

88. Verhoeff, K. 1974. Latent infections by fungi. *Ann. Rev. Phytopathol.* 12:99–110

89. Walker, J. C., Stahmann, M. A. 1955. Chemical nature of disease resistance. *Ann. Rev. Plant Physiol.* 6:351–66

90. Ward, H. M. 1901. *Diseases in Plants.* New York & London: Macmillan. 309 pp.

91. Wargo, P. M. 1972. Defoliation-induced chemical changes in sugar maple roots stimulate growth of *Armillaria mellea. Phytopathology* 62:1278–83

92. Wargo, P. M. 1974. Lysis of the cell wall of *Armillaria mellea* by enzymes from forest trees. *Phytopathology* 64:588 (Abstr.)

93. Wargo, P. M., Houston, D. R. 1973. Infection of defoliated sugar/maple trees by *Armillaria mellea. Phytopathology* 63:209 (Abstr.)

94. Wargo, P. M., Parker, J., Houston, D. R. 1973. Starch content in roots of defoliated sugar maple. *For. Sci.* 18:203–4

95. Weiser, C. J. 1970. Cold resistance and injury in woody plants. *Science* 169:1269–78

96. White, W. C., Weiser, C. J. 1964. The relation of tissue dessication, extreme cold, and rapid temperature fluctuations to winter injury in American arborvitae. *Am. Soc. Hort. Sci. Proc.* 85:554–63

97. Wills, W. H., Moore, L. D. 1968. Heat induced susceptibility of black shank–resistant tobacco to *Phytophthora parasitica* var. *nicotiana. Phytopathology* 58:888 (Abstr.)

98. Wingard, S. A. 1941. The nature of disease resistance in plants I. *Bot. Rev.* 7:59–109

99. Wood, R. K. S. 1967. *Physiological Plant Pathology.* London: Blackwell. 510 pp.

100. Yarwood, C. E. 1959. Predisposition. In *Plant Pathology,* ed. J. G. Horsfall, A. E. Dimond, 1:521–62. New York & London: Academic. 674 pp.

101. Yarwood, C. E., Hooker, A. L. 1966. Heat predisposition to corn rust. *Phytopathology* 56:510–11

102. Zak, B. 1961. Aeration and other soil factors affecting southern pines as related to littleleaf disease. *US Dep. Agric. For. Serv. Tech. Bull.* No. 1248. 30 pp.

PATHOGENESIS OF ❖3620
NEMATODE-INFECTED PLANTS

B. Y. Endo
Plant Protection Institute, Agricultural Research Service,
United States Department of Agriculture, Beltsville, Maryland 20705

INTRODUCTION

In this review, an attempt is made to bring together recent developments in under-standing pathogenesis in nematode-infected plants. As in other disease situations, one must consider factors that affect the attraction between the host and parasite within a given environment. Once in contact with the roots, nematodes may feed either on the surface of roots as ectoparasites or internally after penetration of roots as migratory or sedentary endoparasites. Studies on the feeding processes of ec-toparasitic nematodes have provided valuable information that can be applied to studies of the sedentary endoparasitic forms where direct observations on feeding are difficult. The developmental processes of pathogenesis are related to the way that nematodes adversely affect the host—either as superficial feeders or as endoparasites that induce syncytia or giant cells. Some nematodes cause extreme galling. In many cases, nematodes act alone as primary pathogens, but in other cases they interact with other organisms in disease complexes or serve as vectors for viruses. This review covers some information discussed in other reviews in the specific areas of nematode-induced plant pathogenesis. Studies on the biochemical and physiological aspects of pathogenesis by root-knot nematode have been omitted since they were reviewed recently by Dropkin (22) and Bird (11).

ATTRACTION OF NEMATODES TO THE HOST

Klingler (44), has discussed two hypotheses concerning the attraction of nematodes to host roots: the first is that nematodes move at random and find roots by chance and the second is that nematodes are attracted from some distance away. Agar surfaces have been used to show that nematode movements can be influenced by the host roots. Attraction can be expected to occur in the immediate vicinity of the roots at distances of around 1–2 cm. (44). Longer distance attraction in certain reports were attributed to use of semipermeable membranes that tended to diminish nema-

213

tode movements which were then interpreted as nematode attraction to the host. Klingler concluded that attraction by roots is due to chemical factors, especially CO_2 and certain amino acids.

Pitcher (67) found that high populations of *Trichodorus viruliferus* (half of which were adults) accumulated around the elongating zone of apple roots. If this accumulation was due solely to movement of nematodes to the feeding site rather than to an increase in numbers through reproduction, the attractive force and migratory activity would extend up to an 8 cm radius and effectively evacuate up to a liter of soil. Blake (12) showed that oat roots have a potent attractant which acts locally within a distance of 5 mm on either side of the root. Luc (51) reported that millet roots stimulate movement of *Hemicycliophora paradoxa* for distances up to 40 cm.

Shepherd (82) points out that nematodes are equipped with a neuro-sensory system and that their behavior patterns are influenced by the environment. In terms of root exudates in the soil, quiescent nematodes can be activited and a behavioral pattern set into action which could involve secretory processes of the esophageal glands. The behaviorally activated individuals could then respond to a gradient of stimulants such as root exudates.

In a recent study with potato cyst nematodes on agar plates, Steinbach (83) noted attraction of larvae to tomato roots over a distance of 1 cm. In 4-day-old seedlings, most of the attraction was observed near zones of cell differentiation and elongation. He also found that the attractive force was lost when the roots were disturbed artificially or removed and that a nonhost plant, *Linum usitatissimum*, did not show attraction.

FEEDING SITES AND HOST PENETRATION

Root Surface Feeding

The interaction between plant cells and the parasite determines the damage that is inflicted by nematodes. The site of initial contact varies among ectoparasites. For example, *Tylenchus emarginatus* feed along the entire length of the root with no preferred feeding site (84) whereas *Trichodorus christiei* usually feed at the actively growing root tip. Pitcher & Flegg (68) reported that *T. viruliferus* preferred a feeding site just behind the apical meristem. Colonies of 10 to 100 individuals clustered at a site 1–3 mm from the meristem and rasped the epidermis and hypodermis, causing characteristic browning of the tissues. Although the meristem was not the focus of attack, the affected roots had essentially the same stubby root symptoms observed for *T. christiei* - infected roots. A more detailed account was presented in a later paper by Pitcher (67).

Internal Tissue Feeding

The feeding site of *Hemicycliophora* ranged from 200 to 1000 mm from the root tip with most nematodes at 400–600 mm. Cohn (14) noted that both *Xiphinema brevicolle* and *X. index* fed along the root rather than the root tips. Some nematodes fed at the same site for several days. By comparison, *Longidorus africanus* fed directly on root tips for periods up to 15 min.

Migratory endoparasites such as lesion nematodes generally enter roots at the zones of elongation and root hair formation. However, some species feed on the surface or enter and destroy tissues of the apical meristem (33, 49, 58). Larvae of root-knot nematodes can penetrate anywhere along the root but generally enter at or near the root tips (13). *Heterodera* spp. can penetrate anywhere along the root but according to Mankau & Linford (53), *H. trifolii* generally enters the young piliferous zone. Entry into the central region of the root was observed at apical meristems as well as through cortical tissues separated by emerging lateral rootlets.

Doncaster & Seymour (20) made film observations of 11 tylenchid species and found a similar pattern of behavior leading to feeding. Simple behavior sequences that responded to chemical and physical stimuli led to egg hatching and host penetration.

NEMATODE FEEDING PROCESSES

Tylenchus

In terms of the mechanics of stylet penetration, *T. emarginatus* positioned the anterior one third of the body perpendicular to the root surface prior to insertion of the stylet. The sequence of events leading to nematode feeding consisted of about 20 forward thrusts to puncture the cell, a lapse of 20 seconds, the first signs of median bulb pulsations, and the movement of granular material of the cell toward the stylet. Because no granular material was observed to be drawn up into the stylet, the nematode was assumed to have withdrawn only the fluid portion of the cell (84).

Tylenchorhynchus

Cinemaphotography and interpretation of the penetration and feeding processes of *Tylenchorhynchus dubius* (101) on epidermal cells and root hairs provide further insights in the host-parasite interrelations. After finding a favorable feeding site, the nematode characteristically thrust its stylet irregularly until it punctured the epidermal cell wall. Immediately, the stylet penetrated 2–3 μ into the cell and remained in place during feeding. After stylet insertion, saliva was observed flowing from the dorsal gland ampulla into the esophageal lumen behind the stylet knob. The ampulla was never completely empty and appeared to be resupplied from a source of synthesized material originating in the dorsal esophageal gland. Ingestion of cell contents commenced with the contraction of the muscles in the metacorpus and associated dilation of the metacorporeal pump. During ingestion, the granules near the outlet of the dorsal gland ampulla did not move forward and the reservoir remained distended. The granular constituents of the cytoplasm were not observed to be drawn into the stylet.

The reactions of the host included accumulation of cytoplasm near the stylet region with various degrees of reduction in cytoplasmic flow and subsequent stoppage and apparent coagulation throughout the cell. With root hair feeding, the rapidity of reaction and sensitivity was related to the size and age of the root hairs, with the younger, shorter root hairs being more sensitive.

Trichodorus

In a time-lapse study of root hair feedings of *Nicotiana tabacum* by *Trichodorus similis,* Wyss (102, 103) showed that salivation caused cellular cytoplasm and nuclei to be drawn to the stylet tip (Figure 1). Large portions of cytoplasm and nucleoplasm of pierced cells and nuclei were withdrawn through the nematode stylet. After feeding, the remaining cytoplasm was pulled toward the nematode penetration site where the cytoplasm coagulated. Wyss suggested that virus transmission occurs when cytoplasmic ingestion by the nematode ends before the cells are irreversibly damaged and viruses are in a favorable environment for later replication.

Hemicycliophora

Feeding by a tylenchoid nematode, *Hemicycliophora arenaria* may serve to illustrate the feeding mechanism of other species that are not so readily observable, such as the root-knot and cyst-nematode forms. The similarities may be related at least to the relatively well-developed glandular system of these forms. McElroy & Van Gundy (56) observed that the nematode arched its body to project its stylet perpendicularly to and through the cell wall. Flexibility of the stylet permitted both intra- and intercellular penetration of root tissues, moving to a depth of 2–3 cells. After penetration, nematode salivation occurred for 1–2 hr periods interrupted by slight contractions or pulsations of the muscles in the median bulb. Meanwhile the dorsal gland was moved violently with each pulsation. The dorsal gland ampulla was distended with small globules and as the median bulb pulsated, globules were observed to enter the gland opening and the lumen of the esophagus.

The nematode often fed in the cortical cells bordering the endodermis, in the cells of the endodermis, or upon tissues that would eventually mature to become the endodermis or pericycle. It fed between the root cap and the beginning of the zone of elongation.

A unique feature of the feeding process of *Hemicycliophora* was the formation of an adhesive plug which appears to attach the nematode to the root during feeding. The mechanism of formation of the micro-polysaccharide-like plug was not determined. Tests for the origin and nature of this adhesive material showed periodic acid schiff (PAS) positive reactions for polysaccharides in the esophageal gland, the amphidial pouches, and what appeared to be the stylet sheath. These sites may be the origin of the adhesive material.

The adhesive plug observed with *H. arenaria* was found also while *H. similis* fed on cranberry roots (43). As seen with *H. arenaria,* the adhesive plug associated with *H. similis* had a strong PAS reaction. The plug might be formed by the nematode as a positive adaptation for survival and would help the organism retain the feeding site over long periods.

McElroy & Van Gundy (56) reported that ingestion of host cell contents could be visualized by the concomitant movement of the feed cell contents and membranes with that of the pulsation of the median bulb. Although precise movement of the cell contents could not be observed, feed cells eventually emptied and collapsed under pressure of adjacent tissue.

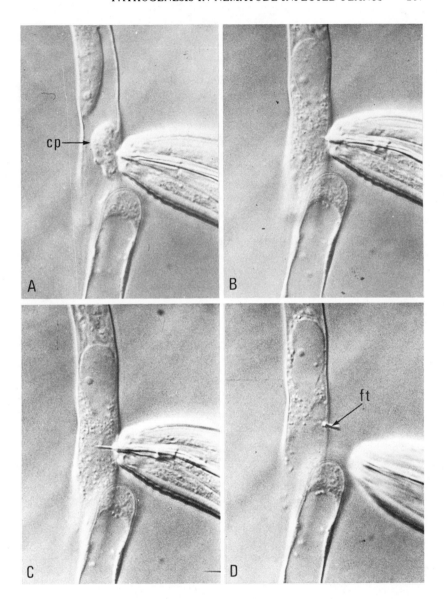

Figure 1   *Trichodorus similis* feeding on *Nicotiana tabacum* root hair: (*a*) cytoplasm (*cp*) drawn to feeding site within 25 sec after cell wall perforation, (*b*) accumulation of cytoplasm 100 sec after wall perforation and presence of an optically empty swollen nucleus, (*c*) initiation of ingestion after deep stylet thrusts into the nucleus and cytoplasm of the cell, 110 sec after perforation, and (*d*) after 15 sec of feeding ingestion, nematode shown to depart from feeding site leaving a feeding tube (*ft*) at the site of feeding. The tube is thought to consist entirely of hardened saliva. (Courtesy of U. Wyss.)

Xiphinema, Longidorus

New observations on virus-transmitting nematodes such as *Xiphinema* and *Longidorus* have stimulated interest in their mechanisms of penetration and feeding. Fisher & Raski (31) described the feeding aspects of *Xiphinema index* and *X. diversicaudatum* on germinating seedlings in petri dish cultures. Although individual cell changes could not be observed because of the depth of the onchiostylet penetration, it was evident that feeding occurred by the contractions and pulsations of the esophageal bulb during each change in depth of the onchiostyle penetration. Fisher & Raski (31) observed some galling of grape roots with *X. index* and with *X. diversicaudatum* on rose, such as that reported by Schindler (80), but the seedling techniques failed to duplicate the development of extensive enlarged galls in plants infected by these nematodes in pot and field experiments.

Sutherland (84) reported that *Xiphinema bakeri*, a parasite of conifers, fed similarly to *X. index* and *X. diversicaudatum*, but noted that *X. bakeri* twisted its onchiostyle when puncturing a cell wall. During muscular contraction of the bulbar portion of the esophagus, a globule, presumably food, was projected into the intestine. As in other species of *Xiphinema*, *X. bakeri* preferred feeding on root tips and caused the characteristic symptoms of galling, darkening of tissues, and inhibiting of lateral root development.

Rotylenchulus

The reniform nematode, *Rotylenchulus reniformis* causes severe root pruning and necrosis in the field and thus dwarfs major crop plants such as cotton and soybeans. Birchfield (4) reported that the nematodes showed no specificity to the region of maximum invasion and infection, but that young roots were infected near the root tips. Recently, significant attention has been drawn to this nematode and its parasitic influence on major crop plants, although the report by Linford & Oliveira (50) emphasized its potential as a pest many years ago. Histological observations showed young females embedded in roots with the anterior portion of the body extending through the epidermis, cortex, and endodermis.

Meloidogyne

Emergence of root-knot larvae from the egg and the initial penetration of plant cell walls has been discussed by Bird (7). He described how the subventral glands were active during the preparasitic phase of development and how the dorsal esophageal gland was involved during the parasitic phase of the nematode. Bird (8) examined the contents of the dorsal gland ampulla and the stylet exudations of living and fixed *Meloidogyne javanica* larvae and histochemically determined that both components consisted of protein. Because Bird (6) was not able to detect enzymes in sterile larval exudates, he suggested that enzymes may not be produced unless a stimulus is received from the host. Thus, the nematode may produce enzymes or materials in the exudate that might influence the host-parasite relationship. The end result is the formation of a giant cell. Similar cytochemical observations and projections for the cyst nematode, not made yet, would prove useful in a comparative study of host-parasite interactions.

DAMAGE OF HOST TISSUE

Ectoparasitic Nematodes

Tissue response to nematode infection is closely related to the region of infection, the feeding habit, and the developmental cycle of the nematode. Within the genus *Trichodorus*, two responses are detectable during the early stages of feeding. Pitcher & Flegg (68) noted that when the nematodes rasped the epidermis and hypodermis of apple roots with their stylets they caused a superficial, but characteristic browning of tissues. Stubby-root symptoms typical of damage by other species of the genus, could be observed at later stages of infection. Whitehead & Hooper (99) reported that *Trichodorus* can cause stubby-root symptoms on sugar beets, as well as on many crop plants including red clover, carrot, wheat, and potato. Tissue damage from *Xiphinema* is reflected in the symptoms of a darkened and thinned-out root system with some sites of cortical breakdown (15). Histological observations (16) of discolored grape roots infected with *Xiphinema index* showed damaged cells in the epidermis and cortex and the development of phellogen tissue three to five cell layers from the injured site. In older roots the injured cells appeared dead and were suberized. No hyperplasia or hypertrophy were observed in the sections.

In contrast to *Xiphinema* infections, hyperplasia and hypertrophy were common reactions in marigold and grape roots infected with *Longidorus africanus*. The swollen root tips resulted from retardation of meristematic activity, hyperplasia of cortical parenchyma, and the maturation of the root tip region, which in some cases included lateral root formation. In a herbaceous host, marigold, the infected root tips became irreversibly nonfunctional whereas in grape, apical growth continued at a reduced rate, depending on the number of secondary invasions by the nematode.

A type of host adaptation to feeding by an ectoparasitic nematode is reflected in the host response by the interaction of *Hemicycliophora arenaria* and tomato (56). Within 6 hr after penetration, protein had accumulated and nucleoli had enlarged in cells. During the ingestion period, the feed cell appeared to enlarge two- to threefold. Feeding during the 24- to 72-hr period removed contents from cells next to the feed cell. When feeding ended, the feed cells were crushed and pushed to the root surface by new tissue formed by the pericycle. Such meristematic tissue continuously supplied cells for nematode feeding. The extensive development of the median bulb, the long salivation process (2 hr), the presence of a stylet plug, and the type of tissue fed on by the nematode indicated an advanced stage of adaptability of a plant parasitic nematode to its host. Such adaptability is limited, however, because *H. arenaria* has a destructive effect on root development and causes reduced growth of 12–37% in citrus and tomato plants (95).

Semiendoparasitic Nematodes

Second-stage larvae of the citrus nematode, *Tylenchulus semipenetrans,* penetrated epidermal and hypodermal cells and fed on the hypodermis and outer layers of cortical parenchyma. Reaction to nematode feeding appeared to be localized or confined to the cells fed on. Larval feeding did not extend beyond the outer layer was situated deeper in the cortex. Young females established a permanent feeding site which consisted of six to ten cortical cells or "nurse cells" which surrounded a cavity in which the nematode head region was located. The cavity was apparently a destroyed cortical cell that allowed the nematode to move from cell to cell as it pierced the surrounding cells during feeding. The heavy staining of cells that were fed upon indicated their increased cytoplasmic density. Other changes such as increased cell size or number did not occur, however. In addition to the increased density of cytoplasm, volume of nucleus and nucleolus of each cell increased. Advanced stages of affected cells were characterized by thickening of the cell walls, very dense cytoplasm, and enlarged amoeboid nuclei. When nematodes fed on small roots, periderm often formed prematurely in the endodermis (93, 94) (Figure 2).

Thus, the mechanism and location of the feeding site of *T. semipenetrans* appears intermediate to the ectoparasitic behavior of *Trichodorus* and the endoparasitic behavior of *Meloidogyne* and *Heterodera* in which syncytia are stimulated. The term "nurse cells" as a feeding site for citrus nematode also has been used to describe other nematode feeding sites including endoparasitic forms. A major difference in the nurse cells induced by *T. semipenetrans*, however, is the apparent absence of multinucleate cells (syncytia), which are common infections caused by both the root-knot and the cyst nematodes.

Sedentary Endoparasitic Nematodes

ROTYLENCHULUS   The pericycle appears to be the primary site of feeding and source of nutrition for the reniform nematode, *Rotylenchulus reniformis.* In one report, Birchfield (4) mentioned that the phloem tissue of cotton roots was the feeding site for *R. reniformis.* In independent reports, however, Oteifa (61) and Rebois et al (76) observed that reniform nematodes cause hypertrophy of the pericycle in cotton and soybeans. They concluded that this tissue alteration was the major factor in the weakening of the root system. Birchfield (5) recently indicated that the pericycle is the preferred feeding site for five dicotyledonous crop plants, but in corn, the nematode fed throughout the cortical parenchyma, with no enlargement of cells. Heald (34) observed slight variations in feeding sites in cantaloupe which was infected with *R. reniformis.* In new roots, the vermiform female penetrated the root, usually perpendicularly to the root surface, and came to rest with its lip region placed against the endodermis. Pericycle cells next to the cells fed on by nematodes hypertrophied, and their cytoplasm increased in density and stainability. All the hypertrophied cells affected by the nematodes were uninucleate and had prominent nucleoli. In older roots, the vermiform female penetrated directly to the pericycle because the endodermis would have been crushed by the normal expansion of the root. Subsequent cell-to-cell changes in the pericycle were similar to those described for young root infections.

Histological (77) and ultrastructural studies of *R. reniformis* on soybean roots emphasized the endodermis and pericycle in the host-parasite interaction. Besides the hypertrophy and increased cytoplasmic density observed by other workers, Rebois reported that the walls between enlarged affected endodermal and pericycle cells ruptured and thus allowed continuity of cytoplasm among several cells. This

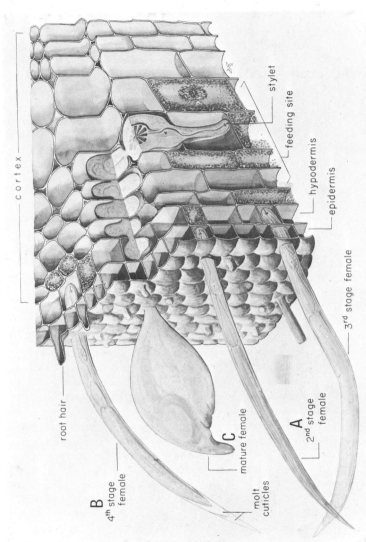

Figure 2 Diagrammatic drawing showing the general cellular arrangement of the citrus nematode, *Tylenchulus semipenetrans*, during penetration and feeding on citrus roots. Three phases of parasitism are depicted: (*a*) larval feeding, (*b*) deeper penetration by young female, and (*c*) establishment of female in a permanent feeding and egg laying position. (Courtesy of Van Gundy & Kirkpatrick.)

separation of cell walls was attributed to partial cell wall lysis and dissolution probably aided by nematode secretions and internal osmotic pressures during the early lysis phase of parasitism. Plasmalemma between affected cells were continuous from cell to cell, and the formation of syncytia was ascribed to the coalescence of cytoplasm between adjoining cells. In the susceptible soybean cultivar, Lee, it was observed that the first cell involved in the formation of the syncytium was from the endodermis. Then within 2 days, portions of the pericycle formed a syncytium. Increased metabolic activity within the affected cells was indicated by the formation of many small vacuoles, an increase in relative numbers of mitochondria, dictyosomes, and plastids. The dense cytoplasm had a high population of ribosomes and rough endoplasmic reticulum; nuclei and nucleoli were obviously hypertrophied. During later stages of nematode development, syncytial walls were modified and developed boundary formations as described for root-knot infections (37). Some thickened cell walls near the infection site had small wall protuberances. Other cell walls near the periphery of the syncytium were about five times the thickness of normal walls. Syncytial and hypertrophied cell walls induced by the reniform nematode were smoother, and had fewer wall protuberances than those described in syncytia induced by root-knot and cyst nematodes (37, 40, 41).

Jones & Dropkin (40) recently emphasized pericycle involvement when soybean roots were infected by the reniform nematode. The authors concurred with Rebois et al (77) that syncytia were formed as a curved sheet of cells that lacked central vacuoles and had dense-staining cytoplasm. Except for the initially stimulated cells near the lip region of the nematode, which are usually two cells wide, the layer of cells appears to arise from pericyclic tissue and may extend halfway or more around the root. Although it was difficult to delineate the effect of a single nematode in roots that had multiple infections, the sheet of cells involved 10–20 cells in longitudinal view and about 11 cells in cross section. Unlike cyst-nematode and root-knot infections, sheets of altered cells rarely involved the vascular elements. Occasional contact between altered cells and xylem elements at protoxylem poles did not induce the formation of clearly defined cell wall ingrowths; however, irregular deposits of wall materials were formed on altered walls contacting xylem elements and on walls between adjacent altered cells.

Parasitism by the reniform nematode appears to involve the induction of a syncytium. Though limited to the initially stimulated cells, it provides an interplay with sheets of stimulated cells of the pericycle. The dense-stained cytoplasm of the altered cells may be comparable to feed cells in the cortex stimulated by *Hemicycliophora* (56) and to the nurse cells of the cortex induced by the citrus nematode, *Tylenchulus semipenetrans*. Neither cell type showed signs of wall dissolution and syncytial formation. Similarly, altered cells induced by *Rotylenchulus* can be similar to initially stimulated cells induced by root-knot nematode (13, 29) and the cells next to developing syncytia of *Heterodera* (25).

HETERODERA   Vascular alterations are obvious during infections by *Heterodera* and *Meloidogyne*. The differences in tissue damage by the two genera have been reviewed and discussed by several workers (21, 26, 46, 59, 81, 97). The mechanism of penetration and establishment of various *Heterodera* spp. in their several hosts can be summarized as follows: second-stage larvae of *Heterodera glycines* can enter the root by direct penetration of cells and traverse the cortex toward the vascular region. If penetration is not direct, the perpendicular orientation is lost. Instead, the larval body is often found lying parallel to the root axis in galleys formed within the cortex during nematode migration. When the nematode becomes sedentary and begins to feed intensively, the anterior portion of the nematode and its lip region is usually directed toward the vascular elements. The second-stage larva may have its lip region in a cortical or endodermal cell from which it can stimulate a syncytium. The multinucleate state can develop within 24 hr after inoculation when cells that lie adjacent to the initially stimulated cell merge with this cell.

Continuity of cytoplasm and ease of movement of nuclei from cell to cell is facilitated by openings in adjacent cell walls. The openings appear to have been caused either by chemical dissolution or physical stresses. Physical stress and chemical action appear to be major factors in cell wall dissolution since cellular hypertrophy is evident during early stages of syncytial development. Such development and increased cytoplasmic density appear to correlate with the developmental stage of the infective larva as it feeds and progresses through the third and fourth stages and eventually becomes an enlarged egg-laying female. Syncytia associated with developing males are limited in size and development and are often limited to the pericycle and cortical tissues. In contrast, syncytia associated with females are extensive and are usually initiated near the protoxylem poles, where vascular tissues are readily incorporated. Syncytia encompass tiers of cells to either side of the nematode laterally and numerous cells longitudinally. Thus large sectors of developing roots are transformed from what would have been vascular tissue into extensive syncytia upon which nematodes feed. The end of feeding and the maturation of the nematode to the cyst stage coincide with the deterioration of syncytial contents and the eventual collapse of the outer wall of the syncytium. Rejuvenative parenchymatous tissue developed between the restricted xylem tissue and the receding syncytial wall (25) (Figure 3).

In an ultrastructural study of the soybean cyst nematode in the susceptible soybean cultivar, Lee, Gipson et al (32) observed that cell walls were perforated in syncytia within 42 hr after inoculation. The edges of the perforations indicated enzymatic breakdown of the walls. The absence of plasmalemma covering these perforations allowed for fusion of protoplasts between adjacent cells. The dense cytoplasm in syncytia indicated in light microscope studies (25) were identified ultrastructurally as plastids and large amounts of endoplasmic reticulum-like materials. The syncytia had many small vacuoles. At later stages of syncytial development (7 days), the syncytium was completely filled with cytoplasm that had an abundance of membranous material resembling smooth endoplasmic reticulum.

The vital nature of the cytoplasm of syncytia, as observed by electron microscopy, was documented also by cytochemical studies that localized several enzymes in syncytia in soybean roots infected with *H. glycines* (28) and *M. incognita acrita* (96).

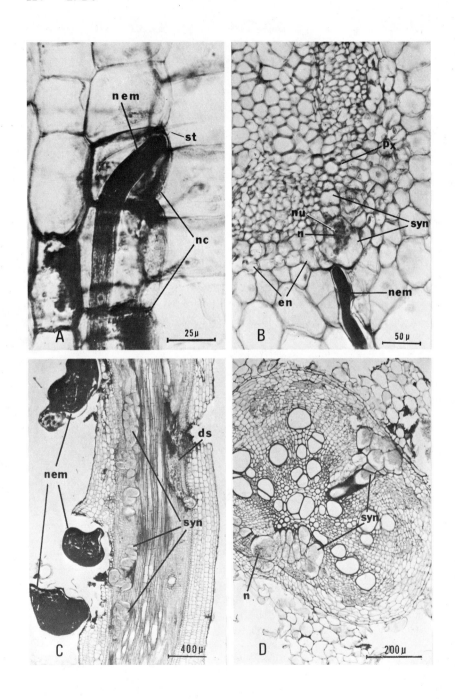

Further documentation of the vital active nature of syncytial cytoplasm is indicated by the ultrastructural observations of syncytia induced by *H. carotae* in carrot roots. Syncytia had many plastids, mitochondria, dictyosomes, rough endoplasmic reticulum, ribosomes, and vesicles. Nuclei were enlarged and multilobed (1).

Attempts were made to determine whether DNA synthesis occurred in the enlarged nuclei of syncytia. Unlike enlarged nuclei in giant cells induced by root-knot nematode (79), the nuclei in syncytia of *H. glycines* were relatively quiescent and showed no extensive localization of tritiated thymidine as an indicator of DNA synthesis. Some label was incorporated in nuclei of syncytia that were in both the cortical and vascular regions of infected roots, however. The most active region in terms of DNA synthesis was in cells along the leading edge of the syncytium. The enlarged nuclei in syncytia induced by the soybean cyst nematode could arise from newly incorporated soybean root cells that had been activated by the presence of the syncytium. These nuclei may have been stimulated to undergo mitosis but failed to enter anaphase. Subsequent incorporation of the nucleus may account for the enlarged state of certain syncytial nuclei (27).

Portions of syncytial walls, especially those adjacent to xylem vessels, are thickened and able to stain heavily with fast green (25). Recent detailed examination of syncytial walls induced by *H. rostochiensis* in potato roots showed the presence of numerous wall protuberances that invaginated the plasma membrane (41). This combination of wall protuberances and the increase in the surface area of the plasmalemma was thought to respond to and facilitate the flow of solutes from xylem to the syncytium. The researchers hypothesized that wall protuberances form as a result of selective flow across the plasmalemma. Similar cell wall ingrowths occur in normal plant systems especially near regions of active growth or secretory tissue and have been termed transfer cells (63, 100) (Figure 4). The function of these cells has been ascribed to intensive selective transport of solutes over short distances. Similarly, Jones & Northcote (41) suggested that the wall ingrowths of syncytia perform a specific function. Also, they indicated that wall protuberances in syncytia formed as a result of feeding demands by nematodes, whereas such outgrowths in transfer cells formed from nutrient demands of actively growing and secretory tissues of plant parts.

Figure 3   Penetration and infection of the soybean cyst nematode, *Heterodera glycines,* in Lee soybean roots, a susceptible reaction: *A*: Intracellular migration of larva (*nem*) through cortex causing necrosis of cells (*nc*); note stylet (*st*) penetration of cortical cell (*c*) wall (1 day after inoculation). *B*: Nematode (*nem*) in contact with early stage of syncytium (*syn*) extending from the endodermis (*en*) into the protoxylem region (*px*). Enlarged nucleus (*n*) surrounds hypertrophied nucleolus (*nu*). Other nuclei are present but not readily distinguishable (2 days after inoculation). *C*: Adult female nematodes (*nem*) adjacent to syncytia (*syn*) in vascular region. After the nematode matures and stops feeding, the syncytium (*ds*) deteriorates (24 days after inoculation). *D*: Vascular damage from syncytial development; syncytia inhibited wide regions of secondary xylem and phloem tissue. Note the cluster of nuclei in a syncytium (27 days after inoculation). (B. Y. Endo, reprinted from *Plant Parasitic Nematodes,* Vol. 2.)

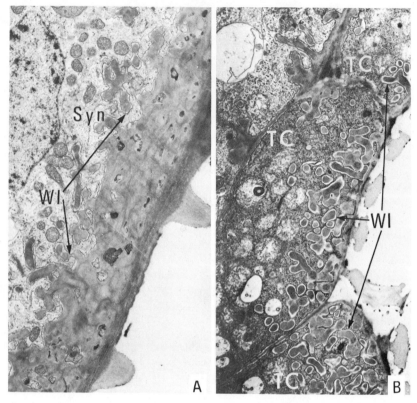

Figure 4   Comparison of cell wall modifications: (*a*) cell wall ingrowths (*WI*) or protuberances within a syncytium (*Syn*) induced in *Impatiens balsamina*, 10 days after inoculation with *Meloidogyne incognita* X 5000 and (*b*) cell wall ingrowths of transfer cells (*TC*) located in the cotyledonary node of *Galium aparine* X 5250. (Courtesy of B. E. S. Gunning and M. G. K. Jones.)

MELOIDOGYNE   An extensive review has recently been completed on the cellular responses to the root-knot nematodes (11). However, some mention should be made of this nematode in comparison to other nematode-host interactions. In contrast to infection by the cyst-nematode species and the reniform nematode, root-knot larvae usually penetrate root tissues by intercellular migration (29) (Figure 5) although larvae also can penetrate intracellularly (13).

The root-knot nematode is unique in its capabilities of inducing syncytia as well as causing a proliferation of tissues that result in well-recognized galls on roots and other plant parts. Larval penetration through the cortex and meristematic tissues of the root tip and the establishment of feeding sites in the central vascular region cause severe alterations in root morphology. Kostoff & Kendall (45) noted that giant cells appeared soon after nematode penetration in the host plant and that the parasite began to eject secretions originating from the salivary glands. In the absence of stylet penetration of cell walls, these authors concluded that the nematode received its nutrition from the host through suction of plant substances that were deposited in the intercellular spaces between stimulated cells. Linford (48) established, however, that root-knot larvae penetrate the walls of *Pisum sativum* with their stylets and feed on induced giant cells. In vitro, the saliva, which apparently flowed from the dorsal gland, was seen to flow forward from the extreme tip of the protruded stylet. Salivary flow into a giant cell was not seen because it was obscured by the dense cytoplasm of the syncytium (48). Bird (6, 8) studied the nature of the stylet exudations and the functional relationships of subventral and dorsal glands to host penetration and feeding behavior. Syncytia induced by root-knot nematodes were characterized as multinucleate units that were derived from plant tissues stimulated by the nematodes. Cell wall dissolution and endomitosis without subsequent cytokinesis were considered the mechanism for syncytial or giant cell formation (23, 45, 47, 62).

In an electron microscope study, Huang & Maggenti (36) did not observe fusion of protoplasts of cells adjacent to a syncytium. They concluded that the multinucleate state of giant cells on *Vicia faba* roots infected by *M. javanica* was derived only from repeated mitosis without cytokinesis of a single stimulated cell. Similarly, Paulson & Webster (64) did not observe cell wall dissolution that would contribute to syncytial formation in their electron microscope study of tomato roots infected with *M. incognita*. On the basis of light microscope observations, they concluded that giant cells or syncytia were formed by cell wall expansion and nuclear duplication without cytokinesis during the early stages of development. Also, Jones & Northcote (42) found that giant cells induced in coleus roots by *M. arenaria* appeared to be formed solely by expansion of single cells with nuclear endomitosis. No cell wall gaps between giant cells were observed with the electron microscope, but gaps were possible especially towards the outside of the complex.

In defense of the cell-wall-dissolution concept in the formation syncytia, Hesse (35) studied the cytology of nematode galls on roots of cacti. He concluded that there are two systems that account for the enlargement of giant cells: disintegration of cell walls and abnormal growth of a single cell. To support the concept that giant cells developed from several smaller cells, he observed the presence of irregularly shaped giant-cell outlines that could extend into several planes. He found short particles of cell wall materials in the middle of the giant cell that had no optical connection to the giant cell wall. The cellulose nature of these particles was verified optically and cytochemically. An abnormal-growth concept of giant cell development is suggested by the presence of small parenchymatous cells that were compressed, hypertrophied, and characteristically arranged around certain giant cells. These curved and malformed cells indicated the directional growth of impinging giant cells. Bird (9) reported that in epoxy-embedded tissues sectioned at 2 μm, syncytia showed close associations with adjacent cells. These cells protruded into the cytoplasm of the syncytium, and at times the wall separating the two units broke down and permitted cytoplasmic contact. He concluded that syncytia are formed partly by incorporation of cells where adjacent cell walls are dissolved. To further

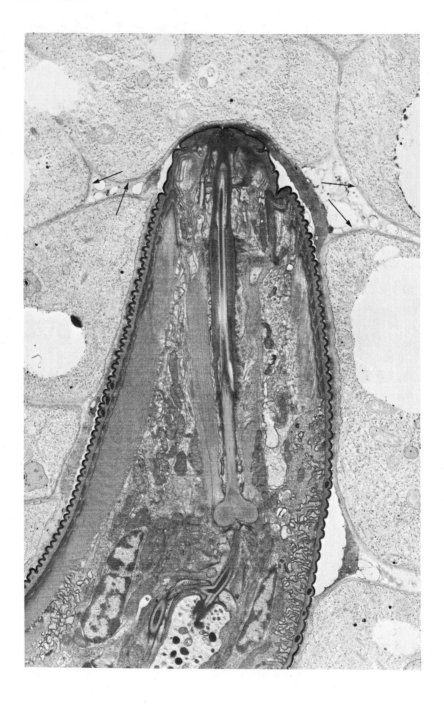

show how nuclei were contributed from several cells rather than a single cell, Bird (10) reported that the ploidy of a syncytium in *Vicia faba* infected with the root-knot nematode did not follow a sequence of 4, 8, 16, 32 N ploidy. This would have been the sequence if a single stimulated cell had developed into a syncytium. In one syncytium, only one nucleus had the 2N complement of chromosomes, whereas others had 4, 6, 8, and 10 N ploidy. Thus, evidence is growing that giant cells induced by the root-knot nematode arise not only from the stimulation of a single cell, but also from the incorporation of adjacent cells through the breakdown of cell walls that separate them.

The thickened syncytial walls mentioned for *Heterodera* species are prominent also in syncytia induced by the root-knot nematode. The role of cell wall protuberances was mentioned by Huang & Maggenti (37) and later by Paulson & Webster (65). Jones & Northcote (42) emphasized proliferation of cell wall protuberances, the resulting increased surface area, and the protuberance-plasma membrane system as a medium through which nutrients could flow. They suggested that syncytia or giant cells induced by nematodes are a multinucleate form of a transfer cell, in which wall protuberances develop in response to the demand of the nematode for nutrients.

INTERACTIONS WITH OTHER PATHOGENS

Several reviews (66, 71–73, 98) have suggested that nematodes interact with and increase or alter the pathogenicity of other organisms.

The ectoparasite, sting nematode, *Belonolaimus longicaudatus,* promoted the development of fusarium wilt in cotton seedlings. This predisposition of the host to fungal infection was similar to that induced by the root-knot nematode. Thus, a question is raised on the mechanism of host susceptibility to an interacting pathogen. If the incidence of fusarium wilt in resistant and susceptible cotton seedlings is related to physiological changes in host tissue, such changes in physiology and susceptibility may be influenced not only by the root-knot nematode, which causes galling and induces giant cells, but also by the sting nematode that feeds on the surface layers of roots (18, 104). It also has been demonstrated that pathogenic fungi in a host can influence nematode populations (19, 60). Significantly larger populations of the stunt nematode *Tylenchorhynchus claytoni* were found on pea roots infected with *Fusarium oxysporum* f. sp. *pisi* race 1 than on previously noninfected plants. Similarly, Yang et al (104) found that among four nematode genera, wilt was most severe in susceptible cotton when the fungus was inoculated with either the sting nematode, *Belonolaimus longicaudatus* or the root-knot nematode. The addition of 100 ppm of cholesterol to the culture medium increased the rate of wilt

---

Figure 5 Longitudinal section through the anterior region of the root-knot nematode, *Meloidogyne incognita,* in roots of red clover, *Trifolium pratense* Kenland. Note the physical pressure exerted by the second-stage larva on the cortical cells which cause separation along their middle lamellae (*arrows*), X 5280. (Endo & Wergin, reprinted from *Protoplasma,* Vol. 78.)

development when *Fusarium* was combined with either of these two nematode genera.

Pratylenchus penetrans, a migratory endoparasitic nematode, can reproduce at a faster rate when associated with a verticillium-infested eggplant and tomato than with plants not in the presence of the fungus (60). In addition to the population effect, the fungus may modify epidermal and cortical tissues of roots to facilitate entry of *P. penetrans* into the root. Similarly, *Pratylenchus minyus* reproduction was higher on verticillium-infected peppermint roots than on noninfected roots. In addition, there was an increase in both the incidence and severity of verticillium wilt symptoms and a reduced incubation period for the fungus in the presence of the nematode (30).

Conroy et al (17) studied the interaction of *Verticillium albo-atrum* with the lesion nematode, *P. penetrans,* using various inoculum levels and controlled environmental conditions. Although their results agree with the general concept that lesion nematodes increase the incidence of verticillium wilt they did not show that nematode populations increased in the presence of the fungus. Conversely, populations of *P. penetrans* on tomato seedlings were smaller when both nematodes and fungus were present than when only nematodes were present. Separation of the nematode and fungus by a split-root technique failed to alter the plants' susceptibility to verticillium.

The interaction of lesion nematodes with a non-wilt–producing fungus is illustrated in the observations of *Pratylenchus brachyurus* and *Phytophthora parasitica* Dast var. *nicotianae* developing on tobacco roots (38). Inoculation with the nematode one week before or during the inoculation with the fungus caused a more rapid development of black-shank symptoms than when only the fungus was used. It was evident that the migration of the nematode through the tissues caused wounds that favored fungus invasion. Unlike fusarium wilt interactions with the root-knot nematode (57), the inoculation of lesion nematodes three weeks before exposure to the black-shank fungus failed to increase the severity of the fungus-induced symptoms. The delayed symptom expression was apparently caused by the nematode through the destruction of cortical cells during migration and the inhibitory effect of necrotic cells on fungal development.

The cyst nematodes that affect the cortex and the vascular system of plant roots interact with fungi in infected plants. The sugar-beet cyst nematode, *Heterodera schachtii,* facilitated penetration and establishment of *Rhizoctonia solani* in sugar beet seedlings (69). The interacting organisms softened tissues throughout the seedlings in 3 days. Giant cells induced by the nematode were especially suitable as a substrate for the fungus. Mycelium entered giant cells soon after they were formed and spread beyond to adjacent root tissues that were not directly affected by the nematode. However, many cell walls broke down before the fungal hyphae advanced.

Interactions of the root-knot nematode with fungi have been implicated in several disease complexes. In a study of the fusarium wilt and root-knot complex in tobacco, Melendez & Powell (57) found vigorous fungal colonization of giant cells and adjacent tissues of both susceptible and resistant plants. Although colonized by the fungus, the giant cells were highly sensitive to the organism. The cytoplasm was destroyed soon after fungal invasion. The effect of the nematode on fungal development was not limited to giant cells, but also involved vascular tissues far removed from the nematode site. Similarly, Porter & Powell (70) showed that root-knot nematode predisposed tobacco plants to fusarium wilt infection. The root-knot infection appeared to affect the physiology of the plant, the fungus infection, and the subsequent development of wilt symptoms. Fungal hyphae were slightly thicker and more extensive on tobacco tissues which previously had been invaded by the root-knot nematode. A 4-week period of predisposition of the host by the root-knot nematode caused more vigorous fungal growth than a 3-week period. This work supported the concept that nematode-induced modification of host tissues and possible altered physiology may be more important than mere nematode-influenced mechanical injury in the interaction between the two types of organisms.

Predisposition of a host to infection was reported earlier for a nonwilt type pathogen, the black-shank fungus, *P. parasitica* var. *nicotianae*. The fungus was attracted to root-knot nematode-induced hypertrophied and hyperplastic tissues. Hyphae progressed rapidly and directly into these tissues. It was apparent that infection by *Meloidogyne incognita acrita* predisposed both black-shank - susceptible and -resistant tobacco plants to become suitable substrates for fungal development (74).

An interaction between the root-knot nematode and an ectomycorrhizae of pine roots was reported by Riffle (78). Roots altered morphologically by mycorrhizal symbionts appeared to be readily invaded by the root-knot nematode and supported giant-cell development. Besides the disruption of normal vascular flow of nutrients and water by giant-cell formation, the fungal mantle formed by the mucorrhiza was ruptured by the developing female. The breakdown of the fungal mantle is important in root exposure to entry by other pathogenic organisms (54). Barham et al (2) found that pine seedling roots with vascular tissues of mycorrhizae can be invaded by the ectoparasitic nematode, *Tylenchorhynchus claytoni* and by *Helicotylenchus dihystera*. In the latter case the spiral nematode, *H. dihystera,* disrupted the structural integrity of the fungal mantle formed by the ectomycorrhiza and provided infection courts for the plant pathogenic fungus, *Phytophthora cinnamomi.* The development of such combinations of organisms demonstrate a three-way interaction in the root tissues.

Interactions of the root-knot nematode, *M. incognita,* and bacteria were described in a comparison of gall development of tomato seedlings grown under aseptic and septic conditions. Plants infected by root-knot nematodes and grown under aseptic conditions had massive galls that were firm and did not disintegrate, whereas galls formed under septic conditions became necrotic (3). Although two genera of fungi, *Fusarium* spp. and *R. solani,* were isolated from necrotic galls, the predominant microflora isolated were bacteria. The influence of the microflora was documented by a 48% reduction in dry weight of roots of septically grown plants. But weight did not differ significantly when plants were infected only with surface-sterilized root-knot nematodes and grown under aseptic conditions (55).

Lucas et al (52) showed the relationship between the root-knot nematode and the Granville wilt bacterium, *Pseudomonas solanacearum,* on tobacco. The increase and intensity of wilt symptoms of plants inoculated with both pathogens were attributed to nematodes that damaged roots and provided ports of entry for bacteria. In a recent histological study, Johnson & Powell (39) found that wilt-susceptible tobacco plants inoculated with root-knot nematode 3 and 4 weeks before inoculation with bacteria developed more extensive wilt symptoms than plants inoculated simultaneously with both organisms. Thus, the root-knot nematode might predispose the plants to invasion and colonization by bacteria as it has been shown with fungi (70).

Moreover, El-Goorani et al (24) observed increased severity of the gladiolus scab when the bacterium *Pseudomonas marginata* was inoculated simultaneously or 1 week after root-knot inoculation.

INTERACTIONS WITH VIRUSES AND VECTOR RELATIONSHIPS

In a recent paper on this subject, Taylor & Robertson (90) divided nematode-transmitted viruses into two groups: (*a*) the nepoviruses that have isometric particles about 30 nm in diameter and are transmitted by species of *Xiphinema* and *Longidorus* and (*b*) tobraviruses, which have straight, tubular particles and are transmitted by species of *Trichodorus* and *Paratrichodorus.*

Teliz et al (92) showed the rapidity at which viruses are acquired from infected plants and transmitted to host plants. They found that *Xiphinema americanum* can acquire tomato ringspot virus from mechanically inoculated cucumber plants within 1 hr and can transfer the inoculum to healthy cucumber plants in 1 hr. The amount of transmission was not proportional to the length of acquisition time up to the first 48 hr. However, in an acquisition time of 4 days or longer, there was 100% transmission among the test plants.

Once the nematodes acquire the virus, retention of the virus within a nematode can be measured. The retention time is determined from when the nematode first contacts the infected plant until the nematode reaches its limits in transmitting the virus to a healthy plant. Viruses appear to be able to persist in nematode vectors for several weeks or more. *X. americanum* stored at 10°C transmitted the virus after 49 weeks. *Xiphinema diversicaudatum* females, males, and juveniles transmitted arabis mosaic virus (AMV) and strawberry latent ringspot virus (SLRV) to cucumber seedlings. In the absence of plants, AMV persisted in nematodes for 112 days and SLRV for 84 days (91).

Although viruses persist in nematode populations for long periods, nematode-transmitted viruses do not persist through a molt or pass through nematode eggs. This feature was found in grape fanleaf virus and its associated vector, *Xiphinema index* (87).

In some cases nematodes can ingest viruses and yet not transmit the viruses. Taylor & Murant (86) found that when *Longidorus elongatus* de Man was allowed to feed on herbaceous hosts infected with each of four viruses, the macerated nematodes could transmit all the viruses on the test plant, *Chenopodium quinoa.* When the nematodes were allowed to feed on test plants, however, only raspberry ringspot (RRSV) and tomato black ring virus (TBRV) were transmitted. It was suggested that transmission was not purely mechanical but a biologically influenced process involving the vector and virus. Other experiments indicated a close biological association between *X. diversicaudatum* and the virus that it transmits (85).

Ultrastructural examinations were made through the anterior region of specimens of *L. elongatus* that had fed on RRSV- and TBRV-infected plants. Virus-like particles were found in the lumen of the buccal capsule and in the space between the stylet and the guiding sheath. When *L. elongatus* was allowed to feed on AMV-infected plants, however, virus particles were not found in the stylet guiding sheath of this nematode. Since AMV is not normally transmitted by the nematode, it was suggested that the association of virus particles with the cuticular guiding sheath could explain the specificity of virus transmission by this nematode (88).

It was suggested that virus particles are selectively and specifically adsorbed onto the cuticular guiding sheath or the lining of the esophagus when the plant sap, with virus, passes from the plant to the nematode intestine. Viruses may (*a*) move directly into the intestine, (*b*) be adsorbed at the retention site, or (*c*) after retention, become dissociated and be released for transmission into a plant host. Specific release mechanisms with ionic changes in the lumen of the esophagus are thought to control the specificity of virus transmission. Furthermore, this specificity appears to be related to the association of protein surface of the virus and the cuticular surface of the nematode in which the virus is retained (89).

Raski et al (75) studied the location and orderly arrangement of virus particles of grapevine fanleaf and yellow mosaic virus in *X. index*. They noted a sharp demarcation of the presence of virus particles in the anterior end of the stylet extension or odontophore and the absence of particles in the odontostyle. The presence and orderly monolayer of virus particles in the odontophore and the esophageal lumen may be due to the cuticular morphology and possible differences in charged receptor sites. The final phase of virus transmission occurs in the host tissues. The nematode can transfer the dissociated virus into a plant cell in a manner that does not kill the cell and allows for virus transfer to deeper lying tissue. Unlike some of the feeding processes described for *Trichodorus* where entire cell contents were consumed (102) only a delicate balance during feeding allows for proper entry and replication of the transmitted virus.

CONCLUSION

This review emphasizes the feeding of certain ectoparasitic nematodes. Many of the feeding mechanisms and the intracellular host responses can contribute information to host-parasite interactions of endoparasitic nematodes that are not readily visualized in vivo. Recent cine-photographic presentations of nematode feeding have emphasized how traumatically a nematode can affect a host cell. However, in the moderation of this feeding process, cells do not always appear to be killed outright, and host cells and tissues are stimulated to provide nutrients for pathogen survival. There appears to be a pattern of sophistication of tissue response to nematode invasion and feeding from the cell-to-cell feeding by an ectoparasite, *Hemicycliophora,* to the nurse cells of *Tylenchulus semipenetrans,* and to the syncytia or giant cells induced by the cyst and root-knot nematodes. The recognition of cell wall protuberances and transfer cell-like functions probably should be studied in other host-parasite relationships. For example, *T. semipenetrans* and the reniform nematode either lack or have restricted boundary formations on cells upon which they feed. Yet, they demand considerable nutrients from host tissues.

The mechanisms of pathogenesis of nematode-infected plants need to be evaluated and studied in relation to both primary pathogens and the other organisms in the environment that are usually associated with the diseased plant.

The challenge of agriculture is to broaden our knowledge of the mechanisms of plant pathogenesis, so we can increase our food and fiber production.

Literature Cited

1. Ambrogioni, L., Porcinai, G. M. 1972. Studio ultrastrutturistico delle cellule giganti prodotte da *Heterodera carotae* Jones, 1950 (Nematoda: Heteroderidae) in radici di carota. *Redia* 53:437–48
2. Barham, R. O., Marx, D. H., Ruehle, J. L. 1974. Infection of ectomycorrhizal and nonmycorrhizal roots of shortleaf pine by nematodes and *Phytophthora cinnamomi. Phytopathology* 64:1260–64
3. Bergeson, G. B., Athow, K. L., Laviolette, F. A., Sister Mary Thomasine 1964. Transmission, movement, and vector relationships of tobacco ringspot virus in soybean. *Phytopathology* 54:723–28
4. Birchfield, W. 1962. Host-parasite relations of *Rotylenchulus reniformis* on *Gossypium hirsutum. Phytopathology* 52:862-65
5. Birchfield, W. 1972. Differences in host-cell responses to the reniform nematode. *Phytopathology* 62:747 (Abstr.)
6. Bird, A. F. 1966. Some observations on exudates from *Meloidogyne* larvae. *Nematologica* 12:471–82
7. Bird, A. F. 1967. Changes associated with parasitism in nematodes. I. Morphology and physiology of preparasitic and parasitic larvae of *Meloidogyne javanica. J. Parasitol.* 53:768–76
8. Bird, A. F. 1968. Changes associated with parasitism in nematodes. IV. Cytochemical studies on the ampulla of the dorsal esophageal gland of *Meloidogyne javanica* and on exudations from the buccal stylet. *J. Parasitol.* 54:879–90
9. Bird, A. F. 1972. Cell wall breakdown during the formation of syncytia induced in plants by root knot nematodes. *Int. J. Parasitol.* 2:431–32
10. Bird, A. F. 1973. Observations on chromosomes and nucleoli in syncytia induced by *Meloidogyne javanica. Physiol. Plant Pathol.* 3:387–91
11. Bird, A. F. 1974. Plant response to root-knot nematode. *Ann. Rev. Phytopathol.* 12:69–85
12. Blake, C. D. 1962. The etiology of tulip-root disease in susceptible and in resistant varieties of oats infested by the stem nematode, *Ditylenchus dipsaci* (Kühn) Filipjev. II. Histopathology of tulip-root and development of the nematode. *Ann. Appl. Biol.* 50:713–22
13. Christie, J. R. 1936. The development of root-knot nematode galls. *Phytopathology* 26:1–22
14. Cohn, E. 1965. On the feeding and histopathology of the citrus nematode. *Nematologica* 11:47–54
15. Cohn, E. 1970. Observations of the feeding and symptomatology of *Xiphinema* and *Longidorus* on selected host roots. *J. Nematol.* 2:167–73
16. Cohn, E., Orion, D. 1970. The pathological effect of representative *Xiphinema* and *Longidorus* species on selected host plants. *Nematologica* 16:423–28
17. Conroy, J. J., Green, R. J. Jr., Ferris, J. M. 1972. Interaction of *Verticillium albo-atrum,* and the root lesion nematode, *Pratylenchus penetrans,* in tomato roots at controlled inoculum densities. *Phytopathology* 62:362–66
18. Cooper, W. E., Brodie, B. B. 1963. A comparison of *Fusarium* wilt indices of cotton varieties with root-knot and sting nematodes as predisposing agents. *Phytopathology* 53:1077–80
19. Davis, R. A., Jenkins, W. R. 1963. Effect of *Meloidogyne* spp. and *Tylenchorhynchus claytoni* on pea wilt incited by *Fusarium oxysporum* f. *pisi* race 1. *Phytopathology* 53:745 (Abstr.)

20. Doncaster, C. C., Seymour, M. K. 1973. Exploration and selection of penetration site by Tylenchida. *Nematologica* 19:137–45

21. Dropkin, V. H. 1969. Cellular responses of plants to nematode infections. *Ann. Rev. Phytopathol.* 7:101–22

22. Dropkin, V. H. 1972. Pathology of *Meloidogyne*-galling giant cell formation, effects of host physiology. *Eur. Plant Protect. Org. Bull.* No. 6:23–32

23. Dropkin, V. H., Nelson, P. E. 1960. The histopathology of root-knot nematode infections in soybeans. *Phytopathology* 50:442–47

24. El-Goorani, M. A., Abo-El-Dahab, M. K., Mehiar, F. F. 1974. Interaction between root-knot nematode and *Pseudomonas marginata* on gladiolus corms. *Phytopathology* 64:271–72

25. Endo, B. Y. 1964. Penetration and development of *Heterodera glycines* in soybean roots and related anatomical changes. *Phytopathology* 54:79–88

26. Endo, B. Y. 1971. Nematode-induced syncytia (giant cells). Host-parasite relationships of Heteroderidae. *Plant Parasitic Nematodes*, ed. B. M. Zuckerman, W. F. Mai, R. A. Rohde, 2:91–117. New York: Academic. 347 pp.

27. Endo, B. Y. 1971. Synthesis of nucleic acids at infection sites of soybean roots parasitized by *Heterodera glycines*. *Phytopathology* 61:395–99

28. Endo, B. Y., Veech, J. A. 1969. The histochemical localization of oxidoreductive enzymes of soybeans infected with the root knot nematode *Meloidogyne incognita acrita*. *Phytopathology* 59:418–25

29. Endo, B. Y., Wergin, W. P. 1973. Ultrastructural investigation of clover roots during early stages of infection by the root-knot nematode, *Meloidogyne incognita*. *Protoplasma* 78:365–79

30. Faulkner, L. R., Skotland, C. B. 1965. Interactions of *Verticillium dahliae* and *Pratylenchus minyus* in *Verticillium* wilt of peppermint. *Phytopathology* 55:583–86

31. Fisher, J. M., Raski, D. J. 1967. Feeding of *Xiphinema index* and *Xiphinema diversicaudatum*. *Proc. Helminthol. Soc. Wash.* 34:68–72

32. Gipson, I., Kim, K. S., Riggs, R. D. 1971. An ultrastructural study of syncytium development in soybean roots infected with *Heterodera glycines*. *Phytopathology* 61:347–53

33. Godfrey, G. H. 1929. A destructive root disease of pineapples and other plants

due to *Tylenchus brachyurus*, n. sp. *Phytopathology* 19:611–29

34. Heald, C. M. 1975. Pathogenicity and histopathology of *Rotylenchulus reniformis* infecting cantaloupe. *J. Nematol.* 7:149–52

35. Hesse, M. 1970. Cytologische untersuchungen an nematodengallen. *Osterr. Bot. Z.* 118:517–41

36. Huang, C. S., Maggenti, A. R. 1969. Mitotic aberrations and nuclear changes of developing giant cells in *Vicia faba* caused by root-knot nematode, *Meloidogyne javanica*. *Phytopathology* 59:447–55

37. Huang, C. S., Maggenti, A. R. 1969. Wall modifications in developing giant cells of *Vicia faba* and *Cucumis sativus* induced by root knot nematode, *Meloidogyne javanica*. *Phytopathology* 59:931–37

38. Inagaki, H., Powell, N. T. 1969. Influence of the root-lesion nematode on black shank symptom development in flue-cured tobacco. *Phytopathology* 59:1350–55

39. Johnson, H. A., Powell, N. T. 1969. Influence of root knot nematode on bacterial wilt development in flue-cured tobacco. *Phytopathology* 59:486–91

40. Jones, M. G. K., Dropkin, V. H. 1975. Cellular alterations induced in soybeans by three endoparasitic nematodes. *Physiol. Plant Pathol.* In press

41. Jones, M. G. K., Northcote, D. H. 1972. Nematode-induced syncytium-A multinucleate transfer cell. *J. Cell Sci.* 10:789–809

42. Jones, M. G. K., Northcote, D. H. 1972. Multinucleate transfer cells induced in coleus roots by the root-knot nematode, *Meloidogyne arenaria*. *Protoplasma* 75:381–95

43. Kisiel, M., Castillo, J., Zuckerman, B. M. 1971. An adhesive plug associated with the feeding of *Hemicycliophora similis* on cranberry. *J. Nematol.* 3:296–98

44. Klingler, J. 1965. On the orientation of plant nematodes and of some other soil animals. *Nematologica* 11:4–18

45. Kostoff, D., Kendall, J. 1930. Cytology of nematode galls on *Nicotiana* roots. *Zentralbl. Bakteriol. Parasitenk. Hyg. Abt. II* 81:86–91

46. Krusberg, L. R. 1963. Host response to nematode infection. *Ann. Rev. Phytopathol.* 1:219–40

47. Krusberg, L. R., Nielsen, L. W. 1958. Pathogenesis of root-knot nematodes to

Porto Rico variety of sweetpotato. *Phytopathology* 48:30–39

48. Linford, M. B. 1937. The feeding of the root-knot nematode in root tissue and nutrient solution. *Phytopathology* 27: 824–35

49. Linford, M. B. 1939. Attractiveness of roots and excised shoot tissues to certain nematodes. *Proc. Helminthol. Soc. Wash.* 6:11–18

50. Linford, M. B., Oliveira, J. M. 1940. *Rotylenchulus reniformis,* nov. gen., n. sp., a nematode parasite of roots. *Proc. Helminthol. Soc. Wash.* 7:35–42

51. Luc, M. 1961. Note préliminaire sur le déplacement de *Hemicycliophora paradoxa* Luc (Nematoda-Criconematidae) dans le sol. *Nematologica* 6:95–106

52. Lucas, G. B., Sasser, J. N., Kelman, A. 1955. The relationship of root-knot nematodes to Granville wilt resistance in tobacco. *Phytopathology* 45:537–40

53. Mankau, R., Linford, M. B. 1960. Host-parasite relationships of the clover cyst nematode, *Heterodera trifolii* Goffart. *Ill. Agric. Exp. Stn. Bull.* 667:1–50

54. Marx, D. H., Davey, C. B. 1969. The influence of ectotrophic mycorrhizal fungi on the resistance of pine roots to pathogenic infections. IV. Resistance of naturally occurring mycorrhizae to infections by *Phytophthora cinnamomi. Phytopathology* 59:559–65

55. Mayol, P. S., Bergeson, G. B. 1969. The role of secondary invaders in premature breakdown of plant roots infected with *Meloidogyne incognita. J. Nematol.* 1:17 (Abstr.)

56. McElroy, F. D., Van Gundy, S. D. 1968. Observations on the feeding processes of *Hemicycliophora arenaria. Phytopathology* 58:1558–65

57. Melendéz, P. L., Powell, N. T. 1967. Histological aspects of the *Fusarium* wilt-root knot complex in flue-cured tobacco. *Phytopathology* 57:286–92

58. Mountain, W. B. 1954. Studies of nematodes in relation to brown root rot of tobacco in Ontario. *Can. J. Bot.* 32:737–59

59. Mountain, W. B. 1965. Pathogenesis by soil nematodes. *Ecology of Soil-borne Plant Pathogens,* ed. K. F. Baker, W. C. Snyder, 285–301. Berkeley, Calif.: Univ. Calif. Press. 571 pp.

60. Mountain, W. B., McKeen, C. D. 1962. Effect of *Verticillium dahliae* on the population of *Pratylenchus penetrans. Nematologica* 7:261–66

61. Oteifa, B. A. 1970. The reniform nematode problem of Egyptian cotton production. *J. Parasitol.* 56:255 (Abstr.)

62. Owens, R. G., Specht, H. N. 1964. Root-knot histogenesis. *Contrib. Boyce Thompson Inst.* 22:471–89

63. Pate, J. S., Gunning, B. E. S. 1969. Vascular transfer cells in angiosperm leaves. A taxonomic and morphological survey. *Protoplasma* 68:135–56

64. Paulson, R. E., Webster, J. M. 1970. The cellular response of a resistant tomato plant to *Meloidogyne incognita. J. Parasitol.* Sect. II, Part 1 56:260 (Abstr.)

65. Paulson, R. E., Webster, J. M. 1972. Ultrastructure of the hypersensitive reaction in roots of tomato, *Lycopersicon esculentum* L., to infection by the root-knot nematode, *Meloidogyne incognita. Physiol. Plant Pathol.* 2:227–34

66. Pitcher, R.S. 1965. Interrelationships of nematodes and other pathogens of plants. *Helminthol. Abstr.* 34:1–17

67. Pitcher, R. S. 1967. The host-parasite relations and ecology of *Trichodorus viruliferus* on apple roots, as observed from an underground laboratory. *Nematologica* 13:547–57

68. Pitcher, R. S., Flegg, J. J. M. 1965. Observation of root feeding by the nematode *Trichodorus viruliferus* Hooper. *Nature London* 207:317

69. Polychronopoulos, A. G., Houston, B. R., Lownsbery, B. F. 1969. Penetration and development of *Rhizoctonia solani* in sugar beet seedlings infected with *Heterodera schachtii. Phytopathology* 59:482–85

70. Porter, D. M., Powell, N. T. 1967. Influence of certain *Meloidogyne* species on *Fusarium* wilt development in flue-cured tobacco. *Phytopathology* 57: 282–85

71. Powell, N. T. 1963. The role of plant-parasitic nematodes in fungus diseases. *Phytopathology* 53:28–35

72. Powell, N. T. 1971. Interaction of plant parasitic nematodes with other disease causing agents. See Ref. 25, 119–36

73. Powell, N. T. 1971. Interactions between nematodes and fungi in disease complexes. *Ann. Rev. Phytopathol.* 9:253–74

74. Powell, N. T., Nusbaum, C. J. 1960. The black shank-root-knot complex in flue-cured tobacco. *Phytopathology* 50:899–906

75. Raski, D. J., Maggenti, A. R., Jones, N. O. 1973. Location of grapevine fanleaf and yellow mosaic virus particles in *Xiphinema index. J. Nematol.* 5:208–11

76. Rebois, R. V., Epps, J. M., Hartwig, E. E. 1970. Correlation of resistance in soybeans to *Heterodera glycines* and *Rotylenchulus reniformis. Phytopathology* 60:695–700

77. Rebois, R. V., Madden, P. A., Eldridge, B. J. 1975. Some ultrastructural changes induced in resistant and susceptible soybean roots following infection by *Rotylenchulus reniformis. J. Nematol.* 7:122–39

78. Riffle, J. W. 1973. Histopathology of *Pinus ponderosa* ectomycorrhizae infected with a *Meloidogyne* species. *Phytopathology* 63:1034–40

79. Rubinstein, J. H., Owens, R. G. 1964. Thymidine and uridine incorporation in relation to the ontogeny of root-knot syncytia. *Contrib. Boyce Thompson Inst.* 22:491–502

80. Schindler, A. F. 1957. Parasitism and pathogenicity of *Xiphinema diversicaudatum,* an ectoparasitic nematode. *Nematologica* 2:25–31

81. Seinhorst, J. W. 1961. Plant-nematode inter-relationships. *Ann. Rev. Microbiol.* 15:177–96

82. Shepherd, A. M. 1970. The influence of root exudates on the activity of some plant-parasitic nematodes. In *Root Diseases and Soil Borne Pathogens, Part Proc. 1st Int. Congr. Plant Pathol. London, 1968,* ed. T. A. Toussoun et al, 134–37. Berkeley & Los Angeles, Calif.: Univ. Calif. Press

83. Steinbach, P. 1972. Studies on the behavior of larvae of the potato root eelworm *Heterodera rostochiensis* on and in the roots of the host plant *Lycopersicon esculentum.* II. The penetration of larvae of the potato root eelworm into the host plant. *Biol. Zentralbl.* 91:743–56

84. Sutherland, J. R. 1967. Parasitism of *Tylenchus emarginatus* on conifer seedling roots and some observations on the biology of the nematode. *Nematologica* 13:191–96

85. Taylor, C. E. 1968. Nematology—association of virus and vectors. *Rept. Scot. Hortic. Res. Inst.* 14:64–68

86. Taylor, C. E., Murant, A. F. 1969. Transmission of strains of raspberry ringspot and tomato black ring viruses by *Longidorus elongatus* (de Man). *Ann. Appl. Biol.* 64:43–48

87. Taylor, C. E., Raski, D. J. 1964. On the transmission of grape fanleaf by *Xiphinema index. Nematologica* 10:489–95

88. Taylor, C. E., Robertson, W. M. 1969. The location of raspberry ringspot and tomato black ring viruses in the nematode vector, *Longidorus elongatus* (de Man). *Ann. Appl. Biol.* 64:233–37

89. Taylor, C. E., Robertson, W. M. 1974. Specific retention and transmission of viruses by nematodes. In *XII Simp. Int. Nematol.,* Granada, Spain, September 1–7, 1974, pp. 98–100 (Abstr.)

90. Taylor, C. E., Robertson, W. M. 1975. Acquisition, retention and transmission of viruses by nematodes. *Nematode Vectors of Plant Viruses,* ed. F. Lamberti, C. E. Taylor, J. W. Seinhorst. New York: Plenum. In press

91. Taylor, C. E., Thomas, P. R. 1968. The association of *Xiphinema diversicaudatum* (Micoletzky) with strawberry latent ringspot and arabis mosaic viruses in a raspberry plantation. *Ann. Appl. Biol.* 62:147–57

92. Teliz, D., Grogan, R. G., Lownsbery, B. F. 1966. Transmission of tomato ringspot, peach yellow bud mosaic, and grape yellow vein viruses by *Xiphinema americanum. Phytopathology* 56:658–63

93. Van Gundy, S. D., Kirkpatrick, J. D. 1964. Nature of resistance in certain citrus rootstocks to citrus nematode. *Phytopathology* 54:419–27

94. Van Gundy, S. D., Kirkpatrick, J. D. 1965. Factors explaining citrus nematode resistance. *Calif. Citrogr.* 50:235–41

95. Van Gundy, S. D., Rackham, R. L. 1961. Studies on the biology and pathogenicity of *Hemicycliophora arenaria. Phytopathology* 51:393–97

96. Veech, J. A., Endo, B. Y. 1969. The histochemical localization of several enzymes of soybeans infected with the root-knot nematode *Meloidogyne incognita acrita. J. Nematol.* 1:265–76

97. Webster, J. M. 1969. The host-parasite relationships of plant-parasitic nematodes. *Adv. Parasitol.* 7:1–40

98. Weischer, B. 1968. Die bedeutung von nematoden für das auftreten und die verbreitung von pflanzenkrankheiten. *8th Symp. Int., Antibes-Juan-Les-Pins, 1965*

99. Whitehead, A. G., Hooper, D. J. 1970. Needle nematodes (*Longidorus* spp.) and stubby-root nematodes (*Trichodorus* spp.) harmful to sugar beet and other field crops in England. *Ann. Appl. Biol.* 65:339–50

100. Wooding, F. B. P., Northcote, D. H. 1965. An anomalous wall thickening

and its possible role in the uptake of stem-fed tritiated glucose by *Pinus pinea. J. Ultrastruct. Res.* 12:463–72

101. Wyss, U. 1973. Feeding of *Tylenchorhynchus dubius. Nematologica* 19: 125–36

102. Wyss, U. 1973. Reaction of the cytoplasm and nucleus within root hairs of *Nicotiana tabacum* to the feeding of *Trichodorus similis. 2nd Int. Congr. Plant Pathol., Minneapolis, Minn.,* p. 1100 (Abstr.)

103. Wyss, U. 1974. *Trichodorus similis* (Nematoda) Reaktion der protoplasten von Wurzelhaaren (*Nicotiana tabacum*) auf den Saugvorgang. Film E 2045. Institut für den Wissenschaftlicken Film, Göttingen, pp. 1–19

104. Yang, H., Powell, N. T., Barker, K. R. 1973. The influence of several nematode species on *Fusarium* wilt of cotton. *Proc. 2nd Int. Congr. Plant Pathol., Minneapolis, Minn.,* p. 0855 (Abstr.)

METEOROLOGICAL FACTORS ✦3621
IN THE EPIDEMIOLOGY
OF RICE BLAST

Hozumi Suzuki
First Agronomy Division, Tohoku National Agricultural Experiment Station,
Akita Prefecture, Japan

Rice blast disease, caused by *Pyricularia oxyzae* Sacc., is one of the most serious and widely distributed diseases of rice. To reduce damage by this disease, most farmers use resistant varieties, apply fungicidal chemicals, and avoid excessive amounts of fertilizer. Despite these precautions, rice blast causes severe losses under certain meteorological conditions. This is due to the interacting influences of weather on both the reproductive ability of the parasite and the resistance of the rice plant. Thus, an outbreak of rice blast in a given paddy field cannot be explained solely by the relation between the parasite and meteorological conditions. Meteorological factors also affect the cultivation and soil conditions, which affect the growth of the rice plant. These conditions also influence the meteorological conditions. Recently, these relationships were made still more complicated by the introduction of new fungicides that influence both the parasite and the rice plant.

In Japan, the primary inoculum for this disease develops after the pathogen has overwintered inside affected rice straw remaining from the previous year. Conidia are formed when the temperature and humidity rise in the spring. After these conidia infect the rice plants, the fungus develops inside the host tissues, and then induces lesions in which new conidia are formed. An epidemic results when this cycle is repeated many times. The intensity of the epidemic is determined mainly by the abundance of conidia, the influence of environmental conditions on infection processes, and the resistance of the host plants. These three factors are influenced not only by meteorological conditions, but also by the varieties and cultivation practices used, so that the development of the disease is very complex.

Epidemics of Rice Blast and Meteorological Factors

According to Kuribayashi et al (21), reproduction and propagation of the rice-blast fungus are greatest at high humidity and comparatively low temperature

239

(21–24°C). The propagating capacity decreases if the temperature is either above or below these temperatures. If temperatures are above 31°C during the growth stage, the rice plants become resistant to disease. During long periods of cloudy, rainy weather both growth and resistance of rice plants are diminished. Accordingly, severe epidemics of rice blast occur in years or in areas in which the above meteorological conditions predominate.

Predicting the years and areas of severe epidemics of rice blast is difficult for the following reasons:

1. The epidemic zones of rice blast are distributed widely from north to south, geographical features of the zones are complex, and the meteorological conditions vary greatly in each zone.

2. Meteorological conditions during rice cultivation periods are subject to change not only on a yearly basis, but even on an hourly basis. It is difficult to determine when and how meteorological conditions influence the outbreak of rice blast and under which type of meteorological conditions the epidemic of rice blast increases.

3. Epidemics of rice blast differ from one paddy field to another even under the same meteorological conditions because the cultivation methods (varieties, amount of fertilizer, etc) are different for each farmer. This human factor increases the difficulty of analyzing the relationship between epidemics of the disease and specific meteorological conditions.

There are many reports about the relation between epidemics of rice blast and meteorological conditions in Japan (2, 5, 8, 19, 21, 22, 24, 29, 30, 50). The most favorable meteorological conditions are as follows: (a) high temperatures with a long period of heavy rain and lack of sunshine before the maximum tillering stage, (b) low temperatures with a long period of heavy rain and lack of sunshine from maximum tillering to the ripening stage, (c) high temperatures with a long period of heavy rain and lack of sunshine after heading time, and (d) low temperatures with a long and heavy rain and lack of sunshine during the entire growth period of the rice plant.

The most common meteorological conditions that favor severe epidemics of rice blast are long periods of drizzling rain, lack of sunshine, slow wind movement, and high humidity.

Yamanaka (50) reported only a poor and highly variable correlation between temperature and amount of rice blast in various districts. But Suzuki (38) reported that temperature was the most important single determinant of disease, and that, in districts or seasons that had favorable temperatures, rain could be a limiting factor. An epidemic of rice blast develops during prolonged periods when the average daily temperatures range between 19 and 28°C and minimum daily temperatures range between 17 and 23°C.

A specific instance of the influence of temperature on an epidemic of rice blast is provided by records of blast in years of cold weather from the Tohoku district in Japan: the first outbreak started as leaf blast in July; however, the activity of fungi soon ceased, and conditions became less favorable for the disease because of high temperatures from late July to early August. The activity of fungi started again during the low temperatures after heading time, and neck- or panicle-blast developed, but then blast ceased further development during the low temperatures that prevailed during the ripening stage of the development of the rice plants. Districts that show such an epidemic are more often in the north than the south, and more often in the mountains and the higher and cooler areas than in the plains of the Tohoku district in Japan. In generally cold years, however, the regions with the most favorable conditions for development of rice blast are in the south and the plains, because of low temperature and high humidity from late July to early August. In these areas, conditions are most favorable for an epidemic of rice blast when leaf blast develops early in the season and the ripening period of the rice plants is prolonged. This provides a maximally long period for development of the disease after initial inoculation of the plants. As a result, panicle blast breaks out severely. In generally cold years, epidemics of rice blast are relatively rare in the northern, higher and cooler areas of the Tohoku district in Japan. This is true because conditions are too cool for development of this disease in these districts which, in warmer years, are the sites of severe epidemics of rice blast.

Rice blast also develops severely after unusual meteorological conditions. Strong winds, floods, cold water, or drought can lower the resistance of rice plants and thus lead to development of epidemics.

METEOROLOGICAL INFLUENCES ON AMOUNT OF INOCULUM

Sporulation

When lesions are placed in a moist chamber, conidiophores begin to emerge in about 6 hr. One hour later the first conidia are produced and reach full size in about 40 min. Soon after, the conidiophores branch just below the point of attachment of the first conidium, and a second conidium is formed at its apex. In the same manner, sporulation continues and 7 to 9 conidia are formed on each conidiophore, each at about 1-hr intervals (43). A maximum rate of sporulation is achieved 3–8 days after the appearance of a lesion (18) although sporulation on a given lesion continues for about 60 days.

Sporulation does not occur below 9°C and over 35°C, the optimum temperature being 25–28°C. Henry et al (14) reported that the mode of sporulation with respect to time varies with temperature. At 28°C the conidia were found to be produced rapidly but production decreased after 9 days, whereas sporulation at 16–24°C tended to increase even after 15 days. The minimum relative humidity for sporulation is 89%, and the higher the humidity rises over 93%, the more sporulation occurs. If enough water is supplied to the lesion from surrounding parts, a lesion can sporulate well despite low humidity. Both conidiophore formation and sporulation are inhibited by poor aeration. Suzuki et al (42) investigated the effect of wavelength and intensity of light between 0 and 2000 lux on sporulation in vitro. He found that sporulation increased as the wavelength became shorter and the light intensity became stronger.

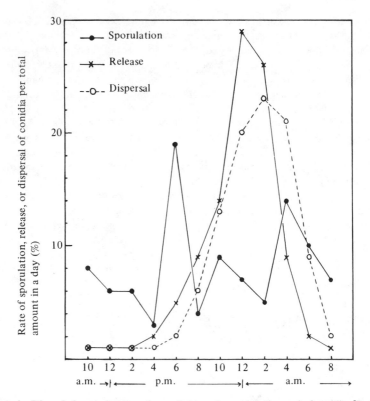

Figure 1   Diurnal changes in rates of sporulation, release, and dispersal of conidia. [Rate of sporulation based on Kuribayashi et al (21).]

As shown in Figure 1, conditions are most favorable for sporulation twice during each 24-hour period—the first peak occurs at 6–7 p.m. and the second between 4 and 5 a.m. (21). Immature conidia formed in the early morning hours stop growing during the less humid daylight hours and resume maturation during the highly humid conditions after sunset. These spores account for the early evening peak of sporulation. Conidia that mature fully during the night are released to form the early morning peak. On cloudy or rainy days, conidia continue to mature during the daylight hours. On a dry, windy day conidia are not formed during night or day.

Release of Conidia

High humidity is one of the most important factors necessary for release of conidia. The effects of temperature on release of conidia are almost the same as on sporulation. Kuribayashi et al (21) report that conidia are released automatically in a saturated humidity. Ingold (17) believes that the conidia are probably discharged violently, although only a very short distance, by the bursting of the hilum attaching them to the conidiophores. According to the electron microscopic study of the conidium by Wu et al (48), the side wall of the hilum consists only of an outer layer. Horino (16) reported that the outer layer of the conidium cell wall is a fragile structure made from a substance secreted from the inner layer.

A diurnal pattern of release of conidia is shown in Figure 1; minimum release occurs between 8 a.m. and 4 p.m., with maximum release between 12 p.m. and 4 a.m. Alternate changes of light and darkness under saturated humidity are essential for release of conidia (4, 33, 35). Continued darkness from time of sporulation inhibits release of conidia. Even if lesions receive light while the conidia are developing, release of conidia is inhibited by continued darkness. Continuous light also inhibits release. Thus, light acts as a temporary stimulus in the release of conidia. From 6 to 8 hr of darkness following a period of light are required for the release of conidia, which then react sensitively to a shift from light to darkness.

Release of conidia is also influenced by water droplets and wind. Although strong winds can cause mechanical release of conidia, winds of less than 3.5 m per sec only dry out the conidiophores. Water drops induce release of conidia by touching the juncture between the conidia and the conidiophore. On a fair day, dew and guttation drops are important in inducing the release of conidia. These several results help explain the increase in the number of conidia dispersed on both rainy and windy days.

Dispersal of Conidia

Released conidia float under the canopy of rice plants and then scatter out into the air above the canopy. Optimum conditions for dispersal of conidia are temperatures of 20.5–21.8°C and relative humidity over 90% for 10 hr or more (21). Dispersed conidia can be detected in the air from the later part of May through the end of November. In northern Japan, maximum dispersal was observed from the end of July through the beginning of August, but in the southwestern and warmer districts of Japan two peaks of conidial dispersion are generally observed—middle of July and middle of September. Although conidia are dispersed for a relatively short period during the growing season in temperate regions, Ou (23) has reported that airborne conidia are present in the tropics throughout the year, with peaks extending from May to June and November to December. This is due both to the temperature and the mode of rice cultivation in the tropics. A diurnal pattern of dispersal of conidia is shown in Figure 1; on a fair day minimum dispersal occurs between 8 a.m. and 5 p.m., with maximum dispersal between 11 p.m. and 4 a.m. Some conidia were detected during most daylight hours. The dispersal curve for cloudy days was similar in shape, but on a rainy day the number of conidia detected was less depending on the time and amount of rainfall.

Several meteorological factors, especially rain, wind, and air currents, strongly affect the number of conidia in the air. Reports concerning the relation between rainfall and conidial dispersion in the air do not coincide with findings of researchers (3, 9). Figure 2 compares the number of conidia dispersed on a day when it started raining at 11 p.m. with the number dispersed on a fair day. As shown in this figure, dispersal decreased rapidly at midnight when precipitation began. The effects of rain on dispersed conidia is most pronounced at the beginning of a rain shower and during heavy rains. Very light rains (0.0–0.1 mm per hr) have no effect on dispersal; rains above 3.5 mm per hr usually decrease dispersal. In a severe outbreak of rice blast, we collected rain drops and counted the number of conidia in them. In a period of maximum outbreak, we found that 1.5 mm of rain falling for 3 hr brought down 8 times as many conidia as were deposited during the same time in a similar area exposed to wind but screened from rain (32). The decrease in numbers of dispersed conidia is due to a combination of capture of airborne conidia by falling rain drops and washing of conidia from sporulating lesions. Under these conditions, diffusion becomes so weak that the conidia stay in the rice-plant canopy, and an epidemic of rice blast frequently results.

Wind is a medium of both long-distance transport and local diffusion of conidia. The concentration of conidia per unit volume of air 2 m above the ground was found to be high at wind velocities below 1 m per sec and to be remarkably low at velocities above 1 m per sec (36). In order to determine whether a relationship exists between epidemics of rice blast and wind velocity, disease ratings were evaluated separately for several successive years and areas where wind velocity differs yearly. Results showed that in plain areas where the wind blew strongly, or in years when it blew strongly, fields were less infected, and infection was uniformly distributed (36). In valley areas or in years where wind velocities were slow, a wide variation in amount of infection was observed, but the epidemic as a whole was "severe." These findings can be explained as follows: under weak wind conditions, dispersal of conidia was limited to a smaller area and most of the conidia stayed within the canopy of the infected field. In such fields, the magnitude of the epidemic was determined mostly by the sensitivity of the plants to infection. Conversely, under strong-wind conditions, conidia were dispersed much more widely and the ensuing disease was uniform and slight.

Diffusion of conidia has been studied by Gregory (9), Waggoner et al (44), Waggoner (45), and Wilson & Baker (46, 47). They surveyed the horizontal distribution of conidia from an infected field or an artificial dispersal source. Their analyses were based on Sutton's diffusion model. The survey of conidial diffusion over an infected field should be a vertical rather than a horizontal distribution of conidial concentration. Inoue (27) has completed an aerodynamic analysis of the vertical distribution of conidia over fields infected with rice blast.

Imagine a horizontal surface at some height, Z, in the atmosphere through which the air particles constantly move upward or downward. The concentration of conidia, C, ascending through a given area per unit time is

$$-K \frac{dC}{dZ}, \qquad\qquad 1.$$

where K is the eddy transfer coefficient. However, the concentration of the conidia descending through this same area per unit time is CW, where W is the terminal velocity of conidia. If we suppose the conidia of the rice-blast fungus to be an ellipsoidal particle, the terminal velocity calculated from Falck's terminal-velocity equation for ellipsoidal particles is 2.4 cm per sec. This value cannot be neglected

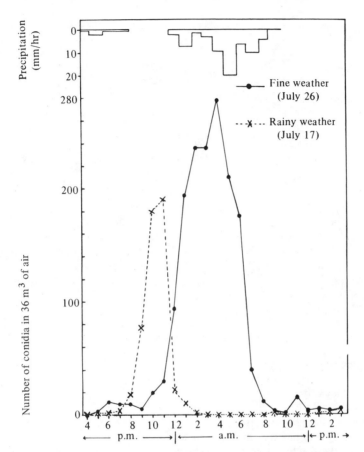

Figure 2 Influence of precipitation on the number of airborne conidia of *Pyricularia oryzae*. Measurements were made with a Hirst spore trap and indicate the number of spores detected per hour.

(9). After a sufficiently long period of time, the concentration of conidia ascending through a given area per unit of time should equal the concentration of the conidia descending through this area per unit of time. This relation is

$$-K\frac{dC}{dZ} = CW. \qquad\qquad 2.$$

Since K is a function of air stability and height, the expression for K is $K = k$ $\cdot(Z - d)\cdot(1 - B\cdot Ri)\cdot V^*$, where, k is Kärman's constant, Ri is Richardson number, B is an empirical constant, about 12, V^* is friction velocity, and d is zero-plane displacement. We obtain the relation between C and Z from equation 2. If the measurement is usually made at height Z from the d plane, equation 2 can be rewritten as

$$- k \cdot Z \cdot V^* \cdot (1 - B \cdot Ri)\frac{dC}{dZ} = CW.$$ 3.

Upon integration we get

$$\frac{C}{C_o} = \left(\frac{Z}{Z_o}\right)^{-n}$$ 4.

where, n is $\dfrac{W}{k \cdot V^* \cdot (1 - B \cdot Ri)}$, and C_o is the number of conidia at the same height, Z_o.

A survey of the vertical distribution of conidia over an infected field was made from the ground surface to a height of 5 m. Figure 3 shows results of such a survey on two different days. On both days many more conidia were present within the plant canopy than above it. The concentration of conidia decreased with increasing height, and the rate of decrease was more rapid within than over the canopy. When the total population over the canopy is decreasing, the logarithm of the number of conidia at a given height gives a straight line with the logarithm of height. This indicates that the vertical distribution of rice-blast conidia can be predicted by equation 4. The slope of the straight line, n, shows the diffusion condition of the conidial cloud, and the slope of the straight line, n, is found to be related to the velocity of wind at 2 m from ground level (U_{200} m per sec). The following equation can be obtained empirically between n and U_{200}:

$$n = 1.80 - 0.33\, U_{200}\ (n > 0).$$ 5.

With this formula, the number of conidia distributed over a given field can be estimated by counting the number of conidia trapped at an arbitrary height and by observing the velocity of wind at 2 m above the ground. In certain instances, the following relationship was observed between disease outbreak and n. When n is nearly equal to 1.8, the disease tends to be severe, and when n is equal to 1.4, the disease is slight (Table 1).

The density of the horizontally distributed conidia within the rice-plant canopy was found to be low on the windward, and high on the leeward sides of infected fields. The tendency is obvious close to the plant height. The vertical distribution of conidia in the rice-plant canopy shows that the maximum number of conidia is between 10 and 15 cm from ground level. The density decreases as a logarithmic curve from the maximum point to the crown of the plant. The decrease in density of conidia above the canopy is especially remarkable at slow wind velocities. The rate of decrease of vertical distribution within or over the canopy is related to wind velocity: the higher the velocity of the wind, the slower the decrease in denisty, and vice versa; this suggests that most conidia are not blown out of the canopy at low wind velocity. The vertical distribution of conidia is also influenced by density of planting. The greater the density of planting, the more of the conidia that stay within

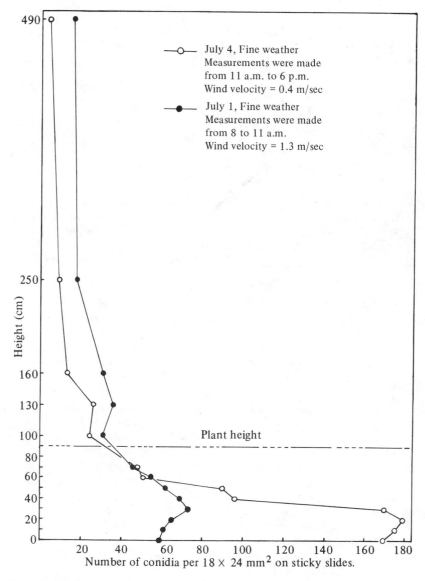

Figure 3 Distribution of conidia above and within the rice canopy on 2 days with fine weather in July. Wind velocity was measured at 2 m above the ground surface.

Table 1   Relationship between vertical distribution of conidia and severity of rice blast

Severity of rice blast	Test locations		
	Yasuzuka	Takada	Kakizaki
	(Number of paddy fields per severity class)		
Light	21	32	34
Moderate	6	2	1
Heavy	7	2	1
Severe	2	0	0
Total	36	36	36
n = vertical distribution of Conidia[a]	1.8	1.6	1.4

[a]
$$n = \frac{W}{k \cdot V^* \, (1 - BR_i)} \, .$$

the canopy and the greater the likelihood of disease. Recently, an aerodynamic survey of vertical distribution of conidia within the infected rice-plant canopy was made by the crop disease section at the Hokuriku Agricultural Experiment Station (28); this permitted an analysis of the distribution of conidia and the distribution of conidial sources within the canopy. This is important from the view of analysis of epidemics of rice blast.

Understanding dissemination of conidia by wind is important for disease control. The following experiment was conducted by Suzuki (36) to determine how rice blast is distributed from the initial inoculum. In fields of about 5 ha where infected rice straw was kept in the center as inoculum, the disease developed widely and rapidly along the wind direction from the inoculum. Infection of hills started rather slowly, but accelerated rapidly after 20–30 hills were infected. Finally the infection spread all over the field and the disease spread from this original field to surrounding fields, depending on the disease intensity in the original field, as well as on the distance from the original field. Variation of the disease in surrounding fields with the distance from the original field was in reciprocal proportion to wind velocity. This dissemination is attributed to dispersal and transport of conidia by the wind. Figure 4 shows how conidia were distributed from the original field to others in relation to wind direction. More conidia were collected on the leeward and less on the windward. Even in the original field, more conidia were collected on marginal areas of the leeward, and less on the marginal areas of windward. Variation in horizontal distribution of transported conidia rapidly declined with distance during 1963. The disease outbreak was severe when the wind was mild, and declined slowly in 1964, when the wind was stronger.

METEOROLOGICAL INFLUENCES ON INFECTION

Deposition and Adherence of Conidia

The ratio of deposition of conidia on rice plants to the number of floating conidia per unit volume of air was investigated (36). This investigation was made with a Hirst type automatic volumetric suction trap positioned at the same height

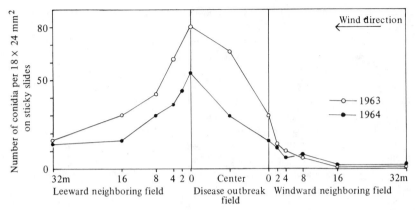

Figure 4   Relation between the horizontal distribution of conidia and distance along the wind direction.

as the plant canopy (15). Approximately 20% of the spores above the canopy were deposited. There is a tendency for the rate of deposition to decrease when wind velocity is high, and vice versa. The amount of deposition varies with the surface of the leaf in question (whether upper or lower surface) and with the position of the leaf on the plant (whether upper or lower on the plant). During winds of low velocity (less than 0.5 m/sec), deposition was generally greatest in the middle of the upper surface of leaves from the middle and lower portions of the plant. More conidia were deposited on horizontal than on inclined leaf surfaces. In a strong wind (about 3 m per sec) many fewer spores were deposited but the spores could be found on the lower surface of leaf and top leaf of the plant. As with a weak wind, more spores were deposited on horizontal than on nonhorizontal plant surfaces (31). The rice-blast fungus usually penetrates the cuticle of motor cells which exist only on the upper surface of the leaves, that is, the parts of penetration coincide with the parts of conidial deposition. The parts of the plant on which lesions are most prevalent coincide with the parts of greatest conidial deposition; also the number of lesions that develop on the rice plant is well correlated with the number of deposited conidia.

As previously mentioned, conidia are captured in rain drops. When such drops roll down a leaf, the conidia adhere to the surface on contact and cannot be freed by rain drops or wind. Conidia floating in the air frequently fall on the surface of drops of dew or guttation fluid from which they adhere to the leaf surfaces on contact. Guttation drops formed at the edge of leaves often roll down the leaf surface. Conidia that fall directly on leaves in daytime covered with a water film at night also adhere to the leaves. This adherence is very important for formation of appressoria.

Conidial Germination and Formation of Appressoria

The conidia formed on rice-blast lesions do not always have high germinability. Many of the old conidia remaining in lesions will not germinate (39), but dispersed conidia usually have high germinability (36).

Conidia germinate in water drops on leaf surfaces and form appressoria. Germination and formation of appressoria occur between 10 and 33°C. The optimum temperature for germination is between 25 and 28°C, and for appressorium formation is between 16 and 25°C (37). Light suppresses germination, but little is known about its effect on formation of appressoria. Even at 100% relative humidity, water drops are essential for germination. Water on leaf surfaces consists of rain, dew, and guttation drops. Formation of guttation and dew drops usually begins just after sunset and about 8 p.m., respectively. Wind and a rise in temperature after sunrise frequently cause these drops to coalesce. After sunrise they usually evaporate and disappear by 6–12 a.m. If the wind blows during the night, the drops may disappear earlier.

Under some climatic conditions, these surface droplets disappear before appressoria are formed; as a result, the protoplasm of the conidia coagulates and dies. Even within the range of temperatures favorable for germination, the germinability of conidia apparently can be diminished by drying, after the spores have been in contact with water drops for about 10 min. This diminution of germinability tends to increase with increasing time of contact with water drops, or increasing time of drying after wetting. This has also been observed when conidia are exposed on glass slides in rice fields. Most conidia can not survive on a leaf surface for more than one day after water drops disappear. For this reason, the time period in which water drops remain on leaf surfaces is a very important factor influencing germination, formation of appressoria, and in turn, epidemics of rice blast. In experiments in which potted plants were inoculated by spraying with conidial suspensions at hourly intervals, the plants inoculated between 8 p.m. and 2 a.m. showed the most severe infection.

It is essential for the establishment of lesions, that the entire process, which consists of germination, appressorium formation, and invasion of the pathogen, must be completed before the water drops evaporate. When glass slides were used to collect conidia in the field at hourly intervals, from 8 p.m. until 7 a.m., maximum germination and formation of appressoria were observed on the slides placed before 2 a.m., and lowest on glasses placed after 4 a.m.

In a test of the influence of planting density on germination of conidia, water drops persisted longer and germination percentages were greater in the more densely planted stands; when rice plants were planted in 120, 60, and 30 hills per 3.3. m^2, the maximum number of lesions formed were 1774, 647, and 521, respectively, while an index of water-drop longevity was 100, 91, and 70, respectively, and germination percentages were 64, 44, and 21%, respectively.

The longevity of dew and guttation drops is influenced not only by cultivation practices such as density of plants, but also by topographic factors. In plain areas where water drops disappeared by 9 a.m., the disease was slight and neck infection was 0.2%. In areas along the foot of a mountain where water drops disappeared by 10 a.m., leaf blast was moderate and neck infection was 6%. In areas between mountains where water drops disappeared about noon, leaf blast was severe and neck infection was 17%. The disease was most severe in the areas between mountains because of the persistence of the water drops.

In paddy fields that extend into a deep valley, the longevity of water drops on leaf surfaces varied with the location of the field in the valley. If the field was deep in a ravine, where there was sunlight only from late morning until early in evening, leaf blast was severe. If a field was located in the eastern part of a valley where the sun shone from late morning until late evening, blast was moderate. If a field was located in the western part of a valley, with sun from early morning until early evening, leaf blast was slight. If a field was located in the center of the valley, with sun from early morning until late evening, the disease was least of all.

Late disappearance of water drops can be important in addition to the total time of their presence. If T is the time of disappearance of water drops in elapsed hours from twelve midnight ($T > 0$), and A is the rate of appressorium formation by adhering conidia ($A > 0$), the following straight-line relationship is found between T and A: $A = 3.4\ T - 19.8$.

Formation and disappearance of dew and guttation drops are influenced by wind velocity. Water drops formed early and lasted longer if the wind velocity at 2 m above the ground remained less than 1 m per sec at night, and even after sunrise; but if the wind blew after sunset at more than 1 m/sec, water drops did not form, or they evaporated after they were formed.

To determine the effect of rain on germination and formation of appressoria, conidia were exposed on glass slides in fields where the rice blast was epidemic. The percentage of appressorium formation on rainy days was 40% compared to 2% on fair days.

Appressoria apparently are formed more frequently when the conidia adhere tightly to plant surfaces. Rain drops loaded with conidia cause the conidia to adhere closely to the rice plant. This is presumed to be one of the reasons for the high rate of appressorium formation on rainy days (34). The relation between the rate of precipitation and both the frequency of appressorium formation and the amount of disease was observed during rains of less than 100 mm/hr. Neither the frequency of appressorium formation nor the number of lesions per hill of rice plant was affected (40).

In addition, the influence of the time and length of rainfall was investigated by exposing conidia on glass slides in paddy fields. In these tests, 37% of the conidia formed appressoria when it rained for 24 hr, 35% when it rained at night, 10% when it rained only in the daytime, and only 2% on cloudy or sunny days. Generally, long periods of drizzle often occur in temperate regions. Ou (23) reported that blast is generally not severe in the tropics. This may be explained by the fact that rains in the tropics usually come as showers that only last a few hours, although when frequent and long periods of rain do occur in tropical areas, rice blast is usually severe.

The relationship between disease severity and the frequency of rainy days is shown in Table 2. In the years between 1961 and 1970, the severity of rice blast was directly related to the number of rainy days during the 30-day period following dispersal of primary inoculum. Severe rice blast occurred in years with more than 22 days of rain.

Table 2 Relationship between the number of rainy days and severity of rice blast

Year	Number of rainy days during the 30 days following primary infection[a]	Disease index[b]
1970	8	3
1969	14	4
1967	17	5
1964	19	6
1961	20	7
1966	21	8
1965	23	9
1962	22	9
1963	23	10

[a] Data obtained from Takada Meteorological Station.

[b] According to this disease index: 3 = light leaf blast, 4 = moderate leaf blast, 5 = heavy leaf blast, 6 = severe leaf blast, 7 = beginning of stunt, 8 = stunt, 9 = severe stunt, 10 = dead.

THE RESISTANCE OF RICE PLANTS

Penetration and Development of the Pathogen in Plant Tissue

Penetration and development of the pathogen inside plant tissue is related closely with resistance of rice plant, as described below. The pathogen penetrates the epidermis of rice leaves by a penetration hypha that develops from the center of the appressorium. The invading hyphae usually swell and fill the epidermal cell within 24 hr after deposition on a leaf; after 48 hr they reach the next layer of cells, and after 72 hr, several dozen cells have been invaded. Relatively few of the penetrating hyphae succeed in inducing lesions. Under conditions considered ideal for the pathogen, Sakamoto (26) recorded that 68% of appressoria penetrated into the epidermal cells; 51% of the penetrating pathogens developed hyphae in the epidermal cell, of which only 17% invaded the surrounding cells. Suzuki (41) considers that successful establishment of a parasitic relationship with the host plant requires growth from the invaded cell into the surrounding cells. The stage of hyphal growth into the surrounding cells is second only in importance to formation of appressoria in the life cycle of this fungus.

Failures of the pathogen to become established in the host tissues are due more to the resistance of the host than to the direct effects of meteorological factors on the growth of the pathogen within the host tissues. The resistance reaction itself, however, can be influenced by meteorological factors such as temperature, humidity, light, and wind.

Invasion of the pathogen can be observed at a minimum of about 10°C, with an optimum at 24°C; it takes only about 6 hr to penetrate the host at the latter temperature. Penetration takes 8 hr at 20 and 28°C, 10 hr at 32°C, and does not occur at 34°C (11). The latent period for formation of lesions is also influenced by temperature; it takes 13–18 days at 9–11°C, 7–9 days at 17–18°C, 5–6 days at

24–25°C, and 4–5 days at 26–28°C. The optimum temperature for developing lesions is 26–28°C (13).

Sunlight also can influence development of the pathogen in host tissue (1, 12). Shading provides more favorable conditions for the pathogen, although the effect of shading varies with the stage of development of the pathogen in the plant tissue.

Resistance Reactions of the Rice Plant

Resistance of the rice plant to blast also varies with its growth stage (10). Younger plants in the leaf-blast stage are quite susceptible, but they become resistant after the booting stage. A protracted susceptible stage and favorable meteorological conditions for the pathogen lead to severe blast. Mean daily temperatures below 19 and above 28°C are limiting for the disease. Although high temperatures are favorable for growth of the pathogen, they also stimulate resistance in the rice plants. Conversely, the disease becomes epidemic at low temperatures because the resistance reaction of the host is decreased even more than the growth of the pathogen is diminished. Decreasing the amount of soluble nitrogen in the plants increases the resistance of the plants to rice blast.

Goto et al (6, 7) have reported on the relation between low temperature and outbreak of rice blast. When rice plants are exposed to low temperatures and then are returned to normal temperatures, their resistance increases just after treatment but they become predisposed to disease after 3 to 6 days. This influence of exposure to low temperatures continues for 1 to 2 weeks. During this time the concentration of nitrogen increases in the plant and their resistance decreases. Yoshino et al (51) observed the same temporary increase in susceptibility to rice blast when plants were shaded. When plants were fertilized with nitrogen, a few days after treatment they became very susceptible to disease and their growth became stunted.

If the period that decreases the resistance of the rice plant becomes prolonged, and the disease-prone stage in turn becomes prolonged, the probability of a severe outbreak of the disease will be increased. In years with very long periods of low temperature, the growth and or set of the heading stage of rice plants is delayed, while in years of high temperature, growth is accelerated and heading is reduced earlier in the season.

In many cases, rice plants are covered with water droplets for half or more of the growing season. There are many studies of the influence of these droplets on the pathogen, but very few studies have been made of the influence on the rice plant itself. In an artificial rain test, rice plant leaves were found to lose their color and become elongated. The resistance of the rice plant at this time should be studied.

Strong wind (e.g. typhoons) injures host surfaces and makes openings for penetration of host tissue (25). Wind also stimulates transpiration of the host, promotes silicification of leaf tissue (20) and accumulation of growth substances in the plant, and strengthens the resistance reaction of the host.

Sunlight attains an intensity of 100,000 lux during some days in the summer. Photosynthesis in rice plants declines when the intensity falls to 40,000-50,000 lux. Intensity is between 10,000 and 20,000 lux on bright cloudy days, and only a few thousand lux on dark cloudy or rainy days (49). During successive rainy or cloudy days, carbon dioxide assimilation and respiration of rice plants decreases, soluble nitrogen accumulates in host tissue, physiological activity of the host decreases, hardness of host tissue decreases, and the resistance to rice blast declines. When the humidity rises without sunshine, transpiration in the host is decreased, silicification of the leaf tissue is inhibited, and, as indicated above, resistance to penetration of the pathogen decreases.

CONCLUSION

Rice blast becomes epidemic when a new race of the pathogen appears or when meteorological conditions activate the pathogen and/or increase the proneness of the plant to disease. Meteorological conditions act directly on the pathogen in the prepenetration stage (before appressorium formation). They also influence formation of appressoria and initial colonization of the host tissues. Further development of the pathogen in host tissues is influenced more by field or genetic resistance of the rice plant than by meteorological conditions; but meteorological conditions then act indirectly on the pathogen through their effects on host resistance. When the host plant is subjected to unfavorable meteorological conditions, its resistance to rice blast decreases as the disease-prone stage is prolonged. Rain is an important factor in outbreaks of rice blast, but wind conditions control the epidemic. If the host is exposed to meteorological conditions that cause the disease to break out early in the growing season, the risk of an epidemic is increased because rice plants have long, susceptible periods to disease from the early growth stage.

So far, meteorological studies on the growth of rice plants and the development of rice blast have not included measurements within the plant canopy. Studies to date have included only the influences of meteorological phenomena measured above plant height (commonly 2 m above the ground). Recently, micrometeorological apparatus and techniques have been developed that will permit quantitative analyses of meteorological factors within the plant canopy. We intend to contribute to this knowledge and apply it so as to diminish the impact of this disease on future crops of rice.

Literature Cited

1. Abe, T. 1931. On the effect of sunlight on infection of rice plants by *Piricularia oryzae. Forsch. Geb. Pflanzenkr.* 1: 46–53
2. Abumiya, S. 1954. Survey and analysis of actual conditions of rice blast development in cool summer, 1953. *Spec. Rept. 1, Proc. Assoc. Plant Prot., North Jpn.* 1–91
3. Asai, G. N. 1960. Intra- and interregional movement of uredospores of black stem rust in the upper Mississippi River Valley. *Phytopathology* 50: 535–41

4. Barksdale, T. H., Asai, G. N. 1961. Diurnal spore release of *Piricularia oryzae* from rice leaves. *Phytopathology* 51:313–17
5. Fujikawa, T., Okatome, Z., Utsunomiya, T. 1955. Severe outbreak of rice blast in Ooita Prefecture 1953. *J. Agric. Meterorol. Tokyo* 11:76
6. Goto, K., Ohata, K. 1960. Influence of low air temperature on susceptibility of rice plants to blast disease. *Ann. Phytopathol. Soc. Jpn.* 25:1 (Abstr.)
7. Goto, K., Ohata, K. 1961. Change of environment and outbreak of rice blast; fluctuation of resistance after low tem-

perature treatment and physiology of rice plants. *Ann. Phytopathol. Soc. Jpn.* 26:215 (Abstr.)

8. Goto, K. 1962. Technical view on the forecasting of blast disease. *Twentieth Anniversary Publication Concerning the Work of Forecasting of Disease and Insect Pests,* 37–52

9. Gregory, P. H. 1973. *The Microbiology of the Atmosphere.* New York: Wiley. 2nd ed. 377 pp.

10. Hashioka, Y. 1950. Studies on the mechanism of prevalence of the rice blast disease in the tropics. *Tech. Bull. Taiwan Agric. Res. Inst.* 8:237

11. Hemmi, T., Abe, T. 1931. On the relation of temperature and period of continuous wetting to the infection of the rice plant by *Piricularia oryzae. Forsch. Geb. Pflanzenkr.* 1:33–45

12. Hemmi, T., Abe, T. 1932. Studies on the rice blast disease. 2. Experimental studies on the relation of environmental factors to the occurrence and severity of blast disease in rice plants. *Mater. Rural Improv. Dep. Agric. For., Jpn.* 47:204

13. Hemmi, T., Abe, T., Ikeya, J., Inoue, Y. 1936. Studies on the rice blast disease. 4. Relation of the environment to the development of blast disease and physiologic specialization in the rice blast fungus. *Mater. Rural Improv. Dep. Agric. For., Jpn.* 105:145

14. Henry, B. W., Andersen, A. L. 1948. Sporulation by *Piricularia oryzae. Phytopathology* 38:265–78

15. Hirst, J. M. 1952. An automatic volumetric spore trap. *Ann. Appl. Biol.* 39:257–65

16. Horino, O. 1971. *Electron-Microscope Study on Rice Plant Infected by Cochliobolus miyabeanus (Ito et Kuribayashi) Drechsler, with Special Reference to Ultrastructural Aspects in Host-Parasite Interaction.* Tsu: Koyo-Insatsu. 136 pp.

17. Ingold, C. T. 1964. Possible spore discharge mechanism in *Pyricularia. Trans. Br. Mycol. Soc.* 47:573–75

18. Kato, H., Sasaki, T., Koshimizu, Y. 1970. Potential for conidium formation of *Pyricularia oryzae* in lesions on leaves and panicles of rice. *Phytopathology* 60:608–12

19. Kobayashi, J. 1965. Survey and analysis of actual conditions of rice blast development in Akita prefecture, 1963. *Spec. Rept. 6 Proc. Assoc. Plant Prot., North Jpn.* 34–48

20. Kumagaya, S., Goto, Y., Hori, C., Matsuoka, M., Nakano, R. 1957. Annual change of silicate absorption and effect of calcium silicate in the rice plant. *Rept. Tokushima Agric. Exp. Stn.* 2: 13–14

21. Kuribayashi, K., Ichikawa, H. 1952. Studies on forecasting of the rice blast disease. *Spec. Rept. Nagano Agric. Exp. Stn.* 13:1–229

22. Nakagawa, K. 1964. Abnormal occurrence of rice blast and composite control method. Comparison between the occurrence in 1953 and 1963 in Fukushima prefecture. *J. Agric. Sci.* 19:160–63, 218–20, 267–69

23. Ou, S. H. 1972. *Rice disease.* Kew, England: Commonwealth Mycol. Inst. 368 pp.

24. Sai, T. 1965. Survey and its analysis on the incidence of rice blast in Miyagi prefecture, *Spec. Rept. 6 Proc. Assoc. Plant. Prot., North Jpn.* 6:54–109

25. Sakamoto, M. 1940. On the facilitated infection of the rice blast fungus, *Piricularia oryzae* Cav. due to the wind. I. *Ann. Phytopathol. Soc. Jpn.* 10:119–26

26. Sakamoto, M. 1968. Studies on the resistance of rice plant to *Piricularia oryzae.* In *Jubilee Publ. Commem. Sixtieth Birthday Prof. Masayuki Sakamoto,* 1–139

27. Section of Crop Disease and Section of Agricultural Meteorology, Division of Environment, Hokuriku Agricultural Experiment Station; Division of Meteorology, National Institute of Agricultural Science. 1962. *Results Res. Dispersion Conidio-Spores 1961.* 1–30

28. Section of Crop Disease, Division of Environment, Hokuriku Agricultural Experiment Station. 1965. *Results Res. Crop. Dis. 1964,* 1–64

29. Section of Plant Protection, Ministry of Agriculture & Forestry. 1965. Epidemic and its countermove of rice blast in 1963. *Spec. Rept. Dis. Insect Outbreak Forecast* 19:213

30. Section of Plant Protection, Ministry of Agriculture & Forestry. 1967. Epidemic and its countermove of rice blast in 1965. *Spec. Rept. Dis. Insect Outbreak Forecast* 21:177

31. Suzuki, H. 1961. Deposition of spores of blast fungus on rice leaves. *Proc. Assoc. Plant Prot., Hokuriku* 9:41–45

32. Suzuki, H. 1965. The falling of *Piricularia oryzae*-spore by rain. *Proc. Assoc. Plant Prot., North Jpn.* 16:135–37

33. Suzuki, H. 1966. The relation between the spore discharge of rice blast fungus and the periodical alteration of light

256   SUZUKI

and darkness. *Proc. Assoc. Plant Prot., Hokuriku* 14:26–30

34. Suzuki, H. 1968. Raindrops, dew and guttation drops related to the spore germination and appressorium formation of rice blast fungus. *Proc. Assoc. Plant Prot., Hokuriku* 16:13–16

35. Suzuki, H. 1969. Diurnal periodicity in spore discharge of rice blast fungus. *Rev. Plant Prot. Res.* 2:64–65

36. Suzuki, H. 1969. Studies on the behavior of rice blast fungus spore and application to outbreak forecast of rice blast disease. *Bull. Hokuriku Agric. Exp. Stn.* 10:118

37. Suzuki, H. 1969. Temperature related to the spore germination and appressorium formation of rice blast fungus. *Proc. Assoc. Plant Prot., Hokuriku* 17:6–9

38. Suzuki, H. 1972. Forecasting method of rice blast with appressorium formation of deposited spore on the slide. *Proc. Assoc. Plant Prot., Hokuriku* 20:9–12

39. Suzuki, H. 1973. Germination ability of the spore on the lesion and the dispersion spore of rice leaf blast. *Proc. Assoc. Plant Prot., Hokuriku* 21:4–5

40. Suzuki, H. 1974. The influence of the precipitation to the appressorium formation of rice blast fungus. *Ann. Phytopathol. Soc. Jpn.* 40:188–89 (Abstr.)

41. Suzuki, N. 1966. Plant Disease. *Biology of Disease. Recent Biology* 8:169–98. Tokyo:Iwanami-Shoten. 198 pp.

42. Suzuki, Y., Yoshimura, S. 1963. Effect of light on sporulation of the rice blast fungus. *Ann. Phytopathol. Soc. Jpn.* 28:62–63 (Abstr.)

43. Toyoda, S., Suzuki, N. 1952. Histochemical studies on the lesions of rice blast caused by *Piricularia oryzae* Cav. I. Some observations on the sporulations on lesions of different types occurring on leaves of the same variety. *Ann. Phytopathol. Soc. Jpn.* 17:1–4

44. Waggoner, P. E., Taylor, G. S. 1958. Dissemination by atmospheric turbulence: spores of *Peronospora tabacina. Phytopathology* 48:46–51

45. Waggoner, P. E. 1962. Weather, space, time and chance of infection. *Phytopathology* 52:1100–8

46. Wilson, E. E., Baker, G. A. 1946. Some aspects of the aerial dissemination of spores, with special reference to conidia of *Sclerotinia laxa. J. Agric. Res.* 72:301–27

47. Wilson, E. E., Baker, G. A. 1946. Some features of the spread of diseases by airborne and insect-borne inoculum. *Phytopathology* 36:418–32

48. Wu, H. K., Tsao, T. H. 1967. The ultrastructure of *Piricularia oryzae* Cav. *Bot. Bull. Acad. Sin., Taipei.* 8 (Spec. No.):353–63

49. Yamada, N. 1955. Crop scientific analysis of rice blast disease. *Nogyo-Kairyo* 5:14–18

50. Yamanaka, I. 1965. Forecasting techniques in warm southern Japan based on weather conditions. *Ann. Phytopathol. Soc. Jpn.* 31:278–82

51. Yoshino, R., Yamaguchi, T. 1974. Influence of sunshine and shade conditions on the occurrence of rice blast caused by *Piricularia oryzae* Cavara. *Bull. Hokuriku Agric. Exp. Stn.* 16:61–119

CHEMICAL CONTROL OF PLANT DISEASES: AN EXCITING FUTURE

❖3622

Ronald J. Sbragia
Ag-Organics Research, Dow Chemical USA, Walnut Creek, California 94598

INTRODUCTION

The past few decades have witnessed very exciting changes in the chemical control of plant diseases. Tremendous advances have been made in fighting the myriad of disease organisms responsible for billions of dollars of production loss annually. This paper is not a conventional review of the state of fungicide research, as this has been covered elsewhere (13, 26, 27, 29, 48). It does represent my own personal approach to research on chemical disease control.

Recently, there have been many developments that will have a profound effect on the future of agriculture and the pesticide industry. We have experienced shortages of agricultural commodities, resulting in traumatic price increases. Restrictions in the supply of petroleum-based products have heightened our awareness of the unrenewable nature of such resources. One possible benefit to be derived from these developments is an increased recognition of the limited nature of many of the products and resources we so often take for granted. Unfortunately, economic development within any society usually brings about a wasteful expenditure of energy and raw materials. At the same time, there is a tendency to overvalue, and thereby limit the full utilization of, its most readily available resource—human labor.

A most important characteristic of agricultural resources is their renewability. Undoubtedly, we will see increased utilization of agricultural products as sources of fuel and synthetic raw materials, and there will be continual pressure to reduce the disturbing spread between average and maximum yields (47). However, the methods used to increase production in the future will be carefully scrutinized, for although yields have continued to increase, the past few decades have seen a trend toward lower net return per unit of input energy (34). The point of diminishing returns has been passed. Nevertheless, it has been estimated that the world's food

257

output must be doubled by the year 2000 in order to feed the expected population (30). Yields must therefore continue to improve, but only through methods not requiring excessive amounts of energy.

Another profound change in agriculture concerns our increased awareness of the environmental impact of chemicals. It is not enough to look only at a chemical's effectiveness; we must be fully aware of the consequences of its use. Some would ban the use of all pesticides, just as others would demand complete freedom from restriction in their use. Reason tells us the path of wisdom lies somewhere in the middle. We must realize that pesticides are economic poisons. They must be considered on a risk-benefit scale and used only when the benefit to be obtained outweighs the risk involved.

The tightening restrictions on chemical usage have greatly increased the difficulty of developing new fungicides; however, they can also be a driving force in the discovery of chemicals both uniquely active and safe in the environment. The discovery of such chemicals will only be achieved through novel test procedures that can reveal unique modes of action. This paper is concerned with the rationale for the development of such procedures.

APPROACH TO FUNGICIDE DISCOVERY

For disease to occur, three factors must be present: a susceptible host, the pathogen, and a favorable environment. Chemicals can control disease by affecting any one of the three. Historically, the goal of fungicide research has been to find chemicals that kill or inhibit the pathogens. Until fairly recently, in vitro methodology has been used in this search. As a result, the chemicals found have, in general, fit that class of compounds called protectant fungicides whose activity and performance depend upon maintaining a protective barrier of toxicant on the host surface.

Following the adoption of in vivo test methodology, the systemic fungicides were born, and with their introduction we have entered a new era of fungicide research. Kirby (27) has outlined the following characteristics that adequately describe the properties of the systemics currently available: 1. They possess direct fungitoxicity and do not control diseases by modifying the host. 2. They have low to very low water solubility. 3. They enter aerial tissues and plant roots and pass into the xylem. 4. They generally move upward in the transpiration stream and may accumulate at the leaf margins. 5. They do not enter organs that do not transpire nor are they reexported to new growth. 6. Downward translocation in the phloem, if any, is of no practical significance. 7. Their activity spectra may be wide or narrow, but none, with one minor exception, control diseases caused by the Phycomycetes.

Stated more succinctly, the term systemic fungicide refers to a compound that gives protection from within. To many, vascular transport is a requirement for being called a systemic. I operate under the premise that movement into the host tissues in amounts sufficient to give control is the only requirement. In order to protect from within, the chemicals must be applied prior to infection; however, several of the compounds possess the additional property of being able to control certain diseases in the postpenetration stage. This type of activity is referred to as curative action and, although generally effective for only a limited time following penetration, can be of major significance in achieving disease control.

Selection of Test Methods

The science of fungicide research depends upon the art of testing methodology. A fungicide evaluation program should strive for uniqueness, and tests should be designed to discover both conventional and unconventional types of activity. In vivo test methodology must be used, for only in this way can one establish the correct interrelationship between the host, pathogen, chemical, and environment necessary to obtain a correct measure of a chemical's effectiveness.

Selection of Test Organisms

If we accept the premise that the closer test methods duplicate field conditions, the more accurate they are, it follows that the choice of test organisms is an important consideration. Optimally, we would normally select the specific disease and host of interest for such tests, but often this is not possible. As an acceptable compromise, we could use a related pathogen on a different host. For example, we could look for compounds that would control Sigatoka disease of bananas by using *Cercospora* leaf spot on peanuts or sugar beets. Most of the major pathogens can be adapted to small-scale tests, and are much preferred over those selected primarily because they are easily manipulated.

I once heard a very appropriate criticism of fungicide researchers worthy of repeating. To paraphrase slightly, it was mentioned that "they have been very successful in discovering a fantastic array of bean rust fungicides." I believe this tongue-in-cheek criticism is well founded. If we are searching for broad-spectrum protectants, the choice of test organisms is less important, but as we strive for increased specificity, the organisms used must be representative of the target disease. Although most of the systemics in use today are effective in conventional tests (17, 28), we are beginning to find compounds so specific that their activity could easily be overlooked. RH-124, 4-n-butyl-1,2,4-triazole, is a prime example of such specificity. As part of a larger study, Rowell measured the activity of this compound as protectant and eradicant leaf sprays and as a soil drench for both stem rust (*Puccinia graminis tritici*) and leaf rust (*P. recondita*) on wheat (40). His results are shown in Table 1. They clearly demonstrate that if wheat stem rust were used as a screening organism, such a compound might not have had sufficient activity to warrant further evaluation, and its unique leaf rust activity might never have been discovered.

Importance of Physical and Biological Properties

Greenhouse tests should give some measure of the properties important for effective field performance. Good in vitro fungicidal activity is quite common, but compounds combining excellent activity with correct physical and biological properties are extremely rare. Today there are many diseases not adequately controlled with chemicals. This is not due to a deficiency of biological activity. The lack of effectiveness stems from an inability of the chemicals to localize in the host at the right place, at the right time, and at sufficiently high concentrations to give effective control. For

Table 1 Dosages of RH–124 giving 50% control (ED_{50}) of infections in wheat seedlings by *Puccinia graminis tritici* and *P. recondita* (40)

Disease pathogen	ED_{50}		
	Protectant (μg/leaf)	Eradicant (μg/leaf)	Soil drench (μg/leaf)
P. graminis tritici	>100	>100	>2000
P. recondita	0.0012	190	3.6

example, we can point to the ineffectiveness of benomyl, methyl-1-(butyl-carbamoyl)-2-benzimidazole carbamate, in controlling verticillium wilt (*V. Albo-atrum*) of cotton in the field (11, 12, 14) when a very high level of activity was found in greenhouse tests (10). Obviously, these greenhouse tests did not correctly measure the properties needed for good field performance. Our primary concern is not that chemicals effective in greenhouse tests may fail in the field, but that improperly designed tests can allow chemicals with desirable properties to pass unnoticed.

PYROXYCHLOR—A CASE HISTORY

I would like to discuss briefly the discovery of pyroxychlor, 2-chloro-6-methoxy-4-(trichloromethyl)pyridine, to emphasize the importance of test methodology in finding unique activity.

Synthetic effort in this area was intensified following the discovery of the nitrification inhibitor, nitrapyrin, 2-chloro-6-(trichloromethyl)pyridine. Certain derivatives possessed excellent fungicidal activity against the Phycomycetes, and a large number were tested as seed treatments for control of *Pythium ultimum*. In these tests, pyroxychlor was definitely a second-best compound and our interest was focused on another, more active derivative.

The chemicals were later tested as soil drenches for control of tobacco black shank (*Phytophthora parasitica* var. *nicotianae*). Activity comparisons were in agreement with the seed treatment data, with the exception of an unusually long residual activity shown only with pyroxychlor. Subsequently, this compound was found to be systemic and effective when applied as a foliage spray.

Looking again at the list of properties possessed by today's systemic fungicides, we can now make some changes. Pyroxychlor is systemically active for several soil diseases caused by Phycomycetes and is translocated downward. Yet, even possessing such unique properties, the compound could have been passed over if only the seed treatment data were used to select a candidate for field trials.

CRITERIA DEFINING EFFECTIVE PERFORMANCE

Protectant Fungicides

To effectively control disease, a chemical must get to the site of action and initiate a biological response. If we consider only protectant fungicides, these properties correspond to the "shaped charge" of Rich & Horsfall (36). We can picture a molecule lying in wait containing an active site that can react with certain cellular constituents of the pathogen. The remainder of the molecule forms the vehicle through which the compound can penetrate the pathogen membrane. The importance of optimum partition coefficient in this ability has been clearly shown (21).

Let us list the three characteristics that typify the protectant fungicides. They are biologically active; their oil:water solubility characteristics allow them to be taken up by the pathogen in toxic amounts; and they have adequate residuality following application. Even though residuality depends on a number of interrelated properties such as resistance to hydrolysis and ultraviolet degradation, tenacity, volatility, and capacity for redistribution, these can be measured independently to derive a reasonable estimate of potential field performance. There is no such thing as a "magic factor" that determines whether a compound will perform in the field. Every failure can be traced to an inadequacy in some measurable property. In vitro test methodology has been used extensively in an attempt to measure these critical properties, but regardless of how sophisticated these methods may be, they have proven adequate only for the development of protectant fungicides (31).

Systemic Fungicides

While relatively few easily measured physical properties can be of overriding importance in determining the field performance of the protectants, the activity of a systemic fungicide is dependent upon a wide range of complex interactions that defy measurement. Byrde (2) proposed a model for systemic action that included five "intrahost" obstacles that must be overcome. These were the host cuticle and subcuticular cells, metabolism in the host, adsorption in the host during translocation, metabolism by the pathogen, and the pathogen membrane. Additionally, the availability and degree of movement of systemic fungicides in plants were found to be highly dependent on the chemical and physical properties of the compound, the nature and growth stage of the host, the method of application, the nature of translocation in the host, and the environment (17, 42, 43). One can now appreciate the difficulty inherent in attempting to measure such parameters independently. The task becomes one of finding compounds that can overcome these obstacles and satisfy all requirements for activity. The most logical approach is to look for the endproduct, control of the clinical symptoms of disease. A good research program should be based on a very simple premise: attempt to bring together the correct combination of host, pathogen, and environment giving a close approximation of the true field situation. The chemicals should then be applied in a manner that gives them the best opportunity for controlling the disease.

Although we discuss protectant and systemic fungicides where the term fungicides implies killing the pathogen, we can easily broaden our thinking to include compounds that control disease by making the host resistant or the environment inhospitable to the organism.

THE SEARCH FOR NEW FUNGICIDES

The search for systemic fungicides is largely based on one of the methods for detecting movement of chemicals as noted by Wain & Carter (44), specifically by reducing disease symptoms at a point removed from the application site. For example, we can treat roots and inoculate foliage or treat one part of the foliage and inoculate roots or a different part of the foliage. Compounds active in these types of tests are considered to be truly systemic, but this implies a definition of high mobility. If the only requirement of a systemic fungicide is that it gives protection from within the plant, the chemical need not move great distances. Quite often the primary goal of finding disease-controlling chemicals has become secondary to the search for highly mobile systemic fungicides. An effective screening program should look for all chemicals that control disease, not just for systemic fungicides.

To best illustrate what I believe to be a logical approach to fungicide screening, I will describe the conceptual development of a test method for verticillium wilt of cotton. This disease is chosen because it has not yet been adequately controlled on a wide scale with chemicals and, even more important, an effective chemical must possess certain properties apparently not present in any of the chemicals available today.

Consider first some information that has considerable bearing on the test methodology selected: 1. Although the organism can infect cotton at any time during the growing season, it is the early season infections that drastically reduce yields (46). 2. A single soil fumigation controlled the disease for two years, but yields were increased the second year only (46). 3. Systemic fungicides such as benomyl and thiabendazole (TBZ), 2-(4-thiazolyl)benzimidazole, provided excellent disease control in greenhouse tests when the plants were inoculated by either stem injection or by pouring inoculum over the root ball (10, 13). 4. In field trials, benomyl was ineffective unless excessively high rates were used (11, 12, 14). This failure may be due to the absence of feeder roots in the treated zone as plants grow older (11). 5. Paraffinic oils augmented the uptake and translocation of benomyl and TBZ in cotton plants in the greenhouse (15), but oils did not improve control of wilt in field trials with either compound (12).

With this as background, we can begin to formulate some generalizations concerning the type of compound we will be looking for. First, a treatment that reduces the inoculum potential in the root zone could possibly reduce early season infection sufficiently to provide beneficial yield responses. This approach, however, has not proven adequate. Although the volatile soil fumigants effectively control the disease, they are expensive. Similarly, with contact fungicides, dosages needed to achieve adequate concentrations and distribution of the toxicant in soil would be economically prohibitive. It is probable that this approach offers little potential for the control of wilt, although it cannot be discounted completely. Second, it seems evident that successful control of this disease is far more dependent on the ability of the chemical to get to the site of action than on the biological activity level of the molecule.

We can now begin to define the possible ways a compound could be effective, exclusive of a host or environmental interaction. One would be as a postemergence foliar treatment effective in either preventing or curing the disease. This would necessitate phloem transport. Another possibility is an at-plant treatment that is selectively accumulated by the roots and not rapidly degraded or translocated, thereby offering an extended period of control.

In the test developed, an attempt would be made to duplicate the actual field situation as closely as possible. This would involve bringing together the chemical, host, and pathogen in soil. The plants must become infected through the roots, and the chemical should be applied to the foliage and to the soil prior to infection. Because the organism is a poor soil inhabitant, the need for uniform infection levels would require some artificial inoculation procedure such as spraying the root ball. Inoculating a few days after treatment is preferred and would also facilitate the discovery of chemicals effective via host or environmental effects.

It has long been recognized that a fungicide with some type of systemic property will be required for controlling verticillium wilt. The impetus for using such procedures as stem injection or double potting (10) has been a desire to separate the chemical from inoculum, thereby eliminating the possibility of a compound being effective through a strictly local effect. This is an example of the primary goal of looking for an effective chemical becoming secondary to the goal of searching for a systemic fungicide. If a chemical moves into and localizes only in the roots or is effective through an interaction at the soil-root interface, it would probably be inactive in a stem injection test, but could be effective in the natural field situation. On the other hand, chemicals like benomyl that are readily taken up by roots but translocate rapidly to the foliage, might be highly active in the stem injection test but ineffective under field conditions. A proper test method must find chemicals that localize in the roots, either through soil or foliar application, not chemicals that translocate readily from the roots to the foliage.

Admittedly, finding novel chemicals with the correct properties is an extremely difficult task. As a result, the "bandwagon" approach to discovery is very common. Since the discovery of the first systemic fungicides, many more have appeared on the scene in rapid succession. In many instances these have been the result of synthesizing analogs and homologues of of previously described active structures. Current examples are the oxathiins and benzimidazoles wherein the initial discoveries were followed by the evolution of several related derivatives.

IMPACT OF CURATIVE FUNGICIDES

There are two approaches to chemotherapy: either protecting plants from infection or curing plants after they become infected. With plants as with humans, the timeworn statement of prevention being the best cure holds true. Too much importance must not be placed on the curative aspect of chemotherapy, although it has obvious value in controlling diseases of foliage and fruit or diseases attacking the roots or scaffold of large trees wherein prevention would be difficult or cost-prohibitive.

A key factor determining the successful use of curative fungicides is an ability to detect infections at an early stage of development. In such instances, it is possible to cure incipient infections without much damage having been done to the host. The curative property allows greater flexibility in spray schedules and reduces the penalty for applications postponed because of inclement weather or other reasons. Spray intervals can be increased or applications eliminated, thereby saving time and money and reducing the total quantity of chemical being applied.

On the other hand, the majority of root diseases pose a much different problem. When symptoms become obvious, a great deal of damage may have already been done, and even if the pathogen is eliminated, the plant may not have sufficient time to recover. A second factor limiting the use of curative fungicides for soil diseases is the increased difficulty of applying them effectively as plants grow larger.

Curative action is not a replacement for a good disease prevention program, nor is it the answer to our unsolved problems. It can be a mechanism through which we can obtain more consistent performance from chemicals. If infections, or the presence of the pathogen, can be identified before irreparable damage is done to the host, curatives can be of tremendous value; if not, their value is greatly diminished.

SHORT-TERM PROSPECTS IN FUNGICIDE RESEARCH

For the immediate future, the discovery of commercially useful compounds effective by methods other than killing the pathogen is unlikely; however, new breakthroughs in activity are possible within the framework of fungicides as we know them today. For example, triarimol, α-(2,4-dichlorophenyl)-α-phenyl-5-pyrimidinemethanol, although withdrawn from development, represented a new dimension in systemic fungicidal activity. Similarly, the control of wheat leaf rust for the entire season with only one application of RH-124 (39) tells us that in the future we may be able to move away from repeat application schedules.

The comparatively large markets found in fungal disease control have tended to limit the work done on bacterial diseases. As a result, the discovery of highly effective bactericides is long overdue. Industry is now beginning to focus more attention on this area and is becoming increasingly aware of the magnitude of the problem and the requirements for effective activity (6, 18). The discovery of highly active compounds will undoubtedly follow.

There is a pressing need for new innovations in chemical control. The not-too-distant future should see systemics and curatives for the downy mildew diseases, good systemic bactericides, and compounds that translocate to new foliage or roots following foliar treatment. As we search for these new parameters of activity, many of the inadequacies found with present chemicals will be eliminated.

Problems Associated with Systemic Fungicides

Since the introduction of the systemics, resistant strains of fungi have begun to develop at an alarming rate. The systemics seem to possess quite specific modes of action, as opposed to the protectants that are "multi-site" inhibitors of cellular processes. This high level of specificity dictates that a one-gene mutation can give rise to a resistant strain (9, 25, 26). Equally important, certain properties of the systemics contribute to the development of resistance (8, 9, 17). Some possible solutions have been discussed (17), but as long as chemicals depend upon killing or inhibiting the pathogen, we must be prepared to live with this problem.

There are other areas of concern with the systemic fungicides. The systemic property increases the potential for high persistence in the crop, and their specificity may alter the ecological balance, thereby increasing diseases caused by nonsensitive organisms (22, 26).

Such problems must not detract from the search for newer and better chemicals. As the systemics brought forth some previously unknown problems, so, likely, will any new group of compounds or concept of control. All new chemicals are merely additions to our arsenal of weapons against plant diseases, and each will differ in its list of advantages and disadvantages.

LONG-RANGE GOALS OF FUNGICIDE RESEARCH

The development of plant chemotherapy will be a true evolutionary process. The protectant fungicides represent the basic ancestral lineage and attained maximum development within the past few decades. We are witnessing the growth of the first major divergence, the systemic fungicides. As usage of the systemics becomes more widespread, the role of the protectants will wane, although it is unlikely they will disappear.

Continual pressure for improved safety with agricultural chemicals will provide a strong impetus for discovering alternative types of activity. Such alternatives reside in chemicals that control diseases through host or environmental effects. These concepts represent new evolutionary lines, and a number of possible modes of action are not without precedent. Several chemicals modify disease levels without affecting the pathogen. While most are purely academic, they represent the beginnings of new, exciting areas of chemistry. Most of these examples have been reviewed elsewhere (16, 22) and receive only superficial coverage here.

Altered Carbohydrate Levels

The interrelationship between sugar levels and disease susceptibility was originally discussed by Horsfall & Dimond (23). A number of chemicals are known to modify disease incidence by altering carbohydrate levels in plants, either directly or indirectly. The most notable example is the use of mineral oil to control Sigatoka. Calpouzos believes the oil reduces sugar levels in the young leaves, making them less susceptible (3). One drawback of this particular mode of action is that there are both high-sugar and low-sugar diseases, and affecting the sugar level to reduce one disease may increase the severity of another.

Stimulation of Host Defenses

Chemicals can trigger existing host defense mechanisms. Production of the phytoalexin pisatin in peas was stimulated by application of copper and mercury salts (7). Similarly, certain chemicals modify the phenolic content or amino acid nutrition of the host, and in doing so can bolster the host's normal disease resistance.

Physical Alteration of the Host

In many cases, diseases could be inhibited by initiating some physical change or disturbance in the host. The response can be straightforward, such as promoting thicker cuticles or affecting stomatal closure, or more complex, such as initiating formation of starch-filled cells or other barriers to invasion.

Maintenance of Mutualistic Symbiosis

An interesting concept was proposed by Futrell (19) following his studies on chemical seed treatments of corn internally infected with *Fusarium moniliforme* and *Helminthosporium maydis*. *F. moniliforme* can live in the host as long as conditions are favorable, but becomes a virulent pathogen only under certain circumstances. The benzimidazoles, benomyl and TBZ, effectively control *F. moniliforme*. Futrell offered the possibility that the chemicals do not kill the pathogen, but keep the host and pathogen in mutualistic symbiosis.

Environmental Effects

A number of reviews discuss the effects of solid, liquid, and gas phase components of soil (4), plant exudates (37, 38, 41, 49), microbial toxins (35, 45), and crop residues (5, 20, 32) on the incidence of soil diseases. Similarly, the effects of host nutrition upon disease susceptibility are well known. Chemical disease control through an environmental effect is very likely linked to an interaction with one or more of these factors.

Nitrogen nutrition has a marked effect upon disease incidence. The interrelationship between nitrogen form and plant disease has been recently reviewed (24), and abundant evidence was presented showing disease levels to be often dependent upon the form of nitrogen present. As regards chemical control opportunities, stabilizing ammonia nitrogen levels with nitrapyrin or other nitrification inhibitors decreased incidence of certain diseases.

Stimulation of Biological Control Mechanisms

An area of research somewhat related to the preceding can be called "chemically induced biological control." Much has been written concerning biological control of soil-borne plant diseases and the subject has been amply reviewed (1). If chemicals could be used to overcome the inherent inconsistency of such control measures, it would be a tremendous step forward. The use of chemicals to aid nature in its normal processes is a sound concept and is worthy of our research efforts. One possible mode of action is to increase the populations of antagonistic organisms in the rhizosphere through a chemically induced alteration in the nature or amount of the host's root exudates.

In summary, the discovery of compounds with such novel modes of action demands that we look for them in the correct manner. If tests are designed to find only fungicidal chemicals, other subtle types of activity can be missed. If tests closely typify field conditions, the probability of finding such activity is greatly improved.

CONCLUSIONS

The development of an agricultural pesticide is an expensive process. Several years and many millions of dollars are spent in bringing a product from discovery to sales.

With such a heavy commitment, market recognition can become a problem. Money spent must be justified, and the justification required is often in terms of dollars spent on control of a particular pest. This approach is shortsighted. To search only for compounds that can carve out a portion of an existing market is not fulfilling our professional responsibility. We must always be sensitive to the growers' needs and must realize that many of today's poorly developed markets hold much potential for the future.

Continued demands for increases in food and energy will heighten man's dependence upon agriculture. Our awareness of the need for conserving nonrenewable resources places much emphasis on agricultural production as a renewable source of energy and raw materials.

The demand for new innovative chemicals is very strong, but chemicals alone are not the answer to reducing disease losses. Integrated programs including sanitation, good cultural practices, resistant varieties, and disease forecasting must go hand-in-hand with judicious application of chemicals.

It has been stated that if pesticides were eliminated, losses would increase only about 7%, from 33.6 to 40.7%, and the additional loss could be compensated for by planting more acres and adopting certain other practices (33). We are negligent in our responsibilities to a hungry world if we adopt such reasoning. Our concern must always be to reduce losses through any means available. We cannot afford to lose billions of dollars worth of potential production every year to the ravages of diseases and pests. However, demands for improved control should not necessitate massive increases in chemical usage. The law of diminishing returns applies to pesticide application. Excessive dosage rates offer ever-decreasing incremental benefits with continually increasing risks of ecological damage. There is a need for better chemicals, better application methods, better sanitation practices, and consistency in control programs. Chemicals are meant to be used in conjunction with good grower practices, and not as alternatives.

The challenge lies before us. We must develop better chemicals, unique types of activity, chemicals that are safe in the environment and that present little hazard to the users. Looking back, we have come a long way, but the road ahead is long and each step becomes more difficult. Yet, every new chemical or technological advance will aid us in an often desperate fight against hunger and misery. For those involved in fungicide research, there can be no greater satisfaction than to know they have made a contribution to this fight.

Literature Cited

1. Baker, R. 1968. Mechanisms of biological control of soil-borne pathogens. *Ann. Rev. Phytopathol.* 6:263–94
2. Byrde, R. J. W. 1969. Systemic fungicides, including recent developments in the agricultural sphere. *Proc. 5th Br. Insectic. Fungic. Conf.* 3:675–79
3. Calpouzos, L. 1966. Action of oil in the control of plant disease. *Ann. Rev. Phytopathol.* 4:369–90
4. Chapman, H. D. 1965. Chemical factors of the soil as they affect soil microorganisms. In *Ecology of Soil-Borne Plant Pathogens*, ed. K. F. Baker, W. C. Snyder, 120–41. Berkeley: Univ. Calif. Press. 571 pp.
5. Cook, R. J., Watson, R. D., Eds. 1969. Nature of the influence of crop residues on fungus-induced root diseases. *Wash. Agric. Exp. Stn. Bull.* 716. 32 pp.
6. Crosse, J. E. 1971. Prospects for the use of bactericides for the control of bacterial diseases. *Proc. 6th Br. Insectic. Fungic. Conf.* 3:694–705
7. Cruickshank, I. A. M. 1963. Phytoalexins. *Ann. Rev. Phytopathol.* 1:351–74
8. Dekker, J. 1969. Acquired resistance to fungicides. *World Rev. Pest Control* 8(2):79–85
9. Dekker, J. 1971. Selective action of fungicides and development of resistance in fungi to fungicides. *Proc. 6th Br. Insectic. Fungic. Conf.* 3:715–23
10. Erwin, D. C. 1969. Methods of determination of the systemic and fungitoxic properties of chemicals applied to plants with emphasis on control of Verticillium wilt with thiabendazole and Benlate. *World Rev. Pest Control* 8:6–22
11. Erwin, D. C. 1970. Progress in the development of systemic fungitoxic chemicals for control of plant diseases. *FAO Plant Prot. Bull.* 18:73–81
12. Erwin, D. C. 1972. Progress report of systemic fungicides for controlling Verticillium wilt. In *Sum. Proc. West. Cotton Prod. Conf.*, 17–20
13. Erwin, D. C. 1973. Systemic fungicides: disease control, translocation, and mode of action. *Ann. Rev. Phytopathol.* 11:389–422
14. Erwin, D. C., Garber, R. H., Carter, L., DeWolfe, T. A. 1969. Studies on thiabendazole and Benlate as systemic fungicides against Verticillium wilt of cotton in the field. *Proc. Beltwide Cotton Prod. Res. Conf.*, p. 29 (Abstr.)
15. Erwin, D. C., Khan, R. A., Buchenauer, H. 1974. Effect of oil emulsions on the uptake of benomyl and thiabendazole in relation to control of Verticillium wilt of cotton. *Phytopathology* 64:485–89
16. Evans, E. 1968. *Plant Diseases and Their Chemical Control.* Oxford: Blackwell. 288 pp.
17. Evans, E. 1971. Systemic fungicides in practice. *Pestic Sci.* 2:192–96
18. Fox, R. T. V. 1971. Development of in vitro and in vivo screening procedures for bactericides. *Proc. 6th Br. Insectic. Fungic. Conf.* 1:187–92
19. Futrell, M. C. 1972. New concepts in chemical seed treatment of agronomic crops. *J. Environ. Qual.* 1:240–43
20. Garrett, S. D. 1970. *Pathogenic Root-Infecting Fungi.* Cambridge: Cambridge Univ. Press. 294 pp.
21. Hansch, C., Fujita, T., 1964. ρ-σ-π-Analysis. A method for the correlation of biological activity and chemical structure. *J. Am. Chem. Soc.* 86:1616–26
22. Horsfall, J. G. 1972. Selective chemicals for plant disease control. In *Pest Control Strategies for the Future*, 216–25. Washington, DC: NAS. 376 pp.
23. Horsfall, J. G., Dimond, A. E. 1957. Interactions of tissue sugar, growth substances, and disease susceptibility. *Z. Pflanzenkr. Pflanzenpathol. Pflanzenschutz* 64:415–21
24. Huber, D. M., Watson, R. D. 1974. Nitrogen form and plant disease. *Ann. Rev. Phytopathol.* 12:139–65
25. Kaars Sijpesteijn, A., Kerk, G. J. M., van der 1969. Biochemical modes of action of fungicides. *Proc. 5th Br. Insectic. Fungic. Conf.* 3:724–33
26. Kerk, G. J. M., van der 1971. Systemic fungicides—new solutions and new problems. *Proc. 6th Br. Insectic. Fungic. Conf.* 3:791–802
27. Kirby, A. H. M. 1972. Progress towards systemic fungicides. *Pestic. Abstr. News Sum. Sect. B. Fungic. Herbic.* 18:1–33
28. Koch, W. 1971. Behaviour of commercial systemic fungicides in conventional (non-systemic) tests. *Pestic. Sci.* 2:207–10
29. Marsh, R. W., Ed. 1972. *Systemic Fungicides.* London: Longman. 321 pp.
30. Miller, P. R. 1970. Crop protecting chemicals must be used to improve food supplies. *Agric. Chem.* 25(12):16–17

31. Neely, D. 1969. The value of in vitro fungicide tests. *Illinois Nat. Hist. Surv. Biol. Notes.* No. 64. 8 pp.

32. Patrick, A. Z., Toussoun, T. A. 1965. Plant residues and organic amendments in relation to biological control. See Ref. 4, 440–59

33. Pimentel, D. 1973. Realities of a pesticide ban. *Environment* 15(2):18–31

34. Pimentel, D. et al 1973. Food production and the energy crisis. *Science* 182:443–49

35. Pringle, R. B., Scheffer, R. P. 1964. Host-specific plant toxins. *Ann. Rev. Phytopathol.* 2:133–56

36. Rich, S., Horsfall, J. G. 1952. The relation between fungitoxicity, permeation, and lipid solubility. *Phytopathology* 42:457–60

37. Rovira, A. D. 1965. Plant root exudates and their influence upon soil microorganisms. See Ref. 4, 170–86

38. Rovira, A. D. 1969. Plant root exudates. *Bot. Rev.* 35:35–57

39. Rowell, J. B. 1971. Wheat (Thatcher): Leaf rust *(Puccinia recondita)*. In *Fungicide and Nematicide Tests,* 26:110. St. Paul, Minn.: Am. Phytopathol. Soc. 196 pp.

40. Rowell, J. B. 1972. Fungicidal management of pathogen populations. *J. Environ. Qual.* 1(3):216–20

41. Schroth, M. N., Hildebrand, D. C. 1964. Influence of plant exudates on root-infecting fungi. *Ann. Rev. Phytopathol.* 2:101–32

42. Shephard, M. C. 1972. Factors limiting the systemic activity of fungicides, In *Herbicides, Fungicides, Formulation Chemistry,* ed. A. S. Tahori, 383–88. New York: Gordon & Breach. 565 pp.

43. Verloop, A. 1972. Factors influencing the availability of systemic fungicides in plants. See Ref. 42, 389–96

44. Wain, R. L., Carter, G. A. 1967. Uptake, translocation, and transformations by higher plants. In *Fungicides, an Advanced Treatise,* ed. D. C. Torgeson, 1:561–611. New York: Academic. 697 pp.

45. Wheeler, H., Luke, H. H. 1963. Microbial toxins in plant disease. *Ann. Rev. Microbiol.* 17:223–42

46. Wilhelm, S., Storkan, R. C., Sagen, J. E., George, A. G., Tietz, H. 1972. Cotton Verticillium wilt control with soil fumigation. *Calif. Agric.* 26(10):4–6

47. Wittwer, S. H. 1974. Maximum production capacity of food crops. *BioScience* 24(4):216–24

48. Woodcock, D. 1971. Chemotherapy of plant disease—progress and problems. *Chem. Br.* 7:415–20, 423

49. Woods, F. W. 1960. Biological antagonisms due to phytotoxic root exudates. *Bot. Rev.* 26:546–69

CRITICAL ANALYSES OF THE PRINCIPLES OF SEED POTATO CERTIFICATION

❖3623

James F. Shepard and Larry E. Claflin
Department of Plant Pathology, Montana State University, Bozeman, Montana 59715

INTRODUCTION

Genotypes of many crop plants including fruit trees, ornamentals, grapes, strawberries, and potatoes are maintained by vegetative propagation. This cultural practice causes certain problems of disease control, which are either nonexistent or of lesser importance in plants that are reproduced from seed. Pathogens already existing in infected propagative materials can continue to develop in the progeny. Cut surfaces of tubers and stem cuttings provide open courts for infection. The high water content of vegetative propagation materials in comparison with true seeds also adds to their vulnerability to infection. Elaborate programs have been devised for many vegetatively propagated plants both to minimize disease incidence in propagative material and to provide relatively disease-free (certified) planting stock to growers. The most extensive and long-lived certification effort has been with potatoes (*Solanum tuberosum* L.) and its success has provided impetus for certification programs in other crops. This article is restricted to an analysis of seed potato certification, and while several of the concepts may also apply to other crops, certain distinctions will be apparent.

HISTORICAL

Potato seed certification was begun in Europe in the early 1900s largely through the efforts of Dr. Otto Appel in Germany (2). During this period, the increase of leafroll and other diseases in North American potatoes stimulated Dr. W. A. Orton of the United States Department of Agriculture to visit Appel (30) and later to present a plan for a certification program patterned after the German program to the First Annual Meeting of the Potato Association of America (1914), the First Official Seed Potato Certification Conference (1914), and the Fifth Annual Meeting of the American Phytopathological Society (1915). In 1914, Appel visited the United States (1)

271

and collaborated with Orton, Dr. William Stuart of the USDA, and W. J. Morse of the University of Maine in proposing a certification program in several potato producing areas (31). Within five years, 12 states and all Canadian provinces were engaged in seed potato certification (37).

Seed certification programs were commonly begun by Cooperative Extension Service personnel primarily to reduce the incidence of virus disease. Later, separate agencies were created to administer and perform the function of seed certification, and the scope of certification was broadened to include tuber-borne diseases caused by other pathogens.

PRINCIPLES OF CERTIFICATION

The basic objectives of seed certification were presented by Orton in 1914 (32) at the First Official Seed Potato Certification Conference: "It [seed certification] presupposes a movement for the betterment of the potato industry through the improvement of seed potatoes, through the development of specialized growers of seed potatoes, and through an organization created to control diseases, to reduce varietal mixture, to improve varietal types, to be stimulated by the inspecting service organized by the state for the purpose of inspecting seed potatoes and granting certificates to such as may be found worthy of such certification through their varietal purity and freedom from disease." Although techniques of certification have changed, these basic goals and principles are still viable today. Precisely how the principles are implemented has, of course, changed and considerably broadened in scope and now probably represents the most extensive disease-control effort for one commodity.

Administrative Organization of Certification Programs

A necessary first step in seed certification is the creation of an agency with a legal base sufficient to administer and coordinate all facets of the program. This is accomplished in a variety of ways. In several countries, seed certification is under the auspices of an agency of the federal government. In the United States, certification may be the responsibility of a state department of agriculture, a land-grant university, or a grower association (see Table 1) with, in frequent cases, cooperative agreements among the three agencies. In general terms, the individual certification agency is responsible for organizing growers who wish to produce seed, formulating and enforcing "rules and regulations," and for conducting field and storage facility inspections, tagging and labeling operations, and in certain instances shipping point inspections.

Funds for agency operation are partially or totally obtained from certified seed growers who pay to register their fields for certification, and for inspection and tags that certify that the seed has met prescribed tolerances.

In most programs, certification does not constitute a warranty by either the certification agency or the grower of certified seed beyond the express representation that the potatoes were produced, inspected, graded, and packed under the regulations of both the certifying agency and an affiliated state or federal regulatory

Table 1 Seed potato certification in North America

Location	Certification started (37)	Certification acreage in 1973 (45)	Name of agency
	1913–1915		
Canada			
Prince Edward		26,231	Canada Department of Agriculture
New Brunswick		17,566	Canada Department of Agriculture
Nova Scotia		331	Canada Department of Agriculture
Quebec		3,273	Canada Department of Agriculture
Ontario		1,164	Canada Department of Agriculture
Manitoba		2,730	Canada Department of Agriculture
Saskatchewan		162	Canada Department of Agriculture
Alberta		1,660	Canada Department of Agriculture
British Columbia		1,087	Canada Department of Agriculture
United States			
Idaho		31,972	Idaho Crop Improvement Association
Maine		52,159	Maine Department of Agriculture
Vermont		169	Vermont Department of Agriculture
Wisconsin		7,714	University of Wisconsin
	1916–1919		
California		3,481	California Department of Agriculture
Colorado		5,735	Colorado State University
Minnesota		25,363	Minnesota Department of Agriculture
Nebraska		2,380	Potato Certification Association of Nebraska
New Hampshire		6	New Hampshire Department of Agriculture
New York		1,727	New York Seed Improvement Cooperative, Inc.
North Dakota		34,145	North Dakota State Seed Department
Oregon		3,006	Oregon State University
	1920–1922		
Michigan		2,356	Michigan Crop Improvement Association
Montana		4,613	Montana Potato Improvement Association
Pennsylvania		148	Pennsylvania Department of Agriculture
South Dakota		593	South Dakota Potato Growers Association
Utah		91	Utah Crop Improvement Association
Washington		2,327	Washington Department of Agriculture
Wyoming		412	Wyoming Seed Certification Service

agency. In the case of unsatisfactory seed, the liability of the certifying agency usually is limited to the value of the seed when sold.

Disease Control

DISEASE TOLERANCES   Profitable "commercial" (i.e. produced for consumption) potato production is possible only when a continued supply of "seed" is available that is totally free from certain diseases and virtually so from most others. It is the responsibility of the individual certification agency to provide the services necessary to assure growers and buyers that the seed in question was produced in compliance with defined standards. The agency must also set minimum necessary standards (rules and regulations) with respect to specific diseases and to a lesser extent other qualifications such as varietal mixtures. The basic principles that are (or should be) followed when regulations are formulated were described by Leach (20): "(1) an accurate knowledge of the extent to which a disease is transmitted in or on seed tubers; (2) the recognition of all other sources of infection; and (3) an accurate evaluation of different sources of infection under various circumstances." For certain diseases insufficient or inaccurate information has resulted in the formulation of disease tolerances that "growers can live with" rather than what may be biologically justified. An example of this is the blackleg disease caused by *Erwinia atroseptica* (Van Hall) Jennison [and possibly by *E. carotovora* (Jones) Holland]. Tuber lenticels of many North American (A. Kelman, personal communication) and European (33) potato varieties are universally infected, but the conditions necessary for the expression of blackleg symptoms in the field are poorly understood. (33). Present regulations that do not include tests for the presence of the bacterium are, therefore, of questionable value.

As shown in Table 2, disease tolerances presently included in the various certification programs in the United States are similar but not completely standardized. Certified seed produced under one program may harbor greater or lesser amounts of infection by certain organisms than that produced in other programs. Nematodes are considered in the certification programs of only seven states and in only two of these is the species of nematode specified. Some differences in the disease tolerances of individual agencies arise from variation in the distribution of pathogenic organisms. It would be wasteful to conduct exhaustive surveys for diseases that do not exist within a given growing area; also, certification agencies differ in their wherewithal to detect or control certain pathogens.

DISEASE DETECTION AND DIAGNOSIS   Most certification standards relate to diseases, and with the exception of giant hill, they do not include such factors as genetic or horticultural differences within a given variety. Hence, establishing levels of disease within fields to be certified as suitable for seed requires an extensive effort on the part of the certification agencies. Disease detection and diagnoses are accomplished by two or three field inspections performed by agency personnel during a growing season. Except for diseases with a zero tolerance, tolerance differentials between the first and last inspection permit growers to rogue out infected plants between inspections. If tolerances are exceeded at the final reading the crop is not eligible for certification. Following harvest, an additional inspection in storage is conducted primarily for the detection of bacterial ring rot, tuber net necrosis, physiological disorders, and occasionally nematodes.

Apart from "latent" virus detection, which will be dealt with later, viral, bacterial, fungal, and mycoplasmal diseases are nearly always diagnosed on the basis of symptoms visible in the field. Exceptions are found in some European programs where routine testing is performed for potato viruses A, Y, and leafroll. Field diagnoses often are difficult, since symptom expression frequently is influenced by the environment and is subject to differences in varietal response to certain pathogens. Certification personnel need extensive field experience before they can become familiar with the symptoms induced by various pathogens in many potato varieties under differing environmental conditions.

Occasionally, laboratory diagnostic techniques are employed to assist in "judgment calls." Such techniques must be of the type in which results are quickly obtainable, as the time interval between inspection and harvest is usually insufficient

Table 2 Maximum allowable tolerances for various seed potato certification programs in North America[a,b]

	Leaf roll	Mild mosaic	Rugose mosaic	Spindle tuber	Total nonlatent	Haywire	Ring rot	Blackleg	Verticillium	Fusarium wilt	Late blight	Fusarium eumartii	Total all other diseases	Giant hill	Varietal mixture	Nematode
California	0.5				1		0	1		5			5		0.25	0.0[c]
Canada					1		0						2			0.1
Colorado	0.5		1.0	0.5	1.5	1	0	excess.			0			0.5	0.25	0.0
Idaho	0.2				1		0	2							0.5	
Maine	1.0	2		2	3		0			1		0	4		0.25	
Michigan	0.5	1	1.0	0.5	1	0.5	0			1					0.1	
Minnesota	0.3	1	0.3	0.1	0.5		0	1		1					0.0	
Montana	0.25	2		1	2		0	0.25			0	0		0.25	0.25	0.0[c]
Nebraska	2	2		1	3		0	2.5	5						0.5	0.0
New York	1	2		1	2		0	1	5	1					0.25	
North Dakota	0.3	1	0.3	0.1	0.5		0	1			0					0.0
Oregon	0.25	1		1	1.5		0			1		0.5			0.25	0.0
South Dakota	1	1.5		1	2	1	0	1						1.0	tr	0.0
Washington	0.2				1		0	2							0.1	
Wisconsin	1			1	3		0							0	0.1	0.0
Wyoming	1	1		1	1.5		0							0	0.1	0.0

[a] Data compiled from rules and regulations published by agencies listed.
[b] All numbers expressed as percentage of maximum allowable for the final inspection of certified grade potatoes.
[c] Root knot.

for the completion of Koch's postulates. This may in turn lead to a reliance on marginally accurate confirmatory techniques for some diseases. In the bacterial ring rot disease caused by *Corynebacterium sepedonicum* (Spieck. & Kotth.) Skapt & Burk., there are two confirmatory techniques presently in use. One is the tomato test (42) which relies on the organism producing defined symptoms when inoculated onto tomato. This procedure requires several weeks for definitive results, however, and even then negative results may be misleading because of poor recovery of the bacterium. The second procedure is the Gram stain, (14, 36) since *C. sepedonicum,* unlike other plant pathogenic bacteria outside of the genus *Corynebacterium,* is gram-positive. Unfortunately, diseased or decaying potato tubers often become invaded with a number of other bacteria some of which may also be gram-positive. Either a positive or a negative Gram stain test may, therefore, be suspect. This point was amply illustrated recently in one lot of Montana-certified seed sold out-of-state. In mid-winter, the buyer complained that the seed lot, which was now in his own storage facility was infected with ring rot since he had obtained a positive Gram stain test through regulatory personnel in his state. Inspection of the seed in question indicated that unknown to the grower, moisture had seeped into the storage facility and soft rot had occurred. The problem was resolved by applying a third test (L. E. Claflin and J. F. Shepard, unpublished) in which antiserum prepared against *C. sepedonicum* was used to test extracts from rotted tubers for the presence of *C. sepedonicum.* The absence of a serological reaction with the gram-positive bacteria in the extracts suggested that *C. sepedonicum* was not responsible for the tuber decay. The seed was planted, no ring rot developed, and a law suit was avoided. This points out the need for additional research on diagnostic tests not only for bacterial ring rot but for other diseases such as leafroll and spindle tuber. Inaccurate results from certain of the tests presently in use may not only be expensive but can unfairly damage the reputation of a careful seed producer. There also should be greater involvement, particularly in the United States, of university or state plant pathologists with the implementation of new techniques in seed certification. The staffs of most certification agencies do not include research plant pathologists. Greater cooperation instead of a mutual "hands-off" policy between certification personnel and state or university pathologists will provide mutually beneficial contacts and encourage efficient translation of research discoveries into pragmatic forms of disease control.

ESTABLISHMENT OF QUARANTINES   The third quarantine program approved in the United States (1912) involved potatoes and was instrumental in preventing an epidemic of potato wart caused by *Synchytrium endobioticum* (Schilb.) Perc. in North America. Spread of the golden cyst nematode, *Heterodera rostochiensis* Wr., from eastern United States to other potato growing regions also was prevented through strict quarantine regulations. At a more local level, barriers have been established by individual certification agencies in growing areas containing both commercial and certified potatoes. As discussed later, certification programs involved in the control of latent virus spread have relied heavily upon isolation and/or quarantine to reduce reinfection rates.

FOUNDATION PLANTINGS   A great deal more is required of seed certification programs than simply the obtention of disease readings. If this were the case, it is probable that there would be no potatoes capable of passing certification standards. Thus, an integral part of disease control includes a number of additional approaches, one of which includes the supervision of or actual provision for foundation plantings.

Certified seed producers must each year have stock available that when replanted conforms to prescribed disease tolerances. This is frequently accomplished by the replanting of certified seedstock. The limitations of this approach become apparent, however, when the basic seed source becomes too heavily infected and the grower must look elsewhere for planting material. As an alternative, foundation blocks may be maintained by the grower that are isolated from certified seedstocks, are of relatively small size, and have more stringent disease tolerances. Foundation blocks because of their small size, may be rogued more intensively to minimize disease problems, and resultant seed that complies with appropriate tolerances may then be entered the following year for certification.

Despite continued vigilance, infections may occur in foundation blocks sufficiently late in the growing season to preclude detection. Hence, an acceptable disease incidence reading may be inaccurate and plantings the following year would be "over-tolerance" at the outset. This, for example, is an all-too-frequent occurrence with infection by potato leafroll virus. It is, therefore, a requirement of all United States certification programs that a "winter index" be performed on seed entered as foundation stock. Soon after harvest, tubers from foundation blocks are sent to a central location and treated with ethylene chlorohydrin (or rindite) or gibberellic acid to break dormancy. The potatoes are then shipped to such southern states as Florida or California for planting in test plots, and during January, February, or March plants are evaluated for disease incidence. A similar approach has been used by other programs that utilize greenhouses for the off-season tests. Limitations on space and disease expression frequently restrict the efficacy of the greenhouse approach, however. As a consequence of the winter index, results of the previous year's field evaluations may be either corroborated or invalidated before spring planting begins, and recommendations are provided for the foundation stock not only with respect to disease incidence but also varietal purity and vigor. The importance of winter indexing is apparent from the fact that all United States programs require it for the foundation seed tag.

Foundation seedstock is produced by many growers as an integral part of their individual seed programs. In other cases, it is made available to growers in certain countries, provinces, or states from isolated foundation seed farms that operate under the auspices of a cooperating certification agency. Foundation farms are responsible for providing an annual source of seedstock for many of their certified seed growers. This service provides assistance to growers whose seedstock is no longer suitable for replanting or who otherwise may wish to change seedstock.

EXTENSION   Participation in seed certification programs is voluntary, and the responsibility to observe and carry out both recommendations and regulations rests with the grower. Hence, all seed growers must continually be apprised of current disease-control recommendations and program regulations. While certain of these functions may be and are performed by cooperative extension service personnel apart from the certification agency, it, nevertheless, is the certification personnel who have the greatest amount of contact with seed growers and hence, maximal opportunity to keep growers informed. In certain regions, for example, potato vines (haulms) must be killed by spraying when populations of the green peach aphid (*Myzus persicae* Sulz.) have attained a critical level (15). This practice effectively reduces late-season leafroll infections. Notice is promptly given to all growers by the certification agency of the final date by which vines must be sprayed.

Varietal Improvement

In addition to the elimination of tuber-transmitted diseases, seed potato quality is enhanced through improvements in the yield and physical appearance of the tubers. Factors generally considered include the grade, type, and maturity of the tubers and organisms that directly affect the outward appearance of tubers. The primary means by which a given variety is improved is through an intensive selection program where such factors as vigor, tuber type, earliness of maturity, and varietal purity are considered. This approach is accomplished through the following procedures and is performed both by individual growers and at certification agency foundation farms.

TUBER INDEXING   Tubers are selected from storage that are uniform in type (i.e. conformation and other physical characteristics), free from detectable diseases, and within an acceptable size range. Each tuber is coded for identification, and an eye, preferably from the stolon end, is removed and planted in the greenhouse. Resultant plants are evaluated for vigor and diseases and those, along with the mother tuber that have undesirable characteristics are destroyed.

TUBER UNIT PLANTING   Tubers kept after indexing are cut into four pieces of nearly equal size and planted in sequence 20 to 25 cm apart to complete the unit. Additional space is provided between tuber units to make each clearly identifiable. There should be four replications in each unit, permiting a more accurate means for selection. Progeny of selected tuber units are placed in coded containers at harvest, and a sample from each is indexed and evaluated during the winter months. Increase from selected units may be used as progeny lines or mixed with other progeny for foundation planting.

HILL SELECTION   This method consists of selecting and labeling promising plants during the growing season. The plants are removed during harvest and the tubers are appropriately evaluated and stored in coded containers. Tubers are then winter indexed, and tubers from desirable hills are increased through tuber unit planting.

MASS SELECTION   This procedure is similar to that of hill selection except that tubers from desirable plants are stored in bulk rather than separately.

FIELD ROGUING   By this procedure, the quality of a given seedstock is maintained by removing all diseased and off-type plants during the growing season. Tubers, root systems (as completely as is practical), and tops are removed from the field and destroyed. Fields are generally subjected to roguing three times during a growing season.

NEW SEED SOURCES   Improved seed sources are commonly made available to growers either in the form of improved varietal selections or new varieties from foundation seed farms of the certification agency. In certain regions where a large number of different varieties are grown and there is an active breeding program, the evaluation and release of improved seedstocks may require an extensive effort on the part of seed-certification personnel.

THE VIRUS-FREE AFFAIR

Viruses known to infect potatoes may conveniently be separated into two general categories: (a) those that normally cause distinct foliar, or in certain instances, tuber symptoms, for example, potato leafroll virus, potato virus Y, etc, and (b) those that frequently do not produce consistent foliar or tuber symptoms, for example, potato virus X, potato virus S, and potato virus M. This second category is often referred to as the latent virus group although the designation is often an inaccurate one depending upon variety, virus strain, and plant-growth conditions.

Certified seed potatoes determined to be completely free of infection by all known viruses of both groups are designated "virus-free." However, concern over the accuracy of the term has led to the emergence of such qualified epithets as "virus-tested," "X-tested," and "X-free" potatoes. These terms also suffer from a lack of precision and have not received wide acceptance. All disease certification programs require that infection by nonlatent viruses be maintained within narrow tolerances (see Table 2). Similar standardization has not been achieved for "latent" virus infections. In many countries including the United States, most seed potatoes are infected with both potato virus X (PVX) and potato virus S (PVS), while potato virus M occurs with considerably less frequency in most varieties.

The first successful virus-free seed potato programs were developed in the Netherlands and in Scotland more than 20 years ago. This stimulated similar interests in the United States. At first, it was anticipated that an improvement in US seed potato quality similar to that achieved in Europe would result from the implementation of European virus technology. Indeed, presentations on the subject of virus-free potatoes are common features of seed potato seminars in many states and "virus-free potato programs" receive annual billing at the certification section meeting of the Potato Association of America. After two decades, however, the concept of virus-free potatoes in North America may still be described in the words of Carl Sagan (40) as one "of muddy surmise, unfettered speculation, stodgy conservatism, and unimaginative disinterest." Technically, there are no latent virus-free potato certification programs in the United States and at presently only three states (chronologically—Nebraska, North Dakota, and Montana) are engaged in large-scale certification for PVX. Certain other states and Canadian provinces are in developmental phases of PVX-free programs or perform limited testing for PVX in foundation seedstocks. The only serious attempt being made in North America to produce PVX- and PVS-free certified seed at the grower level is in British Columbia, Canada.

Why No Virus-Free Potatoes?

In view of the long-term success of European virus-free seed potato programs, it is surprising that none have been established in North America. Cause for this incongruity is at the very least multifaceted and includes a variety of potentialities any one of which may be responsible for precluding the establishment of a latent-virus testing program within a particular growing region.

YIELD INCREASES   Probably the most persistent question that looms over all latent-virus–free attempts in North America is whether increased yields are realized when seed potatoes are freed from all latent virus infection. In this context, unresolved areas include the question of the relative effects on yield of PVX and PVS infection both singly and in combination, and the potential effects of PVX and PVS on plant uniformity, tuber quality, and storage characteristics. It is well established that depending upon the variety and virus strain, each of these points deserves consideration in certain European potato programs, but none has been fully resolved for North American varieties.

Initial attempts at establishing whether or not PVX depressed the yields of North American potato varieties received little acceptance because the PVX-free clones used for comparison were not also reinoculated with the same strain of virus, subsequently increased, and then the effect on yield determined. Instead, progeny of infected and either selected or meristem-derived PVX-free plants were compared. Since clonal variation is the rule in potatoes, results of these studies were difficult to assess. Nevertheless, numerous investigators have reported increased yields when PVX-free stock of several varieties were compared with infected seed sources (16, 29, 43). Probably the most definitive work was recently published by Wright (53) who conducted a systematic study on the effects of PVX- and PVS-simultaneous infections on both total yield and tuber numbers for the Russet Burbank and White Rose varieties. Three US and one Canadian growing region were selected for test plots and results analyzed for each. With one exception, all areas recorded yield increases for latent virus-free stock as compared to infected material of the same clones. Yield increases ranged from 14 to 37%. The one exception was in the plot located in the state of Washington where yield differences were not significant. Subsequently, Kunkel et al (18), in the state of Washington, compared the yields of certified PVX-free Russet Burbank seed and PVX-infected Russet Burbank seed from several growers. In correspondence accompanying the yet unpublished results of further studies, Dr. Kunkel concluded: "When the data for 1972 and 1973 are considered as a whole, the yields of check (PVX- and PVS-infected) tubers . . . were equal to or better than the paired virus tested seed in 8 out of 10 comparisons." The significance of his results may be questioned for the same reasons as were prior studies with different clones, fertility levels, and nonstandardized virus strains.

However, in view of Wright's results (53) in the same region, it is possible that PVX-free Russet Burbank, and perhaps other potato varieties, have a greater yield potential only when grown under certain climatic conditions. If this is true, variations in environmental conditions may contribute to the variation encountered when attempting to establish the depressive effect of PVX and/or PVS infection on yields.

There is a firm scientific basis for believing that environmental conditions may significantly influence the severity of virus disease in plants, and there is little reason to suspect that potatoes are an exception. When potatoes infected with PVX are grown under suboptimal conditions of cool temperatures and low light intensity, many varieties are more severely affected. In the Russet Burbank variety, for example, many PVX strains elicit a pronounced mottle and crinkling of leaves when plants are grown under these conditions. When these varieties are grown at high light intensity with temperatures above 21–27°C neither of these symptoms is readily apparent except for unusually virulent strains of the virus. No studies have been conducted to establish whether environmentally enhanced foliar symptoms in PVX-infected American varieties are accompanied by a corresponding decrease in tuber yields and/or quality.

In conclusion, it has been difficult to establish experimentally that elimination of PVX from North American seedstocks invariably will result in a significant increase in yield. On the part of individual growers, however, far greater concern will continue to be placed on the probable effects of PVX and PVS on average yields over a period of years. Reliable evidence on such long-term effects will be even more difficult to obtain.

Experience with latent-virus–free potatoes in a strict certification program suggests that additional yield increases may occur indirectly. For example, Russet Burbank is a relatively indeterminant variety characterized by luxuriant top growth; if not properly spaced in the row, this variety produces exceptionally large and often knobby tubers. The PVX- and PVS-free Russet Burbank clone presently grown in Montana (originally supplied by R. Stace-Smith, Canada Department of Agriculture) is more vigorous than the normal latent-virus–infected one in terms of plant height, apparent rate of top growth, and average tuber size. When in-the-row spacings were decreased from a maximum of 28 cm to a maximum of 22 cm, a decrease in the number of oversize tubers was observed along with an overall yield increase (our unpublished observations). Thus, an increased plant population effectively enhanced both tuber quality and overall yield. With more determinant varieties such as the Norgold Russet, or under suboptimal growth conditions, a similar effect may occur to only a lesser extent, or perhaps not at all. This should be investigated. In this context, it is important that differences in growth habit between virus-free and latent-virus–infected potatoes should be appreciated fully especially within a variety such as the Russet Burbank. Changes in routine planting and cultivation practices may be required to accommodate increased vigor without a simultaneous reduction in tuber quality.

There is only suggestive evidence that PVX and PVS infection may influence either the susceptibility or resistance of potatoes to other types of diseases (13, 17). There is firm evidence, however, that the stringent sanitation requirements of a

PVX-free program may indirectly decrease the incidence of other types of diseases. A case in point is the blackleg disease caused by *Erwinia atroseptica,* a perennial problem in potatoes. It has recently become established that the causal organism is universally present within the lenticels of potato tubers (33) unless proper steps are taken to eliminate it. One means of eliminating the bacterium from seed sources is to make vegetative cuttings from the stems of potato plants and produce nuclear stock (source material) from them. Meristem tip culture would also be effective. Thereafter, care must be taken through strict sanitation to prevent the recontamination of tubers with *E. atroseptica.* In our experience, the sanitary procedures used in Montana to prevent PVX contamination have also been effective in preventing *E. atroseptica* recontamination. Tests conducted in 1972 and 1973, by A. Kelman of the University of Wisconsin (personal communication) indicated that none of the PVX-free tubers supplied to him by selected growers in Montana contained the blackleg organism even though the seedstock had been in the grower's hands for four years. Also, blackleg has not been observed within the respective PVX-free fields throughout this period. Those growers in Montana who have experienced the black-leg disease in PVX-free stock were also those whose seedstock became recontaminated with PVX during the same time period. Failure to adhere to sanitary procedures is clearly indicated in these examples. Similar logic should pertain to ring rot, a tuber-borne disease caused by *Corynebacterium sepedonicum.*

No convincing evidence has been published regarding the potential effects of PVS infection on North American potato varieties. Studies conducted thus far (N. S. Wright, personal communication) suggest that PVS had no demonstrable effect on the varieties he studied. Detailed investigations with European varieties (39) indicates that the virus depresses yield and quality to greater or lesser degrees depending upon variety and PVS strain.

THE PROBLEM OF VIRUS SPREAD   Potato varieties rendered free of PVX and/or PVS by either selection or meristem-tip culture remain fully susceptible to reinfection by both viruses. In the field, reinfection may occur rapidly if no attempts are made to minimize spread of the virus. And even when serious steps are taken to prevent PVX and PVS reinfection, the numbers of infected plants may nonetheless increase at an alarming rate. The apparent ease by which latent virus-free seedstocks become recontaminated has discouraged in significant measure the extension of latent-virus–free seedstock programs to the grower level.

Potato virus X may be transmitted by natural biological vectors including grasshoppers (*Melanoplus differentialis* Thos.) (52) and the chytrid fungus, *Synchytrium endobioticum* (28). The primary method by which PVX is introduced into virus-free fields is by contaminated equipment, field personnel, and animals (50). The virus may also be introduced into clean seedstock in storage through sprout-to-sprout contact (4, 25) and to some extent by tuber cutting knives (23). Once in a seedstock, PVX may spread by plant-to-plant contact both above (21) and below ground (38). The rate of virus X reinfection within a seedstock protected from external sources of inoculum has been reported to be relatively low (8). However, amounts of infection within a field may increase because of cultural practices in addition to entry of PVX from external sources. Roguing, hilling, and cultivation all are effective, inadvertent means of PVX transmission. In Montana, the incidence of PVX reinfection has been observed to increase from as little as 0.5 to 40% in one year.

A two-step approach is commonly used to control PVX infection. Seedstocks should be produced in growing areas isolated by several miles from sources of PVX inoculum, and strict practices of sanitation must be followed during the production and storage of PVX-free stocks. Decontamination of all implements prior to entry into the field, has been accomplished through the use of appropriate chemicals. Formaldehyde has traditionally been used for this purpose, but in Montana it has been replaced by a 1% solution of pyrrolidine. Pyrrolidine is a highly basic compound which at low concentrations is far less unpleasant to handle than formaldehyde. Potato storage facilities are routinely disinfected with this compound. All persons entering PVX-free fields should wear freshly laundered clothes, disinfect footwear, and avoid contact with plants wherever possible. These two procedures in combination with a genealogical increase (see Table 4) and virus indexing program have generally proven to be successful (26) in the control of PVX reinfection of uninfected seedstocks.

The problem of PVS reinfection is substantially different from that of PVX. Strict sanitation must also be observed to avoid reinfection but in addition, a biological vector also appears to be involved. Despite earlier reports to the contrary (39), at least some strains of PVS can be transmitted by the aphid, *Myzus persicae* Sulz. (6, 22). Aphid transmission is inefficient at best, however, and has been discounted as the primary means of PVS spread in Holland (12). It has been our experience as well as that of others (22) that PVS reinfection has occurred in seedstocks maintained in sufficient isolation from sources of infection to prevent reintroduction of PVX. In some instances, many kilometers have separated PVX- and PVS-free fields from infected ones. The rate of reinfection by PVS in a single year is usually not as high as for PVX, but under certain conditions may be more of a problem than previously believed (12). Aphids and/or any as yet unidentified biological vectors necessitate production of PVS-free potatoes in areas isolated by at least 10 km from any sources of PVS (including home-garden potatoes) if reinfection is to be avoided. Unfortunately, this is a difficult qualification for most North American growing areas, so it may be necessary to concede PVS infection in tolerant potato varieties.

PRIORITIES   The basic question of whether or not to attempt to eliminate latent viruses from seed potatoes can be answered only after consideration of certain crucial questions. These questions relate to the present status and effectiveness of ongoing programs within each state or potato-growing region.

The most successful seed-certification programs require coordinated efforts in extension, research, and disease diagnostic service. They also consider agronomic and horticultural as well as pathological aspects of potato production. When all phases function successfully, optimal production is possible for a given variety grown in a particular region. If this goal has been attained, the time would seem to be right for developing a program to eliminate latent viruses with the aim of further improving seed quality. If on the other hand, production levels are far lower than they should be, improvements within the existing program are necessary before additional features should be added to it. For example, a number of years ago, we visited a growing region in central United States where Russet Burbank potatoes were being grown on sandy soil with essentially no irrigation. Yields were very low in periods when rainfall was inadequate or poorly distributed within the growing season. To compound the problem, the Russet Burbank variety is prone to going "off type" when soil moisture levels fluctuate significantly. Heavy rains falling on fields of this variety resulted in excessive nobbiness of tubers, etc, which in turn reduced the quality of the seed crop. A latent-virus–free program would have little meaning under these conditions. Instead, emphasis here should be placed on improved grower education, including varietal selection and irrigation.

Grower education through extension is very crucial in all seed potato certification programs. Failures in extension usually result in continuing problems in all phases of the program. The challenge of grower education becomes even more acute when attempting to institute a latent-virus–free program. It is first necessary to convince *all* growers that latent-virus–free seed is superior to virus-infected stock. Then their enthusiasm must continually be maintained to ensure that they will practice the necessarily strict sanitary procedures. This task is never easy, but is absolutely necessary if a program is to succeed.

RESOLUTION OF OLD WIVES' TALES AND UNSUBSTANTIATED CLAIMS
Efforts to develop PVX-free potatoes in North America have often been discouraged by many unsubstantiated and otherwise disquieting comments commonly heard within the seed potato industry.

One such claim is that reinfection of PVX-free potatoes by PVX causes "shock symptoms" and a significantly greater decrease in yield than would be experienced if the plants had suffered from tuber-borne PVX infection. Despite repeated attempts, we have been wholly unsuccessful in demonstrating this purported phenomenon (J. F. Shepard and G. A. Secor, unpublished). In our experiments both a mild mottle strain and a more severe strain of PVX were used as inoculum. Much more pronounced foliar symptoms did appear when PVX-free potatoes were inoculated with the more severe strain. This observation has led us to believe that so-called shock symptoms are simply the result of reinfection of PVX-free potato plants with a single relatively severe strain of the virus. In such cases the protective effect of milder PVX strains within the preinfected plant was eliminated.

In this context, another suggestion is sometimes advanced—that it is preferable to continue to utilize the mild strains of PVX already present in most North American varieties to cross-protect against severe strains rather than risk reinfection of PVX-free stock with more virulent strains. This argument has a number of significant shortcomings both practically and theoretically. One factor relates to varietal improvement, selection, and breeding programs. The phenomenon called "potato divergence" was recognized in Europe over 30 years ago. PVX, and in some varieties PVS, may cause a continual divergence in the growth habit of tubers, and foliage from that previously bred or selected in a given variety (39). Hence, intravarietal selection must continually be practiced to minimize virus-induced plant-to-plant variation (i.e. enhance uniformity). Within breeding programs, PVX and PVS may cause even more confusing results because the reaction of F_1 individuals to either virus is unknown. Neither PVX nor PVS are seed transmitted in potato, but because F_1 seedlings are only rarely grown in complete isolation from virus-infected plants reinfection often occurs. Since the effects of the virus(es) on the horticultural characteristics of the selections have not been described, a distinction between genetic and virus-induced effects is difficult to make.

Naturally infected potatoes represent a massive reservoir of PVX and PVS, and this constant source of inoculum results in severe problems when attempts are made to produce PVX-free stock. More virulent strains of the virus may emerge that increase yield losses or negate cross-protection (19, 27). Moreover, PVX is not limited in its host range to potatoes but also infects other commercial crops such as tobacco and tomato as well as solanaceous weed hosts. For this reason, in certain growing areas, it is not advisable to grow potatoes in close proximity to tobacco or tomato fields.

An even more significant deterrent to development of PVX-free programs relates to the amount of PVX in seedstocks certified and sold as "PVX-free." Certain certification agencies in North America test for PVX only in foundation plantings. This may be two or more generations before the certified seed is sold for commercial production. Unfortunately, potatoes often become heavily infected with PVX and PVS when released from agency foundation farms to certified growers. As a result commercial buyers of resultant certified seed receive no data on PVS and PVX infection. Such programs are not only ineffective but also give well-conceived ones a "black eye." In Montana PVX testing is conducted on all seedstocks through the final year of increase and sometimes also in winter indexes. In this way, as much information about PVX infection is provided to the buyer of the seed as for any of the other diseases included within the certification scheme. While current-season infection may increase the amount of PVX in the following year's planting, the problem is not unique to latent virus diseases; this is a basic shortcoming of disease certification in general. When claims of the occurrence of PVX in indexed stock are made by buyers it is often difficult to determine whether contamination occurred prior or subsequent to sale. Insufficient attention to sanitation sometimes applies to certification agencies which may test incoming seed for virus infection. PVX spreads very rapidly when potatoes are grown in greenhouse benches, etc, prior to testing, and this may lead to inflated estimates of infection. Although somewhat less sensitive, tests of tuber sprouts (46) may provide more accurate information in this regard.

Potato leafroll virus is a continual problem in both certified and commercial potato plantings (see Table 3). One of the most recent claims is that PVX-free potato plants are more susceptible to potato leafroll virus infection than PVX-infected ones. To date, this contention has not been scientifically substantiated. It is commonly accepted that PVX-free potatoes are more vigorous in their growth habit than are their PVX-infected counterparts, and it is probably in this that the confusion has

Table 3 Area of potatoes rejected by certification programs in the United States and Canada (1968–1972)[a]

Reason	Number of agencies reporting	Percentage of total rejections	Total area rejected per year (ha)	(acres)	Percentage of total entered for certification
Ring rot	12	59.9	6,536	16,140	5.64 (2.2–13.8)
Leaf roll	10	9.0	985	2,432	.85 (0.0– 3.1)
Varietal mixture	9	7.2	788	1,946	.68 (0.0– 2.7)
Mosaic	9	5.4	591	1,459	.51 (0.0– 3.7)
Blackleg	5	4.5	487	1,202	.42 (0.0– 2.1)
Verticillium wilt	7	3.0	324	801	.28 (0.0– 2.0)
Chemical damage	3	2.8	301	744	.26 (0.0– 2.67)
Isolation	4	2.0	220	544	.19 (0.0– 1.1)
Spindle tuber	7	1.8	197	486	.17 (0.0– 0.9)
Withdrawn	2	1.6	174	429	.15 (0.0– 0.31)
Poor risk	2	1.4	151	372	.13 (0.0– 1.44)
Other	12	1.4	151	372	.13 (0.0– 2.4)

[a] Data compiled from reference 34.

arisen. More vigorous plant growth may reduce the intensity of leafroll virus symptoms at least for a time. Current season infections or infections by mild leafroll strains (55) may thus be more difficult to diagnose visually, and this would result in an apparent increase in the amount of disease during the following season.

THE PROBLEM OF MASS INDEXING   Wherever programs to free seed potatoes of latent viruses have been successful, there has been a concerted team effort between certification personnel and state or university affiliated virologists. The breadth of expertise required to maintain certified seed free of latent viruses is simply too great for a single discipline to handle competently. Improved serological techniques have largely obviated the need for indicator plants as a means of mass indexing potatoes for PVX or PVS infection. Serological testing requires expertise not only in the performance of routine tests but also in virus purification and preparation of antiserum. Qualified personnel must be engaged to evaluate new procedures and incorporate them into the testing programs. Certain of these functions could be accomplished in national laboratories as it is in the Netherlands. Similarly, antisera could be produced at a single location and made available to all testing programs in North America. Annual workshops could be established in which recently developed or improved techniques would be demonstrated to those involved in virus (or other disease) diagnosis. Such a program could also assist certification agencies that do not have affiliated virologists. As an initial step toward this objective, in 1974 the Department of Plant Pathology at Montana State University made available standardized antisera to PVX, PVS, and their respective protein subunits. The former antisera are suitable for microprecipitin or other "whole virus" diagnostic techniques whereas "subunit" antisera are used in immunodiffusion techniques for diagnosis of PVX and PVS (47).

Reliable techniques for large-scale testing are essential to accurately determine the amounts of virus in all categories of seed potatoes. Sensitivity for small amounts of virus is an important consideration with regard to technique reliability, and much emphasis has been placed on the relative sensitivity of different diagnostic techniques for both PVX (41, 48) and PVS (11, 41). Inoculation of *Gomphrena globosa* is probably the most sensitive routine test for PVX. This is true, however, only when plants are grown under the proper environment, several leaves are inoculated, and the strain of PVX present produces optimal numbers of lesions. Some PVX strains produce far fewer lesions per μg virus than do others (J. F. Shepard, unpublished). Growing sufficient numbers of Gomphrena plants under ideal environmental conditions for virus indexing is a formidable and unnecessary task. Serological techniques of far lower sensitivity have been more useful for eliminating PVX infection from seedstocks simply because they can be conducted quickly and in numbers several orders of magnitude greater than Gomphrena tests. One thousand tests per acre with a technique of a slightly lesser sensitivity will provide more information concerning PVX infection rates than 10 idealized Gomphrena tests per acre. Indeed, the relatively insensitive chloroplast agglutination test and the microprecipitin technique have been responsible for the elimination of PVX and PVS from potatoes in Europe (51). Reliable immunodiffusion techniques have also been developed for PVX and PVS diagnosis on a mass scale (47) and have been successfully employed for several years in Montana. In 1973, over 250,000 serological tests were conducted for the two viruses. No PVX-free seed potato program with a long history of success is presently using only the Gomphrena test.

Detection of PVS with indicator hosts is presently not reliable (5). PVS strains vary greatly in the extent of lesion production on *Chenopodium* sp. (11), and *Nicotiana debnyii* in our experience has varied as a systemic host depending upon virus strain and growth conditions. More consistent results have been obtained through properly conducted serological tests of both the microprecipitin and radial-diffusion types.

Whatever the diagnostic technique, successful detection of potato latent viruses is influenced by the pattern of virus distribution in the plant. For PVX, tuber-borne infection commonly, but not always (54), leads to detectable concentrations of the virus throughout most of the foliage. Current-season infection of PVX-free stock, however, may or may not be detectable depending upon the stage of plant development and the virus strain.

In detecting PVS, the problem of virus distribution in chronically infected plants is far more acute than for PVX. Ordinarily, field infections of potato plants with PVS will not lead to detectable amounts of virus in the foliage (10) even though PVS will be detectable in plants grown from tubers of the same plants. Furthermore, the concentration and distribution of PVS in the foliage of systemically infected plants is not consistent throughout the growing season (3). Young as well as very old plants are particularly difficult to index. The general consensus is to sample plants for viruses just before they begin to flower. Because present techniques will only detect PVS infections that developed in the previous year, if in-field spread has occurred, it is difficult to eradicate the virus by roguing.

The Basic Requirements of a Latent-Virus–Free Program

It is beyond the scope of this article to provide a compendium of all past and present approaches to the development of programs for virus-free potatoes. Each has differed in detail and will probably continue to do so. The following list of primary requirements common to several successful programs may be helpful, however.

VIRUS-FREE STARTING MATERIAL   Most commercially important potato seed-stocks are universally infected with PVX and PVS. For this reason, relatively few individuals in a very few varieties can be used to obtain noninfected plants by selection. Usually, meristem tip culture (35) or heat treatment followed by axillary meristem culture (49) is used to eliminate viruses from infected plants. Resultant plants may be rapidly increased in number to provide primary material for further selection. An efficient means of initially increasing meristem-derived or other virus-free starting material is the use of vegetative cuttings (9). This not only permits rapid increase of the plants, but also serves to eliminate other tuber-borne pathogens such as *Erwinia atroseptica*. Many meristem-derived plants are screened initially to eliminate genetic aberrants and to select for improved horticultural characteristics. All phases of increase of virus-free starting material must be conducted in strict isolation from other potatoes or hosts of potato pathogens whether in the greenhouse or in the field. Examples of reinfection by latent viruses or other pathogens, such as the spindle tuber viroid, are far too numerous to assume the contrary. In certain instances, virus-free programs have been set back several years as a result of contamination during the initial greenhouse increase.

SERODIAGNOSTIC CAPABILITY   The diagnosis of latent-virus infections requires laboratory techniques. Test-plant or electron-microscope techniques may be quite efficient for screening initial plants for diseases. However, serological diagnostic methods must be employed during later phases of the program. Commonly, they are conducted at a central or regional laboratory properly equipped to handle a large number of samples. In a few countries, such as Germany, individual growers perform their own tests. Both approaches have been successful over a number of years when a continued supply of antiserum has been assured and sampling procedures are rigidly enforced.

   In Montana, radial-immunodiffusion techniques (47) are used to index plants for PVX and PVS. Antiserum prepared against dissociated (and denatured) viral protein is incorporated into buffered agar and used to test extracts from plants to which a final concentration of 2.5% pyrrolidine has been added. Pyrrolidine serves to dissociate virions into diffusible protein oligomers capable of reaction with homologous antibody preparations. Leaf samples are delivered to the testing laboratory with care being taken to keep the leaves dry and cool. Excessive moisture or humidity leads to rapid deterioration of potato leaves. Juice is expressed from individual leaves with a hydraulic press using a multisample press plate–plunger system. A drop of juice is collected for each sample and placed in a separate 6 X 25 mm test tube. At this point, the juice may either be frozen for later testing or be treated with pyrrolidine and placed in a radial diffusion depot. Extracts to be assayed may not be frozen subsequent to treatment with pyrrolidine. Using these procedures, a laboratory with ten people may test 6000 to 7000 samples per day for PVX and PVS.

GENEALOGICAL PROGENY INCREASE   The process of increasing virus-free starting material through the certified seed stage must continually follow a genealogical selection scheme. This involves the continued entry into the program of recently derived virus-free starting material which will be increased over 4 or 5 growing seasons before marketing. It is well established that replanting of virus-free stock year after year will result in reinfection.

The basic scheme used by all virus-free programs which is slightly modified from Rozendaal & Brust (39) is similar to that shown in Table 4. Nearly all programs differ in one or more specific areas of the scheme, however, particularly within the indexing category. The type and intensity of virus testing is highly variable, and each program must be evaluated separately. Other programs may either add steps to or delete steps from the basic scheme. As indicated in Table 4, the terms used to designate classes of seed potatoes are not uniform. Commonly, the first two or three sequential steps are performed on foundation farms maintained by the state or grower association. Such farms are far from commercial potato plantings, thus minimizing the potential for reinfection by viruses. Foundation farms are not always utilized, however, and in a few programs, for example Montana, growers perform the full sequence of steps.

EFFECTIVENESS OF SEED CERTIFICATION PROGRAMS AND PROSPECTS FOR THE FUTURE

Each seed certification program has merit because seedstocks will be improved if any of the principles of certification are implemented. It is doubtful, however, whether any program has developed to the point where disease control is complete

Table 4   Genealogical scheme for increase of virus-free potato increase[a]

Class	Other common descriptions	Indexing
Mother plant	Pre-elite I, nuclear I	Serologically tested
First year family	Elite I, nuclear II	100% serologically tested plus field inspection
Second year family	Elite II, foundation I	10% serologically tested plus field inspection
Third year family	Elite III, foundation II	2% serologically tested plus field inspection
Fourth year family	Foundation, certified virus-free	30–50 plants/acre tested plus field inspection

[a] After Rozendaal & Brust (39).

or where all existing knowledge has been utilized fully. The value of seed certification in the United States is shown in Figure 1 in which the potato acreage entered for certification was calculated against the number of acres that met prescribed tolerances. During the formative years of 1919–1929, a relatively low percentage of the acreage was found acceptable. A dramatic upswing soon occurred, but was followed by only modest increases in the percentage of acres certified. While these figures have certain limitations, they nevertheless indicate that little overall improvement in the control of the diseases subject to certification standards has occurred during the past several years. Seed certification has been and will continue to be an effective means of improving the quality of seed potatoes, but new approaches and techniques probably will be required before additional improvements are realized.

The science of plant breeding would seem to offer the greatest promise for improving the quality of seed potatoes in North America. Bacterial ring rot, for example, is responsible for more seed rejections in the United States and Canada than all of the other diseases combined (see Table 3). Thus, improved sources of resistance to ring rot would be particularly valuable. Unfortunately, research gains in this area have been limited. Even when increased resistance was obtained, new varieties generally have failed to achieve acceptance by industry. Indeed, the oldest potato variety grown in the United States (i.e. Russet Burbank) commands an ever increasing percentage of the total seed acreage and in 1973 represented 32% of the seed potatoes grown (45). Nevertheless, continued emphasis in potato breeding is necessary, at the very least, to broaden the genetic base of potatoes in North America.

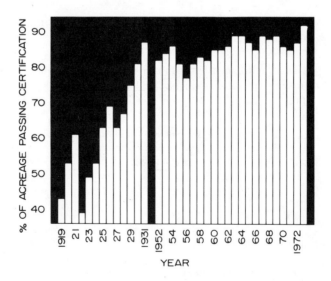

Figure 1   The percentage of seed potato acreage entered for certification that passed certification standards from 1919–1932 (24) and 1952–1973 (44). Data from 1932–1951 could not be obtained.

Another, and as yet untested, approach may be to improve certain varietal characteristics including resistance to diseases through somatic cell selection techniques. Recent advances in the isolation and regeneration of mesophyll cell protoplasts suggest that resistance to disease may be enhanced by these procedures (7) while essential varietal characteristics would be maintained.

Pending new developments in the areas of breeding or cell selection, a great deal may still be done in the control of diseases within existing varieties. Certification programs may be expanded to include the control of latent virus infections, and when maintained under regulations of strict sanitation, may be integrated with programs for blackleg and ring-rot control. Rapid diagnostic techniques for bacterial diseases would be of great value. Serological diagnostic techniques such as immunofluorescence offer great potential for the detection and identification of even small numbers of bacteria in infected tissues. Greater coordination among research agencies and improved methods of diagnosis and disease assessment will be required, however, before these potentialities can be realized.

Literature Cited

1. Appel, O. 1915. Leaf roll diseases of the potato. *Phytopathology* 5:139–48
2. Appel, O. 1934. Vitality and vitality determination in potatoes. *Phytopathology* 24:482–94
3. Arenz, B., Vulic, M., Hunnius, W. 1964. Die Nachweisbarkeit des S-Virus in den Verschiedenen Pflanzenteilen sekundar infizierter Kartoffelpflanzen. *Bayer. Landwirtsch. Jahrb.* 6:683–90
4. Bawden, F. C., Kassanis, B., Roberts, F. M. 1948. Studies on the importance and control of potato virus X. *Ann. Appl. Biol.* 35:250–65
5. Beemster, A. B. R., Rozendaal, A. 1972. Potato viruses: properties and symptoms. In *Viruses of Potatoes and Seed Potato Production*, ed. J. A. de Bokx, 115–43. Wageningen: Centre Agric. Publ. Doc. 233 pp.
6. Bode, O., Weidemann, H. L. 1971. Untersuchungen zur Blattausübertragbarkeit von Kartoffel-M-und-S-Virus. *Potato Res.* 14:119–29
7. Carlson, P. S. 1973. Methionine sulfoximine-resistant mutants of tobacco. *Science* 180:1366–68
8. Cockerham, G. 1958. Observations on the spread of virus X. *Conf. Potato Virus Dis.* 3:144–48
9. Cole, E. F., Wright, N. S. 1967. Propagation of potato by stem cuttings. *Am. Potato J.* 44:301–4
10. de Bokx, J. A. 1968. The translocation of various isolates of potato virus S in potato plants with primary infection. *Meded. Rijksfac. Landbouwwetensch. Gent* 33:1179–85
11. de Bokx, J. A. 1970. Reactions of various plant species to inoculation with potato virus S. *Neth. J. Plant Pathol.* 76:70–78
12. de Bokx, J. A. 1972. Spread of potato virus S. *Potato Res.* 15:67–70
13. Dowley, L. J. 1973. Effects of primary and secondary infection with potato virus X (PVX) on yield, size, chemical composition, blight resistance, and cooking quality of potato variety Kerr's Pink. *Potato Res.* 16:3–9
14. Glick, D. P., Ark, P. A., Racicot, H. N. 1944. Outline of procedure for the diagnosis of bacterial ring rot of potatoes. *Am. Potato J.* 21:311–14
15. Hille Ris Lambers, D. 1972. Aphids: Their life cycles and their role as virus vectors. See Ref. 5, 36–56
16. Hoyman, W. G. 1964. Red Pontiac vine and tuber yields as affected by virus X. *Am. Potato J.* 41:208–11
17. Jones, E. D., Mullen, J. M. 1974. The effect of potato virus X on susceptibility of potato tubers to *Fusarium roseum* 'Avenaceum'. *Am. Potato J.* 51:209–15
18. Kunkel, R., Holstad, N., Butala, H., Thornton, R. E. 1973. Influence of potato viruses X and S on yield of Russet Burbank potatoes. *Am. Potato J.* 50:385 (Abstr.)
19. Ladeburg, R. C., Larson, R. H.,

Walker, J. C. 1950. Origin, interrelation and properties of ringspot strains of virus X in American potato varieties. *Univ. Wis. Res. Bull.* 165. 47 pp.
20. Leach, J. G. 1938. The biological basis for certification of seed potatoes. *Am. Potato J.* 15:117–30
21. Loughnane, J. B., Murphy, P. A. 1938. Dissemination of virus X and F by leaf contact. *Sci. Proc. R. Dublin Soc., NS* 22:1–15
22. MacKinnon, J. P. 1974. Detection, spread, and aphid transmission of potato virus S. *Can. J. Bot.* 52:461–65
23. Mai, W. F. 1947. Virus X in the newer potato varieties and the transmission of the virus by the cutting knife. *Am. Potato J.* 24:341–51
24. Martin, W. H. 1931. Report of the seed potato certification committee. *Proc. 18th Ann. Meet. Potato Assoc. Am.* 115–24
25. McIntosh, T. P. 1944. Potato troubles. *Gard. Chron.* 116:87–88
26. Munro, J. 1954. Maintenance of virus X-free potatoes. *Am. Potato J.* 31:73–82
27. Munro, J. 1961. The importance of potato virus X. *Am. Potato J.* 38:440–47
28. Nienhaus, F., Stille, B. 1965. Übertragung des Kartoffel-X-Virus durch Zoosporen von *Synchytrium endobioticum. Phytopathol. Z.* 54:335–37
29. Ohms, R. et al 1973. Comparison of PVX-free Russet Burbank Canadian source seedstock with regular PVX infected Idaho seedstock. *Am. Potato J.* 50:385–86 (Abstr.)
30. Orton, W. A. 1914. Inspection and certification of potato seedstock. *Phytopathology* 4:39–40 (Abstr.)
31. Orton, W. A. 1914. The potato study trip of 1914. *Phytopathology* 4:412–13 (Abstr.)
32. Orton, W. A. 1914. Improvement of potato seed stocks through official inspection and certification. *Proc. 1st Ann. Meet. Nat. Potato Assoc. Am., Philadelphia* 37–43
33. Pérombelon, M. C. M. 1972. The extent and survival of contamination of potato stocks in Scotland by *Erwinia carotovora* var. *caratovora* and *E. carotovora* var. *atroseptica. Ann. Appl. Biol.* 71:111–17
34. *Proceedings, Certification Sect., Potato Assoc. Am. Meet. Omaha, 1973*
35. Quak, F. 1972. Therapy. See Ref. 5, 158–66
36. Racicot, H. N., Savile, D. B. O., Conners, I. L. 1938. Bacterial wilt and rot of

potatoes—some suggestions for its detection, verification and control. *Am. Potato J.* 15:312–18
37. Rieman, G. H. 1956. Early history of potato seed certification in North America, 1913–1922. *Potato Handbook.* New Brunswick: Potato Assoc. Am. 1:6–10
38. Roberts, F. M. 1946. Underground spread of potato virus X. *Nature London* 158:663
39. Rozendaal, A., Brust, J. H. 1955. The significance of potato virus S in seed potato culture. *Conf. Potato Virus Dis.* 2:120–33
40. Sagan, C. 1973. *The Cosmic Connection: an Extraterrestrial Perspective.* Garden City, NY: Anchor. 274 pp.
41. Sampson, P. J., Taylor, R. H. 1968. A comparison of the electron microscope, microprecipitin tests, and indicator plants for the detection of potato viruses S, X, and Y. *Phytopathology* 58:489–93
42. Savile, D. B. O., Racicot, H. N. 1937. Bacterial wilt and rot of potatoes. *Sci. Agric.* 17:518–22
43. Schultz, E. S., Bonde, R. 1944. The effect of latent mosaic (virus X) on yield of potatoes in Maine. *Am. Potato J.* 21:278–83
44. Seed reports. 1952–1973. *Spudlight.* Certified Seed Editions. Washington: United Fresh Fruit Veg. Assoc.
45. See Ref. 44, 1973 edition
46. Shepard, J. F. 1969. Serodiagnosis of PVX in potato tuber sprouts. *Plant Dis. Reptr.* 53:845–48
47. Shepard, J. F. 1972. Gel-diffusion methods for the serological detection of potato viruses X, S, and M. *Montana Agric. Exp. Stn. Bull.* 662. 72 pp.
48. Shepard, J. F., Secor, G. A. 1969. Detection of potato virus X in infected plant tissue by radial and double-diffusion tests in agar. *Phytopathology* 59:1838–44
49. Stace-Smith, R., Mellor, F. C. 1968. Eradication of potato viruses X and S by thermotherapy and axillary bud culture. *Phytopathology* 58:199–203
50. Todd, J. M. 1958. Spread of potato virus X over a distance. *Conf. Potato Virus Dis.* 3:132–43
51. van Slogteren, E., van Slogteren, D. H. M. 1957. Serological identification of plant viruses and serological diagnosis of virus diseases of plants. *Ann. Rev. Microbiol.* 11:149–64
52. Walters, H. J. 1952. Some relationships of three plant viruses to the differential

grasshopper, *Melanoplus differentialis* (Thos.). *Phytopathology* 42:355–62

53. Wright, N. S. 1970. Combined effects of potato viruses X and S on yield of Netted Gem and White Rose potatoes. *Am. Potato J.* 47:475–78

54. Wright, N. S. 1974. Potato virus X in-

fection of Netted Gem potato. *Am. Potato J.* 51:202–5

55. Wright, N. S., MacCarthy, H. R., Cole, E. F. 1967. Detection and control of mild strains of potato leafroll virus. *Am. Potato J.* 44:245–48

EFFECTS OF SALINITY AND ❖3624
SODICITY ON PLANT GROWTH

Leon Bernstein
US Salinity Laboratory, US Department of Agriculture, Riverside, California 92502

INTRODUCTION

Salinity and sodicity are soil conditions that occur mainly in arid and semiarid regions. Rainfall leaches salts out of soils in humid regions, and salt problems are rare and transitory.

Ions that contribute to soil salinity include Cl^-, SO_4^{2-}, HCO_3^-, Na^+, Ca^{2+}, Mg^{2+}, and, rarely, NO_3^- or K^+. The salts of these ions occur in highly variable concentrations and proportions. They may be indigenous, but more commonly they are brought into an area in the irrigation water or in waters draining from adjacent areas. Natural drainage is often so poorly developed in arid regions that salts collect in inland basins rather than being discharged to the sea.

Saline soils contain soluble salts in quantities that affect plant growth adversely, the lower limit for a saline soil being set conventionally at an electrical conductivity of 4 mmho/cm in the soil saturation extract (75). Actually, sensitive plants are affected at half this salinity, and highly tolerant ones at about twice this salinity (8).

Sodic soils contain excess exchangeable sodium with, by definition, 15% or more of the cation exchange sites of the soil occupied by Na^+ (75). Plants sensitive to sodium are, however, affected at lower exchangeable sodium percentages. Sodic soils may be either nonsaline or saline.

PLANT SYMPTOMS

Plants adversely affected by salinity grow more slowly and are, therefore, stunted. Leaves are smaller, but may be thicker than those of normal plants. Chloride (but not sulfate) increases the elongation of the palisade cells, causing increased succulence (74). Leaves of salt-affected plants are often darker green than those of normal plants, but in some species (e.g. crucifers and some grasses) thicker layers of surface wax cause a bluish-green cast. Stunting of fruits as well as leaves and stems occurs (4, 8). Salt-affected plants may show no distinctive symptoms, and only comparison

295

with normal plants reveals the extent of salt inhibition. Soil salinity measurements, together with salt-tolerance data, aid in diagnosing suspected salt problems in such cases.

Salinity usually varies greatly in a salt-affected field, and may be so high in spots as to inhibit germination completely and cause bare spots. Uneven growth, while suggesting saline conditions, requires confirmatory soil analyses to distinguish salinity from other factors that may cause variable growth.

The foregoing description of plant symptoms relates primarily to annual (non-woody) plants. Trees, vines, and shrubs, in addition to stunting, often show specific leaf injury caused by toxic accumulations of chloride or sodium. When leaves of such woody plants accumulate about 0.5% Cl or 0.2% Na on a dry-weight basis, characteristic marginal or tip burn, or other necrotic leaf symptoms develop (9, 22). Some species, such as bougainvillea, can accumulate chloride without leaf injury (20). Bronzing of leaves, rather than leaf burn, indicates chloride toxicity in privet and citrus (20, 22). Apart from bronzing and some rare effects of cation imbalance, chlorosis is not a characteristic symptom of salt injury. The frequently encountered chloroses in subhumid and arid regions are found on calcareous soils (lime-induced chlorosis) or minor element-deficient soils, rather than on saline ones (22).

Although most nonwoody plants do not develop leaf burn, they do accumulate chloride and sometimes sodium to levels as high or higher than those that injure susceptible woody species. However, sprinkling with brackish waters causes leaf burn in some annual crops (19, 39).

SALT EFFECT MECHANISMS

Over several decades two schools of thought have disputed the relative importance of osmotic and specific-ion effects. The osmotic school (22) claims that most of the adverse effects of salinity are related to the decreased osmotic potential (i.e. increased osmotic pressure) of saline root media, whereas the specific-ion school (30, 74) thinks that the effects are caused mostly by the specific effects of individual ions. The consequences of these opposing views are not trivial. If saline effects depend upon specific ion concentrations, then soil salinity appraisal requires the determination of these individual ions. If, on the other hand, osmotic effects predominate, only osmotic potential or some colligative or related property of solutions needs to be determined—a much simpler task.

The evaluation of soil salinity is complicated by differences in salinity measurement methods. Russian scientists (74) and many ecologists (29) extract all soluble salts and express individual ions or total soluble salts on a dry-soil basis. Salinity of the *soil water,* however, depends not only on the quantity of salt present but also on the volume of water a soil normally holds. Fine-textured soils may retain up to five times as much water as coarse-textured soils. At a given salt content on a dry-soil basis, the coarse-textured soil will have a soil solution concentration five times that of the fine-textured soil (75). Separate salinity criteria must then be established for each soil type, as well as for each type of salinity (74). Such saline soil classifications appear to confirm the specificity of soil salinity effects because plant tolerances vary widely from soil to soil and salt type to salt type. Thus, a prior belief in the specificity of salt effects leads to a method of salinity appraisal that appears to confirm the prior assumption.

Soils may contain amounts of sparingly soluble salts such as gypsum many times greater than can be held in solution in the field water-content range. Gypsiferous soils may appear to be highly saline when exhaustively extracted, but the soil solution may be nonsaline because of the limited solubility of gypsum (the electrical conductivity of a saturated gypsum solution is 2.2 mmho/cm). Soils have even been salinized for salt-tolerance studies without regard to the limited solubility of some of the salts used (10).

Osmotic Effects

CRITERIA AND SUPPORTING EVIDENCE   Osmotic effects can be said to predominate when plant growth is related to the osmotic potentials of root media that contain different salts or combinations of salts. Eaton demonstrated that chloride salinity was twice as inhibitory as sulfate salinity per milliequivalent of salt in his sand-culture solutions (30). However, the osmotic potentials in his media were equal when the chloride solutions had half the concentration of his sulfate solutions on a milliequivalent/liter basis. Thus, response was well correlated with the osmotic potentials of chloride and sulfate media. Similar findings have been reported by others (22).

When isosmotic concentrations of single salts are added to a base nutrient solution, different salts may affect growth quite differently. Thus, $CaCl_2$ is more injurious to bean plants (*Phaseolus vulgaris* L.) than isosmotic concentrations of NaCl (7). Bean plants avidly absorb Ca^{2+} at the expense of K^+ and Mg^{2+} uptake, with resultant nutritional imbalance. Conversely, $CaCl_2$ is tolerated better by maize (*Zea mays* L.) than are isosmotic concentrations of other chloride salts. Maize cannot absorb nutritionally adequate amounts of Ca from a nutrient solution unless added Ca salts contribute to the increased salinity of the media. Salinity in the field is usually caused by mixtures of salts rather than a single salt. With mixed salts, both maize and bean respond to the osmotic potentials of the media with wide tolerances to the proportions of individual salts in the media (7). Thus, under saline field conditions in which Ca^{2+} as well as Na^+ concentrations are high, plant response will be well correlated with the osmotic potential of the soil water. Nonsaline sodic soils that are high in Na and very low in Ca and Mg cause specific nutritional deficiencies, and thus differ from saline soils (see below).

Although the relationship of growth to osmotic potential appears adequate to justify using osmotic potential as a measure of salinity, specific ions may still influence the quality of growth (e.g. the leaf succulence caused by chloride salinity). High Ca^{2+} concentrations cause chlorosis in beans, even in mixed salt solutions that cause no specific effect on growth (7). Specific ions may also influence respiratory pathways, as in pea roots (66).

Difficulty in measuring the osmotic potentials of solutions has only recently been overcome with the availability of commercial osmometers. Since the electrical conductivity (EC) of soil extracts is well correlated with their osmotic potentials and is easily measured, EC has been routinely used as a measure of salinity (75). At a given soil-salt content, EC will vary inversely with fluctuations in soil-water content; for many diagnostic and comparative purposes, EC at a reference soil-water content for a given soil is desirable. Moreover, extraction of soil water from soil in the field water-content range requires special equipment and is time consuming. The EC of the soil saturation-extract has, therefore, been commonly adopted for salinity appraisal. The saturation percentage represents a fairly constant dilution of the soil solution and takes into account the water-holding capacity of the soil (75). Extracts with fixed ratios of soil:water such as 1:2 or 1:5 do not take into account the water-holding capacity of the soil and are also less indicative of the soil-water composition because of the effects of increased dilution (75). Fixed ratio extracts, however, may be used when many analyses are made on a given soil. Conversion to equivalent EC_e values requires determination of the ratio $EC_e:EC_x$ for the soil in question where x denotes the fixed ratio extract (75). Table 1 relates crop response to the electrical conductivity of saturation extracts (EC_e).

Table 1 Crop response to salinity, measured as the electrical conductivity of the soil saturation extract, EC_e

	EC_e range in mmho/cm[a] at 25°C				
	0–2	2–4	4–8	8–16	>16
Crop response	Salinity effects mostly negligible	Yields of very sensitive crops restricted	Yields of most crops restricted	Only tolerant crops yield satisfactorily	Only a few very tolerant crops yield satisfactorily

[a] One mmho/cm is equivalent on the average to 640 mg salts/liter.

OSMOTIC ADJUSTMENT   Although abrupt decreases in osmotic potentials of root media decrease the osmotic gradients for water uptake, such effects do not persist. After a day or two, the osmotic potentials of plant organs decrease to match the decrease in osmotic potential of the medium, and osmotic gradients are therefore not affected (5, 6). If plant resistance to water movement is not changed, water uptake from saline media can equal that from nonsaline media, so that turgor may be unaffected by salinity (34). Some exceptions have been noted (36).

Mechanisms for osmotic adjustment vary among plant species, and are the basis for one major classification of halophytes (29). Those that accumulate salts in effecting osmotic adjustment are known as euhalophytes; those that accumulate organic solutes rather than salts are called glycohalophytes (29). Similar distinctions appear among crop plants. In carrot roots, for example, increased sucrose concentrations under saline conditions provide a major component for osmotic adjustment (13), but in beets salt, rather than sugar, concentration increases. Salt tolerance among species does not appear to be generally correlated with the level of salt accumulation. Some relatively salt-sensitive species, such as maize, restrict salt uptake better than some highly tolerant species such as beets (7).

Short-term osmotic adjustment immediately following salination is mediated by increased cellular K and organic acid concentrations. In a day or two, other cations and anions (e.g. Ca^{2+} or Na^+ and Cl^-) may replace the organic salts of potassium in effecting osmotic adjustment (6).

The growth of salt-affected plants is impaired despite complete osmotic adjustment and the maintenance of full turgor. What, then, causes the decrease in growth? A clue is provided in that osmotic adjustment generally so well matches in magnitude the decreased osmotic potential of the medium. Because growth depends on the maintenance of turgor (40), osmotic overadjustment might increase turgor and accelerate growth, whereas underadjustment would retard growth until additional absorbed solutes (salts) or synthesized solutes (organic compounds) effect the requisite adjustment. Thus, the reduction in growth may be regarded as a consequence of the requirement for osmotic adjustment (5) and the primary mechanism for effecting it. This interpretation is complicated by the different mechanisms for osmotic adjustment. A species may utilize readily absorbable ions when present, but rely on synthesized (organic) osmotica when ions in the saline medium are only slowly absorbable. Chloride and sulfate ions for most species are readily and slowly absorbable, respectively, so Cl^- generally contributes much more to osmotic adjustment than does SO_4^{2-}. Yet the different salts at equal osmotic pressure may cause equal yield decreases. One may infer that the absorption of an equivalent of chloride is metabolically equivalent to the synthesis of an equivalent of organic acid.

It has been observed that salinity may increase respiration (60). This increased respiration is maintained as well after osmotic adjustment as during the course of adjusting osmotically (53). Perhaps as much respiratory activity is needed to retain the solute concentrations required for osmotic adjustment as is needed to increase them initially to effect osmotic adjustment.

Attempts have been made to separate osmotic effects from associated specific ion effects by the use of organic osmotic agents (49, 51, 71, 73). Ideally, these agents should be nonabsorbable so that they cause no secondary effects. Osmotic adjustment may still be achieved by greater accumulation of nutrient ions (e.g. K^+) or of organic solutes (6, 49). The effects of such solutes as polyethylene glycol compared to those of salts, have given inconclusive results. "Inert" osmotica inhibited growth more in some cases than isosmotic salt solutions (73) and less in others (51).

Because saline media invariably include one or more ions that are readily absorbed, they are not, of course, inert osmotic solutions. Even though roots are directly exposed to salinity and absorb and retain more ions than are translocated to the shoots (all roots, for example, take up Na^+), root growth is generally less affected by salinity than is top growth (22). With some salt treatments, such as Na_2SO_4, shoot mineral contents of "Na excluders" such as beans and maize may be affected very little by salinity (7). Specific effects of such salt treatment, if they occur at all, must involve the roots. Because ions, such as Na^+ and Ca^{2+}, that exert quite diverse effects physiologically can produce equal effects on growth, specific ion effects would not appear to be a significant factor in determining growth inhibition. Of course, a degree of physiological balance between Na and Ca must be present, if toxicity due to high concentrations of Na alone is to be avoided (22).

Another possible mechanism whereby effects on roots could be transmitted to shoots without salt transport to the shoots is through altered hormonal status. If the transport of hormones from roots to the shoots is inhibited by salinity (see the following section), shoot growth may be impaired even if no salts accumulate in the shoots. An attempt to supply salinized bean plants with exogenous kinetin to correct for kinetin deficiency only aggravated salt injury (50). In another study, kinetin treatment markedly reduced salt damage to tobacco, but the kinetin treatment itself was more growth inhibitory than were the high salinity levels employed (3). Tolerance of soybeans to abrupt lethal salination was increased by the prior application of growth retardants (56), but growth inhibition of soybeans by *continuous* salinity was not affected by treatment with Amo-1618, a growth retardant (L. Bernstein, unpublished; but see reference 21). In the latter case, the inhibitory influences of salinity and the growth retardant were additive.

HORMONAL CHANGES AND EFFECTS OF RELATIVE HUMIDITY   Salination has been observed to alter the hormone balance in plants. An increase in salinity caused a decreased transport of kinetin from roots to leaves, and an increase in leaf content of abscisic acid (58, 59). Both changes decrease stomatal aperture. These hormonal responses may mediate the adjustment of plants to salinity by reducing stomatal aperture and water loss. These hormonal changes were observed in plants shortly after salination, and their persistence is questionable. Plants growing under saline conditions do not appear kinetin-deficient. Older leaves of salt-affected plants do not senesce earlier and often are retained longer than those of nonsaline controls; nor does decreased stomatal aperture appear to be generally persistent in salt-affected plants (34, 36). Indeed, increased storage of photosynthate under saline conditions in some species indicates that photosynthesis and stomatal aperture are not growth limiting (13). The changes in hormone levels following salination may well be an immediate aftereffect of such treatment, but a rather ephemeral one that may disappear in most cases as soon as osmotic adjustment is effected and water stress is relieved.

Salt-sensitive species are affected less by salinity when relative humidity is high than when it is low (47, 55, 61). This may be the result of reduced transpiration at high relative humidities. Water stress may not be as great when osmotic gradients are reduced, so the check in growth leading to increased solute concentrations is less. Hoffman & Rawlins (47) reported lower leaf osmotic potentials at 45% than at 95% relative humidity. Mizrahi et al (58) found that low relative humidity increased abscisic acid in leaves and suggest therefore that leaf water stress is the primary signal for modifying plant hormone balance.

Water stress induced by salinity or low relative humidity causes a shift in CO_2-fixation from C3 to crassulacean acid metabolism in the halophyte, *Mesembryanthemum crystallinum* (80, 81).

ENZYME LEVELS AND ACTIVITIES   Specific ion effects may be expected because of the well-established influence of specific ions on enzyme activities. Enzymes such as malate dehydrogenase and acetic thiokinase show optimum activity at a salt

(Na^+ or K^+) concentration of 0.04 M, and decreased activity at higher concentrations (45, 46). Growth decreases may, however, occur in organs that do not show increased accumulation of these ions. Moreover, halophilic higher plants, unlike halophilic bacteria (48), do not possess enzymes that require higher salt concentrations for optimum activity (41, 64). The enzymes of saltbushes and bean or maize were equally affected by salinity in vitro, and saline growing conditions did not affect enzyme levels (41, 77). It appears likely that salts are sequestered largely in vacuoles of higher plants where they contribute to the osmotic properties of cells without important effects on enzyme activities.

Nutritional Effects

SALINITY AND CATION NUTRITION   The major nutritional effects of salinity are those associated with cation nutrition. Although imbalances in cation nutrition tend to be corrected when mixed salts are present, imbalances sometimes do occur. Some varieties of carrots contain higher levels of K and lower levels of Ca than others under nonsaline conditions (13). When salinized with sodium and calcium chlorides, all varieties increased in Ca and Na and decreased in K. Those varieties with higher K contents in the nonsaline treatment were less affected by these changes in nutrition, and produced higher yields than the varieties with lower K contents.

Salinity-induced Ca deficiencies cause blossom-end rot of tomato and bell pepper, and blackheart of celery (37). Proportions of Ca in the medium that are adequate under nonsaline conditions become inadequate under saline conditions. To increase available Ca under saline conditions would require heavy applications of Ca salts to the soil, increasing salinity and further inhibiting growth. Spraying the foliage with solutions of $Ca(NO_3)_2$ or $CaCl_2$ provides the Ca required for adequate nutrition through foliar absorption, without increasing soil salinity (38).

SODICITY AND CATION NUTRITION   The nutritional effects of sodicity are not simply related to the exchangeable sodium percentage (ESP) of soils. Plant response is governed by the concentrations of ions in the soil solution, rather than by their proportions on the exchange complex. When soils are nonsaline, total soluble salt concentrations are low. Increases in exchangeable sodium are balanced by decreases in exchangeable Ca and Mg, leading to Ca and/or Mg deficiencies when the concentrations of these elements in solution become deficient (Table 2A). In saline-sodic soils at similar ESPs, the concentrations of all elements in solution are greater and may remain nutritionally adequate (Table 2B). In saline-sodic soils, therefore, salinity effects predominate, and the nutritional effects of sodicity are usually absent (7, 52). The sodium-adsorption ratio (SAR) = $Na^+ / [(Ca^{2+} + Mg^{2+})/2]^{\frac{1}{2}}$ where ionic symbols refer to concentrations in the soil solution in meq/liter. At equal SARs, ESPs are equal, regardless of total salt concentrations (75).

Tolerance of crops to sodicity under nonsaline conditions varies widely. The crops most sensitive to the nutritional effects, such as beans (*Phaseolus vulgaris*), are affected at ESPs of about 10. Most crops are moderately tolerant, and are affected nutritionally at ESPs of about 25. Highly tolerant crops, such as beets (*Beta vulgaris*) and tall wheatgrass (*Agropyron elongatum*), are not affected until ESPs reach

Table 2 Soluble cation concentrations in equilibrium with soils at given SARs and salinities

A. Effect of varying ESP (and SAR) at constant total soluble-cation concentration of 8 meq/liter

ESP:	5	15	25	50
SAR:	4.5	12.8	23.5	69
Na^+, meq/liter	5.25	7.34	7.78	7.97
$Ca^{2+} + Mg^{2+}$, meq/liter	2.75	0.66	0.22	0.03

B. Effect of varying salinity at constant SAR of 36 (ESP = 34)

Total soluble cations:	8 meq/liter	80 meq/liter
Na^+, meq/liter	7.9	72
$Ca^{2+} + Mg^{2+}$, meq/liter	0.1	8

or exceed 50 (24, 65). Tolerance reflects the species' ability to absorb nutritionally adequate levels of Ca and Mg from low concentrations of these elements in the soil solution.

SALINITY-FERTILITY INTERACTIONS   Basic salt tolerance data have been derived primarily under optimal conditions of fertilization (8, 75). In some areas, however, optimum fertilization may be either uneconomic or impossible because of lack of fertilizer. We need to know, therefore, how suboptimal fertilization affects crop response to salinity.

It has been suggested that increased fertilization may overcome some of the inhibitory effects of salinity (44, 67). When salinity induces a nutrient deficiency such as that of calcium, specific corrective treatment will obviously be beneficial. Whether increased applications of N, P, or K over and above the levels required under nonsaline conditions will improve yields on saline soils is quite another matter. Although the effects of salinity on nutrient absorption have been frequently studied (7, 22), only recently have the effects of salinity at different levels of fertility been specifically investigated. Ravikovitch and co-workers (67, 68) have reported apparent increases in salt tolerance when N or P levels under saline conditions were greater than those that were optimum under nonsaline conditions.

In sand-culture studies, Bernstein et al (21) determined the interactive effects of salinity and fertility on yields of grains and vegetable crops. When either nutrient deficiency or salinity was moderate and yields were only moderately depressed (i.e. by about 25%), the effects of salinity and fertility tended to be independent and additive. Strongly inhibitory salinity or nutrient deficiency, however, tended to control yield regardless of the level of the other factor.

Levels of phosphate commonly used in nutrient culture work (0.5–2 mM) increased plant injury under saline conditions, although these phosphate levels were not injurious under nonsaline conditions. The increased injury by high P under saline conditions was attributed to P toxicity (21).

Increasing the level of N (as nitrate) had no significant effect on salt tolerance. Reducing the K level from 2 to 0.35 mM had no effect on salt tolerance of maize.

Although salinity in some cases decreased leaf P when P levels were adequate, it tended to increase leaf P or N when these elements were deficient in the culture solutions. It was concluded that increased nutrient levels would not significantly increase salt tolerance except where salinity induced a nutrient deficiency, or when poor soil physical conditions seriously impaired root development, thereby causing nutrient deficiencies (21).

Physical Effects in Sodic Soils

Nonsaline-sodic soils are characterized by poor physical condition (75). Exchangeable sodium deflocculates such soils, and dispersed soils have low permeability to water and air. They tend to be sticky when wet, and very hard when dry. Black, solubilized organic matter frequently accumulates at the soil surface, giving rise to the name "black alkali." When sodic soils are saline, the high salt concentrations promote flocculation so that permeability is more nearly normal.

Plant growth is adversely affected by the poor physical condition of nonsaline-sodic soils. Even crops that tolerate high levels of exchangeable sodium nutritionally often fail because of the adverse physical condition of sodic soils. Of the crops tested, only rice growing in flooded soils, and therefore insensitive to soil physical conditions, was unaffected by poor soil structure (65). Although it was possible to maintain good soil structure experimentally in sodic soils by means of soil-conditioning agents and to distinguish between the adverse physical and nutritional effects (24, 65), the improvement of sodic soils by these means is not economically practical. Removal of exchangeable sodium by the addition of soil amendments and leaching is the only practical correction for sodic soil conditions (75).

Toxicities

SODIUM AND CHLORIDE   Although the specific effects of an excess of any ion may be regarded as a toxicity, it is useful to distinguish between nutritional effects and direct toxicities (22). Sodium and chloride may be directly toxic and cause characteristic leaf burn in susceptible species (9, 22). The tan or brown necrotic lesions are usually sharply separated from the adjacent healthy green tissue with no symptoms of progressive deterioration in cells adjacent to the lesion (9). Woody plants are generally susceptible, although some exceptions have been noted (9, 20).

In an exceptional case among nonwoody plants, the differential salt tolerance of soybean varieties has been attributed to differential chloride accumulation (2), which appears to be controlled by a single gene with Cl exclusion dominant over Cl accumulation (1). The greater salt tolerance of tideland grass ecotypes compared to upland ecotypes has also been attributed to decreased transport of Na and Cl to the shoots (42).

Species and even varieties within a species among fruit crops may differ in tolerance to chloride (9). The more chloride-tolerant plants absorb Cl^- more slowly, but the leaf-chloride level at which injury occurs tends to be similar for all susceptible plants (22). Tolerance to chloride salinity may be significantly improved by using rootstocks or fruit crop varieties that absorb chloride more slowly. The most chloride-sensitive plants may be injured when chloride in the soil saturation extract exceeds 5 or 10 meq/liter, whereas the most tolerant plants are damaged only at chloride concentrations of about 30 meq/liter or more. Grape rootstocks differ in chloride transport to the leaves by as much as 15-fold (15).

There is generally an overlap between the Cl or Na levels in injured and uninjured leaves of a given plant. Duration of exposure to the accumulated toxic ions may be a factor in the development of injury. Weather conditions are also critical. During the spring, leaves may contain toxic levels of Cl or Na and show no symptoms of injury. Leaf burn appears suddenly following the onset of hot, dry weather in early summer (31). These observations and the similarity of chloride or sodium injury patterns to those of drought injury (20, 22) support the idea that accumulation of these ions may impair the ability of the leaf to regulate water loss under conditions of extreme evaporative demand by perhaps interfering with normal stomatal closure (20).

Toxic ions may accumulate in tissues surrounding vein endings in leaves, that is, at the leaf tip in leaves with parallel venation, or at the leaf margins in leaves with palmate or pinnate venation. Although boron injury appears to follow such patterns of boron accumulation (63), chloride injury does not. Oertli (62) has suggested that leaf injury may be caused by accumulations of excess salts in the cell walls, even to hypertonic concentrations. Leaf cell-wall solute concentrations under saline conditions are, however, less than 10% of the internal cell concentrations (12).

A study of chloride accumulation relative to leaf burn was performed on strawberry plants in culture solutions containing 32 meq $CaCl_2$ per liter (L. Bernstein, unpublished). Leaflets were sampled as soon as incipient injury occurred. The injured marginal leaf tissue was dissected from the leaflet and compared with the remaining central leaf tissue. Major veins were removed. Leaf chloride concentrations were higher in the leaf margins, but the uninjured central leaf tissue of some leaflets had chloride contents as great as those in the injured margins of other leaflets (Table 3). The location of the tissue in the leaf margin, rather than its specific chloride content, thus determined the site of injury. This finding would appear to support the view that water stress caused by uncontrolled transpiration was responsible for the injury rather than localized chloride accumulation. However, the incipient injury symptom indicated toxicity rather than excessive water loss. Instead of a wilt, the marginal leaf tissue at the time of sampling appeared water-soaked, indicating that cell membranes had become leaky, permitting cell sap to fill the intercellular spaces. Irregular lesions that are not marginal (e.g. sodium toxicity in avocado) also imply a mechanism that is not related to water stress, which should cause marginal or tip burn (9). These discordant observations indicate the need for further study of the mechanisms of Cl and Na toxicities in plants.

The increase in sodium concentration of the soil water with increasing sodicity (Table 2A) is a direct hazard to fruit crops specifically sensitive to Na. Toxic accumulations of sodium may occur at relatively low ESPs of about 5 (22) when other nutritional and physical effects of sodicity are absent.

Salts normally are absorbed by the roots and translocated to stems and leaves. When foliage is wetted by sprinkler irrigation, leaves may absorb salts directly. Even sodium, which is not translocated to the leaves from the roots of some species, may

Table 3 Chloride contents in injured margins (ca 1 cm in width) and uninjured central tissues of strawberry leaflets at onset of leaf injury, 5 days after 32 meq $CaCl_2$/liter was added to culture solution

Leaflet[a]	1	2	3	4	5	6
	(meq Cl/100 g dry wt)					
Margin	86.0	75.2	68.2	52.9	61.5	63.5
Central tissue	66.4	60.1	60.6	51.8	58.1	53.4

[a] Leaflet numbers are for identification, and do not refer to leaf age or insertion.

be taken up directly by their leaves and accumulated just as readily as chloride (33). Susceptibility to damage by foliarly absorbed salts depends more on leaf characteristics and rate of foliar absorption than on tolerance to soil salinity. Avocado, which is very salt-sensitive, was not damaged by sprinkling with waters containing Na^+ or Cl^- because foliar absorption rates were very low. Citrus, stone fruits, and almond absorbed salts foliarly and were severely damaged (33). Among other species studied, sugarcane (14) and strawberry (32) have low rates of foliar salt absorption, but grape leaves (L. Bernstein, unpublished) absorb salts readily.

When a species absorbs salts readily through its leaves, tolerance to salinity of sprinkled irrigation waters (specifically Cl^- and Na^+ concentrations) is markedly reduced. As little as 5 meq/liter in the irrigation water wetting foliage in drenching sprays can cause Cl or Na to accumulate to damaging levels in leaves of susceptible fruit crops (33). Intermittent wetting by rotating sprinklers increases the salt concentration of water films on the leaves by evaporation, and as little as 2 or 3 meq of Na or Cl per liter in the irrigation water may cause severe leaf damage (43). Sprinkling during the evening reduces foliar salt absorption by one half compared to daytime sprinkling (33). The symptoms of leaf injury by foliarly absorbed salts are the same as those caused by salts absorbed through the roots.

In comparing foliar- and root-mediated salt uptakes, one must remember that root uptake is continuous but that foliar absorption may occur only during the 10% or less of the time that leaves are wetted by the sprinklers. Thus, foliar absorption per millequivalent chloride in the irrigation water can be about 100 times as great as the rate of root absorption (33). Susceptible species may be sprinkler irrigated by low-head systems that do not wet any significant fraction of the foliage.

Nonwoody species that do not appear specifically sensitive to chloride or sodium when surface irrigated, may be injured by sprinkling with brackish waters containing c 10–20 meq Na or Cl per liter (19). Injury does not occur with waters of 2–3 meq/liter, as in susceptible woody species. Leaf symptoms, including leaf burn, are similar to those observed on woody plants. Even highly salt-tolerant crops, such as cotton, have suffered a 50% decrease in yield when sprinkler irrigated with brackish waters during the day, whereas nighttime sprinkling or furrow irrigation with the same water caused no injury (27).

Salt spray has long been recognized as a source of salt damage to vegetation along sea coasts. Surf throws fine droplets of sea water into the air, which are carried onshore. Deposited on leaves, this cyclic salt causes much of the salt injury to coastal vegetation (35). The salt is probably deposited mostly in dry form, and as such is probably not directly injurious. However, dews common to humid coastal areas may dissolve the salt, permitting osmotic, or toxic effects, or both. Water may also be absorbed from the air by hygroscopic salts.

When salt (NaCl) is used for deicing highways, automobiles may throw salt spray directly onto the foliage of nearby plants (72) or cause injury by aerial salt drift (54). Runoff of saline waters from highways deiced by salt may severely injure roadside trees (28). The symptoms and foliar salt content indicate that injury is similar to that caused by chloride and sodium to susceptible woody species under arid-zone conditions.

Power plants and other industries that use water-cooling towers also cause salt drift. Air passing through the towers picks up fine droplets of water, which, like cyclic salt, is carried to surrounding vegetation. In this case too, salt may be deposited mostly in dry form, but dew or water absorbed from the air by hygroscopic salts may dissolve the salts and cause injury. Although as little as 100–200 lb of salt per acre per year may be deposited downwind from salt-water-cooling towers (78), such salt deposits on leaves dissolved in water films can achieve concentrations approaching or exceeding that of sea water. (One hundred lb of salt per acre per year $= 1.12$ g/1000 cm^2 per year. In a water film 0.10 mm thick the salt concentration will be 112 g/liter, or more than three times the concentration of sea water. Actual duration of cumulative deposition would, of course, depend on frequency of rainfall. The salt deposited in one day would yield a solution of 307 mg salt/liter in a water film 0.10 mm thick, which is equivalent to 5.2 meq/liter for NaCl, a concentration that causes damaging foliar salt absorption in susceptible species.) Because of the low annual rates of salt deposition by such salt drifts, soil salinity would usually be a minor factor compared to foliar salt deposition.

Salt drift from sprinkler-irrigated fields has even injured crops in adjacent fields. The relationships of injury by cyclic salts and salt drift to atmospheric conditions and salt properties (especially hygroscopicity) should be investigated to define the conditions under which such salt injury occurs.

OTHER TOXICITIES   Most sodium and chloride injuries occur on definitely salt-affected soils. Other elements, however, may be toxic at very low concentrations and bear no consistent relationship to salinity. Thus, the essential element boron may be toxic if its concentration is increased only severalfold above the 0.2–0.5 mg/liter that is required for optimum growth (79). Lithium appears to be toxic to citrus at the very low concentration of 0.1 mg/liter (25). Selenium occurs in well waters in Wyoming, and is believed to be toxic at concentrations greater than 0.2 mg/liter (57). Although they do not contribute significantly to soil salinity, these elements (especially boron) are usually included in evaluations of irrigation water quality.

MANAGEMENT PRACTICES IN RELATION TO SALINITY AND SODICITY CONTROL

Reclamation

Before severely salt-affected soils can be cropped, they usually must be reclaimed by applying water to leach excess salts out of the soil (75). If soils are sodic, soil amendments such as gypsum or sulfur may be needed. Land may become saline or sodic during the course of cropping, requiring periodic leaching, or the application of soil amendments, or both. Furrow irrigation salinizes the ridges between the furrows (75), but basin- or flood-irrigated crops in the rotation promote leaching of salts accumulated during a previous period of furrow-irrigation. The need for periodic reclamation may thus influence the cropping pattern for irrigated lands.

For leaching to be effective, soils must have adequate internal drainage, since leaching salts into a shallow water table may only result in a rapid resalination of the root zone. Artificial drainage by tile drains or pumping may be needed to lower the water table (75).

Leaching Requirements and Irrigation Management

Most irrigation waters contain more salts than are removed by the crop, so that continued irrigation without leaching will progressively salinize the land. Water in excess of consumptive use (evapotranspiration) must therefore be applied to carry the residual salts out of the root zone. The fraction of total water applied in irrigation that is needed for leaching is called the leaching requirement (LR): $LR = V_d /V_i$, where V_d and V_i are the required volumes of drainage and irrigation waters, respectively (75). If one assumes salt conservation in the soil water (69, 75), then salt concentration will be inversely proportional to water volume, and $LR = C_i/C_d$ where C_i and C_d are salt concentrations in the irrigation and drainage waters, respectively. If electrical conductivity is used as a measure of salt concentration, then $LR = EC_i/EC_d$. The more saline the irrigation water, the greater the leaching requirement. EC_d in this equation is the maximum EC of the drainage water that permits normal yields. It is greater the more salt-tolerant the crop.

Crops can concentrate the soil water to a greater degree than was previously thought possible, with little or no loss in yield (18). For tolerant crops EC_ds may reach 40–45 mmho/cm, but for sensitive crops only 14 mmho/cm (Table 4). With most irrigation waters and crops, regularity of leaching is not critical. Even when salinities in the lower root zone approximate the tolerable limits for a crop, leaching only every sixth irrigation can be as effective as regular leaching (18).

Soil-water salinity in irrigated fields may range from a low value in the upper root zone that approximates the salinity of the irrigation water, to a value ten or more

Table 4 Crop tolerances to soil salinity (EC_e), and maximum salinities of irrigation waters for full yield potential (EC_i) and of drainage waters for 85–100% of full yield potential (EC_d). All salinities in mmho/cm at 25°C (8, 18)

	EC_i	EC_es at which yields decrease[a]			EC_d
		10%	25%	50%	
Tolerant crops: Bermuda grass, tall wheatgrass, barley, sugar beets, cotton	6.5–8.5	10–13	13–16	15–18	40–45
Moderately tolerant: alfalfa, soybeans, rice, tomato	2–3.5	3–5.5	5–7	8–9	32–35
Sensitive crops: clovers, beans, onions, carrots	1–1.3	1.5–2	2–3.5	3.5–4	14–16

[a] These data are for essentially uniformly salinized root zones.

times greater at the bottom of the root zone. It was previously assumed that in such profiles plants respond to the average soil-water salinity. If this were the case, an increase in salinity of the upper root zone caused by the use of a more saline irrigation water could be compensated for by decreasing the salinity in the lower root zone. However, plants appear more sensitive to changes of salinity in the upper root zone than in the lower root zone. An increase of 1 mmho/cm in EC_i affected yield of alfalfa about as much as a c 20 mmho/cm increase in the lower root zone or EC_d (18). This effect is accounted for by the much greater water uptake from the less saline than from the more saline parts of the root zones.

Salt uptake (Na and Cl) by alfalfa is also relatively unaffected by salinities in the lower root zone. Small changes in irrigation water salinity, however, cause marked changes in Cl and Na contents in alfalfa (18). Thus, an increase in salt uptake by a crop caused by increased Na or Cl content of the irrigation water cannot be compensated for by increasing the leaching fraction to decrease salinity in the lower root zone.

Assessing Irrigation Water Quality

The principal criteria of quality of irrigation water are salinity, sodicity, and toxicity as determined by EC, SAR, and specific ion concentrations, respectively. However, the effects on crops of a given water are not determined solely by these water properties. Even nonhazardous waters may cause problems under adverse conditions of use. Water quality factors should, therefore, be considered in relation to the specific conditions under which the water is to be used (11). However, the *potential* of a given water for crop production can be defined unequivocally as the best attainable result for that water under optimum conditions of use. When a water already contains more salts than the crop in question can tolerate, its potential for crop production is impaired. Tolerable limits for chloride, boron, and other irrigation water components depend on specific crop tolerances (9, 11, 79). Specific limits for salinity can be derived from salt tolerance data [(8,) and Table 4]. Yields of sensitive crops are affected when EC_i exceeds 1 mmho/cm, whereas the most tolerant crops are affected only when EC_i is greater than 6–8 mmho/cm.

Irrigation practices, especially irrigation frequency, influence soil salinity and plant response to it. When crops are irrigated at conventional frequencies, soil-water contents may decrease appreciably during an irrigation cycle, and both decreased matric and osmotic potentials may reduce yields (76). Even when matric potentials remain high (e.g. above −1 bar), the osmotic potential of saline soils may decrease significantly during an irrigation cycle. If water deficits are replaced frequently, roots will have a steady supply of water at the minimum salinity of the irrigation water. This can be achieved by daily small applications of water as by drip (trickle) irrigation (17).

The proposal to irrigate crops in sandy coastal areas with sea water (26) has received much popular attention in recent years. Although halophytic higher plants grow in coastal marshes that are flooded by sea water, no crop plants tolerate sea-water salinities of 35,000 mg/liter. Yields of the most tolerant crops are affected when irrigation water salinities exceed 4000–5000 mg/liter (Table 4). This occurs even in sand cultures through which solutions are flushed twice daily (21). So-called sea-water irrigation may employ coastal or estuarine waters much less saline than oceanic waters (e.g. 7000 mg/liter), and may thus have some potential, especially when used to supplement rainfall in humid or subhumid areas. Brackish ground waters of still lower salinity (e.g. 3000 mg/liter) have considerable potential for the irrigation of salt-tolerant crops even in arid zones, under proper soil and management conditions.

Crop Selection for Salt-Affected Soils

Salinity or sodicity cannot be entirely eliminated when the only available irrigation waters are saline, or when soil properties are marginal. In such situations, only those crops that can tolerate the resultant salinity or sodicity should be grown (8). During the later stages of reclamation, tolerant crops may be grown to yield some economic return. Tolerant grasses and legumes tend to improve soil structure in partially reclaimed soils through the beneficial action of roots, or through their incorporation into the soil as green manures. Salt tolerance data are useful in selecting suitable crops for such conditions (8).

Planting Methods

Plants were thought to be more sensitive to salinity during germination than at later growth stages because germination failures occur so often in salt-affected soils. Salt tolerances during germination and later growth stages, however, are similar for most species, with the notable exception of beets which are much more sensitive during germination than at later stages (22). The *apparent* greater sensitivity during germination in the field is caused, in part, by the tendency for salts to accumulate near the soil surface because of evaporation (17, 19), which also increases the salinity of the residual soil water.

Salt transport is especially pronounced in furrow-irrigated soils. In a single post-planting irrigation, water moving into the planting bed concentrates the salts near the surface at the center of the bed to 5–10 times the salinity of the soil before irrigation. Seeds planted in this location frequently fail to germinate even when the soil salinity prior to irrigation was quite low (23). Salinity decreases from the center to the shoulders of the beds so that double-row plantings near the bed shoulders are at about the same salinity as the original soil before irrigation (16). Sloping seed beds with seeds planted one third of the way below the peak of the bed permit good germination even under extremely high initial soil salinities, because water moving into the bed sweeps salts beyond the plant row into the peak of the bed (16, 23). Sprinkler irrigation is now frequently used to start row crops such as lettuce, because sprinkling washes salts down below the seed depth (70). Sprinkling may also maintain a more favorable soil-water (and sometimes temperature) status, with considerable saving of water.

PATHOGENICITIES AND SALT-AFFECTED SOILS

The effects of salinity and sodicity are physiological, whether nutritional, osmotic, or directly toxic. No pathogens are directly involved, although some secondary pathogenic effects have been noted. Saline soils remain wetter than comparable nonsaline soils under given meteorological conditions, primarily because water use by salt-stunted plants is reduced. Poor soil permeability or drainage, often responsible for salination, also keeps soils wetter. Wet soil conditions may favor phytophthora root rot and other fungus infections (13, 17). None of these diseases is specifically salinity induced. Improving the water regime by better irrigation or drainage may prevent the fungus diseases associated with wetter soil conditions, even without any decrease in soil salinity.

Literature Cited

1. Abel, G. H. 1969. Inheritance of the capacity for chloride inclusion and chloride exclusion by soybeans. *Crop Sci.* 9:697–98

2. Abel, G. H., MacKenzie, A. J. 1964. Salt tolerance of soybean varieties (*Glycine max* L. Merrill) during germination and later growth. *Crop Sci.* 4:157–61

3. Benzioni, A., Mizrahi, Y., Richmond, A. E. 1974. Effect of kinetin on plant response to salinity. *New Phytol.* 73:315–19

4. Bernstein, L. 1959. Salt tolerance of vegetable crops in the West. *US Dep. Agric. Agric. Inf. Bull.* 205. 5 pp.

5. Bernstein, L. 1961. Osmotic adjustment of plants to saline media. I. Steady state. *Am. J. Bot.* 48:909–18

6. Bernstein, L. 1963. Osmotic adjustment of plants to saline media. II. Dynamic phase. *Am. J. Bot.* 50:360–70

7. Bernstein, L. 1964. Effects of salinity on mineral composition and growth of plants. *Plant Anal. Fert. Probl.* 4:25–45

8. Bernstein, L. 1964. Salt tolerance of plants. *US Dep. Agric. Agric. Inf. Bull.* 283. 23 pp.

9. Bernstein, L. 1965. Salt tolerance of fruit crops. *US Dep. Agric Agric. Inf. Bull.* 292. 8 pp.

10. Bernstein, L. 1965. Book Review: Strogonov, B. P. Physiological Basis of Salt Tolerance of Plants. *Soil Sci. Soc. Am. Proc.* 29(2):iv

11. Bernstein, L. 1967. Quantitative assessment of irrigation water quality, pp. 51–65. *Am. Soc. Test. Mater. Spec. Tech. Publ.* 416:1–210

12. Bernstein, L. 1971. Method for determining solutes in the cell walls of leaves. *Plant Physiol.* 47:361–65

13. Bernstein, L., Ayers, A. D. 1953. Salt tolerance of five varieties of carrots. *Proc. Am. Soc. Hortic. Sci.* 61:360–66

14. Bernstein, L., Clark, R. A., Francois, L. E., Derderian, M. D. 1966. Salt toler-

ance of N. Co. varieties of sugar cane. II. Effects of soil salinity and sprinkling on chemical composition. *Agron. J.* 58:503–7

15. Bernstein, L., Ehlig, C. F., Clark, R. A. 1969. Effect of grape rootstocks on chloride accumulation in leaves. *J. Am. Soc. Hortic. Sci.* 94:584–90

16. Bernstein, L., Fireman, M. 1957. Laboratory studies on salt distribution in furrow-irrigated soil with special reference to the pre-emergence period. *Soil Sci.* 83:249–63

17. Bernstein, L., Francois, L. E. 1973. Comparisons of drip, furrow, and sprinkler irrigation. *Soil Sci.* 115:73–86

18. Bernstein, L., Francois, L. E. 1973. Leaching requirement studies: Sensitivity of alfalfa to salinity of irrigation and drainage waters. *Soil Sci. Soc. Am. Proc.* 37:931–43

19. Bernstein, L., Francois, L. E. 1975. Effects of frequency of sprinkling with saline waters compared with daily drip irrigation. *Agron. J.* 67:185–90

20. Bernstein, L., Francois, L. E., Clark, R. A. 1972. Salt tolerance of ornamental shrubs and ground covers. *J. Am. Soc. Hortic. Sci.* 97:550–56

21. Bernstein, L., Francois, L. E., Clark, R. A. 1974. Interactive effects of salinity and fertility on yields of grains and vegetables. *Agron. J.* 66:412–21

22. Bernstein, L., Hayward, H. E. 1958. Physiology of salt tolerance. *Ann. Rev. Plant Physiol.* 9:25–46

23. Bernstein, L., MacKenzie, A. J., Krantz, B. A. 1955. The interaction of salinity and planting practice on the germination of irrigated row crops. *Soil Sci. Soc. Am. Proc.* 19:240–43

24. Bernstein, L., Pearson, G. A. 1956. Influence of exchangeable sodium on the yield and chemical composition of plants. I. Green beans, garden beets, clover, and alfalfa. *Soil Sci.* 82:247–58

25. Bingham, F. T., Bradford, G. R., Page, A. L. 1964. Toxicity of lithium to plants. *Calif. Agric.* 18(9):6–7

26. Boyko, H., Ed. 1966. *Salinity and Aridity.* The Hague: Junk. 408 pp.

27. Busch, C. D., Turner, F. Jr. 1965. Sprinkling cotton with saline water. *Prog. Agric. Ariz.* 17(4): 27–28

28. Button, E. F. 1964. Influence of rock salt used for highway ice control on mature sugar maples at one location in central Connecticut. *Conn. State Highway Dep. Rept. No. 3.* 21 pp.

29. Chapman, V. J. 1960. *Salt Marshes and Salt Deserts of the World.* London: Leonard Hill. 392 pp.

30. Eaton, F. M. 1942. Toxicity and accumulation of chloride and sulfate salts in plants. *J. Agric. Res.* 64:357–99

31. Ehlig, C. F. 1960. Effects of salinity on four varieties of table grapes grown in sand culture. *Proc. Am. Soc. Hortic. Sci.* 76:323–31

32. Ehlig, C. F. 1961. Salt tolerance of strawberries under sprinkler irrigation. *Proc. Am. Soc. Hortic. Sci.* 77:376–79

33. Ehlig, C. F., Bernstein, L. 1959. Foliar absorption of sodium and chloride as a factor in sprinkler irrigation. *Proc. Am. Soc. Hortic. Sci.* 74:661–70

34. Ehlig, C. F., Gardner, W. R., Clark, M. 1968. Effect of soil salinity on water potentials and transpiration in pepper (*Capsicum frutescens*). *Agron. J.* 60: 249–53

35. Ferguson, C. R. 1952. Salt tolerant plants for south Florida. *Proc. Fla. State Hortic. Soc.* 65:306–13

36. Gale, J., Kohl, H. C., Hagan, R. M. 1967. Changes in the water balance and photosynthesis of onion, bean and cotton plants under saline conditions. *Physiol. Plant.* 20:408–20

37. Geraldson, C. M. 1956. Watch nutrient intensity and balance; by checking soil solution soluble salts, we can prevent blossom-end rot of tomatoes and peppers and blackheart of celery. *Sunshine State Fla. Agric. Exp. Stn. Res. Rept.* 1(3):10–11

38. Geraldson, C. M. 1957. Control of blossom-end rot of tomatoes. *Proc. Am. Soc. Hortic. Sci.* 69:309–17

39. Gornat, B., Goldberg, D., Rimon, D., Ben-Asher, J. 1973. The physiological effect of water quality and method of application on tomato, cucumber, and pepper. *J. Am. Soc. Hortic. Sci.* 98: 202–5

40. Green, P. B. 1968. Growth physics in *Nitella:* a method for continuous in vivo analysis of extensibility based on a micro-manometer technique for turgor pressure. *Plant Physiol.* 43:1169–84

41. Greenway, H., Osmond, C. B. 1972. Salt responses of enzymes from species differing in salt tolerance. *Plant Physiol.* 49:256–59

42. Hannon, N. J., Barber, H. N. 1972. The mechanism of salt tolerance in naturally selected populations of grasses. *Search Sydney* 3:259–60

43. Harding, R. B., Miller, M. P., Fireman, M. 1958. Absorption of salts by citrus leaves during sprinkling with water suitable for surface irrigation. *Proc. Am. Soc. Hortic. Sci.* 71:248–56

44. Heimann, H. 1958. Irrigation with saline water and the ionic environment. *Potassium Symp.* 1958:173–220

45. Hiatt, A. J., Evans, H. J. 1960. Influence of salts on activity of malic dehydrogenase from spinach leaves. *Plant Physiol.* 35:662–72

46. Hiatt, A. J., Evans, H. J. 1960. Influence of certain cations on activity of acetic thiokinase from spinach leaves. *Plant Physiol.* 35:673–77

47. Hoffman, G. J., Rawlins, S. L. 1971. Growth and water potential of root crops as influenced by salinity and relative humidity. *Agron. J.* 63:877–80

48. Ingram, M. 1957. Micro-organisms resisting high concentrations of sugars or salts, pp. 90–133. 7th *Symp. Soc. Gen. Microbiol.* Cambridge Univ. Press. 380 pp.

49. Janes, B. E. 1966. Adjustment mechanisms of plants subjected to varied osmotic pressures of nutrient solution. *Soil Sci.* 101:180–88

50. Kirkham, M. B., Gardner, W. R., Gerloff, G. C. 1974. Internal water status of kinetin-treated, salt-stressed plants. *Plant Physiol.* 53:241–43

51. Lagerwerff, J. V., Eagle, H. E. 1961. Osmotic and specific effects of excess salts on beans. *Plant Physiol.* 36:472–77

52. Lagerwerff, J. V., Holland, J. P. 1960. Growth and mineral content of carrots and beans as related to varying osmotic and ionic-composition effects in saline-sodic sand cultures. *Agron. J.* 52:603–8

53. Livne, A., Levin, N. 1967. Tissue respiration and mitochondrial oxidative phosphorylation of NaCl-treated pea seedlings. *Plant Physiol.* 42:407–14

54. Lumis, G. P., Hofstra, G., Hall, R. 1973. Sensitivity of roadside trees and shrubs to aerial drift of deicing salt. *HortScience* 8(6):475–77

55. Magistad, O. C., Ayers, A. D., Wadleigh, C. H., Gauch, H. G. 1943. Effect of salt concentration, kind of salt, and climate on plant growth in sand cultures. *Plant Physiol.* 18:151–66

56. Marth, P. C., Frank, J. R. 1961. Increasing tolerance of soybean plants to some soluble salts through application of plant growth-retardant chemicals. *J. Agric. Food Chem.* 9:359–61

57. Miller, W. M. 1956. Summary of partial analyses of Wyoming waters. Salinity and selenium. *Wyo. Agric. Exp. Stn. Circ.* No. 64:1–11

58. Mizrahi, Y., Blumenfeld, A., Bittner, S., Richmond, A. E. 1971. Abscisic acid and cytokinin contents of leaves in relation to salinity and relative humidity. *Plant Physiol.* 48:752–55

59. Mizrahi, Y., Blumenfeld, A., Richmond, A. E. 1970. Abscisic acid and transpiration in leaves in relation to osmotic root stress. *Plant Physiol.* 46:169–71

60. Nieman, R. H. 1962. Some effects of sodium chloride on growth, photosynthesis, and respiration of twelve crop plants. *Bot. Gaz.* 123:279–85

61. Nieman, R. H., Poulsen, L. L. 1967. Interactive effects of salinity and atmospheric humidity on the growth of bean and cotton plants. *Bot. Gaz.* 128:69–73

62. Oertli, J. J. 1968. Extracellular salt accumulation, a possible mechanism of salt injury in plants. *Agrochimica* 12:461–69

63. Oertli, J. J., Kohl, H. C. 1961. Some considerations about the tolerance of various plant species to excessive supplies of boron. *Soil Sci.* 92:243–47

64. Osmond, C. B., Greenway, H. 1972. Salt responses of carboxylation enzymes from species differing in salt tolerance. *Plant Physiol.* 49:260–63

65. Pearson, G. A., Bernstein, L. 1958. Influence of exchangeable sodium on yield and chemical composition of plants: II. Wheat, barley, oats, rice, tall fescue, and tall wheatgrass. *Soil Sci.* 86:254–61

66. Porath, E., Poljakoff-Mayber, A. 1964. Effect of salinity on metabolic pathways in pea root tips. *Israel J. Bot.* 13:115–21

67. Ravikovitch, S., Porath, A. 1967. The effect of nutrients on the salt tolerance of crops. *Plant Soil* 26:49–71

68. Ravikovitch, S., Yoles, D. 1971. The influence of phosphorus and nitrogen on millet and clover growing in soils

affected by salinity. I. Plant development. *Plant Soil* 35:555–67

69. Rhoades, J. D., Krueger, D. B., Reed, M. J. 1968. The effect of soil-mineral weathering on the sodium hazard of irrigation waters. *Soil Sci. Soc. Am. Proc.* 32:643–47

70. Robinson, F. E., McCoy, O. D. 1965. The effect of sprinkler irrigation with saline water and rates of seeding on germination and growth of lettuce. *Am. Soc. Hortic. Sci. Proc.* 87:318–23

71. Ruf, R. H. Jr., Eckert, R. E. Jr., Gifford, R. O. 1963. Osmotic adjustment of cell sap to increases in root medium osmotic stress. *Soil Sci.* 96:326–30

72. Sauer, G. 1967. Über Schäden an der Bepflanzung der Bundesfernstrassen durch Auftausalze. *Nachrichtenbl. Dtsch. Pflanzenschutzdienstes Braunschweig* 19:81–87

73. Slatyer, R. O. 1961. Effects of several osmotic substrates on the water relationships of tomato. *Aust. J. Biol. Sci.* 14:519–40

74. Strogonov, B. P. 1962. *Physiological Basis of Salt Tolerance of Plants.* Trans. A. Poljakoff-Mayber, A. M. Meyer, Israel Program Sci. Transl. Jerusalem, 1964. 279 pp.

75. United States Salinity Laboratory Staff. 1954. Diagnosis and improvement of saline and alkali soils. *US Dep. Agric. Handb.* 60. 160 pp.

76. Wadleigh, C. H., Ayers, A. D. 1945. Growth and biochemical composition of bean plants as conditioned by soil moisture tension and salt concentration. *Plant Physiol.* 20:106–32

77. Weimberg, R. 1970. Enzyme levels in pea seedlings grown on highly salinized media. *Plant Physiol.* 46:466–70

78. Westinghouse Electric Corp. Environmental Systems Dep. 1973. *The State of the Art of Saltwater Cooling Towers for Steam Electric Generating Plants.* 152 pp. (with appendixes)

79. Wilcox, L. V. 1960. Boron injury to plants. *US Dep. Agric. Agric. Inf. Bull.* 211. 7 pp.

80. Winter, K. 1973. Zum Problem der Ausbildung des Crassulaceensäurestoffwechsels bei *Mesembryanthemum crystallinum* unter NaCl-Einfluss. *Planta* 109:135–45

81. Winter, K. 1973. CO_2-Fixierungsreaktionen bei der Salzpflanze *Mesembryanthemum crystallinum* unter variierten Aussenbedingungen. *Planta* 114:75–85

DISEASE RESISTANCE IN BEANS

♦3625

W. J. Zaumeyer and J. P. Meiners

Plant Protection Institute, Agricultural Research Service, US Department of Agriculture, Beltsville, Maryland 20705

INTRODUCTION

The inefficacy and cost of many of the chemical bean-disease control measures and the susceptibility of many of the important snap and dry bean (*Phaseolus vulgaris* L.) cultivars to damaging diseases stimulated interest in the United States in the development of resistant bean cultivars through breeding. Early epidemics of bean diseases in certain sections of the country and the later discovery of physiological races or strains of many of the causal organisms proved the hopelessness of growing susceptible bean cultivars. None of the American bean cultivars used commercially before 1918 was disease resistant.

The first disease-resistant cultivar, Robust, was developed by F. A. Spragg of the Michigan Agricultural Experiment Station in 1915 (117). A white, pea-bean type, it resisted the common bean mosaic virus. The next resistant cultivar, Great Northern University of Idaho (UI) 1, developed in 1929 by Pierce and Hungerford of the Idaho Agricultural Experiment Station, was resistant to the same virus and was a selection from the mosaic-susceptible common Great Northern.

Wisconsin Refugee and Idaho Refugee introduced in 1934 were the first mosaic-resistant snap bean cultivars developed by hybridization (95). These were followed by the release of United States (US) 5 Refugee in 1935 (124). About 35 disease-resistant dry bean cultivars and 120 snap bean cultivars have been developed in the United States by seed companies, State Experiment Stations, and the US Department of Agriculture.

The greatest progress in breeding for disease resistance in beans has probably been the development of common mosaic-resistant cultivars of both dry and snap bean types. Unless a newly released cultivar is resistant to the type strain of the virus and to a widespread variant strain, generally known as the New York 15 strain, it is not well accepted by growers, the seed trade, and the bean processing industry.

313

IMPORTANCE OF THE CROP

In 1973, about 1.4 million acres of dry beans were harvested in the United States with a production of about 17 million 100-pound bags (122). In order of importance, Michigan, California, Colorado, Idaho, Nebraska, and New York account for almost 90% of the acreage of the United States production. Michigan alone produces about 34% of all dry beans in the country on 42% of the total acreage in beans (122). About 85,000 acres of snap beans were harvested for fresh market and about 292,000 acres for processing in 1973, this acreage producing 302 million pounds and about 1.5 billion pounds, respectively (122).

The farm value of dry beans in the United States in 1973 was approximately $514 million; of snap beans for fresh market, about $53 million; and of snap beans for processing, $77 million. Production of snap bean seed in 1973 was about 52.8 million pounds of the green-podded bush types, 3.8 million pounds of wax-podded bush types, and 2.9 million pounds of pole types. Bush type beans are mainly used for processing and the fresh market; most of the pole beans are used for fresh market and home gardens.

The dry bean is the main edible legume grown and consumed in Latin America and Mexico. In some of these countries it is the second most important crop grown; in many of them it is the principal source of protein in the diet. Brazil leads the world in acreage and total production of dry beans, with approximately 5.9 million acres producing about 53 million 100-pound bags annually. It is followed by Mexico and the United States with acreages of about 3 million and 1.4 million acres, which produce about 9.8 and 17 million 100-pound bags, respectively. The yield per acre for the United States is about 1178 pounds; for Brazil, 578 pounds, and for Mexico, 307 pounds. Argentina, Chile, and Peru have per acre yields of almost 30% less than those of the United States. To a great extent, diseases are responsible for much of these yield differences, and breeding for disease resistance to increase yields is now pursued in many of the Latin American countries.

DISEASE RESISTANCE

Virus Diseases

BEAN COMMON MOSAIC   Although once a major disease, bean common mosaic (BCM) is now almost nonexistent in the United States, because most varieties of both dry and snap beans developed since 1940 are resistant to known strains of the virus (114).

Resistance to BCM in most US snap bean cultivars is derived from Corbett Refugee (136), which was selected in 1929 from Stringless Green Refugee bean, a very popular processing cultivar at the time. Later studies have confirmed that this cultivar is resistant not only to the type strain of the virus, but also to at least nine other strains described in North and South America (64). The same type of resistance is found in Blue Lake type beans. In the United States, the resistance of

Corbett Refugee has not broken down in more than 45 years. However, strains of BCMV have been reported to which neither the dominant Corbett Refugee gene nor the recessive Robust gene confer resistance (92). However, these viruses have not yet been reported to occur in the United States. If and when they are found and become widespread, they could cause severe BCM epidemics.

Ali (3) used the approach-graft inoculation technique developed by Grogan & Walker (67) to distinguish between cultivars with the Corbett Refugee type of resistance and those with the Robust type of resistance. Cultivars in which Corbett Refugee was the resistant parent showed top necrosis, whereas the Robust-type cultivars remained healthy.

Results of mechanical inoculation of the progeny of crosses, Stringless Green Refugee (susceptible) X US 5 Refugee and Idaho Refugee (resistant) indicated that a single dominant gene controlled resistance of the Corbett Refugee type. However, in crosses between Stringless Green Refugee and Robust, he found the resistance of Robust to be controlled by a single recessive gene.

Crosses between Robust and Corbet Refugee gave an F_2 ratio of 13 resistant to 3 susceptible following mechanical inoculation. By use of the approach-graft inoculation the above ratio was broken down into nine necrotic, four healthy, and three mottled. These data suggested two pairs of genes acting with dominant and recessive epistasis. This type of inheritance was confirmed by Rudorf (104). Resistance in Corbett Refugee is based on a dominant inhibitor gene I epistatic to a dominant gene A required for virus infection. Corbett Refugee confers resistance to all strains of BCMV known in the United States and elsewhere and is an excellent example of Van der Plank's (123) horizontal resistance.

Many dry bean varieties derive their resistance from the recessive gene present in Robust. It is effective only against the type strain of BCMV but not effective against the so-called New York 15 strain. Great Northern UI 1 also carries this recessive gene (94), and both varieties, or the resistant varieties developed from them, have been widely used in the development of other mosaic-resistant dry bean types. Both Robust and Great Northern UI 1 have given virus protection for almost 50 years. Some of the most popular mosaic-resistant dry bean varieties are Sanilac, Seafarer, and Gratiot (all navy bean types); also Great Northern UI 59, Great Northern 1140, Pinto UI 111, Red Mexican UI 36, and Big Bend. According to M. J. Silbernagel (unpublished data), resistance depends upon a different pair of recessive alleles for each strain of the virus. Some of the Michigan navy bean cultivars obtained their resistance to the New York 15 virus from the Corbett Refugee source.

BEAN YELLOW MOSAIC   The distribution of bean yellow mosaic virus (BYMV) is found in most areas where beans are grown. The virus and its strains commonly infect *Melilotus alba Medic.*, *Trifolium pratense* L., *T. incarnatum* L., and *Gladiolus* sp. and are transmitted to beans by aphids from these infected hosts. The losses can be very severe if beans are grown close to infected plants of the above species or if viruliferous aphids infest them severely.

Resistance of beans to this virus appears to be conditioned by different genes. Dickson & Natti (57) reviewed the literature on the inheritance of resistance and reported single-factor dominant resistance derived from *Phaseolus coccineus* L. This species and some of its selections resist more strains of BYMV than any variety of *P. vulgaris* (13). Resistance derived from *P. coccineus* has also been reported by Baggett (12) and Baggett & Frazier (14) to be inherited recessively conditioned by two or three major genes with other modifiers affecting the variation in symptom expression. Buishand (28) reported that a single dominant gene from *P. coccineus* conferred resistance to top necrosis caused by a strain of BYMV. Some of the Great Northern bean cultivars such as Great Northern UI 16 and 31 resist one or more strains of the virus (16). Resistance to two severe pod-distorting strains of BYMV in Great Northern UI 31 was found to be governed by three major recessive genes with other modifiers (13). Provvidenti & Schroeder (99) found Great Northern 1140 resistant to 82 isolates of the severe strain of BYMV described by Thomas & Zaumeyer (120) and pea virus 2. They found that resistance in Great Northern 1140 is conditioned by a single recessive gene. Resistance to a pea isolate of BYMV in a Red Kidney X Black Turtle Soup cross was found to be controlled by a single dominant factor (107).

Another useful type of virus resistance in beans is insect nonpreference. Hagel et al (69) showed that aphid nonpreference in the Black Turtle Soup is related to field resistance to BYMV. This is an example of horizontal resistance (123), because it would be effective against all strains of the virus.

CURLY TOP   Curly top, a virus disease of beans and many other cultivated and wild plants (142), was first reported in 1926. The disease is transmitted by the sugar beet leafhopper, *Circulifer tenellus* Baker. The virus is prevalent only in certain arid sections of the United States where this insect thrives, for example, in the Columbia River Basin of eastern Washington, where in most years only bean cultivars immune to the disease survive.

Mackie & Esau (83) were the first to show the feasibility of breeding for resistance to curly top. They recovered resistant plants from segregating populations of crosses between resistant and susceptible cultivars. The United States Department of Agriculture and the University of Idaho have had bean-breeding programs for curly top resistance for over 40 years. These studies have been conducted in the desert regions of eastern Washington and southern Idaho where the sugar beet leafhopper is usually very abundant.

The virus cannot be transmitted mechanically. Segregating populations are exposed to infection with the virus when rows of curly top–susceptible sugar beets are planted about 20 feet apart in bean plots about a month before the beans are planted. The beets attract the viruliferous leafhoppers, which become a virus reservoir. The leafhopper population on the emerging bean seedlings is increased when the leaves are cut off of the beets, thus forcing the leafhoppers to feed on the beans, which are less desirable hosts. Infection during the seedling stage is preferable, because bean plants become more tolerant of the virus with age.

California Pink, Red Mexican, and Burtner Blightless (52) were the sources of resistance for the curly top–resistant bean cultivars grown at present. California Pink and Red Mexican were used to develop the following curly top-resistant dry beans: Big Bend and UI 36, Red Mexican types; Columbia, UI 111 and 114, Pinto cultivars; Royal Red, a dark Red Kidney type; and Great Northern UI 31.

The curly top resistance found in American snap bean cultivars, such as Apollo, Custer, Idelight, Idachief, Jackpot, Rodeo, Valgold, Goldcrop, and Wondergreen, was provided by Burtner Blightless (114). Schultz & Dean (108) reported that resistance was dominant and conditioned by two factors in dominant and recessive epistasis.

Even though a number of strains of the virus have been reported (121) and differences between varieties in degrees of susceptibility have been noted (113), resistance has remained stable in the field in the United States for more than 40 years. Resistance can, however, be broken down in the greenhouse under high inoculum levels and unfavorable host development.

In Australia, Ballantyne (17) found that 56 curly top–resistant bean cultivars in the United States were resistant to Australian summer death virus. This disease, similar to curly top, is caused by a yellow-type, phloem-restricted, leafhopper-transmitted virus. The leafhopper, *Orosius argentatus* (Evans), differs from the species that transmits curly top. Also, many of the curly top, summer death-resistant cultivars have a high degree of resistance to subterranean clover stunt virus, a yellow type, phloem-restricted aphid-transmitted disease (116).

SOUTHERN BEAN MOSAIC   Southern bean mosaic virus (SBMV) was isolated from mottled bean pods originating in Louisiana in 1941 (139). Little is known about the distribution of this virus in the field. Because it is highly infectious, researchers originally assumed that it would become a very serious disease of beans, but it has not. However, because symptoms of SBMV infection are similar to those produced by BCMV or BYMV, it may be more prevalent than is generally recognized. In 1948 it seriously damaged snap beans in southern Illinois. It was also reported from Tennessee and Georgia in 1952 (142). It is not seed borne and its very narrow host range may account for its relative unimportance as a bean-virus disease.

SBMV produces local lesions on some bean cultivars and systemic-mottle symptoms on others (139). No cultivar thus far tested has shown both types of symptoms. Local-lesion-susceptible cultivars can be considered commercially resistant. Most of the snap bean cultivars are susceptible to systemic infection, whereas most of the dry bean cultivars are resistant to it.

Zaumeyer & Harter (138) studied the inheritance of symptom expression to infection by SBMV in bean crosses with nine cultivars. Inheritance was governed by a single gene with local-lesion development dominant to systemic mottling.

POD MOTTLE   Little is known of the distribution of pod mottle virus (PMV). It was first isolated by Zaumeyer & Thomas (141) in 1945 from severely mottled bean pods of the Tendergreen cultivar grown at Charleston, South Carolina. In 1948 and

1949 it was isolated in combination with SBMV from beans grown in southern Illinois, and in 1950 it was found in Florida-grown beans.

Although unrelated to SBMV, PMV also produces local lesions on some bean varieties and systemic mottle symptoms on others. Varieties susceptible to local infection are immune from systemic infection, and those susceptible to systemic infection are immune from local infection.

As with SBMV infection, all cultivars susceptible to local infection can be considered commercially resistant.

Most green-podded snap bean and dry bean cultivars are susceptible to local infection; most of the wax-podded cultivars are susceptible to systemic infection. Practically all of the cultivars resistant to BCMV are also resistant to systemic infection by PMV. Reaction of most of the green-podded, bush, snap bean cultivars to PMV differs from that of SBMV (141).

Thomas & Zaumeyer (119) found that the inheritance of the expression of symptoms of PMV was governed by a single pair of factors. Plants with the dominant factor are susceptible to the local-lesion type of infection whereas those with the recessive factor are susceptible to the systemic mottle type infection.

Fungus Diseases

ANTHRACNOSE   Anthracnose caused by *Colletotrichum lindemuthianum* (Sacc. & Magn.) Briosi & Cavara does not occur in the Rocky Mountain states and in states farther west. Thus, bean seed grown in such regions is free of the disease. Before the use of western-grown seed, bean anthracnose was very serious in the humid areas of the United States (142). The widespread use of western-grown seed has practically eliminated the disease in this country, making it unnecessary to incorporate anthracnose resistance in domestic snap bean breeding programs. In Michigan, however, where the disease is found, and where much acreage of dry bean seed is grown for seed stock, anthracnose resistance has been incorporated into most of the recent navy bean releases.

In Europe, the environment in areas where bean seed is produced is ideal for the development and spread of the anthracnose organism. Hence, the incorporation of resistance to this disease is an important phase of every European bean-breeding program (65). General use of resistant cultivars resulting from these programs has reduced anthracnose to a minor problem. However, in Central and South America, anthracnose resistance is considered one of the primary breeding objectives (114).

Four races of the causal organism are distinguished by their differential pathogenicity to several host cultivars and are designated by alpha, beta, gamma, and delta. Barrus (23) was the first to demonstrate physiologic races of anthracnose—in fact, of any plant pathogen.

Hubbeling (76), Goth & Zaumeyer (66), Walker (126), and Yarnell (133) have reviewed the sources of resistance, differential cultivars, and the inheritance of resistance to bean anthracnose. Emerson No. 847, a bean strain developed at Cornell University but never released to the trade, resists the alpha, beta, and gamma races of the organism. Wells Red Kidney was reported to be resistant to the alpha and beta races (23). Goth & Zaumeyer (66) summarized the status of varietal resistance to the various races. Michelite 62 and Seaway are resistant to beta and gamma; Sanilac, Saginaw, Gratiot, Seafarer, and Manitou, a recent Red Kidney release from Michigan, resist all three races of the organism. They also noted that Plant Introduction (PI) 304110 resisted beta, gamma, and delta races but was susceptible to the alpha race. Cornell 49-242 resists all four races but has undesirable genetic linkages.

In 1918 Burkholder (33) studied anthracnose resistance. His work provided the first contribution on the inheritance of resistance to any bean disease. When segregating progenies of crosses between Wells Red Kidney, resistant to two physiological races of anthracnose (alpha and beta), and Perry Marrow, resistant to only one race (alpha), were inoculated in the F_2 generation with a single race of the organism, resistance was found to be governed by a single dominant gene. Later work by Bredemann (26) confirmed the dominance of resistance and established the presence of several genes for resistance.

Schreiber (105) supplemented these studies using 37 isolates of the organism which he divided into three main groups corresponding to alpha, beta, and gamma races. Reciprocal crosses between Dry Shell No. 22 and Konserva and between Dry Shell No. 22 and Wachs Best von Allen showed a 3:1 ratio with resistance dominant when inoculated in the F_2 generation with only one race of the organism. When progenies of the same crosses were inoculated with two races together, a 9:7 ratio of resistance to susceptibility was noted, indicating two complementary dominant factors. When inoculations were made with all 37 isolates together, a three-factor difference was indicated. If the races used for inoculation were selected from two of these groups, the F_2 hybrids always showed a 9:7 ratio, but if two strains were chosen from the same group, the ratio was always 3:1. Schreiber concluded that each of the three factors for resistance was on a different chromosome.

In 1933 Schreiber (106) reported the inoculation of the same crosses with a mixture of other physiologic races of the pathogen. From these studies he concluded that at least eight dominant genes were responsible for resistance.

Andrus & Wade (5) studied the inheritance of resistance to beta, gamma, and delta races. In crosses of resistant X tolerant and resistant X susceptible parents, resistance was always dominant. In crosses of tolerant X susceptible parents, susceptibility was dominant. Monohybrid and dihybrid ratios were obtained with all three races and trihybrid ratios with two races. A system of ten genes in three allelomorphic series with both duplicate and complementary genes for resistance, one dominant gene for susceptibility, and gene interactions at three points was proposed as the simplest Mendelian hypothesis that would coordinate all the data for the inheritance of reaction to beta and gamma races of anthracnose. Three independent genes were proposed for resistance to the delta race.

Resistance to alpha, gamma, and delta races found in *Phaseolus aborigineus* Burkart gave a 9:7 ratio and was controlled by two complementary dominant factors (103). Resistance to beta was inherited as a simple dominant factor. Hubbeling (77) proposed about eight dominant, allelic genes for resistance to all four races of the organism. Masterbroek (84) recently found a new dominant *ARE* gene in Cornell 49-242, which gives resistance to all four races of anthracnose.

Resistance to alpha, beta, and gamma races was studied by Cardenas-Ramos (35). In eight crosses, resistance to alpha was dominant. Crosses between resistant parents indicated that at least two genes were involved. In three of eight crosses, susceptibility to beta was dominant. Two hypotheses were proposed: (*a*) the existence of duplicate genes, each able to confer resistance and (*b*) complementary action of two genes conferring resistance. One of the duplicate genes is linked with one of the gamma-resistant genes. Thus, cultivars susceptible or resistant to both beta and gamma occur more often than cultivars resistant to beta and susceptible to gamma, or vice versa. In the crosses of gamma-resistant with gamma-susceptible varieties, resistance was always dominant, but F_2 ratios indicated two duplicate genes or two complementary genes, as with alpha.

As many as ten different sites in the chromosomes can govern resistance or susceptibility of the plant to alpha, beta, and gamma races (37). The accumulation of favorable genes at all ten sites (loci) would be a difficult task. No known variety has this combination. Certain PI strains have the single gene *ARE*, which confers resistance to all the races simultaneously.

Resistance to anthracnose in most European cultivars is controlled by the single dominant *ARE* gene derived from Cornell 49-242, which confers resistance to all known European races (82, 84). Mexico 222 and Mexico 227 also are resistant to the European races. These likewise are controlled by a single dominant gene, which differs from the *ARE* gene (20).

The races of anthracnose found in Australia differ from those reported in Europe and the United States. Cornell 49-242 resists all of the Australian isolates collected during the past ten years (B. Ballantyne, unpublished data).

Oliari et al (91) reported the identification of seven physiologic races of *C. lindemuthianum* on *P. vulgaris* in Viscosa, Brazil, and neighboring locations in the state of Minas Gerais. Races BA-1 and BA-2 belong to the alpha group, race BA-3 to the Brazilian group II, races BA-4 and BA-5 to the Brazilian group V, and races BA-6 and BA-7 to the Mexican group II. *Phaseolus aborigineus* 583 and Costa Rica 1031 were used to subdivide the races within groups. Cornell 49-242 was resistant to all seven races.

In Uganda, Leakey (81) reported that the *ARE* gene from a number of French bean accessions appeared to confer virtual immunity to all isolates of *C. lindemuthianum*, except to two races that produced slight anthracnose lesions, but with negligible sporulation. French accessions were used in preference to Cornell 49-242, in which the linkages are very undesirable. The most promising of the French cultivars are Confinel, Peonel, and Verdon.

RUST   *Uromyces phaseoli* (Reben) Wint. has been reported from almost every part of the world (142). Before 1945 it was one of the principal diseases of dry beans in the irrigated areas of Colorado, Western Nebraska, Wyoming, and Montana, but in recent years it has been of minor concern in these areas. It is often important in fall-grown snap bean crops along the Atlantic seaboard and in late fall and winter bean plantings in Florida. Recently, it has been serious in Arkansas and Tennessee. In Central and South America it is considered to be the most important bean disease.

It is of concern also when some of the European cultivars are grown in Africa and other tropical seed-producing areas. In such areas, beans are grown for many months of the year, allowing for a buildup of inoculum.

No varieties thus far have been shown to resist all of the reported races of rust. About 35 races have been described (142) in the United States, and unquestionably many more exist. Fifteen other races of the organism were reported from Mexico by Crispin & Dongo (51). Ballantyne (19) lists 95 additional races from Mexico, Latin America, East Africa, Australia, New Zealand, and the Netherlands, but how many of these duplicate previously described races is unknown.

US Pinto 5 and 14 cultivars (140) resisted more US rust races than any other cultivars at the time of their release in 1946, but are not grown commercially. Quinones (101) developed Luna, a new rust-resistant Pinto bean, for New Mexico. Wingard (131) described ten varieties of pole beans that were resistant to the races prevalent in Virginia. Seminole (132), Florigreen (129), and Dade (38), beans developed in Florida, have shown resistance to races prevalent there. Recently a new race that attacks Florigreen and Dade has appeared (86). Kantzes & Hollis (79) reported that Extender, Wade, Tenderwhite, and Harvester cultivars have a commercial level of tolerance to a new race found in Maryland. In recent field tests in Maryland, the following snap beans are highly resistant: Bush Blue Lake 290, Custer, Mountaineer White Half Runner, and Oregon 1604 (88). Augustin, Coyne & Schuster (10) inoculated 25 varieties with Brazilian rust race B11 and found that only Great Northern 1140 (143) and Kentucky Wonder 765 showed a high degree of resistance.

An earlier report indicated that Westralia was resistant to all races of rust in Australia (7). More recently, reactions of 158 bean lines to natural field infection of rust were assessed by Ballantyne (18, 19) in Australia. She found that many of the green-podded bush and the red kidney types were only slightly infected; this suggested that some of these types showed nonspecific resistance to races of rust. The pole beans and most dry bean types showed either a high level of specific resistance or severe infection; there was no evidence of nonspecific resistance in these groups.

In Mexico, immunity from all but one of the 15 rust races reported there by Crispin & Dongo (51) has been found in Guerrero 6, Guanajuato 10A5, Vera Cruz 10, and Negro 150. N. Vakili in Puerto Rico (unpublished data) has selected 22 lines of dry beans possessing resistance to many races in the western hemisphere.

Meiners (unpublished data) found the following varieties highly resistant to three isolates from Maryland: PI 165426 and 152326, Venezuela 54, Villa Gro, Puerto Rico (PR) 15-R-52, PR 15-R-55, and PR 15-R-57, Aurora, and Cornell 49-242.

Wingard's (130) studies on the inheritance of rust resistance in 1933 before the discovery of a large number of physiologic races of the organism showed resistance to be dependent on a single dominant factor. It is assumed that he worked with only one race.

Zaumeyer & Harter (137) extended the study of inheritance of resistance to six races of rust involving four different crosses of six bean cultivars. Their results showed that resistance to races 1 and 2 in the hybrids was governed by a single factor but that more than one factor was involved in the resistance to races 6, 11, 12, and

17. Resistance was dominant in the hybrids inoculated with races 1, 2, 6, and 12 and incompletely dominant in those inoculated with races 11 and 17. Transgressive segregation was indicated in the progenies inoculated with race 11, since one fourth of the F_2 plants exhibited more resistance than the tolerant parent.

In New Zealand, resistance to US races 10, 17, and 28 was dominant in the F_1 seedlings, but the plants were susceptible when older (134).

Augustin et al (10), in studies with Brazilian race B11, found that in crosses between Great Northern 1140 and four susceptible lines a major gene controlled disease reaction, with resistance being dominant.

Resistance to rust is based on the size of the leaf lesions. At least three systems have been used (51, 53, 70), with ratings from 0 to 5, 1 to 5, and 0 to 10. The lower number denotes immunity, and the higher number extreme susceptibility; intermediate grades denote degrees of resistance. In the 0 to 10 scale, pustule size from 0 to 5 denotes field resistance, whereas larger sizes denote susceptibility (70).

ROOT ROTS   Root rots of bean have long been a problem wherever beans are grown. In the principal bean-growing regions of the United States, fusarium root rot is the most serious, followed by rhizoctonia, pythium, and thielaviopsis root rots. More emphasis has been given to breeding for resistance to fusarium root rot; however, the others can be very destructive when conditions favor the development of the respective organisms.

Fusarium root rot   Breeding for resistance to the root rot caused by *Fusarium solani* (Mart.) Appel & Wr. f. sp. *phaseoli* (Burk.) Snyd. & Hans. has been one of the most difficult problems to solve in the history of the crop. The importance of the disease was first recognized by Burkholder (34) in 1916 in western New York. There as many as 90% of the plants in several counties were infected, with corresponding losses.

Burke (29–32) reported that many cultural and environmental factors affect the severity of the disease. Many attempts have been made to control the disease by the use of organic soil amendments and chemical treatments, but none has succeeded.

Breeding for disease resistance has been hampered by a lack of a high resistance in parental material. Disease tolerance, however, has been found in PI 203958 collected in Mexico by Norvell and in *P. coccineus* (14, 128). Smith & Houston (115) also reported it in PI 165426 and 165435.

Although several bean breeders have measurable root-rot tolerance in advanced breeding lines, the first commercial cultivars resistant to fusarium root rot have recently been released by the US Department of Agriculture and the Washington Agricultural Experiment Station. Deriving their root rot resistance from PI 203958, these include a Red Mexican type, Rufus, and three Pink cultivars, Roza, Viva, and Gloria. The last was released also by the California Agricultural Experiment Station. Besides their root-rot resistance, they are resistant to curly top and to the type and New York 15 strains of BCMV. The cultivars are less resistant to root rot than PI 203958, but more resistant than any known commercial variety. Tests have shown that Rufus yields 15–30% more seed than any other Red Mexican variety in fusarium-infested soil, and the resistant Pink cultivars also are superior in yield under exposure to fusarium root rot.

McRostie (87) was the first to report that resistance to fusarium root rot was recessive to susceptibility and that two factors were involved in the inheritance. In crosses between *P. vulgaris* and *P. coccineus,* Azzam (11) showed that tolerance was recessive and probably controlled by three major genes or two genes with modifying factors. His results indicated no relation between vigor of the root and resistance.

Bravo, Wallace & Wilkinson (25) concluded that resistance to fusarium root rot, whether derived from PI 203958 or scarlet runner, is completely dominant. Additive gene effects are larger than dominant gene effects. Estimates of the number of genes controlling resistance ranged from three to seven, and the effects of individual genes could not be distinguished.

Testing in a greenhouse, Hassan et al (72) found that PI 203958 had four genes for resistance and that breeding line 2114-12, which derived its resistance from *P. coccineus,* had five to six genes for resistance. They concluded that four of the genes from *P. coccineus* are the same as those in PI 203958, and that gene action is mostly additive but that partial dominance of resistance appears in 9- to 13-week-old field-tested plants. Broad-sense heritability was estimated as 62–64% in the greenhouse and as 22% and 79%, respectively, in 5- and 9- to 13-week-old field-tested plants.

Some investigators believe that factors, such as the ability of the seed to germinate in the cold, the ability to develop a large, vigorous root system, and the presence of inhibitory substances in the seed coat and hypocotyls (118) may increase the level of genetic tolerance.

Rhizoctonia root rot   Several PI lines have been reported to resist rhizoctonia root rot caused by *Rhizoctonia solani* Kühn; namely, PI 165426, 165435, 109859, 163583, 174908, 226895, and Venezuela 54 (97, 98). Venezuela 54 and PI 165426 were reported to have the highest degree of resistance (97). Yerkes & Freytag (135) believed that resistance to *Rhizoctonia* in the scarlet runner bean was superior to that found in *P. vulgaris.*

McLean et al (85) reported 12 bean breeding and PI lines among 600 tested as showing some resistance to *R. solani* in artificially infested greenhouse soil. The resistant lines were as follows: PI 165426, 165435, 181954, 318696, 318697, 318699, 318700 and breeding lines B3866, Venezuela 54, and 165426 × Alabama 1. No commercial rhizoctonia root rot–resistant varieties have yet been released whose resistance has been derived from the above sources. However, J. R. Deakin of the U.S. Vegetable Breeding Laboratory, Charleston, South Carolina, released a highly tolerant wax bean cultivar named Goldcoast in 1970 and two breeding lines, B3088 and B3787 (unpublished data).

Deakin & Dukes (54) reported that resistance to *R. solani* is highly heritable, although the precise mode of inheritance is unknown. They found also that resistance is associated with colored seed, and they were unable to obtain resistant white-seeded lines because of epistatic effects.

Black root rot   Resistance to black root rot caused by *Thielaviopsis basicola* (Berk. & Br.) Ferr. was reported by Hassan et al (73) in PI 203958 and a New York State breeding line 2114-12. Both had the same genes for resistance to the organism. Resistance was found to be partially recessive and controlled by about three genes (65). Also, the genes controlling resistance to black root rot differed from those responsible for fusarium root rot resistance and were not linked (74).

PYTHIUM BLIGHT   Resistance to pythium blight caused by five species of *Pythium* was recently found in PI 203958 by Adegbola & Hagedorn (2). They noted it also in Bush Green Pod, a Blue Lake type snap bean.

Kim & Kantzes (80) tested 138 cultivars and lines of *Phaseolus vulgaris* for resistance to *Pythium aphanidermatum* (Edson) Fitzpatrick and found PI 201389 somewhat resistant. They noted PI 203958, previously reported as resistant, susceptible. PI 164893 and 234258 of *Phaseolus lunatus* L., and PI 180466 and 288600 of *Vigna mungo* L. were more resistant than were lines and cultivars of *P. vulgaris.*

Dickson & Abawi (55) found a white-seeded snap bean breeding line 1273 resistant to seed decay and damping-off caused by *Pythium ultimum* Trow in artificially infested soil under growth chamber conditions. Resistance was derived from PI 203958. The results indicated that the association of dark seed-coat color and *Pythium* resistance can be broken.

ANGULAR LEAF SPOT   Angular leaf spot caused by *Isariopsis griseola* Sacc. is of minor importance in the United States, but is an important disease in tropical and subtropical regions. In tests conducted in Colombia, Olave (90) found that the most resistant cultivars were Mexico 11, Mexico 12, and Cauca 27a. In Australia, Brock (27) tested 164 cultivars and found 19 resistant and 11 highly resistant. The latter included Alabama No. 1, Cafe, California Small White, Epicure, Mexico Black, McCaslan, Negro Costa Rica, Scotia, Rojo, Chico, and Case Knife. In Spain, Puerta & Alonso (100) reported the cultivars Boriole and San Fiacre to be resistant.

Cardona-Alvarez (36), in Colombia, reported resistance to be controlled by a single dominant gene. Barros, Cardenosa & Skiles (22) reported resistance to be recessive in some crosses and dominant in others. Resistance of cultivars Decal, Maravilla, and Huila 14 is attributed to three recessive genes (89).

POWDERY MILDEW   Studies by Dundas (58) showed that Pinto, Hungarian, and Pink were resistant to powdery mildew caused by *Erysiphe polygoni* DC. ex St.-Amans. Also, Alabama No. 1, Contender, Logan, Tenderlong 15, Idaho Refugee, Sensation Refugee No. 1066 and 1067, US 5 Refugee, Topcrop, and Wade have been reported to be resistant to one or more races of the fungus (142). Dundas (59–61) found that the inheritance of resistance in several cultivars of field and snap beans to 12 of 14 physiologic races of mildew was controlled by a single dominant factor for resistance. He noted also one dominant factor for tolerance and one for susceptibility during 5 to 7 days after emergence (61). These studies were in part based on the dish-culture method, in which diseased leaflets were floated on a 10% sucrose solution in Petri dishes and inoculated with spores of the powdery mildew fungus.

WHITE MOLD   In a greenhouse screening procedure for resistance to *Sclerotinia sclerotiorum* (Lib.) de By., Adams et al (1) found nine *Phaseolus coccineus* cultivars (Scarlet and White runner types) and *P. vulgaris* lines PI 203958 and 300659, and cultivars Soldier and Steuben Yelloweye tolerant to the organism. In a study of 20 bean cultivars and breeding lines, Anderson et al (4) reported that Black Turtle Soup, Sanilac, Capitol, Aurora, and New York (NY) 6207-2 were tolerant to white mold in western Nebraska in 1973. Whether the tolerance observed in these cultivars resulted from certain physiological or morphological characteristics was not shown.

Nematode Diseases: Root Knot

Barrons (21) studied the inheritance of resistance to root knot caused by *Meloidogyne incognita* (Kofoid & White, 1919) Chitwood, 1949 in a cross between resistant Alabama No. 1 (78) variety and susceptible Kentucky Wonder. He found the inheritance of resistance governed by two recessive genes. Barrons believed that the inheritance is quantitative: all individuals with two or more dominant genes appeared susceptible to root knot and those with one dominant gene appeared intermediate. Later work by Blazey et al (24), using Contender and Cherokee Wax as susceptible parents and Wingard Wonder and Springwater Half Runner as resistant parents, produced a resistant F_1. The F_2 segregated 1 resistant to 15 susceptible plants. They found that bean cultivars as well as *P. coccineus* resistant to *M. incognita* were susceptible to four other species of *Meloidogyne* (24). Certain of these results are considered to be consistent with those of Barrons. Fassuliotis et al (62) reported that PI 165426 and 165435 resisted the root knot nematode.

In crosses between resistant Alabama No. 1 and susceptible Hawaiian Wonder, Hartmann (71) reported that the two-gene hypothesis for resistance reported earlier did not account for the segregation patterns of the F_3 families. Instead, he explained it with a three-gene hypothesis. He found that three pairs of genes equal in action are needed for resistance, but a minimum number of genes for susceptibility is necessary before all resistance is lost.

Bacterial Diseases

HALO BLIGHT   The epidemics of halo blight caused by *Pseudomonas phaseolicola* (Burkh.) Dows. from 1963 to 1970 in the snap bean seed–growing areas of southern Idaho stimulated much interest in research on the control of bacterial diseases of bean (63).

Many dry bean cultivars are highly resistant to halo blight infection in the field (47, 142). In the greenhouse, small necrotic spots develop on inoculated leaves of such cultivars, with no systemic spread of the organism (142). Most United States snap bean cultivars are very susceptible. However, some Bush Blue Lake types show some tolerance.

European bean breeders have devoted more attention to the devleopment of halo blight–resistant snap beans than have breeders in the United States. Many French, German, and Dutch cultivars have much resistance to halo blight. for example,

Chicobel and Colana were found resistant with a rating of 1 on a 1 to 9 scale with 1 being resistant and 9 susceptible. Nine varieties were rated 8, 2 as 7, 14 as 6, and 4 as 4 (9).

In Australia, Richmond Wonder, Clarendon Wonder, Hawksbury Wonder (112), and Windsor Longpod (6) are resistant cultivars developed by hybridization. Sources of resistance include Cornell 49–242, Great Northern UI 59, Great Northern 1140, Great Northern Nebraska No. 1, Scout Pinto, Pinto UI 111, Red Mexican UI 34, Red Kote, California Small White varieties Ferry Morse (FM) No. 51, FM No. 59, Sanilac, Seafarer, Bush Blue Lake, Oregon State University (OSU) 190, OSU 10183, and Oregon 1604, PI 181954, 203958, and 150414 (8, 56, 63, 102, 142). Coyne et al (47) and Baggett & Frazier (15) summarized the sources of resistance in beans and their differential reaction to races 1 and 2 of the halo blight organism.

The general pattern of inheritance of resistance to halo blight is not well understood. In South Australia (125) two genes were found to govern resistance in beans to halo blight. Disease-resistant selections were isolated from a Canadian Wonder × Burnley Selection cross. Schuster (109) crossed the resistant varieties Red Mexican and Ankara Yellow to the susceptible US 5 Refugee and reported resistance due to a single major recessive gene. In the cross Red Mexican with susceptible Asgrow Stringless Green Pod two recessive genes were involved. Using a mixed inoculum of races 1 and 2, Dickson & Natti (56) found resistance in PI 181954 to be due to one or two recessive genes with modifiers that can increase resistant levels. Walker & Patel (127) reported hypersensitive resistance against race 1 in Red Mexican UI 3 to be controlled by one dominant gene whereas tolerance against race 1 and 2 in a selection from PI 150414 was conditioned by one recessive gene when crossed with Tenderwhite (93). Coyne et al (50) reported that tolerance in Great Northern Nebraska No. 1 selection 27 to halo blight race 1 was conditioned mainly by a major dominant gene when crossed with white-seeded Tendergreen. Coyne et al (48) reported that systemic chlorosis of the trifoliolate leaves and watersoaking of the primary leaves caused by race 2 of *Pseudomonas phaseolicola* was heritable. In the cross Gallatin 50 (susceptible to both reactions) × Great Northern Nebraska No. 1 selection 27 (tolerant to both reactions), both reactions were controlled mainly by a single dominant gene. In the cross Gallatin 50 × PI 150414 (tolerant to both reactions), both reactions were controlled mainly by a single recessive gene. In the cross Pinto UI 111 (susceptible only to watersoaked lesions) × Dark Red Kidney (susceptible to both reactions), the systemic chlorosis reaction was governed by duplicate recessive genes. Coupling linkage was detected between the genes controlling each of the reactions.

Hill et al (75) reported that the reaction to pod infection by race 1 of the halo blight organism was controlled by another gene, independent of the genes controlling the two leaf reactions. A fourth major gene independent of the genes described above was found to control the wilting reaction of halo blight–inoculated leaves. This genetic analysis of halo blight reaction shows the importance of selecting bean plants with pods that are tolerant to infection and nonsystemic trifoliolate leaf infection.

COMMON BLIGHT   Unlike halo blight, very few commercial cultivars of *Phaseolus vulgaris* are tolerant to common blight caused by *Xanthomonas phaseoli* (E. F. Sm.) Dows. and fuscous blight caused by *X. phaseoli* var. *fuscans* (Burkh.) Starr and Burkh. Coyne et al (45) tested 1080 PI accessions, cultivars, and breeding lines of *P. vulgaris* for reaction to *X. phaseoli*. None was found free of symptoms after heavy inoculation, and 28 showed slight symptoms. Two accessions of tepary bean, *P. acutifolius* var. *latifolius* Freeman, showed no infection when inoculated in the field. Using Great Northern Nebraska No. 1 selection 27 as a source of resistance to common fuscous blight, Coyne & Schuster (40, 41) developed the Great Northern varieties Tara and Jules. Tara has moderate tolerance and Jules high tolerance to both organisms. Schuster et al (110, 111) reported the strains of *X. phaseoli* to which Great Northern Nebraska No. 1 selection 27 and some other recently discovered sources of resistance (44) were susceptible. PI 207262 gave a tolerant reaction with all isolates they used.

In crosses between Great Northern 1140 and Great Northern Nebraska No. 1 selection 27, Coyne et al (43, 49) found that the latter contributed several genes for tolerance to the hybrid. In advanced self-pollinated or backcross generations, the pattern of segregation suggested polygenic inheritance for resistance.

Pompeu & Crowder (96) reported in crosses between two bean lines resistant to common blight and two susceptible cultivars (Red Kidney and Black Turtle Soup) that resistance was conditioned by several partially dominant genes. This character was quantitative and highly heritable. Transgressive segregation was observed in all of the crosses studied.

BACTERIAL WILT   Although not widespread in the dry bean-growing areas of the United States, bacterial wilt caused by *Corynebacterium flaccumfaciens* (Hedges) Dows. occasionally causes serious losses in western Nebraska. No commercial variety of *P. vulgaris* is resistant to the wilt bacterium. A number of PI lines have shown much tolerance, including PI 165078, 165422, 167399, and 169727. Coyne & Schuster (42) developed the variety Emerson, a great Northern type and the first cultivar tolerant to the bacterial wilt disease.

In the cross Great Northern 1140 X PI 165078 from Turkey, Coyne et al (46) determined that resistance to the organism was inherited quantitatively. In crosses with PI 165078 X Great Northern Nebraska No. 1 selection 27, susceptibility appeared to be conditioned by two complementary dominant genes, with absence of either gene or both genes resulting in tolerance.

BROWN SPOT   Similar to bacterial wilt, brown spot caused by *Pseudomonas syringae* van Hall is generally of minor importance to bean growers, although serious outbreaks of the disease occurred in the mid-1960s in snap beans grown in Wisconsin. Resistance to the disease was reported by Coyne & Schuster (39) in Great Northern 1140 and Tempo. Truegreen was found to be tolerant to one strain of the organism, but susceptible to another. In Wisconsin, Hagedorn et al (68) in field tests reported Earliwax, Michelite, Processor, Puregold, Sanilac, Saginaw, Tempo, Truegreen, and ten PI lines as tolerant to the disease. Greenhouse studies did not substantiate their field results. Inheritance of resistance to this organism has not been studied.

SOURCES OF GERM PLASM

The National Seed Storage Laboratory at Fort Collins, Colorado, keeps stocks of many commercial bean varieties and many older genetic stocks. The US Department of Agriculture Plant Introduction Station at Pullman, Washington, keeps world bean collections of over 8000 items as well as several thousand accessions of related *Phaseolus* species, some of which can be hybridized with *P. vulgaris*. The Norvell Bean Collection of several thousand accessions recently has been purchased by the US Department of Agriculture and will be added to the bean collections at Pullman. This collection consists mostly of bean introductions from Latin America, including wild species and cultivars from all parts of the world. A large segment of the collection is from centers of origin of *Phaseolus* spp. in Mexico. Centro Internacional de Agricultura Tropical (CIAT) at Cali, Colombia, also maintains a world collection of *P. vulgaris* and related species.

GENETIC VULNERABILITY

Genetic vulnerability in snap beans was reported by the Committee on Genetic Vulnerability of Major Crops of the National Research Council (37). In 1970, of 35.5 million pounds of seed of green-podded bush types produced, about 46% had Tendercrop germ plasm in their ancestry, 15% had Blue Lake germ plasm, and about 15% had Harvester germ plasm. Thus, about 76% of the total seed produced of all green-podded bush varieties had germ plasm from only three major sources. The wax-podded bush varieties had more genetic diversity, although the cultivar Earliwax in this class comprised about 20% of the total seed produced.

The production of 80% of the United States' supply of bush snap bean seed in a relatively small area of southern Idaho is dangerous. Potentially the situation is very vulnerable for the users of this seed throughout the United States (37).

In dry beans, according to the Committee (37), genetic vulnerability could likewise be a problem. The most outstanding example is in Michigan where about 500,000 acres are planted annually to four closely related navy bean cultivars derived from the variety Michelite.

The pinto bean rivals the navy bean in total production. The two major pinto cultivars produced in the United States are Pinto UI No. 111 and 114, which are closely related. That about half of the total acreage is in Colorado makes this class of bean potentially vulnerable. However, the pinto bean is grown also in several western states reducing the disease hazard somewhat.

CHALLENGES FOR THE FUTURE

Most of the United States' supply of bush snap bean seed are produced in a relatively small area of southern Idaho. Also the Columbia Basin of central Washington is ideal for bacterial blight- and anthracnose-free seed to grow but the area is subject to serious curly top infection. Therefore, greater effort should be made to develop more curly top–resistant cultivars for seed production in this area. Thus, more of the southern Idaho acreage could be diverted to the Columbia Basin, where thousands of acres are available, to ensure a dependable supply of disease-free bean seed.

Fusarium root rot causes about 6% loss to beans in the United States, if crop rotation is not rigidly adhered to. Although several tolerant dry bean cultivars have recently been released, efforts should be made to incorporate similar resistance into snap bean cultivars.

In some of the Central and South American countries and in parts of Africa, anthracnose, rust, and angular leaf spot cause heavy losses to bean crops. Parental material that resists tne four important races of the anthracnose organism is available, such as Cornell 49–242, Mexico 222, and Mexico 227. Breeding should be begun to incorporate the resistant genes into the important cultivars grown in these countries.

Rust is possibly the most important bean disease in many of the Latin countries, and a number of resistant lines have been developed in several countries. This work should be expanded. When lines or cultivars are found that have a wide base of resistance, efforts should be made to develop, through breeding, resistant cultivars that will be acceptable in many of the countries. To facilitate such breeding for resistance, an International Bean Rust Nursery, similar to the International Cereal Rust Nurseries, has been organized recently. It will be coordinated by CIAT, Palmira, Colombia.

In many tropical areas of Latin America, angular leaf spot often seriously reduces yield during the rainy seasons. Several highly resistant local varieties (27R and Antioquia) were observed several years ago in El Salvador (unpublished data). Others such as Mexico 11, Mexico 12, and Cauca 27a have been reported by Olave (90) from Colombia. Efforts should be made to develop resistant varieties for the areas where the disease is a problem.

Other sources of resistance to common blight and especially fuscous blight are urgently needed. The fuscous blight organism constitutes about 50% or more of the isolates in Michigan (37). The tepary bean is a potential source of resistance to fuscous blight, but the specific barrier to the transfer of genes to *P. vulgaris* remains nearly intact, since a hybrid was obtained from only one of thousands of crosses attempted.

Because many dry bean cultivars resist the halo blight organism, breeders should make every effort to develop resistant snap bean cultivars.

Vertical or specific genes for resistance have been stable for most bean diseases for many years. Nevertheless, breeders should consider using newer methods of breeding to incorporate or retain so-called horizontal or nonspecific resistance to bean diseases. Combination of horizontal resistance with major genes could provide more lasting types of genetic control.

The genetics of resistance to diseases in beans has not been elucidated for many disease-host interactions. These should be studied vigorously, to provide a scientific basis for the practical breeding research.

In recent years, the nature of resistance in crop plants, including beans, has been emphasized. Much information is available on the development of phytoalexins, such as phaseolin and other chemical constituents, that influence disease resistance or susceptibility in bean plants. Such research is aimed, at least in part, toward the development of chemical tests for resistance. The tests would be of value in screening germ plasm and breeding materials for disease resistance. To date, the authors are not aware that such methods are being applied to practical breeding problems, but a certain amount of effort should continue on basic studies that could develop such tests.

Literature Cited

1. Adams, P. B., Tate, C. J., Lumsden, R. D., Meiners, J. P. 1973. Resistance of *Phaseolus* species to *Sclerotinia sclerotiorum*. *Bean Impr. Coop. Ann. Rept.* 16:8–9

2. Adegbola, M. O. K., Hagedorn, D. J. 1970. Host resistance and pathogen virulence in *Pythium* blight of bean. *Phytopathology* 60:1477–79

3. Ali, M. A. 1950. Genetics of resistance to the common bean mosaic virus (bean virus 1) in the bean (*Phaseolus vulgaris*). *Phytopathology* 40:69–79

4. Anderson, F. N., Steadman, J. R., Coyne, D. P., Swartz, H. F. 1973. Tolerance to white mold in *Phaseolus vulgaris* dry edible bean types. *Plant Dis. Reptr.* 58:782–84

5. Andrus, C. F., Wade, B. L. 1942. The factorial interpretation of anthracnose resistance in beans. *US Dep. Agric. Tech. Bull.* 810. 29 pp.

6. Anonymous. 1948. Windsor Longpod—a new French bean. *Agric. Gaz. NSW* 59:468

7. Anonymous. 1954. Plant diseases. Bean rust. *J. Dep. Agric. West. Aust.* 3:439–40

8. Anonymous. 1973. Bush pod bean 1604. *Seed World* 111(5):30

9. Anonymous. 1973. 25 Ste rassenlyst voor groentegewassen, pp. 46–47. Wageningen: Inst. Vered. Tuinbouwgewassen. 219 pp.

10. Augustin, E., Coyne, D. P., Schuster, M. L. 1972. Inheritance of resistance in *Phaseolus vulgaris* to *Uromyces phaseoli typica* Brazilian rust race B 11 and of plant habit. *J. Am. Soc. Hortic. Sci.* 96:526–29

11. Azzam, H. A. 1958. Inheritance of resistance to *Fusarium* root rot in *Phaseolus vulgaris* L., and *Phaseolus coccineus* L. *Diss. Abstr.* 18:32–33

12. Baggett, J. R. 1956. The inheritance of resistance to strains of bean yellow mosaic virus in the interspecific cross *Phaseolus vulgaris* × *P. coccineus*. *Plant Dis. Reptr.* 40:702–7

13. Baggett, J. R., Frazier, W. A. 1957. The inheritance of resistance to bean yellow mosaic virus in *Phaseolus vulgaris*. *J. Am. Soc. Hortic. Sci.* 70:325–33

14. Baggett, J. R., Frazier, W. A. 1959. Disease resistance in the runner bean, *Phaseolus coccineus* L. *Plant Dis. Reptr.* 43:137–43

15. Baggett, J. R., Frazier, W. A. 1967. Sources of resistance to halo blight in *Phaseolus vulgaris*. *Plant Dis. Reptr.* 51:661–65

16. Baggett, J. R., Frazier, W. A., McWhorter, F. P. 1966. Sources of virus resistance in beans. *Plant Dis. Reptr.* 50:532–36

17. Ballantyne, B. 1970. Field reactions of bean varieties to summer death in 1970. *Plant Dis. Reptr.* 54:903–5

18. Ballantyne, B. 1972. Reaction of bean (*Phaseolus vulgaris* L.) varieties to rust (*Uromyces appendiculatus* (Pers.) Unger) in New South Wales. *Hort-Science* 7:345 (Abstr.)

19. Ballantyne, B. 1974. Resistance to rust (*Uromyces appendiculatus*) in beans (*Phaseolus vulgaris*). *Proc. Linn. Soc. NSW* 98:107–21

20. Bannerot, H., Derieux, M., Fouilloux, G. 1971. Mise en evidence d'un second gene de resistance totale a l'anthracnose chez le haricot. *Ann. Amelior. Plantes* 21:1:83–85

21. Barrons, K. C. 1940. Root-knot resistance in beans. *J. Hered.* 31:35–38

22. Barros, O., Cardenosa, R., Skiles, R. L. 1957. The severity and control of angular leaf spot of beans in Colombia. *Phytopathology* 47:3 (Abstr.)

23. Barrus, M. F. 1911. Variation of varieties of beans in their susceptibility to anthracnose. *Phytopathology* 1:190–95

24. Blazey, D. A., Smith, P. G., Gentile, A. G., Miyagawa, S. T. 1964. Nematode resistance in the common bean. *J. Hered.* 55:20–22

25. Bravo, A., Wallace, D. H., Wilkinson, R. E. 1969. Inheritance of resistance to

Fusarium root rot of beans. *Phytopathology* 59:1930–33

26. Bredemann, G., ten Doornkaat-Koolman, H. 1927. Zur immunitätszüchtung bei *Phaseolus vulgaris* gegenüber *Colletotrichum lindemuthianum* und seinen Biotypen. *Z. Pflanzenzuecht.* 12: 209–17

27. Brock, R. D. 1951. Resistance to angular leaf spot among varieties of beans. *J. Aust. Inst. Agric. Sci.* 17:25–30

28. Buishand, T. 1955. Eenige ervaringen met het veredelen van Bonen (I) *Phaseolus* (spp.) A. Het Kruisen van Bonen. B. Ervaringen met het. veredelen tot de F₂ generatie. *Medel Proefstn. Groenteteelt* 1. 48 pp.

29. Burke, D. W. 1965. Plant spacing and Fusarium root rot of beans. *Phytopathology* 55:757–59

30. Burke, D. W. 1965. Fusarium root rot of beans and behavior of the pathogen in different soils. *Phytopathology* 55: 1122–26

31. Burke, D. W. 1966. Predisposition of bean plants to Fusarium root rot. *Phytopathology* 56:872 (Abstr.)

32. Burke, D. W., Miller, D. E., Holmes, L. D., Barker, A. W. 1972. Counteracting bean root rot by loosening the soil. *Phytopathology* 62:306–9

33. Burkholder, W. H. 1918. The production of an anthracnose-resistant white marrow bean. *Phytopathology* 8:353–59

34. Burkholder, W. H. 1919. The dry root rot of the bean. *NY Agric. Exp. Stn. Ithaca Mem.* 26:998–1033

35. Cardenas-Ramos, F. A. 1961. Genetic systems for reaction of field beans (*Phaseolus vulgaris* L.) to infection by three races of *Colletotrichum lindemuthianum*. *Diss. Abstr.* 22:1325

36. Cardona-Alvarez, C. 1956. Herencia de la resistencia a la mancha angular en frijol. *Agric. Trop.* 18:330–31

37. Committee on Genetic Vulnerability of Major Crops. 1972. In *Genetic Vulnerability of Major Crops.* Washington DC: NAS. 307 pp.

38. Conover, R. A., Walter, J. M., Lorz, A. P. 1962. Dade, a rust-resistant pole bean for fresh market. *Fla. Agric. Exp. Stn. Circ.* S-142. 7 pp.

39. Coyne, D. P., Schuster, M. L. 1969. Moderate tolerance of bean varieties to brown spot bacterium (*Pseudomonas syringae*). *Plant Dis. Reptr.* 53:677–80

40. Coyne, D. P., Schuster, M. L. 1969. "Tara," a new Great Northern dry bean variety tolerant to common blight bacterial disease. *Univ. Neb. Agric. Exp. Stn. Bull.* 506. 10 pp.

41. Coyne, D. P., Schuster, M. L. 1970. Jules, a Great Northern dry bean variety tolerant to common blight bacterium (*Xanthomonas phaseoli*). *Plant Dis. Reptr.* 54:557–59

42. Coyne, D. P., Schuster, M. L. 1971. "Emerson"—a new large-seeded Great Northern dry bean variety tolerant to bacterial wilt disease. *Neb. Agric. Exp. Stn.* SB516. 11 pp.

43. Coyne, D. P., Schuster, M. L. 1972. Genetic control of the reaction of the common blight bacterium (*Xanthomonas phaseoli*) in beans (*Phaseolus vulgaris*) in relation to time and stage of plant development. *HortScience* 7:345 (Abstr.)

44. Coyne, D. P., Schuster, M. L. 1973. *Phaseolus* germplasm tolerant to common blight bacterium *Xanthomonas phaseoli*. *Plant Dis. Reptr.* 57:111–14

45. Coyne, D. P., Schuster, M. L., Al-Yasiri, S. 1963. Reaction studies of bean species and varieties to common blight and bacterial wilt. *Plant Dis. Reptr.* 47:534–37

46. Coyne, D. P., Schuster, M. L., Estes, L. W. 1966. Effect of maturity and environment on the genetic control of reaction to bacterial wilt in *Phaseolus vulgaris* L. crosses. *Proc. Am. Soc. Hortic. Sci.* 88:393–99

47. Coyne, D. P., Schuster, M. L., Fast, R. 1967. Sources of tolerance and reaction of beans to races and strains of halo blight bacteria. *Plant Dis. Reptr.* 51:20–24

48. Coyne, D. P., Schuster, M. L., Gallegos, C. C. 1971. Inheritance and linkage of the halo blight systemic chlorosis and leaf watersoaked reaction in *Phaseolus vulgaris* variety crosses. *Plant Dis. Reptr.* 55:203–7

49. Coyne, D. P., Schuster, M. L., Hill, K. 1973. Genetic control of reaction to common blight bacterium in bean (*Phaseolus vulgaris*) as influenced by plant age and bacterial multiplication. *J. Am. Soc. Hortic. Sci.* 98:94–99

50. Coyne, D. P., Schuster, M. L., Shaughnessy, L. 1966. Inheritance of reaction to halo blight and common blight bacteria in a *Phaseolus vulgaris* cross. *Plant Dis. Reptr.* 50:29–32

51. Crispin, A., Dongo, S. 1962. New physiologic races of bean rust, *Uromyces phaseoli typica,* from Mexico. *Plant Dis. Reptr.* 46:411–13

52. Dana, B. F. 1940. Resistance and susceptibility to curly top in varieties of common beans, *Phaseolus vulgaris*. *Phytopathology* 30:786 (Abstr.)

53. Davison, A. D., Vaughan, E. K. 1964. Effect of urediospore concentration on determination of races of *Uromyces phaseoli* var. *phaseoli*. *Phytopathology* 54:336–38

54. Deakin, J. R., Dukes, P. D. 1975. Breeding snap beans for resistance to diseases caused by *Rhizoctonia solani*. *HortScience* 10:269–71

55. Dickson, M. H., Abawi, G. S. 1974. Resistance to *Pythium ultimum* in white-seeded beans (*Phaseolus vulgaris*). *Plant Dis. Reptr.* 58:774–76

56. Dickson, M. H., Natti, J. J. 1966. Breeding for halo blight and virus resistance in snap bean. *Farm Res.* 32:4–5

57. Dickson, M. H., Natti, J. J. 1968. Inheritance of resistance of *Phaseolus vulgaris* to bean yellow mosaic virus. *Phytopathology* 58:1450

58. Dundas, B. 1934. Growing powdery mildew on detached bean leaflets and breeding for resistance. *Phytopathology* 24:1137 (Abstr.)

59. Dundas, B. 1936. Inheritance of resistance to powdery mildew in beans. *Hilgardia* 10:241–53

60. Dundas, B. 1940. A new factor for resistance to powdery mildew (*Erysiphe polygoni*) in beans (*Phaseolus vulgaris*). *Phytopathology* 30:786 (Abstr.)

61. Dundas, B. 1941. Further studies on the inheritance of resistance to powdery mildew of beans. *Hilgardia* 13:551–65

62. Fassuliotis, G., Deakin, J. R., Hoffman, J. C. 1970. Root-knot nematode resistance in snap beans: breeding and nature of resistance. *J. Am. Soc. Hortic. Sci.* 95:640–45

63. Frazier, W. A. 1970. Breeding beans tolerant to several bacterial diseases. *Bean Imp. Coop. Ann. Rept.* 13:12–19

64. Gamez, R. 1973. Los Virus del Frijol en Centro América. III. Razas del virus del mosaico común del frijol de El Salvador y Nicaragua. *Turrialba* 23:475–76

65. Giessen, A. C. v.d., Steinbergen, A. v. 1957. A new method of testing beans for anthracnose. *Euphytica* 6:90–93

66. Goth, R. W., Zaumeyer, W. J. 1965. Reactions of bean varieties to four races of anthracnose. *Plant Dis. Reptr.* 49:815–18

67. Grogan, R. G., Walker, J. C. 1948. The relation of common mosaic to black root of bean. *J. Agric. Res.* 77:315–31

68. Hagedorn, D. J., Rand, R. E., Saad, S. M. 1972. *Phaseolus vulgaris* reaction to *Pseudomonas syringae*. *Plant Dis. Reptr.* 56:325–27

69. Hagel, G. T., Silbernagel, M. J., Burke, D. W. 1972. Resistance to aphids, mites, and thrips in field beans relative to infection by aphid-borne viruses. *US Dep. Agric. Agric. Res. Serv.* 33–139. 4 pp.

70. Harter, L. L., Zaumeyer, W. J. 1941. Differentiation of physiologic races of *Uromyces phaseoli typica* on bean. *J. Agric. Res.* 62:717–32

71. Hartmann, R. W. 1971. Inheritance of resistance to root-knot nematodes (*Meloidogyne incognita*) in beans (*Phaseolus vulgaris* L.) *J. Am. Soc. Hortic. Sci.* 96:344–47

72. Hassan, A. A., Wallace, D. H., Wilkinson, R. E. 1971. Genetics and heritability of resistance to *Fusarium solani* f. *phaseoli* in beans. *J. Am. Soc. Hortic. Sci.* 96:623–27

73. Hassan, A. A., Wilkinson, R. E., Wallace, D. H. 1971. Genetics and heritability of resistance to *Thielaviopsis basicola* in beans. *Proc. Am. Soc. Hortic. Sci.* 96:628–30

74. Hassan, A. A., Wilkinson, R. E., Wallace, D. H. 1971. Relationship between genes controlling resistance to *Fusarium* and *Thielaviopsis* root rots in beans. *J. Am. Soc. Hortic. Sci.* 96:631–32

75. Hill, K., Coyne, D. P., Schuster, M. L. 1972. Leaf, pod and systemic chlorosis reaction in *Phaseolus vulgaris* to halo blight controlled by different genes. *J. Am. Soc. Hortic. Sci.* 97:494–98

76. Hubbeling, N. 1957. New aspects of breeding for disease resistance in beans (*Phaseolus vulgaris* L.). *Euphytica* 6: 111–41

77. Hubbeling, N. 1961. Inheritance and interaction of genes for disease resistance in beans. *Recent Adv. Bot.* 1:438–43

78. Isbell, C. L. 1931. Nematode-resistance studies with pole snap beans. *J. Hered.* 22:191–98

79. Kantzes, J. G., Hollis, W. L. 1962. Susceptibility of snap bean varieties and lines to the Maryland strain of rust. *Bean Imp. Coop. Ann. Rept.* 5:9–11

80. Kim, S. H., Kantzes, J. G. 1972. Species, cultivars and lines of *Phaseolus* resistant to *Pythium aphanidermatum*. *Phytopathology* 62:769 (Abstr.)

81. Leakey, C. L. A. 1970. Anthracnose resistance breeding in Pinto beans in Uganda using the *ARE* gene from Cor-

nell 49–242. *Bean Imp. Coop. Ann. Rept.* 13:60–61
82. Lundin, P., Nilsson, K., Lundin, M. 1971. Nagra Svenska bönsorters reaktion för olika raser av *Colletotrichum lindemuthianum. Agric. Hortic. Genet.* 29(1/4):30–38
83. Mackie, W. W., Esau, K. 1932. A preliminary report on resistance to curly top of sugarbeet in bean hybrids and varieties. *Phytopathology* 22:207–16
84. Masterbroek, C. 1960. A breeding programme for resistance to anthracnose in dry shell haricot beans, based on a new gene. *Euphytica* 9:177–84
85. McLean, D. M., Hoffman, J. C., Brown, G. B. 1968. Greenhouse studies on resistance of snap beans to *Rhizoctonia solani. Plant Dis. Reptr.* 52:486–88
86. McMillan, R. T. 1972. A new race of bean rust on pole beans in Florida. *Plant Dis. Reptr.* 56:759–60
87. McRostie, G. P. 1921. Inheritance of disease resistance in the common bean. *J. Am. Soc. Agron.* 13:15–32
88. Meiners, J. P., Rogers, C. W. 1974. Reaction of bean cultivars to populations of rust in Maryland. *Bean Imp. Coop. Ann. Rept.* 17:52–53
89. Ministerio De Agric., Ofic. De Invest. Espec. 1957. Sexto Informe Anu. 1955–56 *Phaseolus. Rev. Nac. Agric. Bogota* 51:1–89
90. Olave, L. C. A. 1958. Resistancia de algunas variedades y líneas de frijol (*Phaseolus vulgaris* L.) al *Isariopsis griseola* Sacc. *Acta Agron. Palmira* 8:197–219
91. Oliari, L., Vieira, C., Wilkinson, R. E. 1973. Physiologic races of *Colletotrichum lindemuthianum* in the state of Minas Gerais, Brazil. *Plant Dis. Reptr.* 57:870–72
92. Ordosgoitty, A. 1972. Identificacion del mosaico común de la caraota (*Phaseolus vulgaris* L.) en Venezuela. *Agric. Trop.* 22:29–43
93. Patel, P. N., Walker, J. C. 1966. Inheritance of tolerance to halo blight in bean. *Phytopathology* 56:681–82
94. Pierce, W. H. 1934. Resistance to common bean mosaic on the Great Northern field bean. *J. Agric. Res.* 49:183–88
95. Pierce, W. H., Walker, J. C. 1933. The development of mosaic-resistant Refugee beans. *Canner* 77(26):7–9
96. Pompeu, A. S., Crowder, L. V. 1972. Inheritance of resistance of *Phaseolus vulgaris* dry beans to *Xanthomonas phaseoli* common blight. *Cienc. Cult.* 24:1055–63

97. Prasad, K., Weigle, J. L. 1969. Resistance to *Rhizoctonia solani* in *Phaseolus vulgaris* (snap bean). *Plant Dis. Reptr.* 53:350–52
98. Prasad, K., Weigle, J. L. 1970. Screening for resistance to *Rhizoctonia solani* in *Phaseolus vulgaris. Plant Dis. Reptr.* 54:40–44
99. Provvidenti, R., Schroeder, W. T. 1973. Resistance in *Phaseolus vulgaris* to the severe strain of bean yellow mosaic virus. *Phytopathology* 63:196–97
100. Puerta, J., Alonso, A. 1958. Pruebas de resistencia a diferentes enfermedades en diversa variedades de judias. *Bol. Inst. Invest. Agron. Madrid* 18:37–48
101. Quinones, F. A. 1963. Luna, a new high-yielding rust-resistant pinto bean for the Deming area. *NM Agric. Exp. Stn. Bull.* 478. 5 pp.
102. Rudolph, K. 1972. Die Anfälligkeit von Buschbohnen (*Phaseolus vulgaris* var. *nanus* L.) gegen die durch *Pseudomonas phaseolicola* (Burkholder) Dowson hervorgerufene Fettfleckenkrankheit. I. Gewächshausversuche. *Nachrichtenbl. Dtsch. Pflanzenschutzdienst* 24:49–51
103. Rudorf, W. 1958. Genetics of *Phaseolus aborigineus* Burkart. *Proc. X Int. Congr. Genet.* 2:243
104. Rudorf, W. 1958. Ein Beitrag zur genetik der resistanz gegen das Bohnenmosaikvirus 1. *Phytopathol. Z.* 31:371–80
105. Schreiber, F. 1932. Resistenzzuchtung bei *Phaseolus vulgaris. Phytopathol. Z.* 4:415–54
106. Schreiber, F. 1933. Resistenz-Züchtung bei Buschbohnen. *Kuehn-Arch.* 38:287–94
107. Schroeder, W. T., Provvidenti, R. 1968. Resistance of bean (*Phaseolus vulgaris*) to the PV$_2$ strain of bean yellow mosaic virus conditioned by the single dominant gene By. *Phytopathology* 58:1710
108. Schultz, H. K., Dean, L. L. 1947. Inheritance of curly top disease reaction in the bean, *Phaseolus vulgaris. J. Am. Soc. Agron.* 39:47–51
109. Schuster, M. L. 1950. A genetic study of halo blight reaction in *Phaseolus vulgaris. Phytopathology* 40:604–12
110. Schuster, M. L., Coyne, D. P. 1971. New virulent strains of *Xanthomonas phaseoli. Plant Dis. Reptr.* 55:505–6
111. Schuster, M. L., Coyne, D. P., Hoff, B. 1973. Comparative strains of *Xanthomonas phaseoli* from Uganda, Columbia, and Nebraska. *Plant Dis. Reptr.* 57:74–75

112. Shirlow, N. S. 1947. Richmond Wonder French bean. A new, heavy-yielding, disease-resistant variety. *Agric. Gaz. NS W* 58:459–60
113. Silbernagel, M. J. 1965. Differential tolerance to curly top in some snap bean varieties. *Plant Dis. Reptr.* 49:475–77
114. Silbernagel, M. J., Zaumeyer, W. J. 1973. Beans. In *Breeding Plants for Disease Resistance*, ed. R. R. Nelson, 253–69. London & Univ. Park, Pa.: Penn. State Univ. Press. 401 pp.
115. Smith, F. L., Houston, B. R. 1960. Root rot resistance in common beans sought in plant breeding program. *Calif. Agric.* 14(9):8
116. Smith, P. R. 1966. A disease of French beans (*Phaseolus vulgaris* L.) caused by subterranean clover stunt virus. *Aust. J. Agric. Res.* 17:875–83
117. Spragg, F. A., Down, E. E. 1921. The Robust bean. *Mich. Agric. Exp. Stn. Spec. Bull.* 108. 9 pp.
118. Statler, G. D. 1970. Resistance of bean plants to *Fusarium solani* f. *phaseoli.* *Plant Dis. Reptr.* 54:698–99
119. Thomas, H. R., Zaumeyer, W. J. 1950. Inheritance of symptom expression of pod mottle virus. *Phytopathology* 40:1007–10
120. Thomas, H. R., Zaumeyer, W. J. 1953. A strain of yellow bean mosaic virus producing local lesions on tobacco. *Phytopathology* 43:11–15
121. Thomas, P. E. 1969. Five strains of curly top virus. *Phytopathology* 59:1053 (Abstr.)
122. United States Department of Agriculture. 1973. *Agricultural Statistics 1973.* Washington DC: GPO. 617 pp.
123. Van der Plank, J. E. 1968. *Disease Resistance in Plants.* New York: Academic. 206 pp.
124. Wade, B. L., Zaumeyer, W. J. 1938. U.S. No. 5 Refugee, a new mosaic-resistant Refugee bean. *US Dep. Agric. Circ. No.* 500. 12 pp.
125. Waite Agricultural Research Institute. 1943. Report of the Waite Agricultural Research Institute. *Adelaide Univ. Waite Agric. Res. Inst. Rept.* 14:1941–42
126. Walker, J. C. 1965. Disease resistance in vegetable ˚crops III. *Bot. Rev.* 31:331–80
127. Walker, J. C., Patel, P. N. 1964. Inheritance of resistance to halo blight of bean. *Phytopathology* 54:952–54
128. Wallace, D. H., Wilkinson, R. E. 1965. Breeding for Fusarium root rot resistance in beans. *Phytopathology* 55:1227–31
129. Walter, J. M., Lorz, A. P. 1956. Florigreen, a disease-resistant pole bean. *Fla. Agric. Exp. Stn. Circ.* S-92. 8 pp.
130. Wingard, S. A. 1933. The development of rust-resistant beans by hybridization. *Va. Agric. Exp. Stn. Tech. Bull.* 51. 40 pp.
131. Wingard, S. A. 1943. New rust-resistant pole beans of superior quality. *Va. Agric. Exp. Stn. Bull.* 350. 31 pp.
132. Wolf, E. A., Hills, W. A. 1954. Seminole, a new disease resistant, green, round-podded bush bean. *Fla. Agric. Exp. Stn. Circ.* S-73. 6 pp.
133. Yarnell, S. R. 1965. Cytogenetics of the vegetable crops, IV Legumes (cont.). *Bot. Rev.* 31:247–330
134. Yen, D. E., Brien, R. M. 1960. Frenchbean rust (*Uromyces appendiculatus*) studies on resistance and determination of rust races present in New Zealand. *NZ J. Agric. Res.* 3:358–63
135. Yerkes, W. D. Jr., Freytag, G. F. 1956. *Phaseolus coccineus* as a source of root rot resistance for the common bean. *Phytopathology* 46:32 (Abstr.)
136. Zaumeyer, W. J. 1969. The origin of resistance to common bean mosaic in snap beans. *Seed World* 105(4):8–9
137. Zaumeyer, W. J., Harter, L. L. 1941. Inheritance of resistance to six physiologic races of bean rust. *J. Agric. Res.* 63:599–622
138. Zaumeyer, W. J., Harter, L. L. 1943. Inheritance of symptom expression of bean mosaic virus 4. *J. Agric. Res.* 67:295–300
139. Zaumeyer, W. J., Harter, L. L. 1943. Two new virus diseases of beans. *J. Agric. Res.* 67:305–27
140. Zaumeyer, W. J., Harter, L. L. 1946. Pintos 5 and 14, new rust-resistant beans for dry-land areas of the West. *South. Seedsman* 9(8):15, 50, 54
141. Zaumeyer, W. J., Thomas, H. R. 1948. Pod mottle, a virus disease of beans. *J. Agric. Res.* 77:81–96
142. Zaumeyer, W. J., Thomas, H. R. 1957. A monographic study of bean diseases and methods for their control. *US Dep. Agric. Tech. Bull.* No. 868. 255 pp.
143. Zaumeyer, W. J., Thomas, H. R., Afanasiev, M. M. 1960. A new disease-resistant Great Northern bean. *Seed World* 86(11):14, 16

MUTUAL RESPONSIBILITIES ❖3626
OF TAXONOMIC MYCOLOGY AND
PLANT PATHOLOGY

J. Walker

Biology Branch, Biological and Chemical Research Institute, New South Wales Department of Agriculture, Rydalmere, NSW 2116, Australia

INTRODUCTION

From its beginnings in antiquity plant pathology has described and classified diseases and their causes. Its history has been summarized by Keitt (46). Development of the science as we know it began in the nineteenth century, and taxonomic and descriptive work on plant parasitic fungi was of necessity carried out by many pathologists, several of whom made significant contributions to mycology in the process.

Taxonomic mycology has received in recent times little attention in plant pathology, although many authors over the past sixty years have written ·of the place taxonomy should hold in applied biology and of the importance of taxonomic work on fungi in plant pathology (10, 19, 20, 25, 26, 33, 35, 45, 49, 50, 54, 55, 57, 77, 79–81, 84). In 1965, Subramanian (84) commented on the small amount of attention devoted to taxonomy in two important plant pathology texts (41, 42) compared with that given to the many other component·disciplines of present-day pathology, and an examination of the first twelve volumes of *Annual Review of Phytopathology* and of recent texts reveals a similar situation.

The attitude of some pathologists has been frankly apathetic to taxonomic work, part of a wider bias against taxonomy in biology. Rogers (69) has deplored the existence of such a peck order in science, and Davis (27) has pointed out that taxonomy is near the bottom of the priority scale for the support necessary for work to be carried out. The Committee on Systematics in Support of Biological Research (20) commented on a "potentially critical situation" in systematic biology due to its lack of support, and the British Committee (19) stressed the adverse effect that neglect of taxonomic work on fungi can have on plant pathology, and so on agriculture, horticulture, and forestry. In plant pathology, some statements in Christensen's *The Molds and Man* (16) illuminate several areas of antagonism to, and

335

misunderstanding of, taxonomy, and Graham's (35) statement that "taxonomy has too long been the Cinderella of Science" is true in many institutions and research groups.

On the other hand, many taxonomists have often been unaware of the needs and work of applied biologists, including plant pathologists, and some have tended to cut themselves off from the wider fields of biological practice and research. The taxonomist "often leads a cloistered life, protected from the vexations and frustrations of the everyday world and he may well wear blinkers as opaque as any worn by a horse. Living a life of seclusion, . . . he affects an unconcern for the mundane applications of his work" (23). Davis (27) has mentioned the taxonomist's often narrow and static approach to biology, and his disdain for function in some cases.

Not all pathologists and taxonomists fit into these categories, but many do. Both positions are unreasonable and greatly damage taxonomy, pathology, and biology as a whole. This article examines some aspects of taxonomic mycology and plant pathology to show that they have many areas of mutual responsibility and dependence. Having come to taxonomy through plant pathology, and working on taxonomy of fungi within a large group of research and extension plant pathologists, I have been impressed by some factors responsible for misunderstanding. Many of these relate to the everyday investigations of both taxonomist and pathologist, and some to a lack of understanding of what the other's work involves. This article shows how many of the procedures of taxonomy are important to the science and the practice (37) of plant pathology, and how plant pathology has much to offer taxonomic mycology. Many of the points apply to other organisms as well as to fungi.

FUNGI, SPECIMENS, NAMES, AND YOU

In a discussion of the validity of taxonomy and its methods, Rogers (69) recalls that taxonomy, by being the science of classification, "includes not merely the identification of individual organisms, not merely their arrangement in species and genera and more inclusive groups, but the whole basis of comparative morphology and physiology, and the whole framework of phylogenetic hypothesis, on which classification is developed. Ultimately, taxonomy is one sort of synthesis of almost everything that is known about living things." He also comments on the predictive value of a sound classification. These basic attributes of taxonomy were also stressed by the Committee on Systematics in Support of Biological Research (20): "The importance of taxonomic classifications is in (1) keeping track of information about organisms, (2) summarising this information, and (3) predicting the characteristics of incompletely studied organisms."

The most familiar taxa are genus and species. Above genera, fungi are grouped into families, orders, and classes, and below species there are several categories such as variety, form species, and physiologic race (1, 3, 29), of great practical importance to plant pathology. Because of lack of knowledge, "a stable, comprehensive and detailed classification of the fungi is at present impossible" (5). In order to provide a framework within which details of classification can be worked out, while still allowing its use for reference purposes, Ainsworth (5) has proposed a general purpose classification of the classes of fungi, which has been accepted as a basis for the sixth edition of the Dictionary of the Fungi (6) and for a comprehensive review of the biology and taxonomy of all fungi (8, 9). Some of the principles governing taxonomy of fungi have been discussed by Talbot (86), and practical advice on naming and describing them has been published (21, 38).

Some critics of taxonomic mycology have suggested that, as there is no general agreement on classification of fungi, an arbitrary, stable, and unchanging system be set up and universally adopted (16). The difficulties experienced by applied biologists when names of fungi are changed have also been highlighted and indeed are widely known, but a statement such as "Only very seldom are such [name] changes made with the idea of making it easier for students, teachers or research workers to recognize and identify the fungi in question" (16) misses the point and does nothing to help pathology and taxonomy and the relationship that should exist between them.

Taxonomy

There are many ways of classifying a group of things, depending on the opinion of the classifier and the use intended for the classification. For example, words may be classified alphabetically in a dictionary and grouped according to meaning in a thesaurus. Both classifications are valid and useful, though their purposes differ. In biology, some classifications proposed for organisms have been used more widely than others because they seem to relate better to the way the organisms behave in nature, because their predictive value is higher, and because later discovered organisms fit readily into the scheme. Modifications often are necessary, and these should reflect an advance in knowledge, resulting in a more generally applicable classification.

The classification of fungi devised by Saccardo (70) and modified by Clements & Shear (18) has been the one most familiar to plant pathologists. In the past forty years, there have been advances in knowledge in most groups, and the recently published four-volume treatise on The Fungi (8, 9) gives details of these, with comprehensive keys to genus level in the two parts of Volume 4 (8). This work is essential reading for all interested in biology and taxonomy of fungi.

Saccardo used a scheme of spore septation, shape, and color to classify genera, and, while this brought together fungi with a superficial similarity, later work has shown that many of Saccardo's groupings are highly artificial. This can be illustrated by considering some species he placed in the genus *Ophiobolus*. The only feature many of these have in common is long, thin ascospores borne in perithecium-like structures. Investigations of the original collections have shown that the first-described species and generic type, *Ophiobolus disseminans* Riess, has bitunicate asci with long, pale yellowish-brown spores, which break at a central constriction when mature (39, 68). This fungus, and thus the genus *Ophiobolus,* belongs to the order Pleosporales, and this sets the standard for placement of other species in *Ophiobolus* (for type specimens see below under "Nomenclature"). Examination of the original specimens of other species placed in *Ophiobolus* by Saccardo shows that he unknowingly brought together fungi quite distinct from *Ophiobolus,* as judged by its standard, *O. disseminans.* For example, *Ophiobolus graminis* of Saccardo has unitunicate asci and other characters that indicate that it belongs in the order Diaporthales. It is now named *Gaeumannomyces graminis* (91, 92). *O. helicosporus* of Saccardo also differs from *Ophiobolus* and is better placed in the genus *Acanthophiobolus* (92). In Saccardo's concept of *Ophiobolus,* there are at least six quite distinct generic types represented in the original specimens so far studied (J. Walker, unpublished data), and only one of these, *O. disseminans* Riess, the original species of the genus, is correctly placed there.

This is not a unique situation and many similar ones are frequently met with in plant pathology. Continuing taxonomic research is needed to help sort them out. This example also highlights several points of practical importance in everyday identification work and the proper use of names of taxa. These and the difficulties they can cause should be clearly understood by all plant pathologists.

PRECISION IN THE USE OF NAMES   The generic name *Ophiobolus,* used in Saccardo's broad sense, has no precision. It embraces fungi that are completely unrelated and have nothing in common except long, thin ascospores. For precision, a genus must be interpreted in the light of the original specimen and description of its type species.

MISLEADING USE OF NAMES   To use the name *Ophiobolus* for a fungus such as *G. graminis* misleads everyone about its correct relationship to other fungi, and obscures details of its biology. Organisms with the same generic name are expected to show some relationship to each other, just as one expects sodium chloride and sodium sulfate to have properties in common related to their being sodium salts. However, if the name *Ophiobolus* means one thing in the case of *O. disseminans,* a different thing in the case of *O. graminis,* and something else with *O. herpotrichus,* there is no precision in its use and true relationships are obscured.

LITERATURE CONFUSION   *Ophiobolus* in the sense of Riess, the original describer, is a relatively precise and definable genus; *Ophiobolus* in Saccardo's sense is a mixture of several generic concepts with no precision. Yet it is *Ophiobolus* in the latter sense that has been so widely used in the literature of mycology and plant pathology for the last ninety years. In the massive literature on cereal root-rot fungi at least, this has been partly responsible for concealing certain facts about their biology and confusing their identity. It is interesting to speculate on how many different organisms and how much confusion lie in the literature under such common names as *Colletotrichum gloeosporioides, Guignardia citricarpa, Rhizoctonia solani, Sclerotinia sclerotiorum,* and many others. Taxonomic confusion leads to information confusion in biology, and workers may think they are talking about the same organism when really they are not.

SPECIMENS AND THEIR VALUE   Taxonomic work aims at giving precision to concepts of taxa and dispelling confusion. This is only possible if the original specimens on which work is based are kept (see "Herbaria and Culture Collections" below). Without specimens, it is not possible to assess earlier work and named taxa in the light of later knowledge. For example, the importance of unitunicate versus bitunicate asci was not known in Saccardo's day. Today it is used as an important character in Ascomycete classification. If the specimens on which Saccardo's work was based were not kept, it would neither be possible to sort out the earlier classification nor to know the identity of fungi studied earlier. Work can be wasted or severely reduced in information content if specimens that allow it to be checked are not preserved.

NECESSITY FOR NAME CHANGES   Name changes result from the sort of taxonomic investigation described above. They should represent an advance in taxonomic knowledge and a clarification of identity. Wholesale name changing without proper taxonomic study must be avoided, and, as names are based on nomenclatural types, names should not be changed without study of type specimens. The changing of familiar names to new ones is often resented and does require some adjustment. The name, however, reflects the taxonomy; it is better to have a new name with an accurate and more complete information content (see "Nomenclature" below) than to relax with an old, familiar, but misleading friend. It is often argued that, since a name such as *O. graminis* is a label everyone knows and understands, it should stay as it is for the sake of convenience and avoidance of confusion. This, however, can only lead to much greater confusion, for precision requires that names have both a consistency in use and an information content as well as being labels.

NO ARBITRARY SYSTEM OF CLASSIFICATION   The confusion in concepts that can result from the sort of arbitrary classification described above for *Ophiobolus* is one of the strongest arguments against proposals for an arbitrary and unchanging system of classification (16). As work continues in any science, concepts change and rearrangement of knowledge takes place, for example, changes in concepts in atomic physics, elucidation of biochemical pathways, and molecular structure studies resulting in changing of structural formulas. These are all regarded as advances in knowledge and the changes accepted as such. This does not mean uncritical acceptance of every suggested change, but its critical appraisal and testing. As Christensen (16) said in discussing scientific research, it is a matter "of knowledge increased as it is exercised, and of judgement tempered by experience."

INFORMATION CONTENT OF CLASSIFICATIONS   With more homogeneous concepts of taxa, information content of classifications improves. The information content of the various segregate genera is much greater than *Ophiobolus* in Saccardo's sense; for example, the genus *Gaeumannomyces* comprises a group of fungi in the order Diaporthales, parasitic on roots and culms of Gramineae and Cyperaceae, and with hyphopodiate mycelia (92). These form a compact taxonomic-ecologic-pathologic group, and the classification reflects both their morphology and their biology. Taxonomies should, where possible, embody biological as well as morphological data (56, 60), with increase in their information content, predictive value (56), and usefulness for applied biology.

TAXONOMIC WORK   Classifications develop from the informed taxonomic opinions of the mycologists proposing them, and there thus can be several different classifications suggested for one group. While they may be a nuisance to someone wanting information on the identity of an organism, they often result in more critical work being done on the group in an effort to resolve the differences in opinion. Different classifications can be looked upon as different hypotheses put forward for testing and experiment. Many groups of fungi are in this stage of study at present. As Ainsworth (7) pointed out, older descriptions of many fungi are inadequate, and taxonomic work to prepare accurate new descriptions is needed before more workable classifications can be developed. Many examples of such work are found in recent taxonomic literature (28, 32, 43, 47, 85). Such redefinition of older taxa is leading to sounder taxonomic concepts and so to more accurate and rapid identifications.

NEED TO STUDY THE WHOLE GROUP   All members of a group, and not just the plant parasites, must be studied if a proper appreciation of their taxonomy is to be obtained (3). Although plant pathologists sometimes consider time spent by taxonomists on nonparasites as time wasted, it is essential. Moreover, many species have both parasitic and saprophytic phases in their life cycle, and a knowledge of the structures produced in each phase and their function is important in understanding the taxonomy and biology of these organisms. The perfect and conidial stages of many Fungi Imperfecti fall into this category, and the influence of saprophytic states on disease epidemiology is well known.

As the science of classification, taxonomy should attempt to increase the information content of its groupings by incorporating a much wider range of biological data (56, 66). This is considered in "Ecological Plant Pathology and Broadly Based Taxonomy" below.

Nomenclature

While taxonomy deals with classification of organisms and can vary with the opinion of different workers, nomenclature deals with the application of names to described taxa. To avoid confusion, this application of names is governed by rules. The rules for fungi are contained in the International Code of Botanical Nomenclature (44), which is drawn up and modified by successive International Botanical Congresses.

The main standard governing use of a name is the type specimen, the specimen named by the original describer. It is not necessarily typical of the species, because often the species at that time was known only from the few specimens studied by the author. The type specimen is thus the type of a name (the nomenclatural type) and not of a species. Its importance is in fixing the application of a name to a specimen and thus providing a stable reference point that can be examined later (if doubt arises) to determine the essential features that govern the use of the name. The type specimen thus allows the name to be related to other specimens of the same species. However, as Nelson (61) has pointed out, the range of variation of the species can only be determined by extensive studies of many populations, collected and examined from a range of environmental conditions, and studied in various ways.

Because of the need for stability in a type specimen, cultures of living organisms are generally not satisfactory as types, unless accompanied by dried specimens prepared at the start. Changes during the life of cultures can then be assessed against the stable, dead reference specimens. Various techniques for preparing dried herbarium reference cultures have been described (21, 53).

Names are an international code that allow accurate interchange of information about organisms. They are not merely labels, but summarize a set of morphological and biological characters. To call a fungus *Phytophthora* immediately conjures up characters and biological features common to fungi within the genus, and these are understood everywhere. Names are also the key allowing access to the full range of information stored in the literature and in herbaria, and they provide a ready-made code for computer storage of information about living organisms.

For these reasons, incorrect use of names is a serious barrier to information recall and communication. If an incorrect determination is made (i.e. the wrong name is applied to a particular specimen), all future communication about that specimen will be a source of error. When the incorrect name is quoted in the literature, it inserts information of a biological nature about an organism that is not what its name suggests. This can lead astray other workers investigating the literature. When names are used as codes in modern literature data processing methods and information banks, their correct use is essential if the relevant data are to be recalled. These comments apply also to host names.

Sometimes one species has several names (synonyms). Synonyms can be of two sorts: nomenclatural (or obligate) and taxonomic (or facultative). Nomenclatural synonyms are based on the same nomenclatural type specimen (see above) and arise when studies show that a fungus should be placed in a taxon other than that in which it was described or is placed at present. Its name is then transferred to this taxon, and the new and old names are nomenclatural synonyms. On the other hand, taxonomic synonyms arise when two or more different species, each with its own nomenclatural type are studied, considered to be the same, and combined under one name. The various names of the different species have different types, and are thus taxonomic synonyms; they need not, however, be synonyms since they are only considered to be by the worker who combined them. Others may consider some of the species combined under the one name to be distinct and will then not regard their names as synonyms. An example showing both nomenclatural and taxonomic synonyms follows:

(a)   *Pseudocercosporella capsellae* (Ell. & Ev.) Deighton 1973
(b)   ≡ *Cylindrosporium capsellae* Ell. & Ev. 1887
(c)   = *Cercospora albomaculans* Ell. & Ev. 1894
(d)   ≡ *Cercosporella albomaculans* (Ell. & Ev.) Sacc. 1895
      ≡ means nomenclatural (obligate) synonym
      = means taxonomic (facultative) synonym
      [adapted from Deighton (28)].

Because *a* and *b* are based on the Ellis & Everhart type of 1887 they are nomenclatural (obligate) synonyms; *c* and *d* are nomenclatural (obligate) synonyms because both are based on the Ellis & Everhart type of 1894. However, *a* and *b* are taxonomic (facultative) synonyms of *c* and *d,* as each pair is based on a different type.

A knowledge of synonyms is important in taxonomy and nomenclature, and very valuable in plant pathology. References to a species in several years of literature can often be found under any of the several names used for it. Moreover, two or more names for the same organism are sometimes currently in use in the literature, due to differences in taxonomic opinion, varying usage in different areas or countries, or delays in use of a newer name. Facultative synonymy can hide organisms lumped together by one author which may in fact be different in characters of great importance to plant pathologists; for example, over 600 synonyms are given in the literature (90) for *Colletotrichum gloeosporioides* Penz. These are mainly morphologically similar fungi on a range of hosts. Pathologists studying diseases caused by *Colletotrichum* should look at details of the biology, morphology, and taxonomy of the fungus on the hosts being studied, and neither assume that their fungus is the same as *C. gloeosporioides* reported from other hosts nor that information in the literature for *C. gloeosporioides* applies to their disease and fungus. When using the literature, workers should always be aware that wanted information may be hidden under an unfamiliar synonym. Monographs and other taxonomic works are valuable sources where synonymy can be checked before literature searches are started.

The importance of names and their use in biology has been summarized by Ainsworth (2): "Biology has one major advantage over other branches of science in having a built-in, internationally used, basic retrieval code (which also reflects a more or less internationally acceptable classification) in the scientific names applied to living things, especially at the generic and specific levels. The truly international character of the binomial names of species is not always fully appreciated. Such Latin names are frequently the only means by which a Western biologist can ascertain the subject of writings in Russian, Japanese and Chinese. When this is realized, the importance of using names in line with what may sometimes seem to be the arbitrary requirements of the three International Codes of Nomenclature (those for zoology, botany and bacteriology) should be apparent."

Herbaria and Culture Collections

Every biologist knows that the living world is the main study of biology. This is the dynamic ever-changing situation of reality. If many features are to be studied or even detected, however, it is often necessary to sample the living population under study and examine individuals in the static condition. These static and dynamic aspects in biological work are not mutually exclusive but supplement one another to build up a complete picture.

When this is realized, the place of herbaria becomes clear. The herbarium, with its specimens and associated living cultures, is a collection of evidence for biological work. In it the specimens are the unchanging static evidence of what was studied 50, 100, or 200 years ago, or last week, which provide a check on the accuracy of the written word and the material evidence for it. As shown above, they allow work to be reassessed in the light of later knowledge. They allow taxa established many years ago to be reexamined in the light of characters then overlooked or unknown, but now discovered to be of importance in classification. They provide a check on the correct identity of host plants. They enable geographic distributions, host ranges, and seasonal occurrences to be determined on a long-term basis, and every point on distribution maps drawn is checkable against a specimen in the collection. While there may be inaccuracies and error in published descriptions, the specimens will reveal the facts to later workers. They cannot change or evolve; thus, they provide static reference points which help us assess changes in living organisms. With living cultures, preparation of dried cultures as herbarium specimens gives a valuable reference to their original morphology and appearance. Herbarium specimens are not meant to replace genetic, pathological, and biochemical studies, epidemiology, ecology, or any other work carried out in biology with living organisms. Their importance lies in being static samples of the changing world, evidence for work done, material for taxonomic routine and research, and providers of a framework of named reference points to aid those navigating the biological world.

"Biology differs from most other branches of science in that an important part of its information is stored not as literature but as specimens" (2, 4). Access to information stored in herbaria as specimens and their associated records can be difficult. Numbering and accessioning of specimens is not carried out in some institutions, but if it is it provides a valuable basis for improved data-recall systems. Several herbaria and culture collections are developing computer-based recall and indexing programs; a biological data bank built in this way provides rapid access to herbarium-based information (24). For mycological and plant-disease herbaria, such methods can be very valuable. These are double herbaria, as there are two sets of organisms, a host and a parasite, and the cross indexing and multiple access obtained with computer recall systems allows information such as host ranges, disease distributions, regional check lists, quarantine interceptions, and other topics of importance to plant pathologists to be obtained rapidly. Computer-produced index cards, in both host and parasite order, can be produced for easy day-to-day reference (J. Walker, unpublished data).

The following types of specimens (and also an ample host specimen to allow its correct identification) should always be kept in plant pathology:

1. Specimens that are the basis for descriptions of new taxa, or new diseases.
2. New records for countries, regions, and climatic zones, and new hosts for known pathogens. Disease-survey specimens fall into this group.
3. Specimens used as sources of isolates. Both the original material and the cultures obtained from it should be retained. Care should always be taken not to destroy all of the original collection when isolating from it, but a representative portion should be put aside for the herbarium.
4. Specimens used as the basis of pathogenicity studies, or biochemical, physiological, or genetic work.
5. Specimens used in higher-degree studies, and specimens that are the basis of published work. In these cases, specimens should be listed, together with the

herbarium and culture collections where they have been lodged. Standard abbreviations for many of the world's collections are available (52, 62) and these, together with the accession numbers (if any) under which specimens are filed, give unequivocal information on their locality.

6. Specimens of rarely seen diseases or fungi.
7. Specimens showing good sporulation.
8. With root rots and wood rots, where the fungus might not be obvious on the original material, dried cultures of the isolations should be filed with the specimen. This applies to isolations made from any specimens.
9. Microscope slide preparations showing fruiting bodies and spores can also be filed for reference with the specimens, as can prints of photos, drawings, and notes.

When specimens are collected, full details of host, exact locality, date of collection, collector's name (and field collection number if any), disease, and damage caused should be noted for incorporation as label information and notes with the specimen.

If plant pathologists and other biologists fail to keep specimens, they destroy the evidence for their work and make the checking of important aspects of it impossible. They are hindering the progress of their discipline, and of taxonomic work on fungi in general. Clark & Loegering (17) stated: "The scientist who discards the biological materials which were the basis of his published research is neglecting part of his responsibility as a scientist." Blake (13), discussing botanical taxonomy in Australia, wrote, "Inadequacy of material for study either as regards quantity or quality or both is the greatest handicap to systematics." These statements are particularly relevant to plant pathology, where much valuable specimen material has either never been collected, or collected, used, and then discarded. Colleagues returning from field trips have often told me of diseases and fungi observed, but not collected. The excuses given have generally been either, "I didn't think you'd want that" or "I thought you had a specimen of that already." In current work on plant-parasitic fungi in Australia, considerable difficulty has often been experienced because of the apparent nonexistence of specimens on which many published accounts were based.

The reluctance of some plant pathologists to keep specimens may arise partly from a failure to see in the specimen the primary material of their work. They will keep detailed records and field notes, take many photographs of diseases and file them in the photography collection, and, on the way, throw the specimens into the rubbish bin! Far too little emphasis on the specimen as the primary record, the basis for published statements and the only checkable material for the future, has been given in training plant pathologists and applied biologists generally. Specimens in the herbarium need to be seen as working tools, like books and journals in a library. However, in a book only the description of a disease and a fungus can be seen, but in a specimen the disease itself and the fungus itself are there to examine. In most cases one good specimen is worth a thousand words.

Many collections of the same species, collected over a range of host plants, space, time, and season should be kept. It is only in this way that a proper appreciation of a species and its variation can be obtained. Imagine trying to comprehend the species *Homo sapiens* by looking only at the few specimens present in the tea room one morning! Probably the generalizations made about morphology, race, sex ratio, and other characters would bear little relation to reality. One would need to study many thousands of individuals in all parts of the world for a considerable period to gain even a partial picture of the true variation.

There are several references that give lists of herbaria, guidance on setting up and caring for a herbarium collection, and on collecting and sending specimens from the field to the laboratory to avoid deterioration in transit (21, 34, 38, 52, 63, 72, 78). Culture collections and their maintenance have also been described (17, 63). They are best run in conjunction with a herbarium where specimens from which cultures originate can be kept, and where dried herbarium reference cultures can be filed (4). Specimens, and cultures originating from them, can carry the same accesssion number for ease of reference.

In institutions where taxonomic work is carried out or where specimens are filed, administrators must be made aware of the value of the collections and make arrangements for their permanent care and protection. Where a lone taxonomist is working, there is the special danger that collections he accumulates will not be properly cared for later. Several valuable collections have been severely damaged by insects and by misuse, because no curatorial work has been carried out and, in some cases, because of poor storage conditions (4). In a few cases, collections accumulated by workers have been thrown out soon after they retired. Continuity of herbarium work must be maintained if this irreplaceable basic scientific resource is to be kept in a useful and usable condition.

Identification

This is the term commonly used for examination of an unknown specimen leading to its recognition and naming; sometimes the term *determination* is used. The use of these terms is discussed by Talbot (86). Identification of the causal fungus plays an important part in all aspects of routine and research work on plant diseases. Diagnostic and control work depend on accurate identification. Accurate and rapid identifications are essential in quarantine (51), especially in countries like Australia that are free of many exotic diseases. Where several similar fungi occur on the same hosts, critical examination is often needed to distinguish them, for example, the rusts on Gramineae, Leguminosae, *Populus* (94), and many others. Sometimes identification of class, family, or genus can be obtained relatively quickly, but to take it to species or subspecific categories may require much more detailed work. Sometimes a general idea of the fungus group is all that is needed; in other cases, a complete identification may be required.

Identification is saying that one thing is the same as another. The basic and most important standards for identification in biology are specimens and, next in importance, the accurate descriptions and illustrations derived from them. Mason (59) has commented, "But the surest basis of the art of diagnosis is unchanging and is this: the matching of good specimens of the species to be named against good specimens that have already been correctly named." Identification can thus be carried out most effectively where the resources of a comprehensive herbarium and library are available, and many specimens can be adequately identified only with herbarium facilities.

Some of the factors on which identification of a fungus depends are the quality of the specimen being examined, the means used in examining it, the taxonomic state of the group into which the fungus falls, and taxonomic assistance from specialists.

SPECIMEN QUALITY   Quality of the specimen should be high. While much can be done with a poor or sparse specimen (and it may be all that is available), it is often quicker to obtain a better one. Specimens sent to specialists should always have the organism present (in ample quantity if possible), and be large enough for dissection and slide preparation purposes, and retention of portion in the specialist's collection. A duplicate should always be retained for reference in the sender's herbarium; numbering of specimens aids correspondence about them.

MEANS USED   Examination of the specimen should be detailed, with notes and diagrams prepared. Many of the characters needed to identify fungi are small and only seen by detailed microscopic examination. Germ pores in rust urediniospores, types of conidiophores produced by Hyphomycetes, and pycnidial and acervular fungi, spore marking and sculpturing, ascus tip structures, and many other features are of great importance in everyday identification work. They cannot be determined on a slide swimming in mounting medium with the coverslip propped up by the large piece taken for examination. Techniques for microscope preparation have been detailed by various workers. Dring (30) gives details of several methods and many references; Savile (76) gives detailed techniques for rusts. Poor slide preparation is a common barrier to ready identification.

Comparison with published descriptions and illustrations is an essential part of identification. Published descriptions often are old and inadequate for diagnostic purposes or, even more disconcerting, consultation of two or more references shows differences in the descriptions of fungi referred to under the same name. The availability of accurately identified specimens in a herbarium provides a valuable aid to identification, and plant pathologists should develop the habit of using the herbarium to see the organism, as they now use the library to see its description. In a working herbarium, there should be ample collections of material for this purpose, and taxonomists should actively encourage pathologists to make use of it.

Type, old, or sparse specimens are best left alone, as many of them have considerable scientific value and are needed for specialist taxonomic work. While some older type specimens are sparse, it is now common practice to try to have abundant material in type collections, and even to distribute portions of them (isotypes) in different herbaria. Where this is done, great care must be taken to ensure that the same organism is present on every portion.

Host lists and check lists are valuable guides to host, substrate, and geographic ranges of fungi. However, it must not be assumed that all records in these lists are of equal value. Some are based on detailed and proper identification of specimens, whereas others are more doubtful compilations from the literature. Sometimes the identity of both hosts and pathogens listed is dubious (94). Such lists should be used only as a guide and never to jump to conclusions about identity; for example, I have a *Cercospora* on host A, *Cercospora X* occurs on host A in this list, therefore I have *Cercospora X.* While this may give the correct answer, it often gives the wrong one. In New South Wales, the rust of *Medicago* spp. had always been identified as *Uromyces striatus* Schroet. until *U. anthyllidis* (Grev.) Schroet. was also recognized. Reexamination of earlier specimens filed as *U. striatus* showed that several of these were in fact *U. anthyllidis,* which I had misidentified earlier, partly due to placing too much reliance on what had already been recorded on *Medicago* in New South Wales, and letting an assumption that this would be the same replace a sound decision about each specimen on its merits.

Quarantine interception lists are sometimes used to give an indication of the presence of a disease in a country. As with other lists, errors occur and the lists must be used with care; for example, white rust of chrysanthemums, *Puccinia horiana* P. Henn., has been listed (89) as intercepted from Australia and this record has been quoted at least twice since as evidence for the disease here. At the time of writing (October 1974), *P. horiana* has not been recorded in Australia, although it does occur in some neighboring Pacific and Asian countries.

These comments also indicate the need for care when preparing such lists, and distinguishing between the varying authenticity of records listed.

TAXONOMIC STATUS OF THE GROUP   Correct identification is one of the fruits of sound taxonomic research. Difficulty is found in identifying some fungi, including many parasites, because the taxonomy of the group is not well known. In many groups, taxonomic work should be done with the needs of plant pathologists in mind; in the various *Colletotrichum* fungi mentioned above, practicing pathologists are aware of many differences between isolates in morphology, cultural characteristics, and pathology. Much remains to be done on species identification and perfect-imperfect state correlations in genera like *Phytophthora, Sclerospora, Glomerella, Guignardia, Mycosphaerella, Sclerotinia,* and many Fungi Imperfecti. Family, genus, and species concepts in the Pleosporales need a large amount of work. In a country like Australia where perithecia of many powdery mildews are rarely or never seen, identification of species on conidial states alone causes many problems; in countries where perfect and conidial states occur together, the correlation of characters of value in specific identification using conidial states would be of great value. Some northern hemisphere works on the Erysiphales, however, pay little attention to details of conidial state morphology. Many other important fungi in plant pathology need taxonomic attention (see below under "Ecological Plant Pathology and Broadly Based Taxonomy").

Workers should not assign names to specimens when they are inadequate for accurate identification or when no certain identification can be obtained, as this can cause later confusion. Identification should be taken only as far as is possible; it is much sounder practice to record a fungus as *Pythium* sp. with a comment on its possible relationships and a description, than to decide uncritically on *Pythium ultimum* Trow. Parmeter (64) discussed the taxonomy of sterile fungi and made the following statement, applicable to other organisms: "At our present state of knowledge, it appears that informal description of sterile fungi, where necessary, is more desirable than formal description. . . . plant pathologists thus could convey considerable information about sterile fungi without encumbering the literature with formal taxa of dubious value." Under these circumstances, citation of herbarium accession numbers of specimens provides a permanent reference point about them and a quotable reference in the literature, for example, *Stereum* sp. DAR 7865. Later work published on the fungus as a new or already described species can then refer to the same specimen reference. This is also valuable in countries or areas where the proportion of undescribed species is high, as it provides a permanent reference for identification purposes until the necessary taxonomic work leading to formal description can be carried out.

Care is needed when using perfect state names if only a conidial state is present. The connection between perfect and imperfect states of some fungi has been demonstrated only rarely, in specific parts of the world, or only in culture; widespread use of a perfect state name when only conidial state is present may lead to difficulties. This is especially true in genera such as *Fusarium, Drechslera,* and many others where identification to species can sometimes be difficult, and closely related organisms may be confused. In the powdery mildews a similar situation exists. For many years, cucurbit powdery mildew in several countries has been referred to as *Erysiphe cichoracearum* DC. ex Merat when studies indicate that the widespread conidial cucurbit mildew in some countries more often resembles the conidial state of *Sphaerotheca fuliginea* (Schlecht.) Poll. (12). Identification should thus be taken only as far as is possible with the specimens on hand.

TAXONOMIC ASSISTANCE   Specimens are sent to specialists and herbaria for lodging and identification. The importance of lodging specimens regularly, and the information required when specimens are sent, has been dealt with under "Herbaria and Culture Collections." Both identified specimens and unidentified specimens for which identifications are not immediately required can be lodged. When specimens are referred to herbaria and taxonomists for identification, two important points arise and must be considered.

The taxonomist must accept responsibility for making identifications as he has detailed knowledge of the fungi (or knows of specialists to whom unfamiliar organisms can be sent). Some taxonomists are unwilling to help with identifications for plant pathologists or even to provide help with reference specimens and literature. This discourages pathologists and contributes to their poor opinion of taxonomy and their reluctance to keep specimens. At one extreme are institutions such as the Commonwealth Mycological Institute where taxonomic work results in a large flow of essential information for both taxonomy and plant pathology; at the other are taxonomists who do not wish to get involved in anything pathological, and act as a deterrent to the recognition by plant pathologists of taxonomy as an integral part of their science. Taxonomists should try to encourage all applied biologists to retain and lodge specimens, and take a cooperative part in the biological work of others on the taxonomic side. Sometimes the specimens they receive for identification will help their own research too.

The pathologist must show consideration when sending specimens for identification. Sometimes he expects an immediate answer (occasionally showing disgust if it is not provided), with little appreciation of the work needed in many cases to give an accurate identification. Sometimes hundreds of specimens or cultures from soil isolations, surveys, or higher degree projects are referred to taxonomists for routine identification, when much of this work (especially the initial sorting and identification) should be undertaken by the worker himself. Many specialists are willing to help identify reasonable numbers of specimens, but the work can be time consuming and demanding, and delay their taxonomic work to the detriment of taxonomy and applied biology. When a particular species has been identified once or twice for the sender, it is also reasonable to expect that he will be familiar with it and not need to send it again for identification. Pathologists can thus assist taxonomists by care in selection of specimens sent for identification, by providing full details, by giving some idea of the level of identification required, and by appreciating the work needed to identify fungi.

ECOLOGICAL PLANT PATHOLOGY AND BROADLY BASED TAXONOMY

The need for a broader approach to both plant pathology and taxonomy has been stressed by several authors (11, 27, 56, 65, 66, 73, 76, 83). In place of a simple host-parasite concept, disease is now seen as a complex of interacting factors, living and nonliving. Most plants are not damaged by most diseases. Study of ecological factors that sometimes prevent severe disease where it might be expected to occur (and that allow severe disease in other cases) is leading toward practices to minimize disease in agriculture, horticulture, and forestry (11, 65, 67). Moreover, "since the kind of disease epidemic that is familiar to agriculture is not characteristic of undisturbed, natural populations, it seems evident (to me at least) that regulating mechanisms are present in natural systems which operate in such a way as to keep host and parasite populations in balance with each other" (65).

A wide range of chemical, physical, and biological measurements of the environment are made in such studies, and the need for accuracy in making these and understanding the concepts involved has been stressed, for example, for soil pH (82) and soil water–potential studies (22). However, taxonomy of the organisms concerned is usually neglected. Its standard of use may be well below that required in other disciplines and may not be critical enough to detect taxonomic complications hindering the whole project. Incorporation of taxonomy as one of the necessary component disciplines of present-day plant pathology is essential.

A broadly based and experimental approach to taxonomy of fungi will need to supplement herbarium and morphological studies if sound and useful taxonomy is to result. Sutton (85) has pointed out how cultural and pathological work in the genus *Colletotrichum* is leading to a better understanding of species concepts; he stated that "pathogenic, serological, chemical and cultural characteristics of coelomycetes appear to be of increasing value in separation of species and subspecific taxa." The same applies to many other genera. The molecular approach to fungus taxonomy (36), biochemical systematics (88), the use of genetic and pathogenic factors in taxonomy (3, 29, 56, 61), the interrelationships between the taxonomy and biology of fungi and their hosts (1, 40, 56, 71, 74–76, 93, 96), the concept of pathogens as organisms out of place (97), and the dynamic biological equilibrium of natural populations (11, 65) are all areas where characters of value in assisting and supplementing classical taxonomy may be found. As Munk (60) has said, "A modern taxonomist must be prepared to use information from every field of biological science to assist his taxonomic understanding."

Darlington (26) wrote in insect taxonomy about an "increase in the reality of classifications, reached by continual comparison with nature" and of "taxonomy that is correlated with reality and that is becoming increasingly useful as a base for diverse biological studies of living plants and animals." For plant pathology and taxonomic mycology, Luttrell (56) said, "I hope to demonstrate that data on parasitism must be integrated into the taxonomy of fungi and that these data influence concepts of classification and evolution in fungi." Cummins (25) commented that "mycological taxonomists cannot be accused of being non-productive, but we may be, in some degree, criticised for the lack of serviceability of our products." Close cooperative work between taxonomic mycology and plant pathology can help increase the reality of classification in many fungi, and lead to classification with a biological information content in excess of, but often correlated with, the morphological differences seen. In this regard, the following paragraph from Savile (76) dealing with the study of rust fungi is worth repeating, and its message can be applied to many other groups: (The italics are mine.)

> If our taxonomy is to yield maximum information about the rust fungi, and also illuminate the relationships and evolution of the host plants, we must adopt a broadly based and vital approach, using aids from various disciplines. . . . We must also realize that host relationship and ecological specialization, which contribute to genetic isolation, may be as important as morphological distinctions in the assessment of taxa. *The ultrabroad species concept makes it easy to name any rust; but the name is often meaningless and, so far from illuminating the biology of the organism, may actually mask it.* The applied biologist, believing that biological information secured for one member of a complex must apply to all, may be seriously delayed in understanding one that concerns him.

Boerema and his colleagues (14) used on *Phoma* spp. morphological, pathological, cultural, and chemical characters to classify this group, of considerable importance in plant pathology and difficult to identify to species. The results have great relevance for both their taxonomy and biology. In the Hyphomycetes, the classification of genera has been reviewed by Kendrick & Carmichael (48) who often fused into one genus several genera regarded as morphologically indistinguishable. This simplified morphological position could possibly benefit from a taxonomic approach that notes their biology and uses minute but significant morphological differences. In some *Cercospora*-like fungi, Deighton (28) found minute morphological differences in conidium and conidiophore characters that must be considered in both taxonomic and biological work. Talbot (87) has suggested a similar approach to the genera *Drechslera* and *Bipolaris,* considered by some as one genus (*Drechslera*) but by others as two or more distinct genera, differing in their perfect states and in small, but perhaps significant morphological conidial features. The results of two independent studies on susceptibility of a range of fungi to benomyl (methyl-1-(butylcarbamoyl)-2-benzimidazolecarbamate) (15, 31) indicated that classification of Hyphomycetes on morphological differences in conidium ontogeny may be correlated with behavior to this chemical. Martin (58) has stressed the need to study the Fungi Imperfecti as they occur, and Kendrick & Carmichael (48) point to the need for further work by referring to their compilation of the Hyphomycetes "not as an end product, but as providing a new place to begin." Many of the fungi listed earlier could benefit from a broadly based taxonomic approach.

Protesting against the rigid specialization that threatens the research biologist, Savile (73) discussed the interrelationship of plant pathology, taxonomy, and several other biological disciplines. The advantages of a broad biological training for taxonomists has been stressed (10, 19, 56, 57, 73), and the danger in plant pathology, that rigid specialization will cause specialist groups to lose plant pathology has been pointed out (95). One of the great benefits arising from a close association of plant pathology and taxonomy is an overlapping field for broadly based experimental taxonomic work on fungi, leading to greater understanding of their identity, their roles in nature, and how to decrease their damage and increase their benefits.

WHAT CAN BE DONE?

To increase cooperation between pathology and taxonomy, some lines of action are suggested. Possibly many others could also be taken.

Training

The importance of microorganisms as a potent ecological factor should be part of the training of all biologists. The methods of taxonomy, its relevance to problem solving in biology, respect for and the need for preservation of the specimen as the primary record, and taxonomy as an area of specialization all need to be dealt with in training pathologists and applied biologists.

Taxonomists need to be shown the practical relevance of their work in applied biology. In a critique of taxonomy, Raven et al (66) stressed the need for taxonomic work to use a broad spectrum of characters and the need to look for new ways of presenting taxonomic and related information. Too early specialization may be a danger. Moreover, it takes many years to achieve competence as a taxonomist; "it is no exaggeration to say that ten years' work in a herbarium is necessary before a mycologist begins to acquire the general knowledge and sound judgement which is essential to a good taxonomist" (19). At present there is a great shortage of taxonomists, and much more attention to taxonomy as a vital discipline needs to be given in course work.

Location

Taxonomy can only be studied properly in an institution conducive to the work, where the facilities of an active growing herbarium and taxonomic literature are available. Under these conditions, taxonomy and plant pathology could mutually benefit from common facilities of glasshouse and plot areas, media preparation, and library. However, it is essential, in such proximity, that taxonomy not be regarded merely as an answering service for pathology and that the conditions and time necessary for taxonomic research be provided. Taxonomists would be apprised of problem groups of fungi, which could then be considered for more detailed taxonomic work. Pathology can only benefit from having the best possible taxonomy.

Team Work and Planning

Taxonomic advice and cooperation can often prevent problems associated with the organisms under study. Possible confusion with related fungi, synonymy, and perfect-imperfect state correlations are just a few of the areas where taxonomic advice can help. A taxonomist should be included as an advisor or co-worker in pathology and related investigations.

Funding

Funds for taxonomic work are often difficult to get. The relevance of taxonomy is often not obvious to administrators and funding bodies. If it is not supported, both at the training and work stages, and provision made for care of herbaria and purchase of taxonomic literature, the foundation for identification of organisms will be weakened, with severe effects on plant pathology. Taxonomy should be supported, both for its own sake and as an integral part of plant pathology and applied biology as a whole. Funds could perhaps often be obtained by including taxonomic aspects when planning disease projects, and bringing their importance to the notice of administrators. Taxonomy, however, cannot be a stop-start affair since funds are necessary for continued long-term taxonomic research and care of collections of specimens and literature. Only with such support can it continue to make its unique and valuable contribution.

Computers

Computer recall and indexing methods can do much to make available the vast amount of information latent in specimens, labels, and notes of a mycology and plant pathology herbarium (J. Walker, unpublished). This information on identities, host ranges, geographic distributions, host organs attacked, quarantine interceptions, lists of fungi on particular host families, and a wide range of other headings is of great value to the pathologist. Close proximity of taxonomy and pathology can allow ready use of such a system by both groups. Computer-produced index cards and print-outs based on herbarium specimens also provide ready access to the herbarium's resources.

Conclusion

This chapter can be summed up in the sentence written by Stakman (83) when discussing opportunity and obligation in plant pathology, "In diversity there is hope for progress; in unity there is strength to assure it."

Literature Cited

1. Ainsworth, G. C. 1955. Host-parasite relationships. *J. Gen. Microbiol.* 12:352–55
2. Ainsworth, G. C. 1961. Storage and retrieval of biological information. *Nature* 191:12–14
3. Ainsworth, G. C. 1962. Pathogenicity and the taxonomy of fungi. In *Microbial Classification,* ed. G. C. Ainsworth, P. H. A. Sneath. *Symp. Soc. Gen. Microbiol.* 12:249–69. 483 pp.
4. Ainsworth, G. C. 1963. The pattern of mycological information. *Mycologia* 55:65–72
5. Ainsworth, G. C. 1966. A general purpose classification of fungi. *Commonw. Mycol. Inst. Bibliogr. Syst. Mycol.* 1966 (1):1–4
6. Ainsworth, G. C. 1971. *Ainsworth & Bisby's Dictionary of the Fungi.* Kew, England: Commonw. Mycol. Inst. 6th ed. 663 pp.
7. Ainsworth, G. C. 1973. Introduction and keys to higher taxa. In *The Fungi. An Advanced Treatise,* ed. G. C. Ainsworth, F. K. Sparrow, A. S. Sussman, 4A:1–7. New York: Academic. 621 pp.
8. Ainsworth, G. C., Sparrow, F. K., Sussman, A. S., Eds. 1973. *The Fungi. An Advanced Treatise,* Vol. 4A, 621 pp.; Vol. 4B, 504 pp. New York: Academic
9. Ainsworth, G. C., Sussman, A. S., Eds. 1965–1968. *The Fungi. An Advanced Treatise.* Vol. 1, 1965, 748 pp.; Vol. 2, 1966, 805 pp.; Vol. 3, 1968, 738 pp. New York: Academic
10. Baker, A. D. 1963. Some observations on the development of the Nematology Section at Ottawa, and on some of the needs of nematology. *Phytoprotection* 44:5–11
11. Baker, K. F., Cook, R. J. 1974. *Biological Control of Plant Pathogens.* San Francisco: Freeman. 433 pp.
12. Ballantyne, B. 1963. A preliminary note on the identity of cucurbit powdery mildews. *Aust. J. Sci.* 25:360–61
13. Blake, S. T. 1949. The specimen, the species and the botanist. *Aust. J. Sci.* 11:119–23
14. Boerema, G. H. et al 1964–1973. Several papers in *Persoonia,* Vol. 3–7
15. Bollen, G. J., Fuchs, A. 1970. On the specificity of the in vitro and in vivo antifungal activity of benomyl. *Neth. J. Plant Pathol.* 76:299–312
16. Christensen, C. M. 1965. *The Molds and Man.* Minneapolis: Univ. Minn. Press. 284 pp.

17. Clark, W. A., Loegering, W. Q. 1967. Functions and maintenance of a type-culture collection. *Ann. Rev. Phytopathol.* 5:319–42
18. Clements, F. E., Shear, C. L. 1931. *The Genera of Fungi.* New York: Wilson. 496 pp.
19. Committee of the British Mycological Society. 1949. The need for encouraging the study of systematic mycology in England and Wales. *Trans. Br. Mycol. Soc.* 32:104–12
20. Committee on Systematics in Support of Biological Research. 1970. *Systematics in Support of Biological Research.* Prepared for National Science Foundation, Washington DC. 25 pp.
21. Commonwealth Mycological Institute. 1968. *Plant Pathologist's Pocketbook.* Kew, England: Commonw. Mycol. Inst. 267 pp.
22. Cook, R. J., Papendick, R. I. 1972. Influence of water potential of soils and plants on root disease. *Ann. Rev. Phytopathol.* 10:349–74
23. Cowan, S. T. 1965. Principles and practice of bacterial taxonomy—a forward look. *J. Gen. Microbiol.* 39:143–53
24. Crovello, T. J., MacDonald, R. D. 1970. Index of EDP-IR projects in systematics. *Taxon* 19:63–76
25. Cummins, G. B. 1947. Some problems in mycological taxonomy. *Mycologia* 39:627–34
26. Darlington, P. J. Jr. 1971. Modern taxonomy, reality, and usefulness. *Syst. Zool.* 20:341–65
27. Davis, C. C. 1963. Biology is not a totem pole. *Science* 141:308,310
28. Deighton, F. C. 1973. Studies on *Cercospora* and allied genera. IV. *Commonw. Mycol. Inst. Mycol. Pap.* No. 133:1–60
29. Dennis, R. W. G. 1952. Biological races and their taxonomic treatment by mycologists. *Proc. Linn. Soc. London* 16:47–53
30. Dring, D. M. 1971. Techniques for microscopic preparation. In *Methods in Microbiology,* ed. C. Booth, 4:95–111. New York: Academic. 795 pp.
31. Edgington, L. V., Khew, K. L., Barron, G. L. 1971. Fungitoxic spectrum of benzimidazole compounds. *Phytopathology* 61:42–44
32. Ellis, M. B. 1971. *Dematiaceous Hyphomycetes.* Kew, England: Commonw. Mycol. Inst. 608 pp.

33. Emerson, R. 1973. Mycological relevance in the nineteen seventies. *Trans. Br. Mycol. Soc.* 60:363–87
34. Fosberg, F. R., Sachet, M. H. 1965. *Manual for Tropical Herbaria (Regnum Vegetabile*, Vol. 39). Utrecht: Int. Bur. Plant Taxon. Nomencl. 132 pp.
35. Graham, M. W. R. de V. 1972. Taxonomy in distress. *Nature* 238:475
36. Hall, R. 1969. Molecular approaches to taxonomy of fungi. *Bot. Rev.* 35:285–304
37. Harper, F. R. 1969. A profession to deal with the diagnosis and treatment of disease in plants. *BioScience* 19:690–92
38. Hawksworth, D. L. 1974. *Mycologist's Handbook.* Kew, England: Commonw. Mycol. Inst. 231 pp.
39. Holm, L. 1957. Études taxonomiques sur les Pléosporacées. *Symb. Bot. Ups.* 14:1–188
40. Holm, L. 1969. An uredinological approach to some problems in angiosperm taxonomy. *Nytt Mag. Bot.* 16:147–50
41. Holton, C. S., Fischer, G. W., Fulton, R. W., Hart, H., McCallan, S. E. A. 1959. *Plant Pathology Problems and Progress, 1908–1958.* Madison, Wis.: Univ. Wisconsin Press. 588 pp.
42. Horsfall, J. G., Dimond, A. E., Eds. 1959–1960. *Plant Pathology. An Advanced Treatise.* New York: Academic. 3 vols. 2064 pp.
43. Hughes, S. J. 1953. Conidiophores, conidia, and classification. *Can. J. Bot.* 31:577–659
44. *International Code of Botanical Nomenclature.* 1972. ed. F. H. Stafleu et al. Utrecht: Int. Assoc. Plant Taxon. 426 pp.
45. James, H. G. 1947. The importance of taxonomy in biological control work. *Rept. Que. Soc. Prot. Plants* 30:51–53
46. Keitt, G. W. 1959. History of plant pathology. See Ref. 42, 1:61–97
47. Kendrick, B., Ed. 1971. *Taxonomy of Fungi Imperfecti.* Toronto: Toronto Univ. Press. 309 pp.
48. Kendrick, W. B., Carmichael, J. W., 1973. Hyphomycetes. See Ref. 8, 4A:323–509
49. Kern, F. D. 1943. The importance of taxonomic studies of the fungi. *Torreya* 43:65–77
50. Key, K. H. L. 1970. The present status of Australian insect taxonomy. *Search Sydney* 1:69–72
51. Kuiper, K., Boerema, G. H. 1972. Quarantine aspects of the epidemiology of soil-borne pathogens. *Organ. Eur.*

Mediterr. Prot. Plant. Bull. 1972 (7): 61–68
52. Lanjouw, J., Stafleu, F. A. 1964. *Index Herbariorum. Part 1. The Herbaria of the World.* Utrecht: Int. Bur. Plant Taxon. Nomen. 5th ed. 251 pp.
53. Laundon, G. F. 1968. A cold method for preparing dried reference cultures. *Trans. Br. Mycol. Soc.* 51:603–4
54. Linder, D. H. 1941. Mycologists in relation to others. *Mycologia* 33:453–60
55. Luttrell, E. S. 1958. The function of taxonomy in mycology. *Mycologia* 50: 942–44
56. Luttrell, E. S. 1974. Parasitism of fungi on vascular plants. *Mycologia* 66:1–15
57. Martin, G. W. 1958. Mycological taxonomy as related to practical problems in microbiology. *Mycologia* 50:97–102
58. Martin, G. W. 1961. Nomenclature of Deuteromycetes. *Taxon* 10:153–54
59. Mason, E. W. 1940. Presidential address on specimens, species and names. *Trans. Br. Mycol. Soc.* 24:115–25
60. Munk, A. 1962. An approach to an analysis of taxonomic method with main reference to higher fungi. *Taxon* 11:185–90
61. Nelson, R. R. 1965. Assessing biological relationships in the fungi. *Phytopathology* 55:823–26
62. New herbarium abbreviations. 1966. *Taxon* 15:334–36
63. Onions, A. H. S. 1971. Preservation of fungi. See Ref. 30, pp. 113–51
64. Parmeter, J. R. Jr. 1965. The taxonomy of sterile fungi. *Phytopathology* 55:826–28
65. Person, C. O. 1972. Basic biology of rusts and rust disease resistance: moderators summary. *US Dep. Agric. Misc. Publ.* 1221:93–95
66. Raven, P. H., Berlin, B., Breedlove, D. E. 1971. The origins of taxonomy. *Science* 174:1210–13
67. Reichert, I., Palti, J. 1967. Prediction of plant disease occurrence. A pathogeographical approach. *Mycopathol. Mycol. Appl.* 32:337–55
68. Riess, H. 1854. Neue Kernpilze. *Hedwigia* 1:25–28
69. Rogers, D. P. 1958. The philosophy of taxonomy. *Mycologia* 50:326–32
70. Saccardo, P. A. 1899. Tabulae comparativae genérum fungorum omnium. *Sylloge Fungorum Omnium Hucusque Cognitorum* 14:1–62
71. Savile, D. B. O. 1954. The fungi as aids in the taxonomy of the flowering plants. *Science* 120:583–85

72. Savile, D. B. O. 1962. *Collection and Care of Botanical Specimens.* Ottawa, Ontario: Can. Dep. Agric. Res. Br. Publ. 1113:1–124

73. Savile, D. B. O. 1966. Unity from diversity in biological research. *Trans. R. Soc. Can. Ser. 4* 4:245–51

74. Savile, D. B. O. 1971. Coevolution of the rust fungi and their hosts. *Q. Rev. Biol.* 46:211–18

75. Savile, D. B. O. 1971. Co-ordinated studies of parasitic fungi and flowering plants. *Nat. Can.* 98:535–52

76. Savile, D. B. O. 1971. Methods and aims in the study of the rust fungi. *J. Indian Bot. Soc.* 50A:41–51

77. Shaw, C. G. 1965. Taxonomic concepts as applied to the fungi. *Phytopathology* 55:819–21

78. Shaw, C. G. 1966. How to collect, prepare and ship plant disease material for identification. *Wash. Agric. Exp. Stn. Cir.* 132(rev.):1–5

79. Shear, C. L. 1915. Mycology in relation to phytopathology. *Science NS* 41:479–84

80. Shear, C. L. 1929. Mycological nomenclature in relation to plant pathology. *Proc. Int. Congr. Plant Sci.* 2:1225–26

81. Skolko, A. J. 1961. The status of taxonomic mycology in present day plant pathology. *Rept. Sixth Comm. Mycol. Conf., 1960,* 26–29

82. Smiley, R. W., Cook, R. J. 1972. Use and abuse of the soil pH measurement. *Phytopathology* 62:193–94

83. Stakman, E. C. 1964. Opportunity and obligation in plant pathology. *Ann. Rev. Phytopathol.* 2:1–12

84. Subramanian, C. V. 1965. Taxonomy of fungi in relation to plant pathology. In *Advances in Agricultural Sciences and Their Applications,* ed. S. Krish-namurth, 522–27. Coimbatore: Madras Agric. J. 666 pp.

85. Sutton, B. C. 1973. Coelomycetes. See Ref. 8, 4A:513–82

86. Talbot, P. H. B. 1971. *Principles of Fungal Taxonomy.* London: Macmillan. 274 pp.

87. Talbot, P. H. B. 1973. On the genus *Helminthosporium* sensu lato. *Aust. Plant Pathol. Soc. Newsl.* 2 (2):3–7

88. Tyrrell, D. 1969. Biochemical systematics and fungi. *Bot. Rev.* 35:305–16

89. United States Department of Agriculture. 1960. *List of Intercepted Plant Pests, 1959.* Washington DC: U.S. Dep. Agric., Plant Quarantine Div. 86 pp.

90. von Arx, J. A. 1957. Die Arten der Gattung Colletotrichum Cda. *Phytopathol. Z.* 29:413–68

91. von Arx, J. A., Olivier, D. L. 1952. The taxonomy of *Ophiobolus graminis* Sacc. *Trans. Br. Mycol. Soc.* 35:29–33

92. Walker, J. 1972. Type studies on *Gaeumannomyces graminis* and related fungi. *Trans. Br. Mycol. Soc.* 58:427–57

93. Walker, J., Bertus, A. L. 1971. Shoot blight of *Eucalyptus* spp. caused by an undescribed species of *Ramularia. Proc. Linn. Soc. NSW* 96:108–15

94. Walker, J., Hartigan, D., Bertus, A. L. 1974. Poplar rusts in Australia with comments on potential conifer rusts. *Eur. J. For. Pathol.* 4:100–18

95. Walker, J. C. 1963. The future of plant pathology. *Ann. Rev. Phytopathol.* 1:1–4

96. Watson, L. 1972. Smuts on grasses: some general implications of the incidence of Ustilaginales on the genera of Gramineae. *Quart. Rev. Biol.* 47:46–62

97. Yarwood, C. E. 1967. Pathogens as organisms out of place. *Phytopathol. Z.* 58:305–14

CLAY MINERALOGY IN RELATION TO SURVIVAL OF SOIL BACTERIA

❖3627

K. C. Marshall

School of Microbiology, University of New South Wales, Kensington, NSW, Australia

INTRODUCTION

Although biologists are conscious of the role of different soil properties in the development of and interaction between different plant communities, little critical thought has been devoted to the effects of such soil properties on the development of microbial communities within soil habitats. Soils are an important repository for many plant pathogens, and the survival of these pathogens in the absence of their specific host plants depends on a complex interplay between the physical, chemical, and biological factors in the soil. The situation is made even more complex by the tremendous degree of variability in these factors, both within and between different soil types.

Unlike laboratory culture systems, soil is a growth medium consisting predominantly of particulate matter offering a large, but variable, surface area that may drastically modify the growth and interaction between microbial populations. The broad aspects of interactions between microorganisms and soil particles have been the subject of several recent reviews (14, 20, 31, 40, 61). Consequently, this review is restricted to (*a*) the influence of clay minerals on bacterial survival and (*b*) the development of a conceptual basis for more realistic future studies on the role of clay minerals in modifying the survival of bacteria in soils. The successful interpretation of more detailed studies on survival depends on our ability to define the extent and nature of the many solid and other surfaces encountered in soil microhabitats, as well as on our ability to relate such knowledge to the effects of these surfaces on microbial behavior.

CLAY MINERALOGY AND BACTERIAL SURVIVAL

Clay minerals are hydrous aluminosilicates and are the finest fraction of the inorganic component of soils. Different clay minerals vary in their particle size, chemical

357

composition, surface charge properties, cation exchange capacity, and water retention properties. The principal clay minerals found in soils are members of the
kaolinite, montmorillonite, and illite groups. Detailed descriptions of the chemical
and physical properties of these clays are given by Grim (17) and van Olphen (68).

Because the survival of bacteria in soils is influenced by a variety of chemical,
physical, and biological factors, effects of clay minerals in modifying bacterial
survival involve many different mechanisms.

A well-documented example of the relationship between clay mineralogy and
microbial activity in soils concerns the rapid spread of *Fusarium oxysporum* f. sp.
cubense in Central American soils lacking swelling clays of the montmorillonite
type (62). Stotzky (59, 60) and Stotzky & Rem (64, 65) attributed the failure of the
fungus to proliferate in soils containing montmorillonite to the buffering effect of
this clay mineral. Montmorillonite was found to maintain the pH at levels favorable
for bacterial growth, allowing the fungus to be excluded by effective bacterial
competition (33). Similarly, the suggestion of Zeidberg (71) of a relationship between soil type and the occurrence of the human pathogenic fungus *Histoplasma
capsulatum* led to the demonstration by Stotzky & Post (63) that the fungus was
absent in montmorillonite-containing soils.

Although many soil microorganisms are capable of producing antibiotics in the
laboratory, few instances have been recorded of antibiotic production in soils. That
some antibiotics are rapidly sorbed by clay minerals (24, 25, 45, 46, 55, 56) may
provide a partial explanation of this apparent lack of activity. The removal of an
antibiotic from the aqueous phase of the soil may enhance survival of susceptible
bacteria in this phase. Whether such sorbed antibiotics are effective against bacteria
adhering to the same surfaces would depend on the mode of action and the orientation on the surface of the particular antibiotic molecule. Skinner (51) has shown that
inactivation of the antibiotic produced by *Streptomyces albidoflavus* by the addition
of bentonite (predominantly a montmorillonite-type clay) did not prevent the suppression of the growth of *Fusarium culmorum* by the streptomycete. Skinner attributed this suppression to a competitive interaction between the two organisms.
Soulides (54) demonstrated that the induced production of streptomycin in kaoli-
nitic soil increased the tolerance of the bacterial flora to this antibiotic, whereas the
tolerance in the population of a montmorillonitic soil remained unchanged. This
effect was attributed to the adsorption of the antibiotic by the montmorillonitic clay.

The adsorption of extracellular enzymes by soil particulate material (52) may
limit the growth of certain bacteria and lessen their chances of survival in such soils.

Hattori (20) found superior survival of gram-negative, nonsporing bacteria in the
interior of aggregates subjected to desiccation as compared to their survival on the
surface of the aggregates. This was attributed to more favorable water relations in
the interior of the aggregates. This observation may be significant in comparisons
of bacterial survival in sandy soils lacking aggregate formation, and soils of heavier
texture featuring aggregate production. Soil textural differences have been implicated in influencing the survival of root-nodule bacteria in desiccated soils (16, 35,
67, 69). Using sterile sandy soil, Marshall (26) found that fast-growing rhizobia (e.g.
Rhizobium trifolii) were susceptible to desiccation and heating, while slow-growing

species (e.g. *R. japonicum*) were much more resistant to the applied stress. The susceptible bacteria were more resistant to desiccation and heating in soils of heavier texture or in sandy soil amended with montmorillonite or illite, but not kaolinite clays. More recent studies (H. V. A. Bushby & K. C. Marshall, unpublished data) have shown that survival of the slow-growing bacteria is not improved in clay-amended soils exposed to desiccation. Marshall & Roberts (36) reported favorable responses in survival of *R. trifolii* exposed to desiccation and high temperatures in field sandy soils amended with appropriate clays.

Marshall (27–31) has shown that, in aqueous suspensions, colloidal clays form an envelope around bacterial cells as a consequence of electrostatic attraction between charged groups on the bacterial and clay surfaces. It was suggested that the clay envelope might protect fast-growing rhizobia from effects of desiccation by modifying rates of water movement in and out of the bacterial cells during the desiccation or rehydration phases. However, H. V. A. Bushby & K. C. Marshall (submitted for publication) have shown that rates of water movement from fast- and slow-growing rhizobia are so rapid that this mechanism is not significant in explaining the difference in desiccation susceptibility between these groups of bacteria. Evidence obtained from water adsorption isotherms for both groups of bacteria indicates that the greater desiccation susceptibility of the fast-growing rhizobia may be related to their higher state of internal hydration at low relative vapor pressures. The relatively high affinity for water found for montmorillonite suggested that the clay may protect the fast-growing bacteria by reducing their state of hydration at low relative vapor pressures to below some critical level where enzyme activity ceases. The slow-growing bacteria have a lower affinity for water and may be below this critical hydration level. The protection from desiccation injury of aerobic, nonsymbiotic nitrogen-fixing bacteria (11) afforded by kaolinite probably cannot be explained in these terms, because of the larger particle size and lower surface energy of this clay mineral.

Azotobacter chroococcum was protected by montmorillonite from the lethal effects of X rays (41), and the degree of protection of *Klebsiella aerogenes* from UV irradiation by clays and humic acids was related to the light absorption and light-scattering properties of these materials (4).

Clays may influence the survival of bacteria in soils by modifying host-parasite or host-predator interactions. The adsorption of viruses by clays has been reported (13, 39). A direct protection from bacteriophage attack of *Escherichia coli* by sediment colloids, extracted organic matter, and montmorillonite has been reported by Roper & Marshall (49). The evidence presented suggests that the bacterium is protected at low electrolyte levels by an envelope of colloidal materials around the cell surface, whereas at high electrolyte concentrations further protection is provided by adsorption of both bacteria and phages to larger particulates. Preliminary evidence (M. M. Roper & K. C. Marshall, unpublished data) indicates that montmorillonite interferes with the parasitism of *E. coli* by *Bdellovibrio bacteriovorus*.

These examples indicate that clay mineralogy does have a modifying influence on bacterial survival in different soils. To adequately explain these phenomena, however, it is necessary to obtain a clearer understanding of the nature of microhabitats in various soils and to relate this information to the behavior of microorganisms in such microhabitats. In the following sections a conceptual basis is provided for examining the behavior of bacteria at different surfaces (interfaces) by treating bacteria as living colloidal particles, as well as a basis for more closely defining differences in microhabitats within and between various soils.

MICROORGANISMS AND INTERFACES

An interface is the boundary between two phases in a heterogeneous system. Natural microbial habitats such as soils are heterogeneous systems consisting of a variety of solid-liquid, gas-liquid, liquid-liquid, and solid-gas interfaces. These phase boundaries possess physical and chemical properties that differ from those of either bulk phase in a two-phase system. The unique forces at interfaces have a significant effect on the distribution of ions, macromolecules, and colloidal materials near the interface. In many natural habitats of low nutrient status, interfaces serve as sites of nutrient concentration and intensified microbial activity. The extent of the increased activity over that in the bulk phase depends on the overall nutrient status of the ecosystem, the nature of the phase boundaries, and the interfacial area available for microbial colonization.

As a result of their small size, bacteria suspended in an electrolyte exhibit properties characteristic of colloidal systems. Some insight into the behavior of bacteria at interfaces is revealed by the application of certain concepts of classical colloid chemistry to such systems. It must be stressed, however, that bacteria differ from normal colloids in that they are capable of metabolism, growth, and in some instances, independent motion. These biological characteristics do play an important role in the initial attraction of bacteria to interfaces and in the degree of selectivity that is evident in the process of adhesion of bacteria to solid surfaces.

Attraction of Bacteria to Interfaces

To take advantage of the improved nutrient status at interfaces, bacteria must be attracted to interfaces by one or more of the following mechanisms.

CHEMOTAXIS   Motile bacteria are capable of a chemotactic response to very low levels of nutrients introduced into an otherwise nutrient-deficient situation (2). It must be expected that certain motile bacteria in natural habitats will respond to the nutrient gradient resulting from the concentration of nutrients at interfaces. Motile bacteria may be attracted to the gas phase at a gas-liquid interface, and, in fact, Kelman & Hruschka (22) have suggested that aerotaxis provides a means for the selective increase of avirulent strains of *Pseudomonas solanacearum* at the air-liquid interface of static broth cultures.

A negative chemotaxis might occur if inhibitory substances (including hydrogen ions) have accumulated at interfaces (10, 50, 70). Seymour & Doetsch (50) have suggested that negative chemotaxis plays an important biological role in the survival of motile bacteria and that positive chemotaxis, in some instances, may be an incidental or useless response. This would not be true of motile bacteria attracted to nutrients concentrated at interfaces in many natural habitats.

It should be obvious that chemotaxis cannot account for the attraction of non-motile bacteria to interfaces.

ELECTROSTATIC ATTRACTION   Bacteria possess a net negative surface charge at normal physiological pH values (1, 27, 47). The net charge at an interface depends on the nature of the phases involved (in the case of charged solid surfaces) and on the ionic species adsorbed at the interface. Electrophoretic studies of various particulate materials indicate that most possess a net negative surface charge, but this charge may be altered by adsorption to the bacterial surface of proteins and other materials in natural environments (1, 23, 42, 43). The similarity in the charge of the bacteria and most interfaces probably precludes electrostatic phenomena as a means of attraction of bacteria to interfaces.

Evidence has been presented for the role of electrostatic phenomena in the attraction of colloidal clays to bacterial surfaces (28–30). Positively charged edges of colloidal clay particles are attracted to negatively charged sites on the bacterial surface, a process that is prevented when the positive sites on the clays are blocked by using Na hexametaphosphate. In those bacteria possessing some positively charged surface ionogenic groups, the predominantly negatively charged faces of clay particles are attracted to the bacterial surface whether or not the clay has been pretreated with hexametaphosphate.

ELECTRICAL DOUBLE-LAYER PHENOMENA   Cations are attracted to a negatively charged interface, whereas anions tend to be excluded. These cations are not held firmly at the interface, but form a diffuse layer (the so-called electrical double layer) near the surface. The extent of this diffuse double layer depends on the valence and the concentration of the electrolyte used, the double layer being compressed as either the valence or the concentration of the electrolyte is increased. When two negatively charged bodies are in proximity, their cationic double layers overlap and there is a tendency for a mutual repulsion between the bodies. The double-layer repulsion energy is counteracted to some extent by the van der Waals attraction energy of each body. The effective extent of the double-layer repulsion energy is related to the thickness of the electrical double layer and, thus, varies with the electrolyte concentration and valence. However, the extent of the van der Waals attraction energy is independent of the electrolyte composition. At low electrolyte concentrations, the repulsion energy is effective well beyond the zone of influence of the attraction energy (Figure 1), and the two bodies are repelled from each other. At higher electrolyte concentrations, the diffuse double layer may be compressed to a point where van der Waals energies are effective in attracting the bodies to a finite distance from each other (Figure 1). The bodies do not come in direct contact because, if the particle separation is further reduced, the resultant repulsion energy is greater than the attraction energy.

The effects of electrolyte concentration and valence on the attraction of cells of a nonmotile *Achromobacter* sp. to glass surfaces have been examined by Marshall et al (37). The numbers of bacteria attracted to the surface lessened with decreasing electrolyte concentration until a point was reached where all bacteria were repelled from the surface. The concentrations at which this occurred were approximately

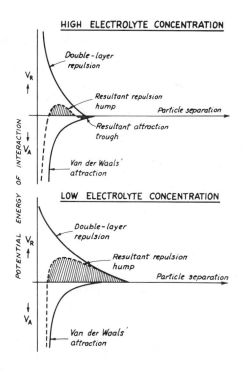

HIGH ELECTROLYTE CONCENTRATION

LOW ELECTROLYTE CONCENTRATION

Figure 1   Idealized curves of the potential energy of interaction between two solid particles as a function of the interparticle distance, shown for both high and low electrolyte concentrations. V_A and V_R = attraction and repulsion energies, respectively.

5×10^{-4} M for a uni-univalent electrolyte (NaCl) and 5×10^{-5} M for a divalent electrolyte ($MgSO_4$). For both electrolytes, these values corresponded to a theoretical double-layer thickness of approximately 200 Å. At higher electrolyte concentrations, where the double-layer thickness was further compressed, the effects of the van der Waals attraction energy was evident in the attraction of bacteria to the surface. This was termed a *reversible* phase of sorption, because the bacteria were readily removed by the shearing force of a jet of water or, in the case of motile bacteria, by their own motive power. It was further shown that the kinetic energy of a motile *Pseudomonas* sp. (5.45×10^{-18} erg) was inadequate to overcome the repulsion energy barrier (approximately 50×10^{-13} erg) adjacent to a glass surface suspended in a 0.2 M NaCl solution.

CELL-SURFACE HYDROPHOBICITY   In any two-phase system involving water, it is feasible that a bacterium possessing a relatively hydrophobic outer surface may be rejected from the aqueous phase. *Mycobacterium tuberculosis* is an excellent example of a bacterium that grows only at the air-water interface in normal liquid media. That it grows as a homogeneous suspension in the presence of a detergent such as Tween 80 ® (12) indicates that its outer surface is hydrophobic. The specific perpendicular orientation of some bacteria at air-water, oil-water, and solid-water interfaces (Figure 2) suggests that one pole of these cells is hydrophobic and is being attracted to the nonaqueous phase. Tween 80 prevented this specific mode of orientation (32, 34).

BUOYANCY   Increased buoyancy of particular bacteria would enable their selective accumulation at air-water interfaces. Possible mechanisms of attaining some degree of buoyancy in bacteria include the possession of stalks, as in *Caulobacter* spp. (48), and gas vacuoles, as in *Ancalomicrobium* and *Prosthecomicrobium* (58).

Firm Adhesion of Bacteria to Solid Surfaces

The existence of a repulsion energy barrier preventing direct contact between bacteria and solid surfaces of similar charge raises the question as to how some bacteria eventually adhere to such surfaces. Displacement of a relatively large body such as a bacterium by Brownian motion is not sufficient to overcome this energy barrier, but displacement of very fine extracellular polymeric fibrils produced by some bacteria may be sufficient to bridge the gap between the surfaces and, hence, anchor the bacterium to the solid surface. The principle involved is the same as that reported for polymeric bridging between adjacent bacterial cells resulting in their flocculation. This effect can be achieved by the addition of anionic or nonionic polymers to normally nonflocculating bacterial cultures (8, 19, 66). The production of extracellular acid polysaccharide by marine bacteria sorbed to glass surfaces has been reported by Corpe (9). Indirect evidence for the role of polymeric bridging in the adhesion of bacteria to surfaces has been reported (21, 37, 44, 73). More direct evidence for this phenomenon has been obtained with bacteria adhering to solid surfaces in a perpendicular manner (32, 34). These studies clearly demonstrated that the bacteria were attached to solid surfaces by either fibrillar or amorphous extracellular materials (Figure 2).

This process of firm adhesion has been termed the *irreversible* phase of sorption (37) and appears to be a highly selective process (15, 38, 72). This selectivity is related to the ability of different bacteria to produce suitable bridging polymers. Zvyagintsev et al (74) have shown that the adhesion forces of a number of microorganisms adsorbed on glass surfaces range from 4×10^{-7} to 4×10^{-4} dyne per cell. They reported that the number of cells firmly attaching to a surface decreases as the age of a culture increases.

CLAY MINERALOGY AND MICROHABITATS IN SOIL

Microhabitats within soils are notable for their remarkable diversity and complexity. Soils may be broadly classified in terms of textural classes to give some indication of the variability in the types and sizes of the major particulate components of the soils (Table 1). With decreasing particle size there is an increase in particle number and in the surface area per gram of soil. It is clear that the interfacial area enlarges with an increase in the proportion of the clay-size fraction and, consequently, the opportunities for sorptive interactions between microorganisms and soil particles should increase. However, knowledge of the gross textural proper-

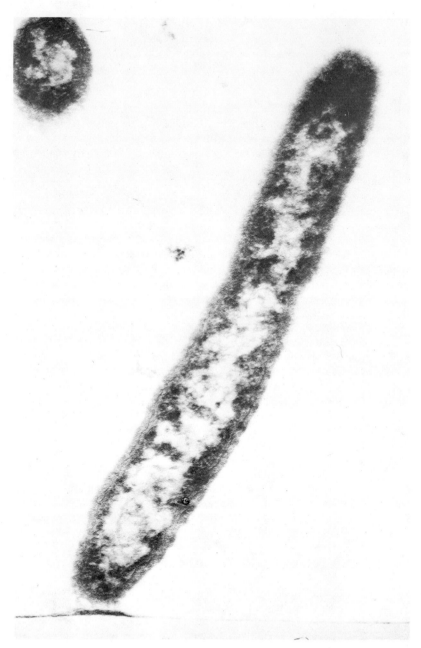

Figure 2   Electron micrograph of a thin section of a reembedded araldite block showing the perpendicular orientation of *Flexibacter* CW7 at the original solid-liquid interface, and polymeric bridging between the bacterial cell and the solid surface (X 52,900).

Table 1   Size distribution and surface area of solid particles found in soils (adapted from Black, 5)

Textural class	Particle diameter (μm)	Approx. no. particles/g[a]	Approx. surface area (cm^2/g)[a]	Mineralogical composition
Coarse sand	2000–200	5.4×10^2	21	Mainly quartz, some rock fragments
Fine sand	200–20	5.4×10^5	210	Mainly quartz and feldspar, some ferromagnesians
Silt	20–2	5.4×10^8	2,100	Mainly quartz and feldspar, some ferromagnesians, mica, and clay minerals
Clay	<2.0	7.2×10^{11}	23,000	Mainly clay minerals, some quartz

[a] Based on an average diameter and a density of 2.65 g/cm^3.

ties or even of the clay mineralogical status of soils is not sufficient for a full understanding of the effects of such properties on the survival of particular bacterial species in different soils. An adequate description of the microhabitats of bacteria in soils requires some recognition of the importance of aggregate formation and of the more detailed organization of the solid phase constituents (fabric) of the soils.

Soil Aggregates

The association of soil components into aggregates or crumbs is a feature of most soils. Hattori (20) has reviewed recent Japanese work on the differentiation of aggregate microhabitats into the outer and inner portions of the aggregates. He has presented a simplified aggregate model to emphasize the advantages of the inner portion of the aggregate as a site for survival of desiccation-sensitive, nonsporing, gram-negative bacteria. Spore-forming bacteria, actinomycetes, and fungi predominate in the outer portion of partly dried aggregates but, on remoistening, the numbers of gram-negative bacteria on the outer portion increase rapidly until they become dominant. The preparation of surface films from soil aggregates (18) provides a method for the direct observation of those microorganisms at the solid-liquid interface of the aggregate outer surface. The technique could be adapted to an examination of inner portions of aggregates by the use of freshly broken crumbs.

Because of problems of oxygen diffusion into the inner portion of aggregates under conditions favoring very active microbial growth, anaerobic conditions often may exist in such regions. Relevant to this and to the activity of aerobes in both regions of aggregates is the possible control of the activities of aerobes by microbial ethylene production in the anaerobic portion of the aggregate (53). Decreasing aggregate size, with concomitant increase in surface area, should favor more rapid oxygen diffusion into the inner aggregate sites. Seifert (quoted by Hattori, 20) has found that the rate of the aerobic process of nitrification increased as the size of soil aggregates decreased.

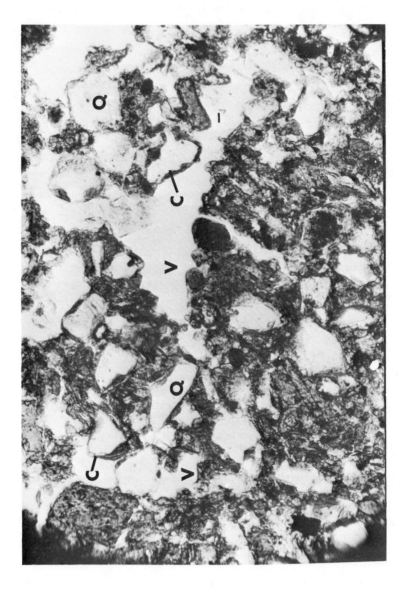

Figure 3   Thin section of a red-earth soil (X 400). *Above,* Bright-field illumination, *opposite,* crossed nicols—revealing darkened voids (*V*), bright quartz grains (*Q*), and birefringent, oriented, clay-iron oxide cutans (*C*).

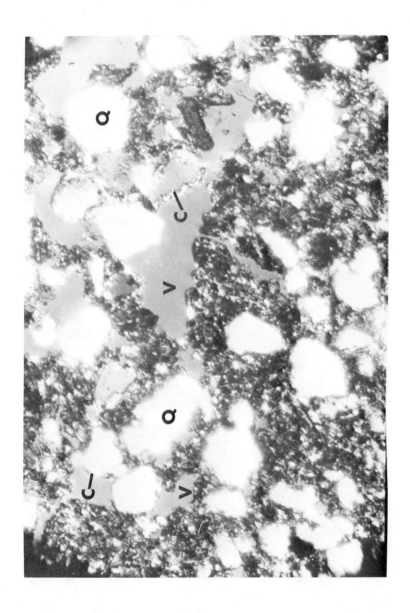

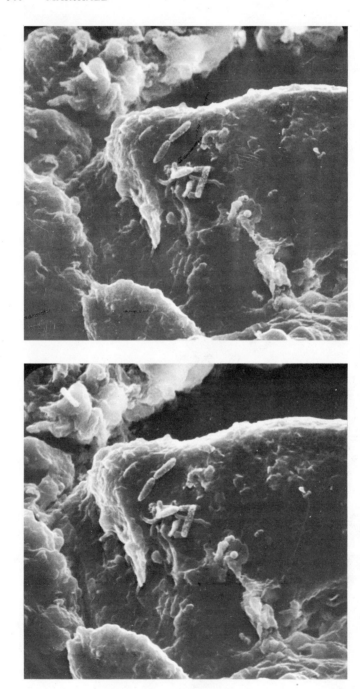

Figure 4   Stereo-pair scanning electron micrographs of a red-earth soil (X 2070). *Left*, Specimen tilt 39°, *right*, tilt 45°. Note the bacteria and their orientation on the surfaces of soil particles. These surfaces probably are clay cutans covering quartz grains (see Figure 3).

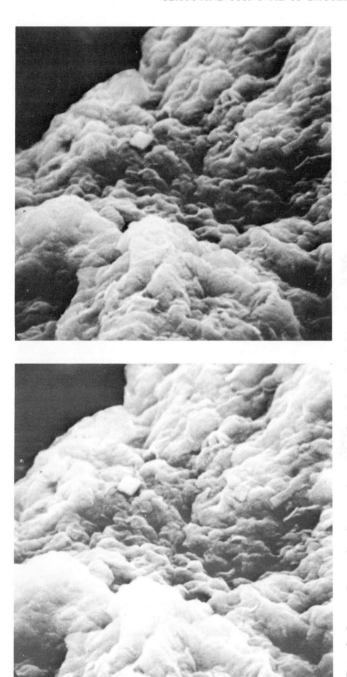

Figure 5 Stereo-pair scanning electron micrographs of a black-earth soil (× 5520). *Left*, Specimen tilt 39°, *right*, tilt 45°. Bacterial cells are not obvious even in the three-dimensional image obtained under the stereo viewer. It is assumed that the vague bacteria-like humps are, in fact, bacteria heavily coated by colloidal clay and organic matter.

The Soil Fabric

The mode of organization of the particulate components on a microscale determines the types of surfaces and the available surface area of the solid-liquid interfaces in soils. *Soil fabric* has been defined by Brewer (6) as "the physical constitution of a soil material as expressed by the spatial arrangement of the solid particles and associated voids." Voids or pore spaces are regarded as distinct entities surrounded by the solid phase. It is in the aqueous phase and at the solid- or gas-water interfaces within the voids that soil microorganisms normally function. Brewer & Sleeman (7) have divided the solid part of soils into either skeleton grains or plasma on the basis of their respective physical and physicochemical properties. Skeleton grains are relatively stable and are not readily translocated, concentrated, or reorganized by soil-forming processes. They include mineral grains and resistant siliceous and organic bodies larger than colloidal size, and are capable of being weathered to form plasma. Plasma is that part of the soil that has been or is capable of being moved, reorganized, or concentrated by the processes of soil formation. It includes all materials, both mineral and organic, of colloidal size. The manner in which the skeleton grains and plasma are organized in different soils determines the microhabitats wherein various microbial species function in the soils.

The detailed concepts and language of the soil fabric analyst are beyond the scope of this review, and the reader is referred to the text by Brewer (6) on this topic.

An example of the important detail revealed by soil fabric analysis is the presence of clay skins (cutans) around quartz grains of the red-earth soil (3), shown in Figure 3. The birefringence of iron-rich clay cutans (or ferriargillans) indicates that these cutans are highly oriented around the quartz grains (Figure 3B). The influence of the argillan-water interface on microbial behavior would be very different from that of interfaces in soils where quartz grains remain free of any clay coating. Microbial colonization of an argillan-water interface in this red-earth soil is shown in Figure 4. Microorganisms at such an interface retain direct access to the aqueous phase, a condition probably conducive to high levels of microbial activity and to rapid organic matter decomposition. This must represent a very different microhabitat to that at the outer surface of an aggregate of a black-earth soil (57), where the components of the soil are organized in a much more amorphous form, and where the presence of microorganisms at solid-void interfaces is not revealed by examination under the scanning electron microscope (Figure 5). It is suggested that most bacteria in this soil are masked by a substantial envelope of colloidal clay and organic matter, a condition in which microbial activity would be greatly reduced and in which organic matter may readily accumulate. These brief and obviously crude descriptions of some soil microhabitats serve to indicate our lack of understanding of such microhabitats, and our present inability to relate the composition and organization of soil materials to the factors affecting the survival of certain bacteria in soils.

The interpretation of the relationships between microorganisms and the solid components of soils as seen under the scanning electron microscope requires careful examination of stereo-pair photographs under a stereo viewer (see Figures 4 and 5).

This technique provides a three-dimensional view and allows a more valid interpretation of the position and orientation of microorganisms with respect to the solid-phase components.

CONCLUDING REMARKS

There is no doubt that the clay mineralogical status of a soil has a profound effect on the survival of microorganisms in the soil. Clays modify the physicochemical status of the microhabitat which, in turn, modifies the microbial balance within the ecosystem. Although broad comparisons of bacterial survival have been made in soils of different clay status, the exact effects of clays on the behavior of the microorganisms must still be explained in detail. This can only be achieved by an appreciation of the variable effects of interfaces on bacteria at the colloidal level, of the effects of bacteria on the properties of the interfaces, and of the detailed microorganization of soil components as they affect the nature of the interfaces "seen" by soil bacteria.

ACKNOWLEDGMENTS

Unpublished original work mentioned in this review has been supported from grants by the Australian Research Grants Committee, the Rural Credits Development Fund of the Reserve Bank of Australia, and the University of Tasmania.

Literature Cited

1. Abramson, H. A., Moyer, L. S., Gorin, M. H. 1942. *Electrophoresis of Proteins and the Chemistry of Cell Surfaces.* New York: Reinhold. 341 pp.
2. Adler, J. 1969. Chemoreceptors in bacteria. *Science* 166:1588–97
3. Beattie, J. A. 1970. Peculiar features of soil development in parna deposits in the Eastern Riverina, N.S.W. *Aust. J. Soil Res.* 8:145–56
4. Bitton, G., Henis, Y., Lahav, N. 1972. Effect of several clay minerals and humic acid on the survival of *Klebsiella aerogenes* exposed to ultraviolet irradiation. *Appl. Microbiol.* 23:870–74
5. Black, C. A. 1957. *Soil-Plant Relationships.* New York: Wiley. 332 pp.
6. Brewer, R. 1964. *Fabric and Mineral Analysis of Soils.* New York: Wiley. 470 pp.
7. Brewer, R., Sleeman, J. R. 1960. Soil structure and fabric; their definition and description. *J. Soil Sci.* 11:172–85
8. Busch, P. L., Stumm, W. 1968. Chemical interactions in the aggregation of bacteria bioflocculation in waste treatment. *Environ. Sci. Technol.* 2:49–53
9. Corpe, W. A. 1970. An acid polysaccharide produced by a primary film-

forming marine bacterium. *Dev. Ind. Microbiol.* 11:402–12
10. Doetsch, R. N., Seymour, W. F. K. 1970. Negative chemotaxis in bacteria. *Life Sci.* 9 (Part II):1029–37
11. Dommergues, Y. 1964. Étude de quelques facteurs influant le comportement de la microflore du sol au cours de la dessiccation. *Sci. Sol* [2]:141–55
12. Dubos, R. J., Middlebrook, G. 1948. The effect of wetting agents on the growth of tubercle bacilli. *J. Exp. Med.* 88:81–88
13. Fildes, P., Kay, D. 1963. The conditions which govern the adsorption of a tryptophan-dependent bacteriophage to kaolin and bacteria. *J. Gen. Microbiol.* 30:183–91
14. Filip, Z. 1973. Clay minerals as a factor influencing the biochemical activity of soil microorganisms. *Folia Microbiol.* 18:56–74
15. Gibbons, R. J., van Houte, J. 1971. Selective bacterial adherence to oral epithelial surfaces and its role as an ecological determinant. *Infect. Immun.* 3:567–73
16. Giltner, W., Langworthy, H. V. 1916. Some factors influencing the longevity of soil micro-organisms subjected to

desiccation, with special reference to soil solution. *J. Agric. Res.* 5:927–42

17. Grim, R. E. 1968. *Clay Mineralogy.* New York: McGraw-Hill. 2nd ed. 596 pp.

18. Harris, P. J. 1972. Micro-organisms in surface films from soil crumbs. *Soil Biol. Biochem.* 4:105–6

19. Harris, R. H., Mitchell, R. 1973. The role of polymers in microbial aggregation. *Ann. Rev. Microbiol.* 27:27–50

20. Hattori, T. 1973. *Microbial Life in the Soil; An Introduction.* New York: Dekker. 427 pp.

21. Hirsch, P., Pankratz, St. H. 1970. Studies of bacterial populations in natural environments by use of submerged electron microscope grids. *Z. Allg. Mikrobiol.* 10:589–605

22. Kelman, A., Hruschka, J. 1973. The role of motility and aerotaxis in the selective increase of avirulent bacteria in still broth cultures of *Pseudomonas solanacearum. J. Gen. Microbiol.* 76: 177–88

23. Kiremidjian, L., Stotzky, G. 1973. Effects of natural microbial preparations on the electrokinetic potential of bacterial cells and clay minerals. *Appl. Microbiol.* 25:964–71

24. Krüger, W. 1961. The activity of antibiotics in soil. I. Adsorption of antibiotics by soils. *S. Afr. J. Agric. Sci.* 4:171–84

25. Krüger, W. 1961. The activity of antibiotics in soil. II. Movement, stability and biological activity of antibiotics in soils and their uptake by tomato plants. *S. Afr. J. Agric. Sci.* 4:301–15

26. Marshall, K. C. 1964. Survival of root-nodule bacteria in dry soils exposed to high temperatures. *Aust. J. Agric. Res.* 15:273–81

27. Marshall, K. C. 1967. Electrophoretic properties of fast- and slow-growing species of Rhizobium. *Aust. J. Biol. Sci.* 20:429–38

28. Marshall, K. C. 1968. Interaction between colloidal montmorillonite and cells of Rhizobium species with different ionogenic surfaces. *Biochim. Biophys. Acta* 156:179–86

29. Marshall, K. C. 1969. Studies by microelectrophoretic and microscopic techniques of the sorption of illite and montmorillonite to rhizobia. *J. Gen. Microbiol.* 56:301–6

30. Marshall, K. C. 1969. Orientation of clay particles sorbed on bacteria possessing different ionogenic surfaces. *Biochim. Biophys. Acta* 193:472–74

31. Marshall, K. C. 1971. Sorptive interactions between soil particles and microorganisms. In *Soil Biochemistry,* ed. A. D. McLaren, J. Skujins. 2:409–45. New York: Dekker. 527 pp.

32. Marshall, K. C. 1973. Mechanism of adhesion of marine bacteria to surfaces. *Proc. Int. Congr. Marine Corros. Fouling, 3rd.* 625–32

33. Marshall, K. C., Alexander, M. 1960. Competition between soil bacteria and Fusarium. *Plant Soil* 12:143–53

34. Marshall, K. C., Cruickshank, R. H. 1973. Cell surface hydrophobicity and the orientation of certain bacteria at interfaces. *Arch. Mikrobiol.* 91:29–40

35. Marshall, K. C. Mulcahy, M. J., Chowdhury, M. S. 1963. Second-year clover mortality in Western Australia—A microbiological problem. *J. Aust. Inst. Agric. Sci.* 29:160–64

36. Marshall, K. C., Roberts, F. J. 1963. Influence of fine particle materials on survival of *Rhizobium trifolii* in sandy soils. *Nature London* 198:410–11

37. Marshall, K. C., Stout, R., Mitchell, R. 1971. Mechanism of the initial events in the sorption of marine bacteria to surfaces. *J. Gen. Microbiol.* 68:337–48

38. Marshall, K. C., Stout, R., Mitchell, R. 1971. Selective sorption of bacteria from seawater. *Can. J. Microbiol.* 17:1413–16

39. Miyamoto, Y. 1959. Further evidence for the longevity of soil-borne plant viruses adsorbed on soil particles. *Virology* 9:290–91

40. Müller, G., Hickish, B. 1970. Die Adsorption von Bodenbakterien an Substrate. *Zentralbl. Bakteriol. Parasitenk. Infektionskr. Hyg Abt. 2* 124:271–78

41. Müller, H. P., Schmidt, L. 1966. Kontinuierliche Atmungsmessungen an *Azotobacter chroococcum* Beij. in Montmorillonit unter chronischer Röntgenbestrahlung. *Arch. Mikrobiol.* 54:70–79

42. Neihof, R. A., Loeb, G. I. 1972. The surface charge of particulate matter in seawater. *Limnol. Oceanogr.* 17:7–16

43. Neihof, R. A., Loeb, G. 1974. Dissolved organic matter in seawater and the electric charge of immersed surfaces. *J. Mar. Res.* 32:5–12

44. Pertsovskaya, A. F., Duda, V. I., Zvyagintsev, D. G. 1972. Surface ultrastructures of adsorbed microorganisms. *Soviet Soil Sci.* 4:684–89

45. Pinck, L. A., Holton, W. F., Allison, F. E. 1961. Antibiotics in soils: 1. Physico-chemical studies of antibiotic-clay complexes. *Soil Sci.* 91:22–28

46. Pinck, L. A., Soulides, D. A., Allison, F. E. 1962. Antibiotics in soils: 4. Polypeptides and macrolides. *Soil Sci.* 94:129–31

47. Plummer, D. T., James, A. M., Gooder, H., Maxted, W. R. 1962. Some physical investigations of the behaviour of bacterial surfaces. V. The variation of the surface structure of *Streptococcus pyogenes* during growth. *Biochim. Biophys. Acta* 60:595–603

48. Poindexter, J. S. 1964. Biological properties and classification of the *Caulobacter* group. *Bacteriol. Rev.* 28:231–95

49. Roper, M. M., Marshall, K. C. 1974. Modification of the interaction between *Escherichia coli* and bacteriophage in saline sediment. *Microb. Ecol.* 1:1–13

50. Seymour, W. F. K., Doetsch, R. N. 1973. Chemotactic responses by motile bacteria. *J. Gen. Microbiol.* 78:287–96

51. Skinner, F. A. 1956. The effect of adding clays to mixed cultures of *Streptomyces albidoflavus* and *Fusarium culmorum*. *J. Gen. Microbiol.* 14:393–405

52. Skujins, J. J. 1967. Enzymes in soil. In *Soil Biochemistry,* ed. A. D. McLaren, G. H. Peterson, 1:371–414. New York: Dekker. 509 pp.

53. Smith, A. M. 1973. Ethylene as a cause of soil fungistasis. *Nature London* 246:311–13

54. Soulides, D. A. 1969. Antibiotic tolerance of the soil microflora in relation to type of clay minerals. *Soil Sci.* 107:105–7

55. Soulides, D. A., Pinck, L. A., Allison, F. E. 1961. Antibiotics in soils: 3. Further studies on release of antibiotics from clays. *Soil Sci.* 92:90–93

56. Soulides, D. A., Pinck, L. A., Allison, F. E. 1962. Antibiotics in soils: V. Stability and release of soil-adsorbed antibiotics. *Soil Sci.* 94:239–44

57. Stace, H. C. T. et al 1968. *A Handbook of Australian Soils,* 117–26. Glenside, South Australia: Rellim Tech. Publ. 435 pp.

58. Staley, J. T. 1968. *Prosthecomicrobium* and *Ancalomicrobium:* New prosthecate freshwater bacteria. *J. Bacteriol.* 95:1921–42

59. Stotzky, G. 1966. Influence of clay minerals on microorganisms. II. Effect of various clay species, homoionic clays, and other particles on bacteria. *Can. J. Microbiol.* 12:831–48

60. Stotzky, G. 1966. Influence of clay minerals on microorganisms. III. Effect of

particle size, cation exchange capacity, and surface area on bacteria. *Can. J. Microbiol.* 12:1235–46

61. Stotzky, G. 1967. Clay minerals and microbial ecology. *Trans. NY Acad. Sci.* Ser. II, 30:11–21

62. Stotzky, G., Martin, R. T. 1963. Soil mineralogy in relation to the spread of Fusarium wilt of banana in Central America. *Plant Soil* 18:317–38

63. Stotzky, G., Post, A. H. 1967. Soil mineralogy as a possible factor in geographic distribution of *Histoplasma capsulatum*. *Can. J. Microbiol.* 13:1–15

64. Stotzky, G., Rem, L. T. 1966. Influence of clay minerals on microorganisms. I. Montmorillonite and kaolinite on bacteria. *Can. J. Microbiol.* 12:547–63

65. Stotzky, G., Rem, L. T. 1967. Influence of clay minerals on microorganisms. IV. Montmorillonite and kaolinite on fungi. *Can. J. Microbiol.* 13:1535–50

66. Tenney, M. W., Stumm, W. 1965. Chemical flocculation of microorganisms in biological waste treatment. *J. Water Pollut. Control Fed.* 37:1370–88

67. Vandecaveye, S. C. 1927. Effect of moisture, temperature, and other climatic conditions on *R. leguminosarum* in the soil. *Soil Sci.* 23:355–62

68. van Olphen, H. 1963. *An Introduction to Clay Colloid Chemistry.* New York: Interscience. 470 pp.

69. Vyas, S. R., Prasad, N. 1960. Investigations on the failure of peas in 'Goradu' soils of Gujarat. *Proc. Indian Acad. Sci. Sect. B* 51:242–48

70. Young, L. Y., Mitchell, R. 1973. Negative chemotaxis of marine bacteria to toxic chemicals. *Appl. Microbiol.* 25:972–75

71. Zeidberg, L. D. 1954. A theory to explain the geographic variations in the prevalence of histoplasmin sensitivity. *Am. J. Trop. Med. Hyg.* 3:1057–65

72. Zvyagintsev, D. G. 1959. Adsorption of microorganisms by glass surfaces. *Microbiology USSR* 28:104–8

73. Zvyagintsev, D. G., Pertsovskaya, A. F., Duda, V. I., Nikitin, D. I. 1969. Electron-microscopic study of the adsorption of microorganisms on soil and minerals. *Microbiology USSR* 38:937–42

74. Zvyagintsev, D. G., Pertsovskaya, A. F., Yakhnin, E. D., Averbach, É. I. 1971. Adhesion value of microorganism cells to solid surfaces. *Microbiology USSR* 40:889–93

PRESENT STATUS OF COFFEE RUST IN SOUTH AMERICA[1]

❖3628

Eugenio Schieber

Post Office Box 226, Antigua, Guatemala

INTRODUCTION

In early 1970 coffee rust incited by *Hemileia vastatrix* Berk & Br. was discovered in Brazil for the first time in the western hemisphere. Since then, the fungus has spread considerably in Brazil and has invaded Paraguay and Argentina. In Argentina the region of Misiones close to Paraná (Brazil), a very small coffee region, is now affected. By 1974 coffee rust covered three Latin American countries. There is no doubt that wind played an important role in its wide-range dissemination, as discussed in this review.

Since the discovery of rust in 1970, many developments in research and control of the disease have come about in dynamic form in Brazil (8, 11, 13, 17, 21), and necessary changes have been effected in the technology of coffee growing.

Since early 1970, the disease situation in South America has changed greatly; this article covers only some of the more important aspects of coffee rust that have produced these changes: the present race situation, aspects of rust spread, the new planting technology in relation to disease control, changes in chemical control, and changes in the production of resistant varieties.

PRESENT RACE SITUATION

When rust was discovered in 1970, race II of *H. vastatrix* was immediately identified by D'Oliveira in Portugal. Because Race II is the most common and prevalent race in the eastern hemisphere and of wide occurrence in the African continent (as well as for other reasons) it may have been involved in the establishment of rust on the American continent (22–24). Race II (Figure 1) also reached Angola in West Africa in 1966 as the rust spread from east to west Africa.

In 1971 race II also invaded the state of São Paulo from Minas Gerais, later reaching the coffee regions in Paraná and the state of Santa Catarina. However, in 1972 the race status changed for Brazil, for race XV was also identified in planta-

---

[1]This review is a supplement to "Economic Impact of Coffee Rust in Latin America," published in 1972 by *Annual Review of Phytopathology.*

Figure 1   Young uredopustules produced by race II of *H. vastatrix.*

tions there. The first identification of race XV was made by the Coffee Rust Research Center (Oeiras, Portugal) from a collection made in Espirito Santo (20). Race XV carries the V_4 and V_5 virulence genes, in contrast to race II, which has the factor V_5.

Race XV has also been reported from Angola in west Africa. In addition, Race XV has been reported from Ceylon and Portuguese Timor in Asia, and from Kenya, Island of São Tomé, Ethiopia, and Angola in Africa.

In March 1973, a new race, race III, again changed the race status in Brazil. Found recently in the state of São Paulo (Jaú region and Campinas), this race also crossed the African continent, finally reaching West Africa in 1966. Race III is found in India and Portuguese Timor in Asia, and in Tanzania, Island of São Tomé, and Ethiopia in Africa.

It is significant that these three races are found in Ethiopia, where coffee originated, and two of these (II and XV) in Ceylon, Kenya, and Angola. Two races, II and XV, that appeared in Brazil in the last four years possibly came to Bahia from Angola, West Africa, with trade winds across the Atlantic in 5–7 days (3, 6). How then did race III, not found in the northern coffee regions of Brazil but farther to the south in the state of São Paulo reach São Paulo if it came from Africa? Was it developed in São Paulo by heterokaryosis? Why is it not found in the northern states of Minas Gerais and Espirito Santo? All of these questions arise when analyzing the presence of race III in Brazil.

As noted in Table 1, the appearance of new races in Brazil has been almost continuous. With the vast coffee regions and the vast substrata for the pathogen to develop, mutate, or produce new races by heterokaryosis, the occurrence of new races would be expected in Brazil. However, there may be other factors involved in the change of race status.

Table 1   Present status of races of *H. vastatrix* in Brazil

1970	Race II
1972	Race II and XV
1973	Race II, XV, and III
1974	Race II, XV, III, and I

In late 1974, race I appeared in Brazil, further complicating the race and resistant variety situation. Race I is reported from the Congo but not from Angola.

The question of the origin of these races will not be answered; the important task for Brazil is the search for resistance.

Research in Portugal and Brazil has detected these races rapidly and in different areas within the different coffee regions. It is of prime importance to follow the appearance of new rust races in South America, because breeding programs must focus on new trends in the production of resistant varieties.

SPREAD OF COFFEE RUST IN SOUTH AMERICA

Since coffee rust was discovered early in 1970 in Bahia, Brazil, it has spread in a southwesterly direction, covering most of the coffee regions of this country and reaching coffee plantations in Paraguay and in Misiones in Argentina. In four years, the rust has spread in South America over an area equivalent to Central America, following the direction of prevailing winds.

Until late 1974 the coffee regions north of Bahia, Brazil (Pernambuco and Ceará) were still free from rust, apparently because of the prevailing wind currents. Late in 1974, rust was discovered in the economically unimportant region of Acre in northwestern Brazil. This is a potentially very serious situation because of the proximity of this region to Bolivia and Peru.

Referring to spread of the rust from Africa to Ceylon, Rayner (19) in 1960, speculated on spread by wind of the uredospores of *H. vastatrix*.

Schieber (22) in 1970 noted that the rapid dissemination of coffee rust in Brazil suggested wind as an important factor. He observed that wind currents in the affected region were from north to south and that *Hemileia* had been disseminated in this direction. In his review article (23) in 1972 he wrote, "It is of interest that the Hemileia rust was found in the state of São Paulo during January 1971, as expected from the direction of the prevailing winds." Figure 2, modified from that article, shows the air movement in Brazil in January, in relation to the recent discoveries of rust in Paraguay, in Misiones in Argentina, and in the region of Acre.

Trapping tests by Brazilian investigators (Martinez and collaborators) in 1971 over the state of São Paulo showed uredospores at high altitudes over the state of Paraná when this state was still free from rust and when rust was spreading rapidly over the state of São Paulo (12, 23, 24). Rust invaded São Paulo plantations during January 1971, and in October of the same year, the rust was found in the first plantations in Paraná. Later the rust was found in Santa Catarina and the western coffee region of Mato Grosso.

🖤 Coffee regions in Brazil

Figure 2   Air movement during January in Brazil.

Firman (10) in 1972, describing the spread in the state of São Paulo, wrote that "there was a pattern of progressive spread in a southwesterly direction, giving rise to speculation about the possibility of the dispersal of the uredospores by air currents along the path of prevailing winds."

Figure 2, which records the air movement during January over the coffee regions in Brazil, reveals why Paraguay and Misiones in Argentina have been also invaded by the rust. There is no question that wind currents carried the uredospores of *H. vastatrix* in the way uredospores of cereal rusts are disseminated throughout the world.

The Department of Meteorology in Brazil is studying the effect that the polar mass movement might have on further spread of the rust in South America. Lessman in 1973 (14) studied air movement from Brazil to Colombia, concluding that uredospores cannot be transported in that direction.

At very high altitudes (5000 m), temperatures have a relatively wide variation, and temperatures are commonly as low as 4°C; this may help preserve uredospore viability. It is common practice at rust laboratories in many countries, including the Coffee Rust Research Center at Oeiras, Portugal, to maintain viability of uredospores by storing them at temperatures from 4–8°C.

NEW PLANTING TECHNOLOGY

Rust disease has led to the development of new planting systems in Brazil, among these, different spacing and pruning techniques. Changes in spacing have been effected to make chemical control feasible and to change the microecology at the plantation level. Before rust was discovered in Brazil, coffee plantation rows were closely spaced. The resultant heavy foliage affected the amount of light reaching the plants, and an appropriate substrata for the rust pathogen was created which favored infection, disease development, and inoculum buildup.

The new spacing technology in Brazil involves the planting of coffee trees (with no shade) between 2–2.5 m in the row, and 3.5–4 m between rows, depending on the variety to be planted (Figure 3). In the rows, the system of cova (planting a number of coffee seedlings in one site) has been discarded, and only two plants were used in each site. This new spacing not only provides the room to enter with spraying

Figure 3 New technology in growing coffee in Brazil. Wider spacing between rows is needed for rust disease control.

equipment, but also changes the microecology at the plantation level, affecting rust establishment and development. In Kenya, it has been observed that less plant density favors more aeration and thus also decreases rust incidence (22).

In Central America and Mexico, coffee is now placed at higher density, including replantings in older plantations. Spacing methods should be studied in Central America and Mexico to prepare for rust, should it invade the area as expected.

CHANGES IN CHEMICAL CONTROL

Since 1970, the Instituto Brasileiro do Café has investigated the efficiency of fungicides, dosage, timing of applications, and application equipment from knapsack to airplane (4, 8).

Different copper fungicides have been tested, together with organic fungicides and new systemics (1, 2, 8, 13, 15, 18). Under Brazilian conditions, copper fungicides as well as some organics including maneb (Dithane M-45®) are effective. Tests in Brazil show that new systemics (including Pyracarbolid) have shown a "curative effect" on the uredopustules. One aspect of interest has been the effective control recently obtained in the field by alternating copper and organic fungicides.

Research in Brazil has defined the right timing in the use of fungicides; applications must be made between October and April.

The Grupo Errad. Café (GERCA) division of the Instituto Brasileiro do Café, has been carrying on most of these studies under field conditions in different ecological regions. Also the University at Viçosa, Minas Gerais has been active in research on chemical control.

In recent years Brazil has made significant advances in the field of chemical control against rust, and the use of fungicides by coffee growers has increased significantly.

CHANGES IN THE PRODUCTION OF RESISTANT VARIETIES

Since 1952, Brazil has produced varieties resistant to *H. vastatrix*. Carvalho, a Brazilian investigator, has carried out an excellent breeding program for many years. He also studied resistance to race II of the rust pathogen, even before the rust was found in 1970. The program has been directed to find the ideal combination of resistance, good agronomic adaptation, and quality (5, 7, 9).

The Coffee Rust Research Center in Portugal has collaborated with the Brazilian program for many years in research on rust resistance. In 1973, Brazil had already distributed seed of some of these new resistant varieties to farmers.

The natural hybrid of *C. canephora* with *C. arabica* Híbrido de Timor with factors for resistance that are dominant has been crossed with Arabica coffees like Catuaí, Caturra, Mundo Novo, and Geisha. The cross with Caturra was made by D'Oliveira and collaborators in Portugal (Figure 4). The breeding program for rust resistance in Brazil has been directed toward finding resistance to race II, the first race found in South America. The presence of race XV has not altered the breeding program significantly, because race XV only attacks coffee lines that have the SH_4 and SH_5 factors.

Figure 4   A cross between Caturra and Híbrido de Timor, in search for resistance toward *H. vastatrix*. (Coffee Rust Research Center, Oeiras, Portugal.)

As pointed out in the section on the race status in South America, the appearance of new races in the extensive coffee regions of Brazil represents a problem in the production of rust-resistant varieties. For example, the variety Geisha, which has been introduced in almost all coffee-producing countries in Latin America, is resistant to race II, but susceptible to race III, recently found in the state of São Paulo.

Schieber wrote in 1970 and 1972 (22, 23), "Vertical resistance to race II of the pathogen now present in the Western Hemisphere is what Brazil needs in the immediate future; however horizontal resistance (found already in plants of *C. arabica* in Angola and Brazil) would be desirable if it is possible to utilize it." The new races XV, III, and I had not been discovered in South America at that time.

Brazil is now searching for "horizontal resistance" (25). This search will be of great value to other coffee-producing countries in the Americas.

ACKNOWLEDGMENT

The author is indebted to Dr. George A. Zentmyer, Professor at the University of California, Riverside for suggestions in the writing of this manuscript.

Literature Cited

1. Almeida, S. R., Matiello, J. B., Mansk, Z. Andrade, I. P. R., Abreu, R. G. 1973. Dosage X season interaction in cupric fungicide spray for coffee rust control in South Minas Gerais. *Proc. I Congr. Brasil. Sobre Pragas e Doencas do Cafeeiro,* Vitória, Brasil
2. Andrade, I. P. R., Mansk, Z., Chaves, G. M., Matiello, J. B. 1972. Efeito de fungicidas cúpricos, sistêmicos e orgânicos no contrôle da ferrugem do cafeeiro (*Hemileia vastatrix* Berk. et Br.) e na produçao de café. 15 pp. In *Novos resultados de contrôle quimico da ferrugem do cafeeiro no Brasil.* Tech. Bull. Rio de Janeiro: Inst. Brasil. do Café. 55 pp.
3. Anonymous. 1970. Influéncia dos fatores meteorológicos na ocorrência da *Hemileia vastatrix. Dep. Nac. Meteorol., Tech. Bul.,* Brasilia, Brazil
4. Anonymous. 1973. Agricultural aviation. *Granja* 20(301):9–10, 12–14, 16–18, 20–21
5. Bettencourt, A. J., Carvalho, A. 1968. Melhoramento visando a resistência do cafeeiro â ferrugem. *Bragantia* 27:35–68
6. Bowden, J., Gregory, P. H., Johnson, C. G. 1971. Possible wind transport of coffee leaf rust across the Atlantic Ocean. *Nature* 229:500–1
7. Carvalho, A., Monaco, L. C. 1972. Melhoramento do cafeeiro visando a resistencia a ferrugem alaranjada. *Cien. Cult. Sao Paulo* 23(2):141–46
8. Chaves, G. M. 1972. *Chemical control of Hemileia vastatrix Berk. & Br.* Presented at Symp. Coffee Rust, Am. Phytopathol. Soc. 64th Ann. Meet.
9. Centro de Investigacao das Ferrugens do Cafeeiro. 1971. Junta de Investigaçoes do Ultramar. Oeiras, Portugal. 29 pp.
10. Firman, I. D. 1972. Coffee leaf rust in Brazil. *FAO Plant Prot. Bull.* 20:121–26
11. Instituto Brasileiro do Café. 1970. A ferrugem do cafeeiro no Brasil. *Tech. Bul. Inst. Bras. Café- Grupo Errad. Café-GERCA,* Rio de Janeiro, Brazil. 75 pp.
12. Instituto Brasileiro do Café. 1971. Vento carrega ferrugem. *Inform. Grupo Errad. Café–GERCA, Inst. Bras. Café.* I (4) Rio de Janeiro, Brazil. 7 pp.
13. Instituto Brasileiro do Café. 1971. Relatório das pesquisas sobre *Hemileia vastatrix* Berk. et Br. *Proc. I Reuniao*

Anual Inst. Brasil. do Café. Inst. Bras. Café–GERCA. Rio de Janeiro, Brazil
14. Lessman, H. 1973. Puede propagarse la roya del café desde el Brasil hasta Colombia? *OMM. Rept. Expert Organ. Mundial Meteorol.*
15. Matiello, J. B., Almeida, S. R., Andrade, I. P. R., Abreu, R. G., Mansk, Z. 1973. Effect of organic and cupric fungicides in isolated mixed or alternating spray, at high and low volume, on coffee rust control. *Proc. I Congr. Brasil. Sobre Pragas e Doenças do Cafeeiro,* Vitória, Brasil
16. Nutman, F. J., Roberts, F. M. 1971. Spread of coffee leaf rust. *Pest Artic. News Summ.* 17:385–86
17. Ortolani, A. A., Vianna, A. C. C., Abreu, R. G. 1971. *Hemileia vastatrix* Berk. et Br. Estudos e observaçoes em regioes da Africa e sugestoes a cafeicultura do Brasil, Rep. Inst. Brasil. do Café, p. 193. Rio de Janeiro, Brasil
18. Paiva, F. A., Chaves, G. M. 1972. Uso de fungicidas sistémicos no contróle da ferrugem do cafeeiro (*Hemileia vastatrix* Berk & Br.) In *Proc. V Reuniao Soc. Brasil. Fitopatol.,* Fortaleza, Ceará
19. Rayner, R. W. 1960. Rust disease of coffee. II. Spread of the disease. *World Crops* 12:222–24
20. Ribeiro, I. J. A., Monaco, L. C., Filho, O. T., Sugimori, M. H., Scali, M. H. 1973. Strain bearing the factors $V_4 V_5$ in Sao Paulo. *Proc. X Congr. Brasil. Sobre Pragas e Doenças do Cafeeiro,* Vitória, Brasil
21. Sebastiao, J. M. J. 1970. El problema de la roya del cafeto en el Brasil. Enfoque del Instituto Brasileiro del Café. Mesa Redonda sobre Roya del Cafeto. *Proc. VII Reunión Latinoam. Fitotec.* (*ALAF*), Bogotá, Colombia
22. Schieber, E. 1970. Viaje al Brasil y el Africa para estudiar y observar el problema de la herrumbre del café. *Rep. Org. Int. Reg. San. Agric.* 109 pp.
23. Schieber, E. 1972. Economic impact of coffee rust in Latin America. *Ann. Rev. Phytopathol.* 10:491–510
24. Schieber, E. 1972. Comparative Review of Coffee Rust in the Eastern and Western Hemispheres. Presented at Symp. Coffee Rust, Am. Phytopathol. Soc. 64th Ann. Meet.
25. Van der Plank, J. E. 1963. *Plant Diseases. Epidemics and Control.* New York: Academic. 349 pp.

AUTHOR INDEX

SUBJECT INDEX

CUMULATIVE INDEXES

CONTRIBUTING AUTHORS VOLUMES 9-13

CHAPTER TITLES VOLUMES 9-13